Springer Textbooks in Earth Sciences, Geography and Environment

The Springer Textbooks series publishes a broad portfolio of textbooks on Earth Sciences, Geography and Environmental Science. Springer textbooks provide comprehensive introductions as well as in-depth knowledge for advanced studies. A clear, reader-friendly layout and features such as end-of-chapter summaries, work examples, exercises, and glossaries help the reader to access the subject. Springer textbooks are essential for students, researchers and applied scientists.

Norbert de Lange

Geoinformatics in Theory and Practice

An Integrated Approach to Geoinformation Systems, Remote Sensing and Digital Image Processing

Springer

Norbert de Lange
Universität Osnabrück
Institut für Informatik
Osnabrück, Niedersachsen
Germany

ISSN 2510-1307 ISSN 2510-1315 (electronic)
Springer Textbooks in Earth Sciences, Geography and Environment
ISBN 978-3-662-65757-7 ISBN 978-3-662-65758-4 (eBook)
https://doi.org/10.1007/978-3-662-65758-4

This book is a translation of the original German edition „Geoinformatik in Theorie und Praxis“ by de Lange, Norbert, published by Springer-Verlag GmbH, DE in 2020. The translation was done with the help of artificial intelligence (machine translation by the service DeepL.com). A subsequent human revision was done primarily in terms of content, so that the book will read stylistically differently from a conventional translation. Springer Nature works continuously to further the development of tools for the production of books and on the related technologies to support the authors.

This Springer imprint is published by the registered company Springer-Verlag GmbH, DE, part of Springer Nature.
The registered company address is: Heidelberger Platz 3, 14197 Berlin, Germany

Foreword

Since the first edition of this textbook in Germany in 2002, Geoinformatics has progressed tremendously under the influence of technical changes and innovations and has been realigned in many aspects. Particularly noteworthy are the introduction of the smartphone in 2007 as well as the rapid increase in the importance of the Internet and its resulting changes in user behaviour. Access to diverse and rapidly growing Internet options has also led to new applications in Geoinformatics: e.g. OGC web services, the development of geodata infrastructures, mobile geoinformation systems and information systems about geoobjects on the Internet, which are often named Web-GIS. Data are now ubiquitously available to many users on mobile devices. These have become indispensable in everyday life and in particular in applications in Geoinformatics. The process of digitalisation which has diverse spatial implications and thus also affects Geoinformatics will shape the 2020s.

This book was translated from the German 4th edition with the help of artificial intelligence (machine translation by the service DeepL.com) first and then significantly revised with regard to technical items and special topics of Geoinformatics.

This new edition would not have been possible without the support of several colleagues. Therefore, the author would like to thank everyone who has given him valuable comments and suggestions. In particular, I would like to thank my colleague M.Sc. M. Storch for critically reading all the chapters. Furthermore, I would like to thank the editors of Springer-Verlag who offered and created the opportunity to publish the book for the international market. My thanks include Mrs. Charu Pancholi and her team who edited the book in India as well as Mr. Shah-Rohlfs for the good cooperation.

Contents

Introduction

1

1.1 Geoinformatics: Approach and Tasks

Geoinformatics is an interdisciplinary field that bridges the gap between computer science, geographic information technologies and geosciences or other spatial sciences (Fig. 1.1):

The emergence of geoinformatics as an independent discipline has been completed. Scientific institutions and professorships, courses of study, textbooks and journals, conferences and associations are clear indicators of this development, as are trade fairs for geoinformatics or public acceptance and use of the term as a component in company names. A broader science-theoretical discussion about the contents of geoinformatics has only gradually begun in Germany (cf. Bill and Hahn 2007 and Ehlers 2006), but it has come to a standstill.

In contrast to the more recent term geoinformatics, the very dazzling term *GIS* has been introduced for some time, which in the narrower sense only stands for *geoinformation systems*, but is often equated with the new field of work and research in geoinformatics, without, however, covering it. The older designation *geographic information systems* for GIS points to its origin or to its earlier self-conception as a tool of Geography. The dominance of "GIS", among others, in the spatially oriented disciplines is mainly due to the rapidly advancing software development of GIS technology and the accelerating, very broad application of these technologies. It must be noted that a methodical approach has existed in the field of Geography for a long time, which overlaps with geoinformatics, also with regard to the persons involved: geospatial analysis, spatial data analysis or just spatial analysis (cf. above all widely established textbooks by Longley and Batty 1997, Fotheringham, Brunsdon and Charlton 2000, Longley 2003, de Smith, Goodchild and Longley 2018).

The history of the development of *geoinformation science* can be divided into several periods (extended cf. Bartelme 2005 p. 10):

N. de Lange, *Geoinformatics in Theory and Practice*, Springer Textbooks in Earth Sciences, Geography and Environment,
https://doi.org/10.1007/978-3-662-65758-4_1

- 1955–1975: period of the pioneers (self-programmed software, hardly any data, hardly any hardware support)
- 1970–1985: period of the authorities (rearrangement of the administration of geodata to computers, limited data and hardware support)
- 1982–1990: period of the companies (appearance of commercial GIS products like Arc/Info of the software producer ESRI (Environmental Systems Research Institute))
- from 1988: period of the users (user-specific solutions, data structuring, special applications, networks)
- since 1995: period of the open market of geoinformation (internet-enabled products, online data exchange, spatial data infrastructures, free geoinformation systems)
- since 2010: period of mobile use of geoinformation by geo-apps on smartphones and tablet computers (navigation systems and location-based services), which enable ubiquitous availability of and mobile work with geodata.

This systematic approach focuses on geoinformation systems. In some cases, the interests and perspectives of the dominant players differ considerably. To this day, the development and applications of and with GIS have played "the" central role in the development of geoinformatics. In contrast, Table 1.1 compiles *milestones* that are independent of GIS or software. Thus, among the significant events are: 1978 the introduction of the software package ERDAS (now ERDAS-IMAGINE) as a market leader in raster data processing and digital image processing, 1982 the start of ARC/Info (in the meantime further development of the software Arc/GIS to a GIS market leader) and also the increased emergence of free geoinformation systems (e.g. development of GRASS – Geographic Resources Analysis Support System) by the US Army Construction Engineering Research Laboratories, meanwhile a project of the Open Source Geospatial Foundation, since 1999 under the GNU General Public License, and e.g. development of QGIS (formerly Quantum GIS, 2002) as a professional GIS application based on open source software.

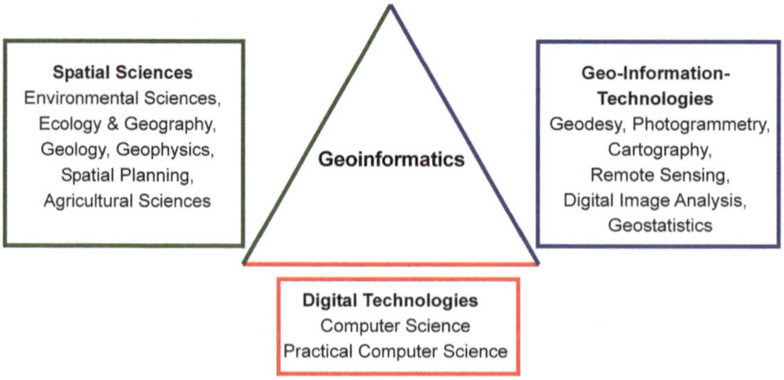

Fig. 1.1 Relationships of geoinformatics to other disciplines

Table 1.1 Milestones in geoinformatics

1964	Founding of Harvard Laboratory for Computer Graphics and Spatial Analysis, where pioneering software systems such as SYMAP (a first raster GIS) or ODYSSEY (a first vector GIS) were developed
1971	Operational use of Canada Geographic Information System (CGIS) by Roger Tomlinson ("father" of GIS)
1972	Launch of first Landsat satellite (original name ERTS-1)
1973	Publication of target concept for ALK (Automated Real Estate Cadastre) as the basis for the property database by the Working Group of the Surveying Authorities of the Federal States of the Federal Republic of Germany
1983	Development of Map Algebra in the PhD dissertation by C. Dana Tomlin as a set of basic operators for manipulating raster data
1984	1st International Spatial Data Handling Symposium in Zurich with trend-setting lectures among others by Tomlinson and Goodchild
1985	Operational availability of NAVSTAR GPS (Global Positioning System), the world's most important satellite-based positioning system
1988	Founding of National Centre for Geographic Information and Analysis (NCGIA), a research network for the further developments in theory, methods and techniques for the analysis of geographic information with GIS
1989	Publication of ATKIS overall documentation (Official Topographic Cartographic Information System) by the Working Group of the Surveying Authorities of the Federal States of the Federal Republic of Germany
1989	First AGIT symposium in Salzburg (originally Applied Geographic Information Technology), since then annual symposia for applied geoinformatics, as the most important in the German-speaking world
1994	Founding of Open Geospatial Consortium (OGC) as the Open GIS Consortium, a non-profit organisation for the development of standards for spatial information processing (especially geodata) to ensure interoperability
1999	Civilian availability of high-resolution satellite images of up to less than 1 m ground resolution (earth observation satellite IKONOS-2)
2000	End of so-called selective availability of GPS signals and start of a broad civil use of GPS (navigation systems and location-based services)
2004	Start of Open Street Map project, the development of freely usable geodata (open data) and a world map similar to the concept of Wikipedia
2005	Geohype due to increasing use of Google Earth
2006	Founding of Open Source Geospatial Foundation (OSGeo), a non-profit organisation for the development and use of free and open source GIS
2007	Enforcement of INSPIRE Directive of the EU (Infrastructure for Spatial Information in Europe) for the establishment of a European spatial data infrastructure
2007	Launch of iPhone and expansion of navigation systems as well as location-based services for smartphones and tablet computers that use device-specific sensor technology for geo-applications
2009	Federal Act on Access to Digital Geodata, including regulating the provision of Federal geodata and geodata services free of charge

(continued)

Table 1.1 (continued)

2014	Launch of the first Sentinel satellite of the Copernicus programme for complex Earth observation for environment, transport, economy and security policy
2016	General availability of GALILEO, ESA's first global satellite navigation and positioning system under civilian control
2016	Virtual and augmented reality (popularised by geo-games like Pokémon Go)

The Core Curriculum in GIS presented by the National Center for Geographic Information and Analysis (NCGIA) in 1990 (cf. NCGIA 1990) had a great influence on the development of standards for geoinformation systems and on the development of geoinformatics in general. The abundance of international readers dealing with the topic of GIS cannot be estimated. Reference should only be made to the two early standard works from the 1990s (cf. Burrough and McDonell 1998 and meanwhile as current readers Longley et al. 2005, 2015) and to two more modern readers in open textbook libraries (Campbell and Shin 2011 and Olaya 2018). In the meantime, with the works by Bartelme (4th ed. 2005), Bill (6th ed. 2016), Dickmann u. Zehner (2nd ed. 2001), Ehlers u. Schiewe (2012), Hennermann (2nd ed. 2014), Kappas (2nd ed. 2012), de Lange (4th ed. 2020), Penzkofer (2017) and Zimmermann (2012), established as well as newer German-language textbooks have become available.

While "GIS" in the sense of geoinformation systems established itself in the German-speaking world in the 1990s, an international discussion about "GIS" as a science began. This reorientation began with the fourth "International Symposium on Spatial Data Handling" in 1990, at which Goodchild demanded in his programmatic keynote "GIS" as *Geographic Information Science* (see also the later article of Goodchild 1992):

> We need to move from system to science, to establish GIS as the intersection between a group of disciplines with common interests, supported by a toolbox of technology, and in turn supporting the technology through its basic research. (Goodchild 1990 p. 11)

Since the mid-1990s, the term *geoinformatics* has appeared in the German-speaking world. In a first approach to the term, "geoinformatics is understood as those aspects of spatial information processing that deal with formal and theoretical foundations and are analysed using methods of computer science" (translation of Kainz 1993 p. 19). The revised new edition of the book "GIS-Technologie" (Bartelme 1989), combined with renaming to "Geoinformatik" (Bartelme 1995), clearly marks the reorientation, but still the relationship to geoinformation systems is tight.

However, geoinformatics is more than a geoinformation system (GIS). Undoubtedly, geoinformation systems are the most important tools of geoinformatics, which has just led to the fact that the new discipline itself was equated with GIS and a discussion "tool or science" started (cf. Blaschke 2003). However, remote sensing and digital image processing are inseparable parts of geoinformatics. They record and provide spatial data or geoinformation and analyse them with regard to spatially relevant issues. Likewise, in geoinformation systems, remote sensing data are often and increasingly processed and analysed together with other geodata. Ehlers in particular has already addressed the close relationship between GIS and remote sensing in several publications (cf. Ehlers et al. 1989, 1991, Ehlers and Amer 1991, Ehlers 1993, Ehlers 2000).

Here, geoinformatics is to be defined in very general terms. The proximity to computer science as well as the special nature of the information to be processed in geoinformatics should become clear:

Geoinformatics is dedicated to the development and application of methods and concepts of computer science to solve spatial problems, with special consideration of the spatial reference of information. Geoinformatics deals with recording or sourcing, modeling, processing and above all with analysing as well as with representing and disseminating geodata.

The following example illustrates the implementation of this definition: It is to be shown for a city how the residential population is served by public transport. Methodological approaches to the task can be evaluation of passenger numbers, surveys of passengers or the residential population as well as assessments of the accessibility or equipment of a stop. Geoinformatics, however, examines the spatially related problem, taking into account the spatial reference of the information, i.e. the locations of the stops and the households as well as the routes: Residents of how many households can reach the next bus stop within a maximum of ten minutes on foot? Based on this, the number of trips per direction and day or the number of available seats a bus line can be examined.

The first and simple approach to the problem is to determine the catchment areas with a radius of 550 m from a bus stop (cf. Fig. 1.2). The traditional circular impact method is operationalised here. The complex implementation takes into account the real, existing network of roads and paths and, starting from a bus stop, determines all the shortest routes of a certain length that is approached 10 minutes on foot. This leads to a spider web of lines which, with regard to the target group, comprises individual streets (but no expressways), one-way streets or footpaths (cf. Fig. 9.26). Connecting the ends of this spider delineates the catchment area of the bus stop at the centre of the spider. The network is modeled as a weighted graph using computer science methods. Path algorithms provide the shortest paths. The convex hull of the terminal nodes defines the catchment area. The implementation in a map or the graphical presentation in a geoinformation system already illustrates the catchment areas and the coverage in an urban area very well.

As a further step, the buildings, which are modeled as areal objects and to which attributes such as building age or number of resident households belong, are determined within the catchment areas (so-called spatial intersection of two data layers by the tools of a geoinformation system). The evaluation of this joint spatial average finally provides the number of households sought. It is precisely this intersection and the calculation of a coverage rate as a proportion of households in the catchment area of all households that provide the added value compared to a purely visual representation (for the intersection of two data layers cf. Sect. 9.4.4).

Geoinformatics is to be understood in particular as the science behind the technologies according to Goodchild (1990, 1997) who differentiates the three large groups "Global

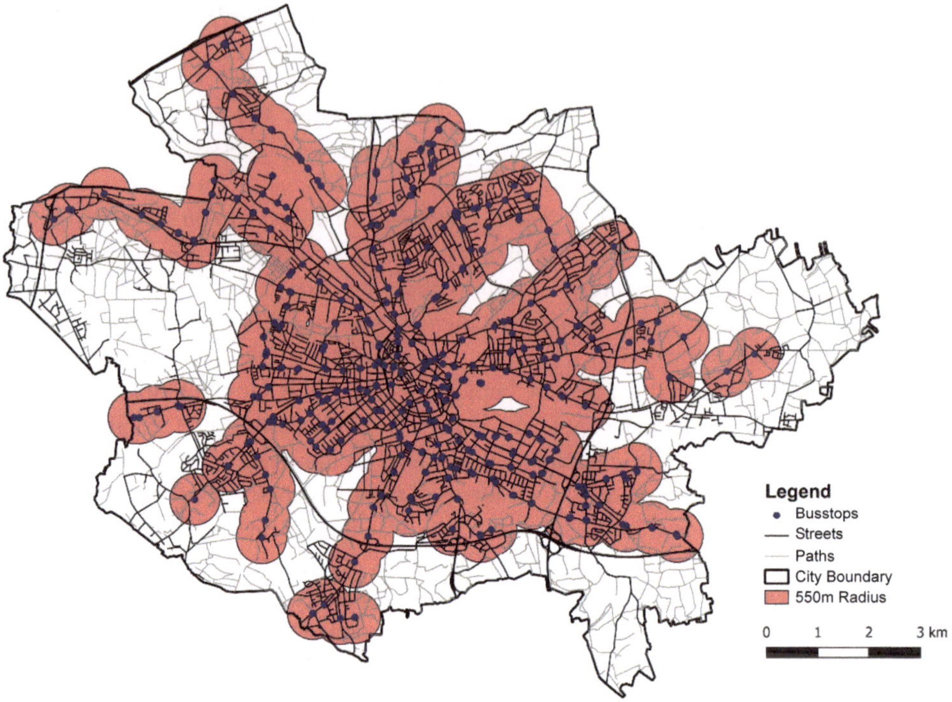

Fig. 1.2 Catchment areas of bus stops in the urban area of Osnabrück

Positioning System", "Geoinformation Systems" and "Remote Sensing" among the geographic information technologies. Knowledge of the geometric-topological modeling of geoobjects, the representation possibilities of geoobjects in coordinate systems and map network designs as well as geodetic basics are necessary prerequisites for the application of the technologies of geoinformatics (for the definition of geoobjects cf. Sect. 4.1). In general, methods of geodata acquisition as well as geodata in general including data quality and metadata as well as official basic geodata (in German: Geobasisdaten, cf. Sect. 5.5) must be addressed. The management of geodata requires database management systems, whereby knowledge of the conceptual modeling of geoobjects is necessary for the modeling of data structures. The representation and presentation of geodata require knowledge of graphic or cartographic presentation, for which the field of cartography must be extended with regard to the use of the new digital visualisation tools. In particular, basic knowledge of computer science is a prerequisite of geoinformatics. Accordingly, Fig. 1.3 shows the internal view of geoinformatics and names areas of study. Above all, it should become clear that computer science, i.e. not geography, geodesy or cartography, is the basis and that remote sensing and digital image processing are integral components of geoinformatics.

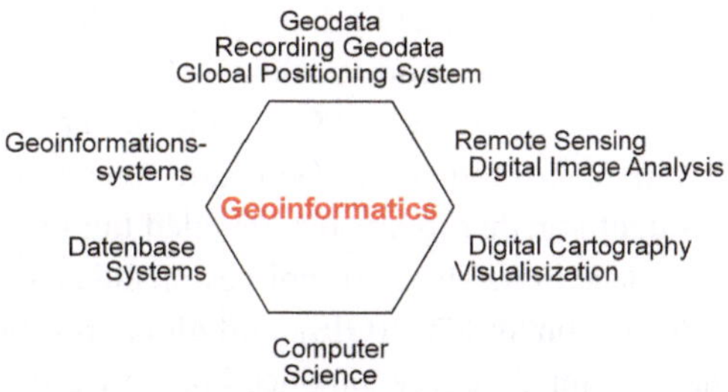

Fig. 1.3 Structure and central branches of geoinformatics

Although geoinformatics is a relatively young science, it is possible to clearly identify the main areas of work and research. The development has not been linear, but has taken several innovative leaps triggering the keywords "world wide web", "smartphone", "cloud" and "autonomous driving". A good overview of the changing issues and tasks of geoinformatics is provided by the contributions to the annual International Conference on Geographic Information Science (first AGIT now GIScience, cf. more recently Strobl et al. 2023). Current topics are:

- geoinformation systems (GIS) and derived specialist information systems
- 3D/4D Geovisualisation, Augmented Reality, Virtual Reality
- web-GIS and web service, mobile GIS
- business geomatics, location based services
- open source GI Software
- geogovernment and challenges of local government, development of spatial data infrastructures (SDI) and georeferenced services
- spatial data science and big data analytics
- digital earth – from real world to the digital twin
- free geodata, volunteered geographic information (VGI)
- geo-it in mobility, traffic management, logistics
- building information modeling (BIM), geoinformation technology (GeoIT) and industry 4.0
- climate change and climate impact, energy transition and geoinformation
- sustainable spatial planning, smart city concepts
- satellite based navigation, indoor navigation, positioning techniques
- analysis of high-resolution remote sensing data, laser scanner data and data from unmanned aerial vehicles

If the contents of the new discipline can be outlined quite clearly, two ambiguities must be clearly named, which concern the international terminology and the (German-speaking) courses of study:

Geoinformatics, Geoinformatics, Géomatique, Geomatik or Geographic Information Science all describe the same discipline. The name *Geomatics* goes back to the French photogrammeter Dubuisson, who describes the fourth chapter in his textbook (1975) as:

"Photo-Géomatique: La Cartographie Photogrammétrique Automatique". Already in 1990, the definition of geomatics by Gagnon and Coleman (1990 p. 378) anticipated the definition of geoinformatics: "Geomatics is the science and technology of gathering and using geographic information. Geomatics encompasses a broad range of disciplines that can be brought together to create a detailed but understandable picture of the physical world and our place in it. These disciplines include surveying, mapping, remote sensing, geographic information systems (GIS), and global positioning system (GPS)." However, this concept has remained largely unnoticed in the German-speaking world. Rather, geomatics is often understood there in a very limited way only as an intersection of geodesy and computer science. Accordingly, the introduction of the geomatics technician in Germany since 2010 is to be understood as a state-approved profession that continues the predecessor profession of surveying technician and cartographer. Against this background, geoinformatics has to be regarded as a comprehensive scientific discipline.

In the meantime, new Bachelor's and Master's degree courses have been created at technical colleges and universities in German-speaking countries, which have the terms "geoinformatics" or "geoinformation" in their names. It can be seen,

- that mainly courses of study from geodesy or surveying technology have adopted the name "geoinformatics",
- that departments of geography at universities are hardly involved in the creation of their own courses of study in "geoinformatics",
- that only very few independent geoinformatics courses of study exist at universities at all.

The contents and requirements of the studies and ultimately the performance characteristics of a geoinformatician are correspondingly heterogeneous. It must be noted that the situation in other countries is somewhat similar. But some research centers or faculties use GIS as geoinformation science directly in their names (e.g. Faculty of Geo-Information Science and Earth Observation, University of Twente, Department of Geography and Geoinformation Science, George Mason University, Fairfax Va, Department of Geography and Geographic Information Science, University of Illinois Urbana, Texas Center for Geographic Information Science, Department of Geography and Environmental Studies, Texas State University, San Marcos Tx) or far more often as part of a name (e.g. GIScience and Geoinformatics at the Dep. of Geography UC Santa Barbary Geography or the Geoinformation Science Lab at the Geography Department of the Humboldt University Berlin). Geoinformatics is currently still in line with traditions and methodologies of, above all, computer science and the already long-established spatial sciences such as geodesy and geography. In particular, the frequently cited spatial reference of information seems to be a constituent feature, although a strange ambivalent situation is now emerging that goes back to a rather simple use of this term:

On the one hand, due to recent web technology and development of the information society, a decline in the importance of spatial reference or location dependency can be observed. All information (including tools for navigation and orientation) is available anytime and anywhere, which has a considerable influence on the development of web services in particular. On the other hand, the importance of spatial reference is increasing significantly, there is an increasing geo-referencing of information and retrieval of "neighboring" information, e.g. via smartphones. The buzzword term is "connecting through location". Geoinformatics must address these new challenges. The value of geoinformation is recognised. The raw material "geodata" is the basis for spatial decisions in many areas such as real estate, radio mast planning, vehicle navigation, global environmental monitoring or spatial planning in general. Geodata and their valuation open up a high economic potential.

Sometimes the extent to which geoinformatics should be considered as part of computer science is discussed. It can be generally defined as follows:

Computer science is the science of systematic processing, storage, transmission and representation of information, especially automatic processing with the help of information-processing machines.

In this context, *systematic processing* is understood to be the planned, i.e. formalised and targeted solution of issues with the help of algorithms (cf. Sect. 2.4.1).

Due to the historical development, the diverse content outside of computer science, and especially due to the strong geo-components of this discipline, geoinformatics has established itself as an independent science, even if the relation to computer science is undisputed. It is not for nothing that "computer science" is at the base of the hexagon in Fig. 1.3.

1.2 Computer Science and Geoinformatics: Ethical Challenges

Information technology has a key role in recent developments that are affecting all areas of economy and society. This is particularly noticeable in automotive engineering. Whereas in the past development was primarily determined by engineering services, in future computer science will take their place. Hardware such as electric engines, which make gearboxes or exhaust systems superfluxes, will be available as standardised components. Software will control the drive, take over navigation and regulate autonomous driving. In general, mobility of the future will be determined by information technology.

Information technology will bring enormous changes that will also affect traditional personal interactions (increased growth in digital advisory services, e.g. in the banking and insurance sectors, e-learning). Endless new possibilities are still beyond our today's imagination, especially of so-called artificial intelligence. *Computer science* will make work considerably easier and also trigger great rationalisation effects. In addition to a reduction of jobs, new jobs will be created, so that it can be assumed that the use of these new technologies will have stabilising effects on economic and social developments. Particular reference should be made to the opportunities created by the new technologies associated with flexible working from home ("home office").

Computer science already has a profound impact on our society. Against this background, in 2018 the Presidium of the German Society for Computer Science (GI) adopted Ethical Guidelines in 12 articles, replacing the Guidelines first set up in 1994 and revised in 2004. Especially article 10 (social responsibility) aims at the ethical responsibility of computer scientists:

"The GI member shall contribute to the improvement of local and global living conditions through the design, production, operation and use of IT-systems. The GI member bears responsibility for the social and societal impacts of its work. It shall contribute to the socially acceptable and sustainable use of IT-systems through its influence on their positioning, marketing and further development" (translated Gesellschaft für Informatik 2023). A wide range of topics concern ethics in computer science:

- Cyberwar
- Cybercrime (i.e. computer crime and cybercrime)
- Computer games and gambling addiction
- Rights to own data and security of own data
- Copyright and ownership of software
- Computers and education
- eHealth and digital health record
- Privacy and anonymity

Geoinformatics must also face *ethical questions*. However, here too, it is impossible to overlook all the implications.

The concept of "*Digital Earth*", a digital replica of the entire planet, first appeared in Gore 1992 and was further developed in a speech at the opening of the California Science Center in 1998 (cf. Gore 1998). Already, the *Google Earth Engine* public data archive exemplifies this vision by providing vast amounts of georeferenced information. The archive includes more than forty years of historical imagery and scientific datasets that are updated and expanded daily (cf. Google Earth Engine 2023). It is already technically possible to envision that, through ubiquitous access to data, mobile end devices such as tablet computers directly on the spot will be able to complement the real view of, for example, a street with adjacent buildings with data and views such as the underground and, in real terms, invisible pipeline system (augmented reality).

The mobility app of the future, whereby the future has already begun in Vilnius, navigates, pays for the electronic train or bus ticket, calls and pays for the shared taxi, unlocks the rental car or the rental bicycle. This is certainly only a preliminary stage of future mobility concepts, which are only dimly recognizable by the keyword "autonomous driving".

Due to the exploding spread of smartphones and geo-apps, special ethical challenges arise for geoinformatics from a spatial perspective. Smartphones enable the constant localisation of its owner and (re)tracking of its spatial fingerprint. Google automatically detects the computer's location based on its IP address, location history if enabled, and recent reached locations. This technique has become popular in localizing traffic jams (cf. current traffic situation in Google Maps, after – consciously or unconsciously – activating location detection on Android devices). The integration of Android Auto by Google or CarPlay by Apple into so-called infotainment systems of motor vehicles was widely introduced in 2019.

Challenges to data protection and ethics arise from access to spatially detailed information. In China, it is already possible to identify persons on the street using cameras and facial recognition, and to automatically issue fine notices in the event of traffic violations. High-resolution satellite images, which are already available but are also planned for civilian use, can also be used to identify small objects on Earth. By recognizing faces or license plates and comparing them with databases, it will be technically possible to have knowledge of people and to store who is where and when. Dobson and Fisher (2003) labelled this vision “geoslavery”.

However, not everyone wants to share personal data. At the very least, there is a call for sensitive handling of these personal geodata. In particular, there is a need for a legal clarification of which data must be protected (cf. data protection problems with Google Street View, making faces and number plates unrecognizable, controversial discussion about the recording of single-family homes or smaller apartment buildings or farmsteads, recording of WLAN parameters during image recording).

References

Bartelme, N. (1989): GIS Technologie. Geoinformationssysteme, Landinformationssysteme und ihre Grundlagen. Berlin: Springer.

Bartelme, N. (1995): Geoinformatik-Modelle, Strukturen, Funktionen. Berlin: Springer.

Bartelme, N. (2005): Geoinformatik: Modelle, Strukturen, Funktionen. Berlin: Springer. 4. Ed.

Bill, R. (2016): Grundlagen der Geo-Informationssysteme. Berlin: Wichmann. 6. Ed.

Bill, R. u. M. Hahn (2007): Akkreditierung von GI-Studiengängen – eine neue Qualität in der Hochschulausbildung?. In: GIS-Zeitschrift für Geoinformatik 2007-4, S. 8–15.

Blaschke, T. (2003): Geographische Informationssysteme: Vom Werkzeug zur Methode. In: Geograph. Zeitschr. 91, S. 95–114.

Burrough, P.A. a. R.A. McDonell (1998): Principles of Geographical Information Systems. Oxford: University Press.

Campbell, J. and M. Shin (2011): Essentials of Geographic Information Systems. Open Textbook Library. https://open.umn.edu/opentextbooks/textbooks/67 (14.04.2023).

de Lange, N. (2020): Geoinformatik in Theorie und Praxis. Grundlagen von Geoinformationssystemen, Fernerkundung und digitaler Bilbverarbeitung. Berlin: Springer Spektrum.

de Smith, M., Goodchild M. a. P. A. Longley (2018): Geospatial Analysis: A Comprehensive Guide to Principles Techniques and Software Tools. London: Winchelsea Press 2021. 6ed. (cf. for the update and free online book 2022) (14.04.2023).

Dickmann, F. a. K. Zehner (2001): Computerkartographie und GIS. Braunschweig: Westermann. 2. Ed.

Dobson J.E. u. P.F. Fisher (2003) Geoslavery. In: IEEE Technology and Society Magazine: S. 47–52. https://pdfs.semanticscholar.org/c0e1/0fa50dfb89b571e7e9dd1817f165d50f4a0a.pdf (14.04.2023).

Dubuisson, B. (1975): Practique de la photogrammétrie et de moyens cartographiques derivés des ordinateurs, Editions Eyrolles. Paris.

Ehlers, M. (1993): Integration of GIS, Remote Sensing, Photogrammetry and Cartography: The Geoinformatics Approch. In: Geo-Informations-Systeme 6 (5), S. 18–23.

Ehlers, M. (2000): Fernerkundung und Geographische Informationssysteme: von der Datenintegration zur integrierten Analyse. In: Blotevogel, H.H. u.a. (Hrsg.): Lokal verankert – weltweit vernetzt. Tagungsbericht und wiss. Abhandlungen 52. Deutscher Geographentag Hamburg. Stuttgart, S. 586–591.

Ehlers, M. (2006): Geoinformatik: Wissenschaftliche Disziplin oder alter Wein in neuen Schläuchen? In: GIS 11, S. 20–26.

Ehlers, M. a. S. Amer (1991): Geoinformatics: An Integrated Approach to Acquisition. In: Processing and Production of Geo-Data. Proceedings, EGIS '91, Brüssel, S. 306–312.

Ehlers, M., Edwards, G. a. Y. Bédard (1989): Integration of Remote Sensing with Geographic Information Systems: A Necessary Evolution. In: Photogrammetric Engineering and Remote Sensing 55, S. 1619–1627.

Ehlers, M., D. Greenlee, Smith T. a. J. Star (1991): Integration of Remote Sensing and GIS: Data and Data Access. In: Photogrammetric Engineering and Remote Sensing 57, S. 669–675.

Ehlers, M. a. J. Schiewe (2012): Geoinformatik. Darmstadt: Wiss. Buchgesellschaft.

Gagnon, P. a. D.J. Coleman (1990): Geomatics: An Integrated, Systemic Approach to Meet the Needs for Spatial Information. In: Canadian Institute of Surveying and Mapping Journal 44-4, S. 377–382.

Gesellschaft für Informatik 2018 (2023): Ethische Leitlinien. https://gi.de/ueber-uns/organisation/unsere-ethischen-leitlinien (14.04.2023).

Fotheringham, A. S., C. Brunsdon a. M. Charlton (2000): Quantitative Geography: Perspectives on Spatial Data Analysis, London: Sage.

Goodchild, M. F. (1990): Spatial Information Science. Keynote Adress, 4th. Int. Symposium on Spatial Data Handling. Proceedings, Vol. 1, S. 3–12. Zürich: Dep. of Geogr.

Goodchild, M.F. (1992): Geographical Information Science. In: International Journal of Geographical Information Systems 6, S. 31–45.

Goodchild, M. F. (1997): What is Geographic Information Science? NCGIA Core Curriculum in GIScience, https://escholarship.org/content/qt5k52c3kc/qt5k52c3kc_noSplash_6e9f98376fef4ac654ecb7e66635df098.pdf (14.04.2023).

Google Earth Engine (2023): A planetary-scale platform for Earth science data and analysis. https://earthengine.google.com (14.04.2023).

Gore, A. (1998): The Digital Earth: Understanding our planet in the 21st Century. http://portal.opengeospatial.org/files/?artifact id=6210 (14.04.2023).

Gore A (1992) Earth in the Balance: Ecology and the Human Spirit. Boston: Houghton Mifflin.

Hennermann, K. (2014): Kartographie und GIS. Eine Einführung. Darmstadt: Wiss. Buchgesellschaft. 2. Ed.

Kainz, W. (1993): Grundlagen der Geoinformatik. In: Kartographie und Geo-Informationssysteme.

Grundlagen, Entwicklungsstand und Trends. Kartographische Schriften 1. S. 19–22. Bonn: Kirschbaum.

Kappas, M. (2012): Geographische Informationssysteme. Braunschweig: Westermann = Das Geographische Seminar. 2. Ed.

Longley, P.A., Goodchild, M.F., Maguire, D.J. u. D.W. Rhind (2005, Ed.): Geographical Information Systems: Principles, Techniques, Management and Applications. West Sussex: John Wiley & Sons. 2nd Edition, Abridged.

Longley, P.A., Goodchild, M.F., Maguire, D.J. u. D.W. Rhind (2015): Geographic Information Science and Systems, 4th Edition Hoboken, NJ: Wiley

Longley, P.A. (2003): Advanced Spatial Analysis. The CASA Book of GIS. Redlands. CA: ESRI Press.

Longley P A, Batty M, eds. (1997) Spatial analysis: Modeling in a GIS environment. J Wiley and Sons, New.

NCGIA (1990): Core Curriculum-Geographic Information Systems (1990). https://escholarship.org/uc/spatial_ucsb_ncgia_cc (14.04.2023).

Olaya, V. (2018) Introduction to GIS. This book is distributed under a Creative Commons Attribution license. https://volaya.github.io/gis-book/en/gisbook.pdf (14.04.2023).

Penzkofer, M. (2017): Geoinformatik: Von den Grundlagen zum Fachwissen. München: C.H. Beck. 2. Ed.

Strobl, J., Zagel, B., Griesebner, G. a. T. Blaschke (2023, Hrsg.): AGIT. Journal für Angewandte Geoinformatik 8-2022. Berlin: Wichmann.

Zimmermann, A. (2012): Basismodelle der Geoinformatik. Strukturen, Algorithmen und Programmierbeispiele in Java. München: Hanser.

Information Processing: Basic Concepts 2

2.1 Information, Message, Signals, Data

In computer science, "information" is of central importance. Likewise, in computer science, the terms message, signal, sign and date are associated with clearly defined contents that do not coincide with every day notions:

Information includes a message along with its meaning to the receiver.
Messages are a finite sequence of signals including their spatial and temporal arrangement, i.e. composed according to given rules.
Signals are elementary recognizable changes such as a sound, a facial expression, a flash of light, a change in color, a movement or an electrical impulse. A distinction is made between analog signals, which have a temporally/spatially continuous course (e.g. sound waves), and digital signals, which are short in time and can only assume a limited number of values, i.e. discrete values. While in analog signals the information is encoded by means of signal height and duration, in digital signals the information is encoded by means of signal number, signal distance and possibly signal duration. In a digital computer data is processed on the basis of discrete number representations, which are represented by two discrete and clearly separable signals (0 and 1) (cf. Sect. 2.5).

The messages listed in Fig. 2.1, which are composed of a structured sequence of signals, i.e. in this case changes in brightness on a white paper, initially are meaningless for the reader and receiver. It is only through the processing of this message by the receiver, which may include decoding, calculation and interpretation, that the message acquires meaning and becomes information for the receiver.

The Japanese character stands for the syllable "dai" meaning "big".

N. de Lange, *Geoinformatics in Theory and Practice*, Springer Textbooks in Earth Sciences, Geography and Environment,
https://doi.org/10.1007/978-3-662-65758-4_2

大 13 • • • ▬ ▬ ▬ • • •

Fig. 2.1 Examples of messages

The number 13 only has a meaning for the receiver if for example temperature data are to be transmitted in degrees Celsius (not 13 degrees Fahrenheit, 13 years or 13 dollars).

The last character sequence represents Morse signals. Then the signal sequence can be decoded as the letter sequence SOS. This message only becomes information for the recipient, if he knows the internationally recognised meaning of this letter sequence (call for help, "Save Our Souls").

For the representation of information, *characters* are mostly used, which are understood as an element from a finite set of objects (the character set) agreed upon for the representation of information. A linearly ordered character set is generally referred to as an *alphabet. Alphanumeric characters* originate from a character set consisting of digits, letters, and special characters (e.g., point, comma, parentheses). *Numeric characters originate* from a character set consisting of digits and those supplementary letters and special characters (+, −, decimal point) that are required for such a number representation. Numeric characters can be understood as a subset of alphanumeric characters. The sequence 49076 of alphanumeric characters is understood as a name with which no arithmetic operations can be associated. The sequence 49076 of numeric characters denotes, for example, the population size of a city, which can be added to another population size. These different *semantics* of characters is implemented in computer science or in programming languages by the data type concept (cf. Sect. 3.2).

Data are compositions of characters or continuous functions that represent information based on conventions. They are primarily used for processing or as its result, whereby processing includes the performance of mathematical, transforming, transferring and storing operations. Digital data consist only of characters, analog data only of continuous functions. According to areas of application, several classifications of data arise. A distinction must be made between at least:

- *Input data* provides the information necessary to solve a task, while *output data* describes the solution to the task.
- *Active data* such as program instructions control and monitor a work process, while *passive data* such as input data are processed in a work process.
- *Numeric data* includes digits and certain special characters such as the sign, while *alphanumeric data* is composed of any characters in the character set (i.e. letters, digits, special characters).

The basic terms of information processing, which include the terms mentioned here, have been defined in the nine parts of DIN 44300. With this terminology standard, the German Institute for Standardisation has attempted a German-language paraphrase or definition of central terms, which, however, sometimes appear quite long-winded. DIN 44300 has since been withdrawn and replaced by the ISO/IEC 2382 standard.

2.2 Machine, Computer, Program, Hardware, Software

A *automaton (automatic machine)* is generally understood to be a machine (i.e. a technical or mechanical device) that receives an input and produces an output depending on the input and the current state of the machine. A finite automaton only has a finite number of input options and states, and the terms automaton and finite automaton are usually used interchangeably. Well-known examples of automatons in everyday life are beverage vending machines or card dispensers (cf. Fig. 2.2). According to the above definition, they expect an input (insert money, select the desired product by pressing a selection button, press money return button), have different states (amount of money has been sufficiently entered, vending machine is ready) and generate an output (output of the product, return of the money, output of a signal tone).

Such vending machines, like all automatons, i.e. also like the computers and automatons for information processing to be discussed in more detail here, operate according to a central principle which describes the basic sequence of the technical functions of an automaton "input", "processing", "output" and is referred to as the *IPO principle* (cf. Sect. 2.3). Compared to these still illustrative examples, in computer science automatons are understood to be abstract mathematical models of devices that process information.

A *computer* is a machine that is controlled by programs. The use of different programs to solve different tasks makes a computer universally usable. Just the free, arbitrary programmability characterises a fundamental feature of a computer.

Computers, computer systems or digital computing devices can be divided into different categories based on performance or cost. The classic distinction is between microcomputers, minicomputers, mainframes and supercomputers. Among these, the class of microcomputers is called *personal computers* (PCs) and that of minicomputers is called workstations. *It* is these two classes of computer systems that are used for

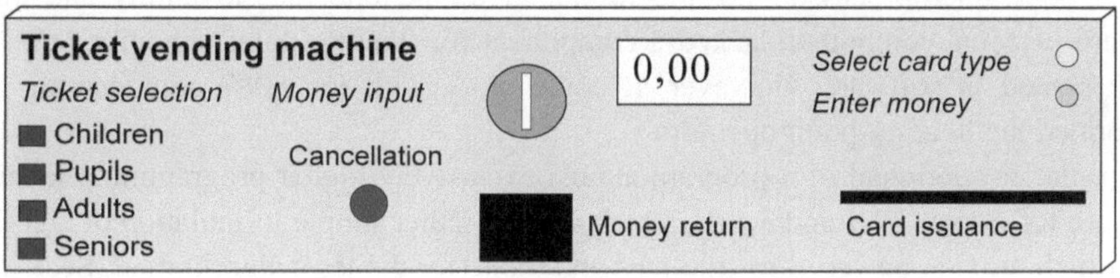

Fig. 2.2 Ticket vending machine for a swimming pool

applications in geoinformatics. However, the differences are not rigid and the transitions are not distinct.

A *program* consists of a sequence of instructions or execution rules in a syntax defined according to the rules of the programming language used for the input, processing and output of information (cf. Sect. 3.1). A program converts the steps of a generally formulated instruction into a programming language (for the term algorithm cf. Sects. 2.4 and 3.3). The individual steps of a program are usually executed one after the other (i.e. sequentially), whereby repetitions, i.e. so-called loops, or jumps can occur. Meanwhile, special programs also allow parallel processing of program steps.

Depending on the external conditions, especially on the inputs and the states of the computer, the dynamics of the program execution can be different for an identical program. Therefore, the program execution, i.e. the program together with the associated inputs (i.e. data), is defined as one *process.*

Software comprises all or part of the programs that can be used on a computer system. The programs enable the operation of a computer system and the solution of tasks with the aid of a computer system. Accordingly, a rough distinction must be made between system software and application software (cf. in more detail Sect. 2.7).

System software includes all programs that are required for the correct operation of a computer system (e.g. operating systems) and that perform program generation. *Application software* is the task-related and subject-specific software for solving specific user tasks (e.g. for word processing, accounting, simulation). The prefix "soft" is intended to make it clear that the software is an easily changeable components of a computer system.

The *hardware* comprises all or parts of the components and technical devices of a computer system (cf. Sect. 2.6). This includes above all the so-called processor, which executes the processes (i.e. the programs with the associated data). Hardware also includes the (internal and external) memories, the peripheral devices (e.g. printer, scanner) for input and output as well as the components of the network. The prefix "hard" is intended to make it clear that hardware is the physical material part of a computer system that are unchangeable (apart from the replacement of individual components).

Frequently, computer functions can be realised by both software and hardware (e.g. complex arithmetic operations or zoom functions on the screen). Here, hardware realisation is generally faster, whereas software realisation is more flexible. For example, Intel's 8086 microprocessor, i.e. the forefather of the Intel 80*86 family and thus of modern personal computers, lacked a component for floating point operations that were implemented in software. However, it could work with the 8087 coprocessor, which performed the floating-point operations.

For the development of a program almost exclusively higher programming languages are used (cf. Sects. 3.1.1 and 3.1.3), which allow a rather simple formulation of algorithms and which provide powerful instructions and tools (for the term algorithm cf. Sect. 2.4.1). Before being executed in the computer, these instructions must be translated into *machine code*, i.e. into binary-coded machine instructions that make up the so-called machine program (cf. in contrast the programming language Java, cf. Sect. 3.1.1). The execution of an instruction in machine code is ultimately carried out by several elementary operations in the hardware (e.g. circuits). These elementary operations are controlled by microinstructions, each of which forms a microprogram.

In micro-programmable computing devices, *firmware* refers to the set of all microprograms implemented in a processor. Thus, the common processors of personal computers have copy instructions consisting of microprograms with which data are copied from one memory cell to another memory cell. The prefix "firm" is intended to make clear that the microprograms can in principle be changed, but remain fixed over a longer period of time. As a rule, only the manufacturer of computers makes changes to the firmware. The firmware is counted neither to the hard nor to the software. It stands between the devices and the programs.

Overall, there is a distinct *software-hardware hierarchy in* terms of universality and user relevance.

2.3 Information Processing: IPO Principle

A computer converts input data into output data according to certain rules. These rules and instructions are communicated to the computer by a program. In doing so, the program implements the algorithm to be executed (e.g. calculating of the property size with known corner coordinates of the property, cf. Fig. 3.11) so that the working steps can be interpreted by the computer and processed step by step.

Figure 2.3 describes the *IPO principle of* information processing, which applies to all computer classes. Primarily for data backup and simplification of data input, the storage of data and programs on external storage devices is added as a further technical function in

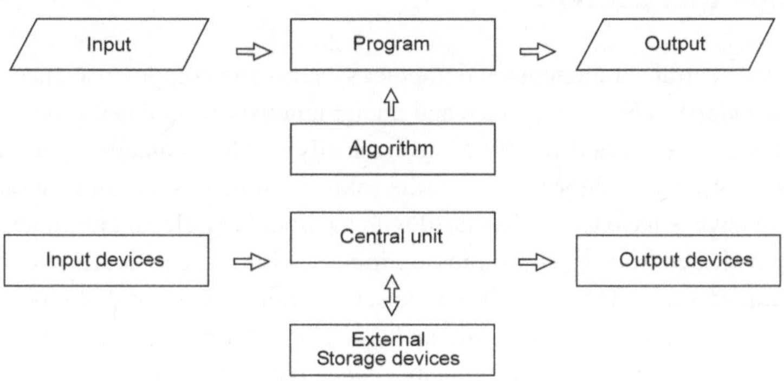

Fig. 2.3 IPO principle of information processing and schematic representation of a computer system

addition to input, processing and output. Data processing also includes simple read access to stored data in order to present it in information systems, for example.

The *central processing unit (cpu) of* a computer system always performs the data processing. However, the term central processing unit is often understood ambiguously (cf. Sect. 2.6 for more information on the structure of a computer system). For example, all functional units that are required for the interpretation and execution of (machine) instructions are referred to as the central processing unit or computer core. The terms central *processing* unit, processor core or *central processing unit* (*cpu*) are used interchangeably for the processor. The *central processing unit* is often used to refer to the processor together with the main memory.

While the processing of the data always takes place in the central unit, the input and output as well as the storage of the data is done with very different devices:

- devices for the input of data (e.g. keyboards, scanners, microphones),
- devices for the output of data (e.g. monitors, printers, loudspeakers),
- devices for storing data (e.g. hard disks, DVD drives).

Here it becomes clear that the term "computer", which is mostly used colloquially, hardly corresponds to the abstract definition of a digital computing system. In view of the diversity of components and the technical system structure, we should speak of a *computer system* that consists of many individual components, but always of the three main groups of central processing unit, input and output devices, and storage devices (cf. also Sect. 2.6). The term computer could then at best be used in a simplified or shortened form for all system components.

2.4 Computer Systems: Algorithms and Programs

2.4.1 Algorithm Concept

Algorithms are central components of computer systems and computer science, in which an *algorithm* is understood to be a precise and unambiguously formulated processing instruction so that it can be carried out by a mechanically or electronically operating machine (cf. also Sect. 3.4). Thus, algorithms describe solution strategies for application tasks. This term does not always have to be associated with a particularly elegant arithmetic operation, such as sophisticated sorting algorithms optimised with respect to arithmetic speed and necessary storage space. However, it is precisely computer science that has the important task of developing optimal algorithms and associated forms of evaluation. In a much simpler and more generally understandable way, the term algorithm encompasses the formalised description of a solution, which is then implemented by a program in a form that is understandable and executable by a computer. Thus, a program can be defined as the realisation of an algorithm (cf. Fig. 2.3).

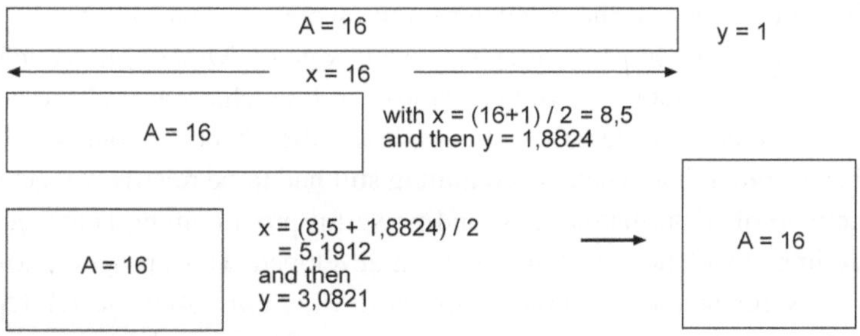

Fig. 2.4 Geometric illustration of the Heron process

A classic example is the *Heron method* for calculating an approximation of the square root of a number (cf. Herrmann 1992 pp. 36). The Heron method is a special case of *Newton's method*, which is generally used to calculate zeros of polynomials (cf. Schwarz u. Köckler 2011 pp. 192; here: calculation of the zero of a quadratic function $f(x) = x^2 - a$, cf. also Sect. 3.3.3.1 for the solution of a nonlinear system of equations). The iteration rule is:

$$x_{n+1} = \frac{\left(x_n + \frac{a}{x_n}\right)}{2}$$

The Heron method provides a simple geometric illustration (cf. Fig. 2.4). The starting point is a rectangle with the area a and the side lengths $y_0 = a$ and $x_0 = 1$. Now the side lengths are successively changed so that a square with the area a is approximated. The arithmetic mean of the side lengths x_0 and y_0 is taken as the improved value of the new side length x_1, from which the new side length $y_1 = a/x_1$ is then calculated. The procedure is continued until x_n and x_{n+1} no longer differ significantly, i.e. until: $|x_n - x_{n+1}| < e$ or $(x_n \cdot x_{n+1} - a) < e$ (e given lower bound).

In addition to the Heron or Newton method, numerical mathematics has other algorithms for determining the root or the zeros of functions. Accordingly, usually several algorithms are available for solving a task, but their efficiency varies. Computer science has developed methods to evaluate the efficiency of an algorithm (for the so-called complexity of an algorithm cf. Sect. 3.3.4).

When reproducing the instructions of the Heron procedure, a notation is used which already resembles a computer program (cf. Fig. 2.5).

2.4.2 Program Flow-Charts and Structure Diagrams

An algorithm that is to be developed for processing in a computer must be prepared in a way that is suitable for the computer. First of all, a rough concept of the algorithm can also be designed on paper or – in the case of less complex algorithms – immediately interactively with the help of the computer on the screen. In this case, the flow-chart of a smaller algorithm can be shown by graphs (so-called *flow-charts* or *program flow-charts),* which use standardised graphical symbols. However, flow-charts are only suitable for short algorithms. With increasing length and complexity they become confusing. Such flow-charts originate from a time when programming still had to be relatively machine-oriented and with the help of jump instructions, and powerful programming languages that allow more elegant implementations had not yet been developed. Thus, program flow-charts are used today only for smaller programs. *Structure diagrams* offer clearer representation possibilities (so-called *Nassi-Shneiderman diagrams*). These graphic means of expression belong to the so-called structured programming (cf. Sect. 3.1.4.1). In this illustration, a program is broken down into several structural blocks (cf. Fig. 2.5).

```
Input: a
x := a
y := 1
e := 0.00000001
if x < 1
        swap x and y
repeat
  x := (x + y) / 2
  y := a / x
as long as (x – y) >
e
root (a) = x
output (x)
```

	x	y
step 0	16	1
step 1	8,5	1,8824
step 2	5,1912	3,0821
step 3	4,1367	3,8678
step 4	4,0023	3,9977
step 5	4	4

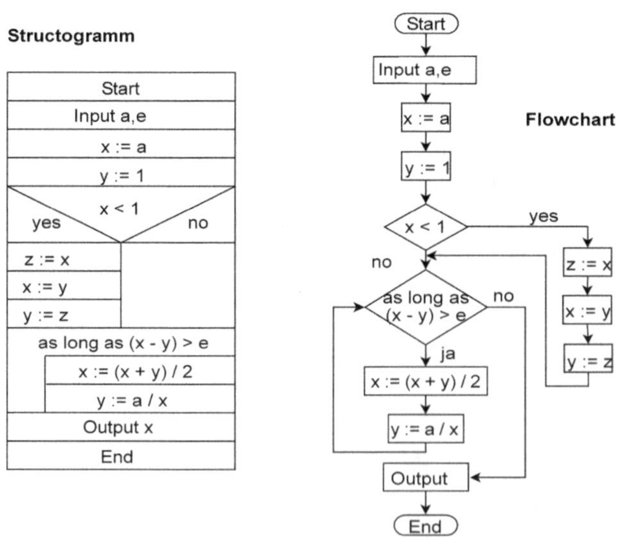

Fig. 2.5 Heron method as a structure diagram and a flow-chart

In computer science, such graphic tools can be used to make central processing rules and program sequences accessible to a wide range of users without revealing program details or requiring knowledge of a programming language. They can also be used to formalize and clarify general solutions and solution strategies and thus have a general significance for the formulation of research procedures and processes.

2.4.3 Levels of Algorithm Execution in a Computer System

After formulating the algorithm in a notation that is as formal as possible, the instructions for action must be converted into execution instructions of a program that can be processed by a computer. The algorithm is converted into a *program* by the process of *programming.* As a rule, a so-called higher-level or problem-oriented programming language is used (cf. Sect. 3.1.4), which is independent of the technical equipment of the computer, is (relatively) easy for the programmer to understand and has language elements from everyday language (e.g. "repeat" for a request to repeat).

However, the computer is not able to understand this language and to convert it into actions such as arithmetic steps. The program written in the so-called *source code,* i.e. in the comprehensible formulation by the higher programming language, must be translated into a form suitable for the computer and into computer instructions (cf. in more detail Sect. 3.1). The processor has its own instructions into which the program must first be transformed. Processors have different *instruction sets* depending on the manufacturer. It is true that programming could be done in *machine code* as in the early days of computer science. But this programming in binary code is difficult and error-prone. Thus, programs are now created in a higher programming language, which then have to be translated into programs in machine code. This process is carried out with the help of special programs. The program translated into machine code is then executed by the processor. The programs, which translate the source code into machine code, the compilers (cf. Sect. 3.1.2), are processor-dependent, since they are adapted to the instruction set of the respective processor. Part of this process is the "language check". Among other things, it is checked whether the source code has the correct structure. The result of a successful compilation is an executable file. Thus, machine code created for certain processors cannot be used on other processors (cf. machine programs for processors from Intel or for Apple computers or for smartphones). The great success of Intel's 80*86 processor family is due, among other things, to the fact that programs developed for an older type of processor can still be run on a younger type of processor.

2.5 Computer System: Presentation of Information by Bit Sequences

The entire Sect. 2.5 aims to clarify the basic principle: digitalisation of all information of the real (physical) world and processing of digital information in the digital world.

2.5.1 Digital World, Digitalisation

The automated processing of information in a computer system requires that the real world is completely mapped into the *digital world.* All information must be represented by digital data, which is ultimately composed only of two discrete and clearly separable signals (0 and 1). Technically these signals can be represented, for example, by the fact that a certain voltage is present in an electric circuit or not, or a certain location of a magnetisable layer on a rotating aluminium disk (as in classical hard disks) being magnetised or not. The basic approach of technical computer science, i.e. the technical realisation of information processing and data storage, consists in tracing back all information of the real (physical) world (e.g. logical values, texts, numbers, pictures, sounds) to the two basic states of electronic components by means of suitable coding. This transfer or transformation of the analog information of the real world into the digital world is called *digitalisation.*

Computer science provides methods for transforming logical values, numbers, extensive texts or images and graphics or complex audio-visual information, which is ultimately traced back to numerical values, into a digital world (cf. Sect. 2.5.3). Geoinformatics has the special task of mapping spatial data such as roads with information, for example about their length, into the digital world (cf. Chap. 5). In order for this digital world to be complete, it must be possible to process the digital signals, i.e. to "compute" them. With the so-called dual number arithmetic, mathematics provides the theoretical foundation of a closed and complete digital world of computation, which corresponds to our descriptive arithmetic world. Here, using the example of whole numbers, it is shown how the basic arithmetic operations can be carried out only with digital integers, i.e. with dual numbers, to which all complex mathematical processes can be traced (cf. Sect. 2.5.9). Finally, this digital arithmetic world, i.e. arithmetic with dual numbers, can be traced back to the linking of statements with the truth content "true" or "false" with the help of Boolean Algebra, so that ultimately mathematics also provides the basis for logic circuits (cf. Broy 1998 pp. 291).

2.5.2 Bit and Bit Sequences

The smallest possible unit of information is a *bit* (abbreviation for binary digit). A bit identifies the amount of information in an answer to a question that allows only two possibilities, such as yes or no, true or false, left or right. Such answers, which include only two possibilities, can be easily coded by two characters. Mostly the characters 0 and

1 are used. Mostly the answer to a question allows more than one bit of information. If four answers (north, east, south, west) are possible to the question of the prevailing wind direction, the information content of the answer is nevertheless only 2 bits. The original question can be broken down into two other questions, each allowing only two answers (yes = 1, no = 0). However, the intuitive conversion into the two questions does not lead to uniqueness:

- Is the prevailing wind direction north or east (yes/no)?
- Is the prevailing wind direction south or west (yes/no)?

	N	E	S	W
North or East	1	1	0	0
South or West	0	0	1	1

Only the translation into the following two questions leads to uniqueness:

- Is the prevailing wind direction north or east (yes/no)?
- Is the prevailing wind direction east or west (yes/no)?

	N	E	S	W
North or East	1	1	0	0
East or West	0	1	0	1

With 3 bits $2^3 = 8$ possibilities can be coded. Such codings are obvious if all possible combinations are considered from the outset: 110 (N), 100 (O), 010 (S), 000 (W), 111 (NO), 101 (SO), 011 (SW), 001 (NW). Thus, each additional bit doubles the number of possible bit sequences. In general, exactly 2^n possible bit sequences of length n exist.

With 5 bits presenting $2^5 = 32$ possibilities, already the 26 letters of the alphabet can be represented without distinction of upper and lower case. In a computer, however, far more characters are needed, i.e. text characters such as "+", "<", number signs such as "1" or small and capital letters, so that 7 or 8 bits are used in a computer to encode text characters. However, it is a waste of memory to represent a number like 123456 by using 7 or 8 bits for each number sign. So numbers and also logical values are expressed by special bit sequences.

In a computer system, large quantities of bit sequences are always processed, always taking groups of bits, either 8 bits, 16 bits, 32 bits or 64 bits. Always the length of a bit block is a multiple of 8. Therefore a group of 8 bits is also called 1 *byte*. The abbreviation 1 B is used for 1 byte. Binary addressing results in storage capacities of 2^n bytes. Since there were no unit prefixes for powers of two until international standardisation, the decimal prefixes with the factor $2^{10} = 1024$ instead of 1000 had become common (e.g. 1 kilobyte = 1024 bytes). The ISO standard IEC 8000013:2008 now provides new binary prefixes for the designation of powers of two and recommends to use the (usual) decimal prefixes also only in the decimal meaning. However, the new binary prefixes are still quite unusual:

Decimal prefixes			Binary prefixes		
Name	Symbol		IEC name	IEC symbol	
Kilobyte	kB	10^3 byte	Kibibyte	KiB	2^{10} byte
Megabyte	MB	10^6 byte	Mebibyte	MiB	2^{20} byte
Gigabyte	GB	10^9 byte	Gibibyte	GiB	2^{30} byte
Terabyte	TB	10^{12} byte	Tebibyte	TiB	2^{40} byte
Petabyte	PB	10^{15} byte	Pebibyte	PiB	2^{50} byte

2.5.3 Logical Values

The values “false” and “true” are logical values. They can be represented by means of exactly one bit.

2.5.4 Numbers

Numbers can be represented by bit sequences by converting them into the dual number system (from the Latin “duo” for “two”, also known as the binary system). This is based on a mathematical formalism that allows numbers to be represented in different *positional numeral systems* (place-value notation). In general, a system for representing numbers by digits is called a *positional numeral system,* if the value of a digit depends on the position at which it is written within the number. In this formalisation it is not essential, how many different digits exist at all. In general, all (positive) integers z can be represented in positional numeral systems in the so-called *radix notation* with B as the base and the following digits z_j (i.e. notated from the right):

$$z = z_0 \cdot B^0 + z_1 \cdot B^1 + z_2 \cdot B^2 + \ldots$$

The best known positional numeral system is the *decimal system,* i.e. the positional numeral system to the base 10. In the case of the number 135, the value of the number “3” results from the position within the sequence of digits, where each digit is the power of the number 10, i.e. the number of different digits.

$$135_{10} = 5 \cdot 10^0 + 3 \cdot 10^1 + 1 \cdot 10^2 + \ldots$$

Accordingly the *hexadecimal system* is defined, i.e. the positional numeral system to the base 16. Now with 16 digits: 0, 1, 2, 3, 4, 5, 6, 7, 8, 9, A, B, C, D, E, F. The figure 135 and 1DE (in the hexadecimal system) represent in the decimal system:

$$135_{16} = 5 \cdot 16^0 + 3 \cdot 16^1 + 1 \cdot 16^2 = 309_{10}$$

$$1DE_{16} = 14 \cdot 16^0 + 13 \cdot 16^1 + 1 \cdot 16^2 = 478_{10} \ldots$$

The *dual number system*, i.e. the positional numeral system with base 2, has exactly the two characters 0 and 1 and is suitable for representing numbers as bit sequences:

$$101_2 = 1 \cdot 2^0 + 0 \cdot 2^1 + 1 \cdot 2^2 = 5_{10}$$

$$1111011_2 = 1 \cdot 2^0 + 1 \cdot 2^1 + 0 \cdot 2^2 + 1 \cdot 2^3 + 1 \cdot 2^4 + 1 \cdot 2^5 + 1 \cdot 2^6 = 123_{10}$$

In the dual number system, both whole and fractional decimal numbers can be represented, so that the entire range of decimal numbers can be converted:

$$0.101_2 = 1 \cdot 2^{-1} + 0 \cdot 2^{-2} + 1 \cdot 2^{-3} = 0.625_{10}$$

Suitable conversion methods and algorithms exist for transformations between dual number and decimal numbers (cf. e.g. Gumm and Sommer 2013 pp. 17). Table 2.1 shows some examples.

The last line in Table 2.1 proves that the decimal number 0.1 can only be represented as an infinitely long, periodic, fractional dual number. Since in a computer system each digit of a dual number is stored in a memory cell and only a finite number of memory cells are

Table 2.1 Examples of dual numbers with their values in the decimal system

Dual numbers	Corresponding decimal numbers
0	0
1	1
10	2
101	5
11110101101	1965
0.1	0.5
0.01	0.25
111.111	7.875
11010.101	26.625
10101010.10011001	170.59765625
0.00011001100110011...	0.1

available, the number 0.1, which is very easy to represent in the decimal system, is stored "inaccurately" in computer systems. Thus, a computer calculates 0.1 * 0.3 "inaccurately" because the infinite decimal value cannot be represented. For example, technically often only 32 or 64 bits are available for a number. Even an extension to e.g. 256 bits does not solve the basic problem that in the end only finitely long dual numbers can be processed! However, not every task requires arbitrarily large numbers, and often different precision requirements exist for different calculations. Therefore in computer science different representations of numbers with bit sequences of different lengths (or their storages) are common and technically realised. Computer science provides the concept of *data types* for this purpose (cf. Sect. 3.2).

2.5.5 Text

Characters (text) are represented in a computer by encoding the alphabet and punctuation marks in bit sequences. Decimal digits, letters and special characters are individually specified by a bit sequence of fixed length. For the representation of all characters, only 7 bits are needed, which make up 128 different possibilities. 26 lowercase and uppercase letters, punctuation marks, special characters such as & and non-printable formatting characters such as line breaks make up about 100 characters of a typewriter keyboard (i.e. a standard computer keyboard without special characters). Several so-called *character sets* are in use, which define an encoding (cf. Table 2.2):

Table 2.2 Encoding of selected characters in ASCII and ANSI code

	ASCII		ANSI	
	Decimal	Dual	Decimal	Dual
1	49	0011 0001	49	0011 0001
9	57	0011 1001	57	0011 1001
@	64	0100 0000	64	0100 0000
A	65	0100 0001	65	0100 0001
a	97	0110 0001	97	0110 0001
Z	90	0101 1010	90	0101 1010
z	122	0111 1010	122	0111 1010
Ü			220	1101 1100
ü			252	1111 1100
ê			234	1110 1010
£			163	1010 0011

- the *ASCII code* (*American Standard Code for Information Interchange*, 7 bits per character, a total of 128 different characters, 95 of which are printable characters),
- the *ANSI code* (*American National Standards Institute, 8 bits per character*),
- the *EBCDI code* (*Extended Binary Coded Decimal Interchange Code*, 8 bits per character, used on mainframes, i.e. almost exclusively on IBM mainframes),
- *UTF*-8 (8-bit UCS (*Universal Character Set*) *Transformation Format*).

To represent and store e.g. the number 123 as text, i.e. by alphanumeric characters, 24 bits are required. However, a bit sequence of length 7 is sufficient as a binary number ($123_{10} = 1111011_2$). This example shows that numbers can be represented and stored more efficiently in the dual number system than in ASCII code (for internal storage of numbers cf. Sect. 3.2.2).

Unicode is an international standard that is intended to define a code for encoding all characters of all scripts (32 bits are possible per character, western character sets manage with 8 bits). Maintained by the Unicode Consortium, the standard defines 144,697 characters and covers 159 modern and historical scripts, as well as symbols and non-visual control and formatting codes. Unicode must also be translated into bit code. The ASCII-compatible Unicode Transformation Format UTF-8 is the most widely used encoding.

The *Universal Character Set* (UCS), the character encoding defined by ISO10646, is almost identical to Unicode. Characters encoded in the value range from 0 to 127 correspond exactly to the encoding in ASCII format.

It should be noted that the 8-bit character sets of the ISO/IEC 8859 family of standards are of central importance in the World Wide Web. The encoding is specified, for example, in the document header of an HTML page (specification charset = iso-8859-1, Latin-1 character set Western European e.g. with special characters like ä and ö, cf. Fig. 2.13).

2.5.6 Spatial Information

The digital representation of spatial information is a central concern of geoinformatics (cf. Sect. 5.2). Two fundamentally different approaches exist:

In the *vector model,* spatial information is resolved by points that are recorded in the form of coordinates, i.e. mathematically by vectors. The representation of point-like objects such as trees is unambiguous. Linear objects such as road side lines or road centre lines as well as very abstract units of investigation such as municipal boundaries are decomposed into sequences of vectors. The contour lines of spatial objects such as buildings or lakes are recorded (cf. Fig. 2.6). A curvilinear line is usually approximated by a polyline of straight line segments. In surveying, however, the connection is often also described by curves (e.g. specification of a curve radius, e.g. in the representation of road curves). The start and end points of these line segments are always *coordinates* in a reference system, i.e. so-called vectors *(vector model, vector graphics).* These coordinates are ultimately

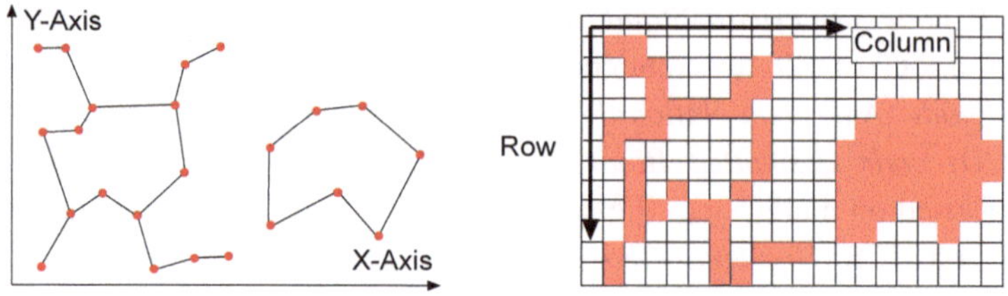

Fig. 2.6 Spatial information in vector and raster models

encoded as numerical values. The type of connection, i.e. color, width or shape of the line or also the design as a straight line or curve with a certain radius, is also encoded by numerical values.

In the *raster model,* spatial information is decomposed by a fine but rigid grid of small areas. The study area is represented as a matrix of so-called *pixels (pixel graphics* or *raster graphics).* Each individual spatial element is uniquely identifiable by specifying the number of rows and columns (i.e. "row and column coordinates"). Thus, B(7,27) identifies the pixel in the seventh row and 27th column. The information of each pixel is then also encoded by numerical values. Finally, as with the vector model, the numerical values are easily digitised.

In the technical transformation into the digital world a distinction must be made between recording of spatial information directly in the field on the one hand and from analog sources and templates (e.g. maps or aerial photographs) on the other:

In the first case, for example, (2D or 3D) position coordinates are available in digital form through classical geodesy procedures, through terrestrial or airborne laser scanning or through GPS-supported coordinate recording mostly via the recording devices. The data can also be entered manually using the keyboard. Digital aerial and satellite images provide spatial information as already digitally resolved raster data.

In the second case, spatial information from analog drawings or images must be transferred into the digital world. Recording (colloquially "digitising") with the help of a *graphics tablet,* a *digitising tablet* or more simply a *digitiser* is now obsolete (cf. Fig. 2.7), but this technique illustrates very well the underlying representation principle of spatial information in the vector model, i.e. the recording of (x, y) coordinates (cf. Sect. 5.2.1 and Fig. 5.3). With the help of a so-called digitising magnifier a specific point such as a house corner is identified in a map. A pulse is transmitted from the digitising magnifier to a grid system integrated in the tablet and two electrical conduction paths are activated, which provide the corresponding coordinates. This data input is associated with a high degree of accuracy, so that it was used in surveying, among other fields, and was fundamental to the digitalisation of cadastral maps.

In geoinformatics, *scanners* for the recording of raster data are of particular importance. In the low-cost office scanners based on the so-called CIS technology (CIS = Contact Image Sensor), many sensors arranged in a line next to each other on a carriage are passed directly under the document (cf. Fig. 2.8). For each line, the individual detectors measure

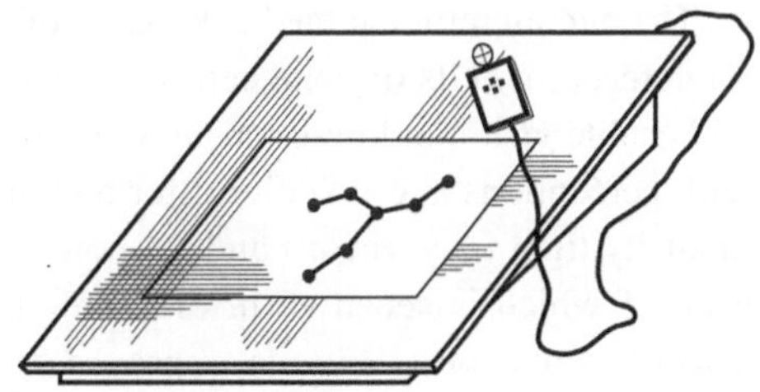

Fig. 2.7 Recording principle of coordinates with a digitising tablet

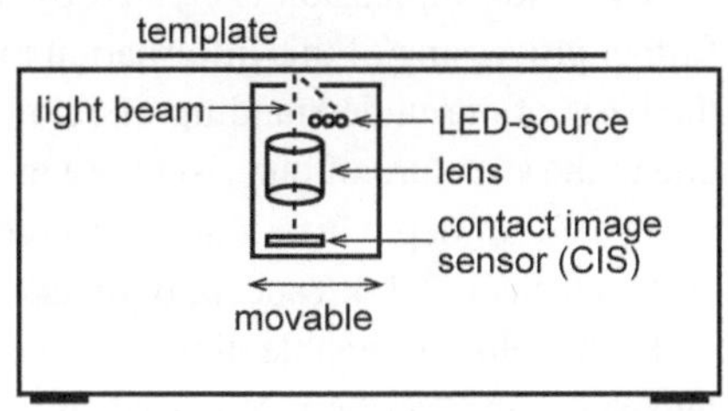

Fig. 2.8 Operating principle of a scanner

the intensity of the incoming light, i.e. the light reflected by the document, at a tiny point. The radiation detector converts the brightness of the light signal into voltage values, which are converted into digital signals by the processing electronics. In this process, a sensor can only differentiate between the different forms of brightness (monochrome sensors). In office scanners, three rows of LEDs are installed, which emit red, green and blue light one after the other at extremely short intervals. For each pixel, a sensor successively measures the intensity of the light reflected from the original or arriving at the sensor, i.e. blue, red and green light. The basis for this is the composition of visible light from three different wavelength ranges of the electromagnetic spectrum (cf. Sect. 2.5.7). A scanner with a device resolution of 600 dpi (dots per inch, approximately 240 dots per centimeter) requires almost 5000 sensors for a width of a DIN-A-4 page.

2.5.7 Color Information

Colors have a special meaning in geoinformatics. They are used for the visualisation of graphic information on monitors or printers, they are an essential component and information carrier in digital or analog presentations. However, the colors visible to the human eye are only a rather small part of the electromagnetic spectrum. This so-called visible light covers the wavelength range between approx. 400 nm and 800 nm (for the structure of the electromagnetic spectrum cf. Sect. 10.3.1). In remote sensing, on the other hand, there are a variety of sensors that can receive areas of the electromagnetic spectrum far beyond this range (cf. Sect. 10.4). However, the understanding of colors and the digitalisation of colors and visible light is fundamental to geoinformatics and in particular to remote sensing and image processing:

The human retina at the back inside of the eye is covered by tiny rod- and cone-shaped photoreceptor cells or cone cells or cones. While the rods have a much higher sensitivity and enable vision in low-intensity light (twilight and night vision), but only evaluate light-dark perceptions and no color information, the cones are responsible for vision in higher-intensity light (day vision) and for color perception. Three different types of cones exist, each of which is sensitive in different spectral ranges. They have a maximum spectral sensitivity of approx. 420 nm, approx. 530 nm and approx. 560 nm respectively. According to the maximum color sensitivity, the cones are also called blue, green and red receptors.

The color impression is created by different stimulation of these three sensor types and further processing of this information in the brain. This so-called three-component theory is the basis of our understanding of color vision and explains *additive color mixing,* which is due to the structure of our visual organ (additive color mixing as a physiological process in the eye; for in-depth information on color perception and its physical principles cf. e.g. Lang 1993, cf. Sect. 7.7.4, especially on additive and subtractive color mixing).

Each color of visible light can be decomposed into three primary colors and their intensities and can be reproduced by additive (or subtractive) color mixing of the primary colors (cf. Sect. 7.7.4). Thus, each color can be represented formally unambiguously as a vector in a three-dimensional color space spanned by the three primary colors (cf. color cube in Fig. 2.9). The *RGB color model* uses a three-dimensional Cartesian coordinate system whose axes represent the primary colors red, green and blue. Colors are each encoded by their three coordinates. Typically, each axis (i.e., each primary color) is scaled by an integer from 0 (no color) to the value corresponding to the strongest intensity. For example, brightest saturated blue is written in the different RGB notations as: (0,0,1) or (0%, 0%, 100%) or (0,0,255, i.e. within the range 0 to 255, 8-bit per primary color). With a scaling of 0 to 255 per primary color, a total of combinations $256 \cdot 256 \cdot 256 = 2^{24} = 16.777.216$ or colors can then be encoded and stored. Thus, a true-to-life reproduction is given, which explains the name true color (or truecolor). Overall, in computer science, colors are represented by numerical values and then by bit sequences:

- For (approximate) true color representation, each pixel is represented by 3 bytes, encoding 2^{24} color combinations.
- To represent grades of a single color, 1 byte is needed, so $2^8 = 256$ grades per pixel are possible.
- Only 1 bit per pixel is required to represent a black and white image, since only two states are encoded.

Fig. 2.9 RGB color model

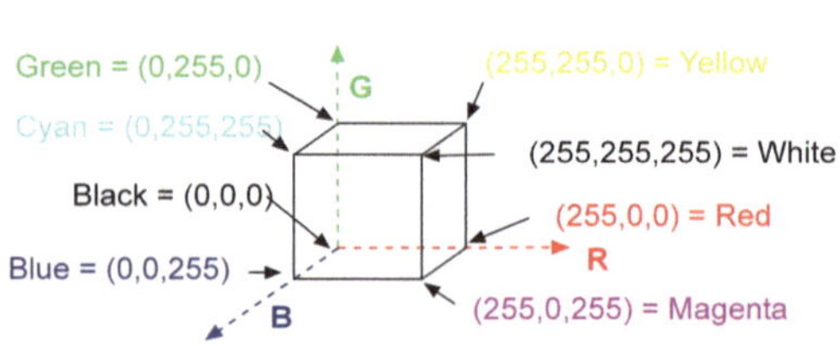

According to this grading, the representation accuracy of a color is reflected by the so-called *color depth,* which is understood as the maximum number of displayable colors. The color depth is given as an exponent of the number 2. Often 8-bit color and 24-bit color have the same meaning. The 8-bit refers to the grading of each red, green, and blue primary color, while 24-bit means all three $2^{24} = 2^8 \cdot 2^8 \cdot 2^8 = 16.777.216$ color combinations.

2.5.8 Sensor Data

Measurement technology has developed many sensors in order to automatically record very different characteristics and properties and to digitally display the analog measurement results. For example, in the widely used three-cup anemometer, the wind speed is measured with the help of three horizontally aligned spherical cups, which are mainly fixed crosswise on the vertical axis of the cup anemometer. The revolutions of this cup star per second or minute are converted into a speed value, which is usually represented as length per unit of time (e.g. m/s). Accordingly, many sensors exist that detect different conditions, such as humidity or temperature, gas concentrations, pH values, the sound level or radiation values, and ultimately provide numbers that can then be easily digitised. The basic principle, i.e. converting analog signals from a sensor into digital numerical values *(analog-to-digital conversion),* can be found in almost countless forms.

2.5.9 Dual Numbers Arithmetic

The arithmetic of dual numbers is fundamental to understanding how to process information encoded by bit sequences. A very elementary calculation is presented, which is the basis of all calculation operations in computer systems (cf. Gumm and Sommer 2013 pp. 17). These computing steps are implemented in hardware, in particular through the technical implementation of transistor circuits.

The addition is defined for the simplest case of the two dual numbers 0 and 1 by:

$$\begin{aligned} 0 + 0 &= 0 \\ 0 + 1 &= 1 \\ 1 + 0 &= 1 \\ 1 + 1 &= 10 \end{aligned}$$

The *addition of* the two dual numbers is done in the same way as the addition of decimal numbers. A carry over at one digit position is added to the next higher digit position.

			1	0	1	0	1	0	42
	+	1	1	0	1	1	1	1	1111
Carry	1	1		1	1	1			
Total	1	0	0	1	1	0	0	1	153

Somewhat more complex than addition is the *subtraction of* two dual numbers. Here, the subtraction is traced back to the addition: The subtrahend is not subtracted, but the negative subtrahend is added. In the dual number system, negative dual numbers are required for subtraction, which are represented by complement formation. In an intermediate step, the so-called *ones' complement* is formed first. The positive dual numbers remain unchanged, while the ones complement of a negative number is formed by exchanging the digits 0 and 1 in the corresponding positive number for their complementary digits 1 and 0.

Decimal	Binary	8-Bit representation	Ones' complement
13	1101	00001101	11110010
43	101011	00101011	11010100
56	111000	00111000	11000111

However, the representation of whole dual numbers in ones' complement has the great disadvantage that there are two dual numbers for the number zero: 0000 0000 (+0) and 1111 1111 (−0). For this reason, the so-called *two's complement* is introduced. The positive dual numbers remain unchanged here too, while a negative number is first transformed as a dual number into its ones' complement and then 1 is added.

Decimal	Binary	Ones' complement	Two's complement
13	1101	11110010	11110011
43	101011	11010100	11010101
56	111000	11000111	11001000

Subtraction is then performed by adding the two's complement. The calculated number must be further calculated depending on the type of overflow. In the first example, which calculates 56 + (−13), an overflow occurs. The final result is then a positive number, namely the dual number without overflow (here: 0010 1011):

		0	0	1	1	1	0	0	0	56
	+	1	1	1	1	0	0	1	1	−13
Total	1	0	0	1	0	1	0	1	1	43

In the second example, which calculates 13 + (−56), no overflow occurs. The end result will be a negative number:

		0	0	0	0	1	1	0	1	13
	+	1	1	0	0	1	0	0	0	−56
Total	0	1	1	0	1	0	1	0	1	−43

The final result is not yet available. The calculated two's complement 1101 0101 must be processed further. By "taking back" the addition of 1, a dual number is produced which is to be regarded as a ones' complement: 1101 0100. Then the complement calculation is to be taken back: 0010 1011. The decimal number (without sign) is then: 43. Since it is already known that the final result is a negative number, the result is: −43.

For the *multiplication* and *division* of multi-digit dual numbers there are quite simple calculation procedures, which are analog to the procedures in the decimal system. The multiplication of a dual number by a 1, which is in the ith position of a multi-digit dual number, is equivalent to a "shifting" of the dual number by (i-1)-positions and "filling up" with 0.

```
110  •  1011 (i.e. 6 • 11)      1000010  :  110  =  1011
      110000                     110
           0                      1001
        1100                       110         i.e. 66 : 6 = 11
         110                        110
     1000010   = 66                 110
```

In computer systems multiplication is also performed as repeated addition. The continued subtraction, which however can also be traced back to an addition, replaces the division of whole numbers. Thus, concerning the technical realisation, the arithmetic unit of a processor can be designed very simple, as it only needs to be able to perform additions. As shown here, all arithmetic operations can essentially be traced back to the addition of dual numbers. However, modern processors have other arithmetic calculators besides an adder (a digital circuit that performs addition of numbers).

The techniques of dual number arithmetic explained here for integers can be transferred accordingly to fractional dual numbers, which are stored in normalised representation with *mantissa* and *exponent* (cf. Sect. 3.2.2.3). Mantissa and exponent are to be considered separately. For example, before an addition, the magnitude of the exponents must be compared, and the mantissa belonging to the number with the smaller exponent must be adjusted accordingly. The following example in the decimal system illustrates (with a calculation accuracy of three digits!) the calculation principle:

$$0.123 \cdot 10^3 + 0.456 \cdot 10^5 = 0.001 \cdot 10^5 + 0.456 \cdot 10^5 = (0.001 + 0.456) \cdot 10^5 = 0.457 \cdot 10^5$$

The arithmetic, which at first seems complex and cumbersome, has considerable advantages when it is carried out with an automatic calculator:

The numbers are represented as bit sequences that are technically easy to implement.

The arithmetic operations are reduced to basic operations that are technically simple to implement, such as addition and forming complements.

2.6 Structure of a Computer System: Hardware

A computer system is made up of very different components that implement the *IPO principle* of information processing (cf. Fig. 2.3). The basic components are always input and output devices, memory devices and (in the case of personal computers or workstations) above all a motherboard on which various electronic components are located, which are connected by several electrical conduction paths. Many components are permanently installed on the motherboard. In addition, some slots connect other components to the motherboard. This structure is basically the same for personal computers or workstations, notebooks, tablet computers or smartphones (cf. Fig. 2.10).

The *processor (cpu, central processing unit)* is the central processing component of a computer system, as it can execute programs. In addition to the so-called control unit, a processor has the components required for information processing, such as the arithmetic logic unit (alu), which must performs at least addition, negation and conjunction (so-called logical "and" operation), since all arithmetic and logical functions can be traced back to these minimal operations.

The *central memory* is used for temporary storage of the program and intermediate results (RAM = Random Access Memory). When the computer is switched off, this

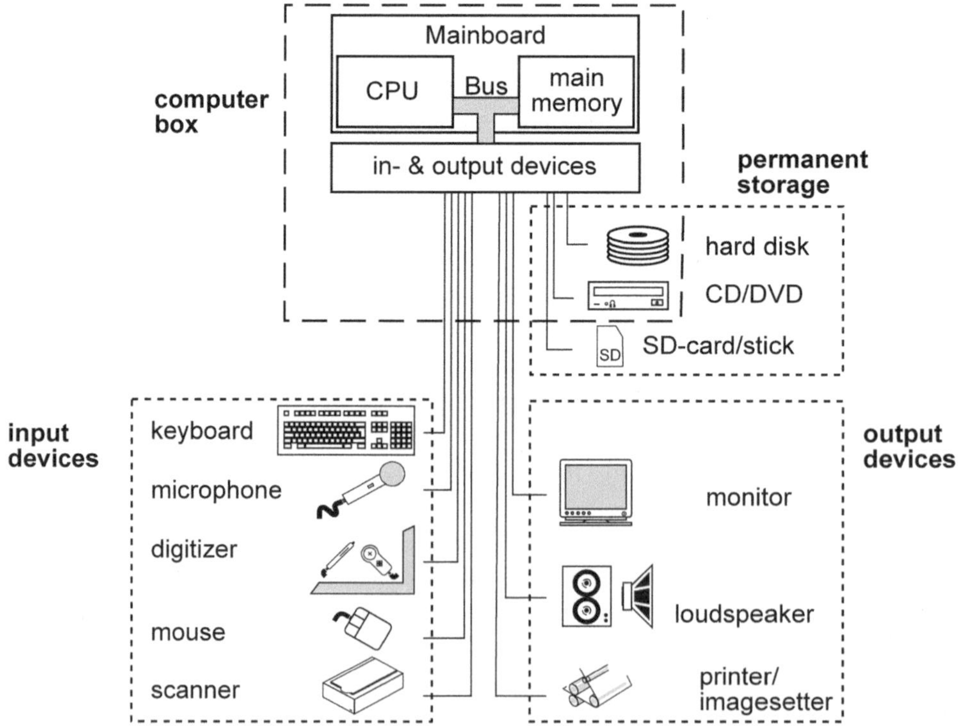

Fig. 2.10 Structure and selected components of a computer system

information is lost. It must therefore be stored or backed up outside the main memory in permanent memories. In contrast to permanent memories, which are used for the permanent storage of programs and data, the main memory has extremely short access times. The memory cells can be accessed directly.

The operation of *permanent storages* such as hard disk drives or DVD drives is controlled with the help of special processors (so-called input/output controllers, i/o controllers for short), which control the functions of these mass memories and establish the connection to the processor. In the meantime, so-called solid-state drives (SSD), i.e. non-volatile memories that are implemented by semiconductor components, are used. They serve as a replacement for the previously common hard disk drives, which are based on magnetisable disk technology. The name "drive" is misleading, as because an SSD has no moving parts.

Information is transported via electrical conduction paths that connect the various components on the motherboard. These lines are commonly called a *bus* or *bus system.*

Monitor and printer belong to the standard output devices of a computer system. Geoinformatics also uses these standard devices to output spatial, i.e. two-dimensional information such as graphics, images or maps. For the output of large-format graphics, such as construction drawings or maps, so-called plotters play an important role, which can be roughly divided into vector and raster plotters. Vector plotters represent the output device corresponding to digitisers. Here, drawing pens are moved across the paper so that lines appear as solid lines rather than being broken up into several raster dots. However, most modern plotters, which are used for tasks in geoinformatics, work on the principle of inkjet printers (besides laser and photo plotters). The individual characters are made up from a grid of dots, which can be very small and close together and determine the resolution and sharpness of the text output or the graphic representation.

2.7 Structure of a Computer System: Software

2.7.1 System Software and System-Related Software

The operation of a computer system is only made possible through the software, which is divided into system and application software. The main component of the *system software* is the *operating system,* which, depending on its efficiency, enables various operating modes and forms of use of the computer system and which primarily regulates the execution of the user programs. The system software also includes the programs of the programming languages, which translate the programs, which are usually written in a higher programming language, into machine code to be executed by the computer system. The *system-related software* includes the area between system and application software that cannot be clearly separated. System-related denotes the proximity to the processor architecture and the interfaces of the operating system. This can include programs that perform administrative, development and monitoring tasks:

- programs to support software development (i.e. integrated development environments, cf. Fig. 3.3), programming tools and program libraries (collection of programs to be integrated into user programs),
- programs for (statistical) recording of computer utilisation for resource optimizing,
- programs for monitoring and, if necessary, remove of so-called computer viruses.

It is controversial whether so-called browsers (cf. Sect. 2.8.3 and Fig. 2.13) belong to the operating system or to the system-related software. Compilers can be counted among the operating systems, since their performance depends on the operating system, or they can also be assigned to the software development tools (and thus to the system-related software).

The operating system of a computer comprises all the software required to perform the entire computer operation and thus, in particular, to execute the application programs and to control the most important peripheral devices such as keyboard, monitor, memory drives and printer. Among other things, the operating system forms the central interface between the user and the computer system. In addition to other tasks it regulates communication between users (e.g. input of control commands, starting of programs) and the computer (e.g. output of the operating status, of results or error messages during program execution).

2.7.2 Business, Individual and Standard Software, Apps

The software ultimately makes a computer system universally applicable. The applications software is used to solve user and task-specific issues. In general, a distinction can be made between business software and individual and standard software. *Business software* is tailored to the special requirements of individual lines of business such as construction, trade, commerce, banking or tax consultancy with mostly very specific tasks. *Individual software* is specially created for a specific application and is based on the specific needs of a user. In most cases, special software solutions are programmed according to the specifications of a user or after the development of a requirements specification. However, user-specific adaptation of standard software can also be carried out, which is usually less costly than very complex individual programming. *Standard software* is (largely) independent of the requirements of a particular business. This includes so-called office packages, which include programs for word processing, spreadsheets or presentations.

The advantages of standard software compared to individual programming are primarily cost advantages, immediate availability and direct use, a greater possibility of use for different areas of application in a company or an authority (synergy effects mainly due to widespread program knowledge), simpler operation compared to special programs and a large range of training opportunities. User forums on the WWW, which only arise with standard software, offer a diverse exchange of experiences between several users. Disadvantages of standard software compared to individual software are above all conformity problems with the special requirements of a user, less optimal operational behaviour for an individual application (e.g. more cumbersome operation, longer computing or reaction times, superfluous program modules at the expense of clarity), adaptation and interface problems to individual program systems. However, the advantages of individual software mentioned do not always have to exist. For example, standard software which is aimed at a wide range of users can have a generally understandable, simple user interface and operation that does not require any specialists. Individual programming, however, can very well achieve these goals as well or even better.

The so-called *software risk* and *software quality* have not yet been listed. The risk of having to manage alone and without the support of the manufacturer in the event of problems or requests for changes or enhancements tends to be lower with standard software. Software maintenance by the software manufacturer is more intensive with standard software, but this can also lead to a series of rapid (chargeable) updates that a user does not (always necessarily) need, but still has to implement in order not to lose support or not to buy expensive new ones. Due to the greater experience of the programmers of a manufacturer of standard software, a higher quality tends to be expected. However, this tendency does not always have to apply in individual cases. Custom software can usually be tested in more detail with regard to the specific application.

The increased use of mobile devices such as tablet computers or smartphones has led to the explosive spread of associated application software, which is referred to as *mobile app* (or only *app*) and which works with or without access to the WWW outside a browser (cf. Sect. 3.1.8). Platform-dependent apps (so-called native apps) only work under one operating system and are (usually) specially adapted to this platform through access to platform-specific hardware and software functions (e.g. camera, microphone, GPS or acceleration sensor). Almost all conceivable areas of use are covered, ranging from office, fitness and leisure applications to any fun applications. These apps are mostly available in the respective app stores for free or for little money.

2.7.3 Application Software in Geoinformatics

Geoinformatics, like any other information processing, uses standard software and often adapts it to user-specific tasks. Furthermore, the software specific to applications and tasks in geoinformatics essentially falls into four major categories:

- cartography, presentation and visualisation systems,
- database management systems,
- geoinformation systems,
- software systems for remote sensing and digital image processing.

Of central importance are geoinformation systems which also enable database management and presentation. Geoinformation systems are usually standard programs that can be used for very different tasks such as the management of parcels of a real estate cadastre or in geomarketing to optimize the sales area. Especially the user-specific adaptation of geoinformation systems in the environmental sector is an important task of geoinformatics.

2.7.4 Proprietary, Open Source and Free Software

Proprietary software is usually created by a software company and passed on to the user in the form of a paid license, who does not acquire any software but is merely granted rights of use. These rights are sometimes severely restricted. In most cases, only one single backup copy may be made. Resale is not legally permissible or impossible, since the user does not own the software. The user cannot make changes to the software because the source code is not open. He often does not know which specific algorithm is being used. Proprietary software generally offers user-friendly and simple installation, good documentation and, as part of its warranty, high reliability, professional support and regular updates.

Free software has gained considerably in importance in recent years. A large number of free alternatives to the established proprietary software systems are available via the WWW.

The open source and free software movements is oriented towards "freely" accessible software. It should be emphasised in advance that *"open source software"*and *"free software*"are not synonymous and, above all, that the word "free" should not be used in the sense of "free of charge". Thus, "open source" stands rather for technical and economic aspects and emphasizes the developer's point of view, since open source enables further developments by disclosing the source code and the underlying algorithms and should ultimately lead to better as well as less expensive proprietary software. In contrast, the free software movement, in particular the *Free Software Foundation*, generally rejects proprietary software for social-ethical reasons, since in their opinion software should be transparent, verifiable, free of charge and freely accessible, i.e. common property.

A first step towards the development of free software was the GNU project in the early 1980s, which aimed to develop a free, UNIX-like operating system (GNU as a recursive acronym for "GNU's not UNIX"). In 1985 the Free Software Foundation (FSF) was founded to give the GNU project a formal framework. It defines four freedoms, which are basically freedoms of users and developers, respectively, concerning the development of software. The freedom (cf. GNU 2023a).

1. to run the program for any purpose,
2. to study how the software works and adapt it to your own needs,
3. to redistribute copies of the program,
4. to improve the program and to publish these improvements.

Free access to the source code is a necessary prerequisite for freedoms (2) and (4). Thus, according to the Free Software Foundation, free software encompasses more than public accessibility of the source code, which does not yet have to include the possibility of using, modifying or redistributing the source code.

In contrast to the previous fuzzy use of the term "open source", the open source definition of the *Open Source Initiative (OSI)* provides a set of criteria for software licenses that is very similar to free software, but avoids misinterpretations of the word "free". According to the OSI's open source definition, a license for open source software requires:

1. free redistribution of the software, i.e. the license shall not restrict the redistribution of the software
2. available source code, i.e. the source code must be freely available to all users
3. derived works, i.e. new software developed from the source software and its distribution under the same license as the base software
4. integrity of the author's source code, i.e., the license may restrict source-code from being distributed in modified form only if the license allows the distribution of "patch files" with the source code for the purpose of modifying the program at build time. The license must permit distribution of software built from modified original source code. The license may require derived works to have a new name or version number.
5. no discrimination of persons or groups, i.e. use must discriminate against any person or group of persons
6. no discrimination against fields of endeavor, i.e. the licence must not restrict use to individual purposes
7. distribution of license, i.e. the license may be used immediately without purchasing other licenses
8. license must not be specific to a product, i.e. the license must not refer to a specific software distribution
9. license must not restrict other software, i.e. the license must not, for example, insist that all other programs distributed on the same medium must be open-source software
10. technology-neutrality, i.e. the provision of the licence must not exclude any technology or form of distribution.

The OSI's open source definition is a guideline for evaluating software licenses (cf. Open Source Initiative 2023).

In order to implement the free software idea several formal legal software license models have been developed that are intended to ensure the freedoms of a software or its type of use. The most widespread is GNU-GPL, the GNU General Public License, which primarily includes the so-called copyleft principle. This regulation (also called "share alike") guarantees that free software always remains free software and thus, freedoms of distribution may not be restricted. This means that programs developed from software under the GPL may also only be distributed under the terms of the GPL. This licensing model leads to conflicts when GPL-licensed software is to be integrated into proprietary programs, e.g. as a program library. Against this background, the GNU-LGPL, GNU Lesser General Public License, was developed, which allows the use of LGPL-licensed,

Table 2.3 Creative Commons licences (cf. Creative Commons 2023)

	Short form	The name of the author must be given	The work must not be changed	The work must not be used commercially	The work must be passed on under the same license after changes
	by	x			
	by-sa	x			x
	by-nd	x	x		
	by-nc	x		x	
	by-nc-sa	x		x	x
	by-nc-nd	x	x	x	

free programs also in proprietary programs (for more GNU licensing models cf. GNU 2023b).

In addition to the GNU licensing models similar models exist, of which those of the Creative Commons are of greater importance (cf. cf. Table 2.3). *Creative Commons (CC)* is a non-profit organisation dedicated to supporting the publication and distributing of digital media content in general, i.e. not only to support software, through various standard licence agreements. Currently, six different licenses are offered, which graduate the freedoms to share and modify (cf. Creative Commons 2023):

The simplified view that the use of free software is also completely free of charge must be corrected. The implementation of free software in a municipality or in a company is usually carried out by a service provider who very often has to make individual software adjustments or further developments on the basis of free software so that the free software can be integrated into the existing software environment. Business models in the context of free software aim at training and general customer support in addition to technical adaptation and maintenance. Thus, switching from formerly proprietary to free software can be more costly than extending a license agreement, at least in an initial phase. Cost savings can result in the medium term, since any number of licenses can now be used and (annual) license fees are no longer apply.

2.8 Networks and Connectivity

2.8.1 Definition and Distinguishing Features

A computer network is the totality of conductor cables, exchanges and subscriber facilities, all of which are used for data communication. Several independent computer stations or workstations are connected and can jointly access databases, exchange data and use resources such as printers or backup devices.

With regard to the hardware, a computer network consists at least of the conductor system, which can usually be constructed from coaxial cable, also from fibre optic cable or from a wireless connection e.g. via radio (e.g. Bluetooth), infrared or satellite, as well as the connected computers, which must have suitable network adapters. A common network type for local networks standardised by the IEEE (Institute of Electrical and Electronics Engineers) is the so-called *Ethernet,* which offers transmission speeds of 10 MBit/sec. or 100 MBit/sec. (Fast Ethernet) or 1000 MBit/sec. (GBit-Ethernet) up to 10 GBit/sec. can be achieved. With regard to software, a computer network consists of special control software or a suitable operating system that controls access to the computer network and manages the resources (including connected devices, but also software). Networks can be classified according to several aspects:

According to the size of the network or physical distance of the computers:

In a *local area network (LAN)*, the computers are integrated at one location (building, company premises). In *wide area networks (WAN)*, the computers are located far away from each other. Wide area networks can also connect several local computer networks. With regard to local computer networks, a distinction must be made between wired networks and networks in which data is transmitted wirelessly (*WLAN, wireless local area network*). So-called WLAN routers are very often used, which are connected to a local area network or another wired data network (e.g. telephone network) and, as so-called wireless access points, establish a radio connection e.g. to mobile end devices such as notebooks. The data transfer rates for WLANs are generally significantly lower than for a LAN. However, there are considerable difficulties in comparing the data transfer rates, since different technical standards apply for both WLANs and LANs (for LANs Ethernet, Fast Ethernet, Gigabit Ethernet), whereby the cabling (copper cable according to various categories such as CAT 5 or fiber optic cable) and the hardware components (e.g. hubs, switches, Gigabit Ethernet switches) are of great importance. For example, an optimally configured WLAN based on the widely used IEEE 802.11ac standard can be faster than an old LAN. Current routers achieve transmission rates of up to 1300 Mbps. In buildings with solid and thick walls the connection quality is sometimes unfavorable, which, together with the often time-consuming handling of a router and speculation about health hazards from electromagnetic WLAN fields, make PowerLAN a sensible alternative. In this case, data exchange takes place via the in-house power supply network, whereby, according to the IEEE 1901.FFT standard, ranges of up to 300 m and data transmission rates of up to 2000 MBit/sec. should be possible, but not via different circuits that are protected by their own fault circuit.

According to the *network topology:*

The network topology or network structure is the arrangement of how the individual computers are connected to each other. The most important network topologies are, whereby the logical structure of a network outlined here does not have to correspond to the physical structure:

bus topology: All stations are connected to each other via a common cable (bus). If a station fails, the transmission is not affected.

ring topology: All subscribers are connected to each other in a ring from station to station. If one station fails, this results in a total failure of the entire network.

star topology: The participants are connected in a star configuration to a central computer (so-called hub). If one station fails, the transmission is not affected.

According to the type of distributed processing:

In the so-called *peer-to-peer network* several fully equipped personal computers are usually integrated, which can also be operated independently, i.e. not networked. The individual computers have the same priority and each take on the administration and control tasks, without a fixed server. With a peer-to-peer network data can be exchanged (joint or mutual access to data carriers, sending of messages) and peripheral devices such as printers can be used jointly. This form is suitable for smaller networks with up to approx. 15 computers.

The so-called *client-server networks* represent the most important modern form of a network (cf. Fig. 2.11). Here, the computers involved do not have equal rights. There is a clear separation of functions between powerful computers that offer services as servers and other computers that request these services as clients. According to the services, a distinction is made between data and program servers (so-called file and application servers), which provide data and/or programs, print servers, which process temporarily stored print jobs, and communication servers. Different computers do not have to be used. One powerful computer can take over several tasks. The distribution of functions is ultimately regulated by software, which also determines the performance of the network in terms of response time and processing speed of the programs.

A computer can be both a server and a client. Passive servers only provide data or programs. In contrast, active servers execute programs themselves (distributed processing).

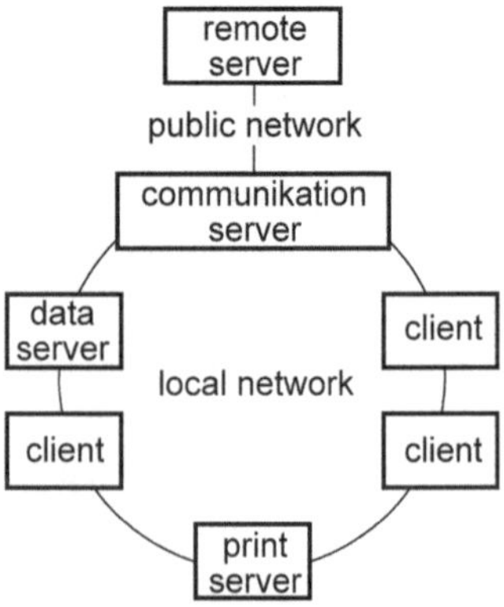

Fig. 2.11 Principle of a client-server network

2.8.2 Internet

The International Network, abbreviation *Internet,* is a worldwide connection of different networks, with no network consisting of permanent connections, but the join is based on agreements on forms of communication. The exchange of information between very different computer platforms takes place via the TCP/IP protocol, which was specified until about 1982. The research institution ARPA (Advanced Research Projects Agency), established by the US Department of Defense, initiated the so-called *ARPANET in* 1969, which from the early 1970s connected universities and research institutions that cooperated with the US Department of Defense. The actual birth of the Internet can be dated to around 1983, when ARPANET was converted to the TCP/IP communications protocol. The former objective has been changed to a non-commercial science network (NSFNET) with support from the National Science Foundation (NSF). Research institutions and universities were involved, and in the 1980s research networks in Europe were also connected. The term "Internet" was born.

Three dates mark significant steps in development: In 1989, the *World Wide Web* network service was developed at CERN (European Organisation for Nuclear Research, originally Conseil Européen pour la Recherche Nucléaire). In 1993, the first freely available webbrowser Mosaic was released, which enabled the display of graphics and text. In the following years, the National Science Foundation supported privatisation until finally, in 1998, the net access points and administrative functions were commercialised (cf. National Science Foundation 2003). After that, exponential growth began. In the meantime, every scientific and public institution, every interest group as well as every commercial enterprise and also private individuals are represented on the Internet or World Wide Web. As a mass medium, the Internet has achieved enormous economic importance, among other things, through electronic commerce, so-called e-commerce. Characteristic features are therefore: open system, decentralised control, heterogeneous hardware and software systems.

The Internet provides many services that are implemented by various protocols based on the TCP/IP protocol. Some of the services are offered as separate programs or functions in operating systems or in various webbrowsers. The currently most important or most frequently used services are:

- *Electronic Mail* or *e-mail (Simple Mail Transfer Protocol SMTP):* Exchange of messages in the form of data. In most cases, a message, i.e. an e-mail, is temporarily stored on an e-mail server to which the user has access authorisation (indirect information exchange). Files with any content can be attached to an e-mail (so-called attachments). In addition to the Internet e-mails can be sent via so-called online services or mailbox networks.
- *Usenet News (Network News Transfer Protocol NNTP):* Discussion forums on Usenet, a system of providers and users of international newsgroups. The news server or mailbox

of an Internet provider provides information and discussion forums that can be subscribed to or viewed.

- *Terminal Emulation* or *Telnet* (*Telnet Protocol*): Program execution on a computer on the Internet. This enables remote access to computers on the Internet, for which a user must have access authorisation and on which he can execute programs online.
- *File Transfer (File Transfer Protocol FTP):* Based on the File Transfer Protocol and a FTP program a simple file transfer and data exchange between different computer worlds is given.
- *World Wide Web*, WWW (*Hyper Text Transfer Protocol HTTP*): Worldwide information system (also abbreviated to: Web). Next to e-mail the World Wide Web now has the greatest importance and is often (incorrectly) equated with the Internet. The communication standard HTTP is the basis for the transfer of information between a WWW server and a WWW client. Central tasks in communication are performed by so-called *browsers* (cf. Fig. 2.13). Information is accessed via a standardised address (so-called *URL*, *uniform resource locator*), which consists of the service or protocol (e.g. http://) and specifying the location of the resource (e.g. www.uni-osnabrueck.de). With *HTTPS (Hyper Text Transfer Protocol Secure)* data can be transmitted securely (including encryption and authentication, i.e. verification of the identity of the connection partners).

2.8.3 Web Technologies

The communication structure for the transmission of text pages on the Internet is based on a fairly simple principle: A webbrowser is used to establish a connection from a client in a network to a webserver (cf. Fig. 2.12). A webbrowser is a program that can be used to display HTML web pages or, more generally, documents such as images. HTML (*Hypertext Markup Language*) is a text-based description language used to structure texts, images and hyperlinks in documents (cf. Fig. 2.13 and Sect. 3.1.7).

To request a specific page, a query is generated from various instructions and sent to the server. This request consists of several so-called methods (e.g. get or post), which are defined in the Hyper Text Transfer Protocol. With a so-called response, the server sends back instructions and, above all, data. The connection is then terminated (apart from the relatively rare exception of a so-called keep-alive connection). The information transferred

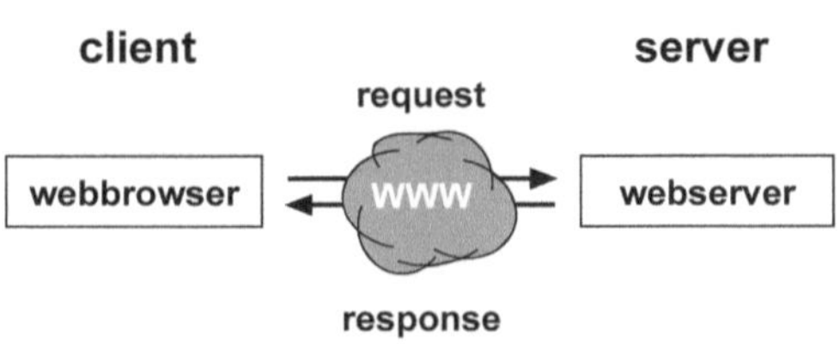

Fig. 2.12 Principle of a client-server application

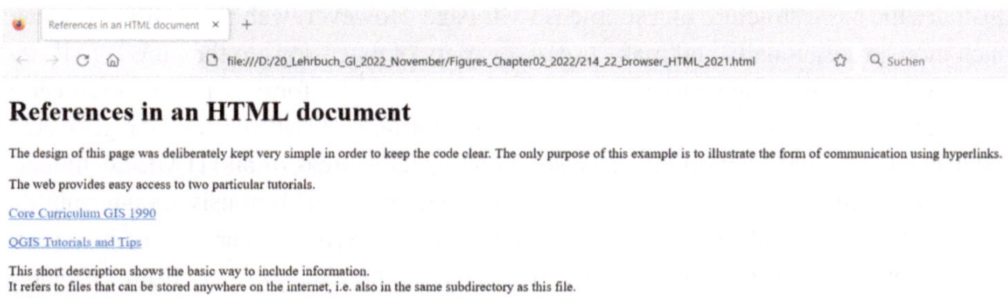


```
<!DOCTYPE HTML PUBLIC "-//W3C//DTD HTML 4.01//EN">
<html>
<head>
<title>References in an HTML document</title>
<meta http-equiv="Content-Type" content="text/html">
</head>
<body>
    <h1> References in an HTML document </h1>
    <p>The design of this page was deliberately kept very simple in order to keep the code clear.
    The only purpose of this example is to illustrate the form of communication using hyperlinks.
    <br><br>The web provides easy access to two particular tutorials.
    </p>
    <p><a href="https://escholarship.org/uc/spatial_ucsb_ncgia_cc/">Core Curriculum GIS 1990</a></p>
    <p><a href="http://www.qgistutorials.com/en/">QGIS Tutorials and Tips</a></p>
    <p>This short description shows the basic way to include information.<br>
    It refers to files that can be stored anywhere on the internet, i.e. also in the same subdirectory as this file.
    </p>
</body>
</html>
```

Fig. 2.13 Webbrowser and hypertext

and temporarily stored (cached) in the browser's own cache on the hard disk of the client is displayed by the browser.

In addition to text, graphics as well as sound and video sequences can be integrated in the text documents, i.e. in the HTML pages, which can be played depending on the hardware equipment. In this way, multimedia objects, i.e. any files consisting of text, graphics, sound or video sequences, are logically integrated into a document (so-called hypertext) in a special form of distributed processing. The actual localisation of the data sources is unimportant. Special markings (connection symbols such as text underlining as a standard form) in text documents, which are displayed in *webbrowsers*, indicate connections (so-called *hyperlinks* or *links* for short) to other documents on any servers on the World Wide Web. By clicking on such a marking with the mouse, the browser establishes a connection to the document identified by the marking and transfers it to the user's own computer. Figure 2.13 (with associated HTML text) illustrates this principle with a very simple example.

The integration of information, which can be located anywhere in the World Wide Web, is realised by browsers as information organisers. The example shown in Fig. 2.13

illustrates the basic structure of a simple HTML page. However, web pages are now created much more professionally and make use of a variety of extensions to the pure HTML text. With the help of so-called cascading style sheets (CSS), the form of presentation can be separated from the content of a structured document (e.g. HTML or XML pages). *AJAX* (*Asynchronous JavaScript and XML*) also allows specific areas of an HTML document to be updated without having to reload the entire page by asynchronously transferring data between the server and browser. For example, when entering a search term in a search mask, answer suggestions are already displayed in parallel (known from Google or Wikipedia). This can save time and volume for the redundant transmission of unnecessary or already transmitted data.

In this form of data transmission, which has long been the most common form of data transmission on the Internet, information is only transmitted statically. The user receives information from a data server, which is then processed offline on the client (so-called downloading or downloading of data). In this case, no processing of information takes place via the Internet apart from the information retrieval. In contrast, programs or parts of programs that are integrated in HTML pages can be transferred to the client and executed on the client, can extend the functionality of the client and carry out work on the client. This form of distributed working on the Internet can be implemented by several variants:

Plug-ins are programs that insert themselves into the browser (or other programs) to provide additional functions. They are transmitted via the Internet, for example, and installed on the client. This can significantly increase the functionality of the browser, shifting a lot of computing activity to the client, ensuring fast transmission and reducing the load on the network, especially for interactive graphics applications. An example is the Acrobat Reader plug-in, which enables files in PDF format to be viewed in a browser.

Furthermore, the performance of browsers can be increased e.g. with *Java applets* or with *JavaScript* (cf. Sect. 3.1.7). While plug-ins remain on the client and are also available for further applications, web applets must always be retransmitted (with the HTML pages). Web applets are computer programs that run directly in the browser on the client side without transferring data from a server. The user interacts with the program directly through the browser. However, their use is critical for security reasons, as applets transferred from an unknown external source can potentially cause damage (e.g. delete or lock files on the local hard drive). Due to this security risk, Java applets without a certificate are blocked on many browsers (for application software on mobile devices cf. Sect. 3.1.8).

On the server side a web server software, often simplified as *webserver*, controls the communication with other programs on the server (cf. Fig. 2.14). In the simplest form of distributed working on the Internet, static files, e.g. unchangeable HTML or image files, are transferred from a server to a client for further processing. An important application in geoinformatics is the download of a map. Zooming or panning of a section of the map is done on the client (including reloading tiles after panning). Very often, dynamically generated files are provided to the user or his browser, which are created individually according to the user's requirements and requests. In order to display a complete web page, the user receives the HTML page itself, a file with the instructions for designing the page,

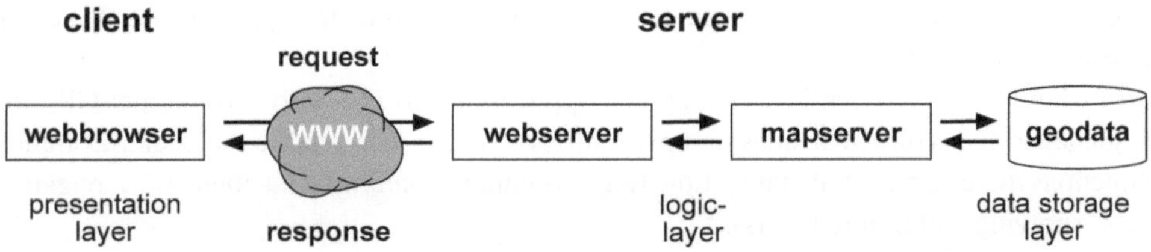

Fig. 2.14 Application of the client-server model in geoinformatics

and several image files individually. The browser makes a separate request to the web server in each case, so that sometimes hundreds of requests and server responses are necessary for a complex web page.

Dynamic web pages, for which the scripting language PHP is often used, are compiled from various sources and transferred to the client. The PHP code is located on the server. The client transmits which PHP code or which PHP file is to be executed. Only the result of this server-side processing is sent back to the browser and then displayed to the user. The special feature of this technique is that any number of HTML lines can be created depending on data processing on the server. For example, a database query can create a new table which is sent back to the client as an HTML page.

Applications in geoinformatics specify the general client-server architecture with regard to specific functions on the server side. As in the general model, the user calls up functions in the webbrowser, which serves as a client. These functions are processed on an external server (hardware). Via a web server working as software (e.g. the free Apache Tomcat web server) on the external server, the requests are forwarded to a *mapserver* (software, e.g. the free GeoServer), which processes the request with access to the geodata (cf. Sects. 7.2.2, 7.2.3 and 6.4.3). The result is then usually sent back to the client in the form of a map from the *mapserver* via the *webserver* (cf. the (OGC) geodata services in Sect. 6.4.2).

On the software side it is a three-layer architecture. The *presentation layer* is responsible for the representation of data, for user input and generally for the user interface. The *data storage layer* contains geodata, possibly in a database, it is responsible for storing and loading data. The *logic layer* contains the processing mechanisms and the application logic. It is precisely the scope of performance of this logic layer that ultimately determines the alignment of the entire system. It ranges from displaying maps with only simple navigation functions to web mapping (cf. Sect. 7.2.1) to Web-GIS (cf. Sect. 9.2). The layer model shown in Fig. 2.14 should not be confused with the so-called ISO/OSI reference model (cf. Schreiner 2019 pp. 3). This model, published by the International Organisation for Standardisation (ISO), describes the architecture of communication protocols in computer networks (OSI = Open Systems Interconnection) built up in a kind of layer structure. The client-server communication shown in Fig. 2.14 refers here primarily to the upper layers of the ISO/OSI reference model (representation and presentation of the transmitted data, application-oriented). The lower layers of the ISO/OSI reference model (transport protocols, transmission of individual bits over the network, transport-oriented) are not shown in Fig. 2.14.

It should be noted that in web mapping applications, as in a web and Internet-GIS, the logic layer that typically resides on the server side may also extend to the client. Common functions here are spatial navigation (e.g. zooming, panning) and thematic navigation (e.g. showing and hiding layers).

Server-based programs offer several advantages. The user only needs a webbrowser as part of the standard configuration of a computer system in order to get the desired information. No additional software is required for the client. It should be particularly emphasised that no training in another, possibly very complex software is necessary. The operation is only possible with the knowledge of web browsing. It is precisely this advantage that helps to open up large user groups.

2.8.4 Web 2.0

The World Wide Web owes its breakthrough to the countless free information offers that can be easily accessed by anyone and almost from anywhere via browsers. The beginnings are characterised by the fact that the user was offered a variety of information that he could view and evaluate (only) through query functions. Information was provided exclusively by the providers. This generation is represented by the now countless web portals. On the one hand they are used for self-presentation e.g. of enterprises and public institutions as well as the dissemination of information in order to reduce the volume of standard inquiries. On the other hand they offer sales products. These websites have neither been displaced nor replaced by the new forms of Web 2.0. They continue to make up a large, if not the vast majority, of the information on the WWW. No name exists for this generation. The term Web 1.0 is only occasionally used from the perspective of Web 2.0.

The term *Web 2.0* goes back to the article by O'Reilly "What is Web 2.0?" (cf. O'Reilly 2005). This identifies and names a new generation of the WWW, which is characterised by a new level of interactivity and in which the user (not the information provider) plays an increasingly important role. New information is generated, modified and presented by the users (user generated content). Even non-specialists can use simple programming techniques and tools, such as wikis (hypertext system for websites for the collaborative creation of texts e.g. Wikipedia), blogs (a (diary) kept on websites and mostly public) as well as podcasts (audio and video files), or create profiles with personal data in social networks (exchange of information and networking with other users, transfer of personal information with or Instagram).

Web 2.0 uses many techniques developed as early as the second half of the 1990s, but which only became generally accessible with the greater spread of broadband Internet access. Typical techniques are those that enable web applications to behave like desktop applications (cf. Sect. 2.8.3) and subscription services (e.g. in RSS format) that exchange information (especially news reports) between web pages (e.g. RSS feed or newsfeed). An essential element of web technology is that many Internet services provide programming interfaces (so-called *APIs*, *application programming interface*). An API allows access to certain functions, data structures or variables of a provider. For example, Google Maps offers many APIs which can be used to embed Google Maps functions can into one's own web pages (e.g. creating a route by calling the routing functions of Google Maps, cf. Sect. 7.2.4).

2.8.5 Cloud Computing

With increasing connectivity and the development of powerful servers and external storage a new IT approach has emerged, which is known as *cloud computing*. In simplified terms, part of the hardware and software, i.e. primarily storage space, network and user programs, is no longer stored locally on the computer at a desk, but is used as services by one or more providers. The providers are located in an undefined part of the Internet, in a figurative sene, somewhere in a "cloud". Thus, "cloud computing" includes on-demand infrastructure (computing and storage capacity, networks) on the one hand and on-demand software (operating system, user programs) on the other. Both services are billed according to use and scope and, above all, are requested depending on the respective operational requirements.

The pioneer was the company Amazon, which set up very large server parks, but they were not consistently utilised. So the idea of offering computing power was obvious. Amazon Web Services (AWS) were launched in 2006, with Amazon Elastic Compute Cloud (Amazon EC2), for example, providing computing capacity.

The National Institute of Standards and Technology (NIST), a US federal authority, published a now widely accepted definition back in 2011 (cf. Mell and Grance 2011). Important (technical) characteristics are:

on-demand self-service: automatic allocation of computer capacity on demand
broad network access: availability via the Internet, access to so-called thin or thick clients such as notebooks, PDAs or smartphones
resource pooling: bundling of the provider's computer capacities independent of the physical localisation
rapid elasticity: fast and flexible provision of computer capacities, easy scalability
measured service: automatic control and optimisation of the use of resources.

Furthermore, NIST identifies three service models for cloud computing:

SaaS (*Software as a Service*): Access to software collections and (special) application programs that are charged on a per-use basis, run in a browser and are usually platform-independent.

PaaS (*Platform as a Service*): Provision of (virtual) runtime environments in which user-generated or other programs run, as well as development environments with access to programming languages and test tools.

IaaS (*Infrastructure as a Service*): (almost arbitrary) access to resources such as computers, networks and storage and independent composition of virtual computer hardware by the users (mostly storage capacities and virtual servers with associated software).

In addition, the NIST definition approach lists three (or four) deployment models:

private cloud: Operation of the cloud infrastructure solely for a company, regardless of the location of the provider.

community cloud: shared use of cloud infrastructure by multiple companies or organisations

public cloud: general public access to the cloud infrastructure

hybrid cloud: composition of two or three deployment models

The service models are particular relevant for users:

With the IaaS model, the user is freed from the questions and problems of purchasing and updating hardware or preventive data backup. However, the user is independently responsible for the functioning of his software on this virtual computer environment.

The PaaS model is primarily dedicated to developers, who receive a complete virtual working or development environment. The "Google App Engine" as the most important or best-known example, with which web applications can even be developed free of charge under certain quantity restrictions and made available on the Internet, provides development environments for Java and Python, among others.

In the SaaS model, the provider ensures that the software works within the user's infrastructure and provides the service requested by the user, which can be an invaluable advantage in everyday life if consistent functionality of the software is desired or even essential. In the simplest case, the management of e-mails, appointments and addresses on the web can be attributed to these services, whereby these services are offered as part of the cost of access to the web (mostly free) and do not generate any additional profits. In contrast, large software companies, which previously provided their enterprise software by default via licensing models, also offer their services as part of cloud computing. For example, the cloud-based solutions "SAP Business By-Design" or "DATEV-Unternehmen online" map business processes. However, this often seems to be just a new term for the already long existing services of running office applications on a server system of a software provider. Specific geoinformatics software solutions are now also being offered as part of cloud computing. For example, ArcGIS online is a cloud-based GIS mapping software (cf. Sect. 9.1.5). Smaller IT companies also offer cloud-based GIS applications (i.e. Software as a Service, cf. PlexMap by Geoplex).

The service model BaaS or mBaaS (backend or mobile backend as a service) is not part of the NIST definition. BaaS has only been formalised as an abstraction in recent years to meet the increasing spread of apps on mobile devices. A BaaS is used to link mobile apps to a backend cloud. In addition, the backend can be managed e.g. via a webbrowser (e.g. user management, push notifications).

References

Broy, M. (1998): Informatik Eine grundlegende Einführung. Band 1: Programmierung und Rechnerstrukturen. Berlin: Springer. 2. Ed.

Creative Commons (2023): Mehr über Lizenzen. https://creativecommons.org/licenses/ (14.04.2023).

GNU (2023a): Free Software. http://www.gnu.de/free-software/index.en.html (14.04.2023).

GNU (2023b): Documents. http://www.gnu.de/documents/index.en.html (14.04.2023).

Gumm, H.-P. u. M. Sommer (2013): Einführung in die Informatik. München: Oldenbourg. 10. Ed.

Herrmann, D. (1992): Algorithmen Arbeitsbuch. Bonn: Addison-Wesley.

Lang, H. (1993): Farbmetrik. In: Niedrig, H. (Hrsg.): Optik. Bergmann Schaefer Lehrbuch der Experimentalphysik Bd. 3. Berlin: de Gruyter. 9. Ed.

Mell, P. u. T. Grance (2011): The NIST Definition of Cloud Computing. Recommendations of the National Institute of Standards and Technology. Special Publication 800-145. Gaithersburg, MD. https://nvlpubs.nist.gov/nistpubs/Legacy/SP/nistspecialpublication800-145.pdf (14.04.2023).

National Science Foundation (2003): A Brief History of NSF and the Internet. https://www.nsf.gov/news/news_summ.jsp?cntn_id=103050 (14.04.2023).

Open Source Initiative (2023): The Open Source Definition. https://opensource.org/docs/osd (14.04.2023).

O'Reilly, T. (2005): What Is Web 2.0. Design Patterns and Business Models for the Next Generation of Software. https://www.oreilly.com/pub/a/web2/archive/what-is-web-20.html (14.04.2023).

Schreiner, R. (2019): Computernetzwerke: Von den Grundlagen zur Funktion und Anwendung. München: Hanser. 7. Ed.

Schwarz, H.R. u. N. Köckler (2011): Numerische Mathematik. Wiesbaden: Vieweg + Teubner. 8. Ed.

Basic Concepts of Computer Science

3

3.1 Programming Computer Systems

3.1.1 Programming Levels

A computer is a universally usable automatic machine (automaton) which is controlled by a program (cf. Sect. 2.2). The entirety of all programs, which can be used by a computer, is called software (cf. Sect. 2.7). It should be emphasised that both the system and the application software are created with the help of programming languages, i.e. by "*programming*". These (artificial) languages play a central role in communication between humans and machine, i.e. between user and computer system. In principle, the programming of a computer can take place at different levels (cf. Fig. 3.1):

Machine codes, which are specific to each type of processor, describe an algorithm as a sequence of binary coded instructions (e.g. machine code 010100001010) so that the processor can execute this program immediately. However, programming is very time consuming, difficult and error prone. A program in machine code cannot be transferred to another computer model or processor model because it consists of processor-specific instructions.

Assembly languages are machine-oriented programming languages which express instructions and operations by easily understandable symbols. Here, mnemonic (i.e. memory-aiding) terms such as ADD or SUB serve to abbreviate the instructions in an understandable way. Assembler programs are efficient and require little memory. They allow quite flexible and fast program execution. However, program creation is (still) very laborious, the programs are relatively difficult to read and confusing. Assembler programs depend on the respective hardware and can hardly be transferred to another computer model. But these programs cannot be directly understood by the computer. The assembler

N. de Lange, *Geoinformatics in Theory and Practice*, Springer Textbooks in Earth Sciences, Geography and Environment,
https://doi.org/10.1007/978-3-662-65758-4_3

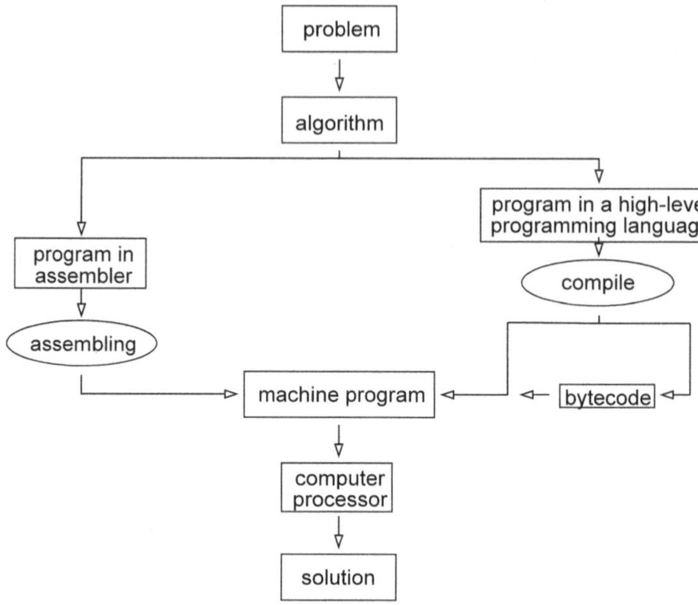

Fig. 3.1 From problem to solution: programming a computer system

program has to be translated, i.e. assembled, into the machine code with the help of a program in order to be executed. The program and data addresses are no longer specified by absolute machine addresses. Instead, a symbolic representation is chosen, where, for example, (variable) names stand for addresses. Also, constants such as 47 are specified in the usual notation rather than in a binary form. Through assembling, which again is done with the help of special programs, programs in assembly code are then automatically translated into machine code. Programs in assembly code are still written today when short processing times are required or only little memory is available. In particular, programs in assembly code are included as subroutines in programs of a higher programming language when a particularly efficient fast code must be generated, e.g., for a frequently repeated arithmetic operation.

Problem-oriented or *high-level programming languages* realize algorithms using simple forms adapted to the problem, regardless of a specific processor or computer system. They are close to the problem, usually convenient and less time-consuming to create, as well as easier to read, better structured and less error-prone. However, they are memory intensive and slower than equivalent programs in assembly code. Programs written in a problem-oriented programming language cannot be directly understood and executed by the computer. By compilation, which like assembling is done with the help of special programs (so-called compilers), programs are translated into an assembly code or already directly into machine code (cf. Sect. 3.1.2). The programs in a high-level programming language are transferable to different types of computer systems, a new, processor-dependent compilation may be necessary.

In contrast to this principle, the Java programming language does not generate machine code directly from a source text during compilation, but an intermediate code. This so-called *Java bytecode,* which represents the machine code of a virtual machine (VM), i.e. a machine that does not consist of hardware but is simulated by software, is only translated and executed during runtime by the so-called *Java Virtual Machine (JVM)* into the corresponding machine instructions of the processor. Since the bytecode is independent of the real hardware, a so-called platform independence is achieved. On the one hand, this is an advantage; on the other hand, Java programs are relatively slow due to the lack of hardware proximity. Furthermore, the virtual machine must be provided for each computer platform on which the program or the bytecode is to be executed. The Java Virtual Machine is part of the *Java Run-time Environment (JRE)*, which exists for every computer platform and is available for free as GNU General Public License. In addition to Java, the .NET languages and the programs developed from them also use bytecode. The term .NET covers several software platforms published by Microsoft, such as Visual C#.NET, C++. NET or Visual Basic.NET.

3.1.2 Creating and Executing Programs

Several programs are involved in the creation and execution of programs (cf. Fig. 3.2). An editor, i.e. a program for entering and editing texts, numbers and special characters, is used to enter the program instructions. This is preceded by the analysis of the problem, the development of a solution strategy, a first or rough program sketch or even the arrangement of all program instructions (for software development cf. Sect. 3.5). Such a program written in a high-level programming language is called a source program, which contains the so-called *source code*. Since such source programs cannot be executed directly by the computer or processor, a compiler translates the programs into the machine program.

In general, a *compiler* is a program that reads a program written in a high-level programming language and then translates it into a program of a target language. An essential sub-task of a compiler is to report syntactic errors contained in the source program. Compilers are therefore important software tools. Their reliability and manageability (e.g. error analysis) have a significant impact on the productivity of program development and user software. The *compilation* of a program must be understood as a complex process, which is divided into several phases:

The source program is broken down into its components. The lexical analysis (scanning) checks, among other things, whether the characters and symbols contained in the source program are permitted and included in the vocabulary, i.e. in the alphabet of the programming language.

Subsequently, the syntax analysis (parsing) checks whether the words or instructions used or the formal structure of the instructions comply with the rules of the language. Thus, syntax is generally understood to mean the sentence structure or the formal structure of the sentences or words of a language. Thus, for example, it is checked whether the statement

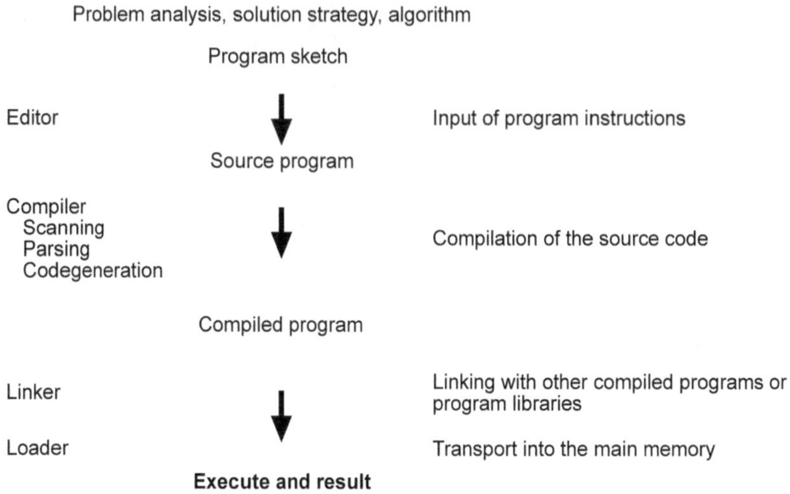

Fig. 3.2 Creating and executing programs with a computer system

"statik" belongs to the selected programming language. Sometimes, especially in a simple software development environment a detailed error analysis may be missing, so that the programmer is not made aware that the correct instruction must be "static" when programming in Java.

The next step is the semantic analysis. In general, semantics characterises the meaning of the content of a language. So here the program is checked for meaning errors. Code generation is prepared.

In the last step of code generation, the actual target program is created. The machine instructions are generated. The code generation often takes place in several steps. Thus, initially only an intermediate code is generated. Then, during code optimisation, an attempt is made to improve the intermediate code to create more efficient machine code. Often, instead of direct machine code, compilers first generate assembly code that must be translated by an assembler into machine code. Since assemblers, like compilers, generate machine programs, i.e. processor-specific instructions, these tools depend on the respective processor (or its instruction set).

After translation, the compiled program is linked together with other programs that already exist in machine code to form a single program. These programs can come from several different translations. In this way, a large program can be broken down into individual parts, which are tested and compiled one after the other. In almost every case, system-provided library functions, such as standard input and output routines or mathematical functions such as sin(x), are added during linking. The loader loads the program into main memory and executes it. Often, these steps are performed automatically in a comfortable software development environment. Ideally, the computer system responds without error message and delivers the desired result. If not, the debugging starts, with which many programmers spend most of their time.

Instead of compilers, *interpreters* translate a program step by step and execute the instruction directly afterwards. This means that not the entire program is first translated into another language. Interpreters are mainly used for programming languages that are intended for dialog operation. The advantage of interpreters is that after making changes to the source code, the program can be executed immediately, that is, without compiling. If an error occurs in the program, the programmer can try to change the relevant line of code directly and then continue the program again. This makes error handling and the creation of error-free programs much easier. However, interpreters have much longer computing times. For example, for an assignment (e.g. streetlength = 123) the addresses of all used variables have to be searched for based on their names, whereas the addresses are calculated once in the compiler. Python is a modern interpreter language that is quite easy to learn and supports several programming paradigms such as functional or object-oriented programming (for the importance of Python in geoinformatics cf. Sect. 3.1.5).

The *editor,* the *linker* and the *loader* are components of the operating system. Together with a compiler, which depends on the programming language used and the *instruction set* of the processor, these programs form the minimum equipment necessary for the creation of programs. Mostly, however, the (simple) editor of the operating system is not very suitable for the development of programs. Typing errors, i.e. language-dependent typing errors, cannot be recognised during the input. Above all, no help can be offered here that can explain commands or error messages.

While such simple programming conditions characterised the beginning of programming, today an interactive way of working dominates with a graphical user interface and intuitive user guidance (integrated development environment, IDE). Figure 3.3 shows the (graphical) software development environment Eclipse, a freely available programming tool for developing software for many programming languages such as Java, C/C++, PHP or Python (cf. Eclipse Foundation 2023). In several windows, different programming tools are provided: In the code window, the source code is entered, with e.g. syntax checking and autocompletion of code lines that are already carried out during the input. Statements, variables and comments are marked in different colors, structure blocks are (automatically) indented. In the software development environment C, for example, so-called makefiles can be generated automatically from a collection of elements, so that a simpler and faster program creation can take place. The software development environment also includes compilers and test aids, e.g. for diagnosing and finding errors (debuggers). These tools, which represent individual programs, are integrated into the development environment and do not have to be called individually by the programmer.

In the context of creating programs two concepts are of great practical importance:

- Programming interfaces are often offered for programming languages or for existing software. Such an *API (Application User Interface)* enables software developers to access additional resources so that they can program (individual) extensions. Examples

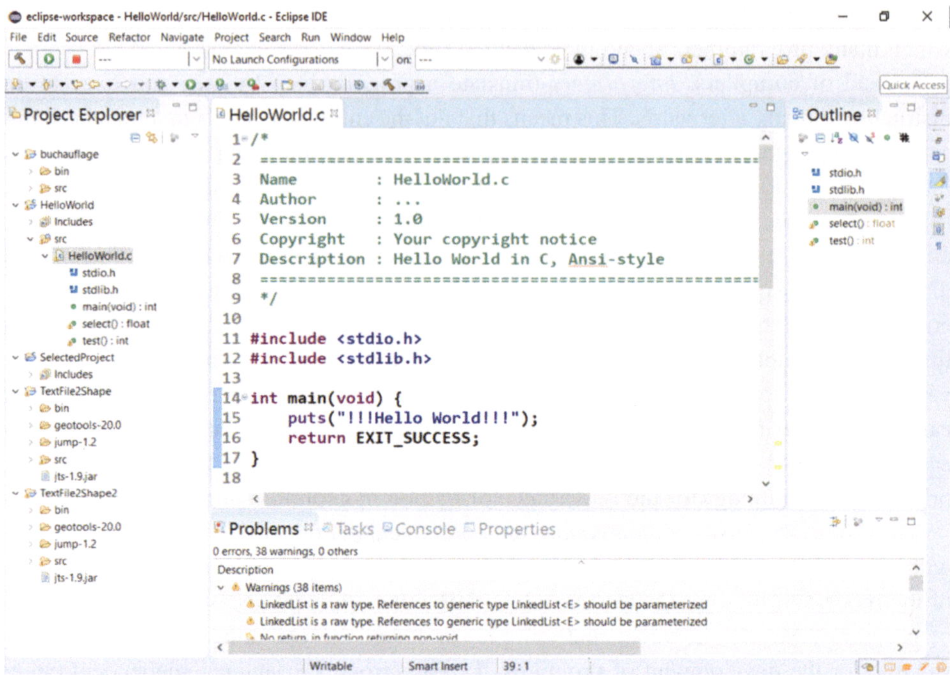

Fig. 3.3 Eclipse software development environment for C

are the Google Maps API in the context of developing graphical applications for smartphones (cf. Sect. 7.2.4) or the ArcPy and PyQGIS programming interfaces (cf. Sect. 3.1.5).

– A *plug-in* is a "ready-made" software component (often synonymous with "add-on") that extends a piece of software. In addition to plug-ins for webbrowsers (cf. Sect. 2. 8.3), extensions to geoinformation systems are particularly important in geoinformatics. Plug-ins can be installed by the user for special applications and integrated during the runtime of the program (cf. the abundance of plug-ins for QGIS cf. QGIS 2023a and QGIS 2023b).

3.1.3 Programming Languages

Machine and assembler codes are not very suitable for programming even simple arithmetic instructions. Thus, it would be absolutely necessary for a mathematical-technical programming language to write just exactly this instruction for the computation of $y = x^2$. If a problem solution (i.e. an algorithm) with problem-adapted operations is transferred to a computer, this is referred to as programming in a high-level programming language. Before being executed by the computer, however, these instructions must first be translated into the machine code (cf. Sect. 3.1.2). Programming languages are often classified as follows:

The languages of the so-called *first generation languages* were true binary languages. Until the mid-1950s, computers were programmed exclusively in machine code. This programming was very cumbersome and overall a hindrance to computer work.

The *second generation languages* formed assembler languages, which also still worked very close to the machine (i.e. the cpu) and therefore required an extensive understanding of the hardware. The instructions from the instruction set of the Intel 8086/88 processor:

```
MOV eax, 100h   (load 100hex into the register eax)
ADD eax, 70h    (add 70hex to eax)
```

run the addition 256 + 112, where ADD eax stands for a machine code in the form "1010 0001 0000 0010 1010" and 100hex stands for 256 or 70hex for 112 in the hexadecimal number system. Assembler languages are no longer important because of the power of high-level languages.

The *third generation languages* comprise the high-level or problem-oriented languages that are remote to the machine (i.e. the cpu) but have language elements of the human language. They ensure a better structuring of the tasks to be solved, make them easier to read and above all are less error-prone. The current high-level languages such as C, C++ or Java belong to this generation.

The user languages are the languages of the *fourth generation languages*, in which, for example, databases were linked to programming languages. This includes above all the Structured Query Language (SQL, cf. Sect. 8.4.3). In general, this generation also includes macro languages and scripting languages, which can be used, for example, to customize and control word processing or spreadsheet programs.

The *fifth generation languages* include the so-called AI languages (AI = artificial intelligence). The programs developed in these languages are intended to mimic human thinking. Solutions are worked out with the help of certain rules. These include logical languages such as Lisp or Prolog.

Languages for creating web pages or for exchanging data between computer systems cannot be classified in this series of generations (cf. e.g. HTML and XML). Strictly speaking, these are not programming languages that generate machine code by compiling instructions. The *Extensible Markup Language (XML)* is a language format for outlining and formatting text. The text file in this interchange format is readable by both humans and machines. Properties, affiliations and presentation forms of sections of a text (e.g. characters, words, paragraphs) are marked and structured with so-called opening and closing tags (cf. the XML structure with the tags in angle brackets in the SVG file, cf. Sect. 3.1.7).

The variety of programming languages has become almost immense. However, of the more than 1000 programming languages used today, only about 20 are more widely used (for geoinformatics cf. Table 3.1), and these can be roughly divided into four groups. These

Table 3.1 Programming languages relevant in geoinformatics

C C++ C#	Imperative language developed at the beginning of the 1970s initially only for the UNIX operating system (UNIX is essentially based on C), combination of structural elements of high-level programming languages such as data types, blocks, functions or loops with relatively machine-oriented constructions (e.g. register operations), better computer utilisation, relatively small language scope. C++ is the further development of C with regard to object-oriented programming. C# (c sharp) is the programming language developed by Microsoft for .NET Framework, which continues concepts from C++, Java and Delphi, among others
BASIC Visual BASIC	Easy-to-learn language (Beginners All Purpose Symbolic Instruction Code) developed at the beginning of the 1960s on the basis of FORTRAN, which is particularly widespread in the field of personal computers (latest version Visual Basic .NET fully object-oriented)
Java	One of the newest programming languages that is completely object-oriented. Java is quite straightforward and easy to learn, yet very powerful, compact and versatile. The Open Java Development Kit (OpenJDK) is licensed under the GNU General Public License GPL-2.0-only and is therefore freely available as a free, open source implementation of the Java platform, Standard Edition. Extensive free program libraries are available for Java, among which "GeoTools" is of great importance for applications in geoinformatics
Javascript	JavaScript is a scripting language primarily used for web applications that is executed on the client side. Despite the similarity in name, it has only little in common with Java
PHP	Script language developed in 1995, frequently used language in the WWW, executed on client and server side, database support possible, high functionality
Python	Object-oriented programming language developed in 1991, also supports functional programming; often used as a scripting language; easy to read; structured without brackets but by indentation; open development model with a large community; extendable by a large number of modules; modules of other programming languages such as C can also be integrated
R	1992 developed programming language for calculation of statistics and their graphical implementation

groups each have fundamentally different approaches. Imperative and increasingly object-oriented programming languages are the most widespread today and dominate programming:

The *imperative programming languages* (*command languages*) represent the oldest high-level programming languages. Here, a program is composed of a sequence of commands to the computer such as: "assign the value 3.3 to variable z", "jump to location y in the program". Here the variable concept and especially the data type concept (cf. Sect. 3.2) are realised: The input values are assigned to variables and no longer directly to memory cells. However, the definition of data types was still limited to the standard data types in the older languages such as FORTRAN or BASIC, whereas the complex data types of modern languages are suitable to optimally model the information structures of reality. Examples are: Algol, BASIC, C, COBOL, FORTRAN, Pascal or Modula (cf. programs in Sects. 3.1.4.1 and 3.1.4.2). *Procedural programming* is a part of *imperative programming,* in which a more complex computer program is built up from smaller subtasks, the procedures. *Modular programming* goes one step further by grouping procedures and data into logical units. Finally, *object-oriented programming* completes this development, in which data and functions that can or should (only) be applied to this data are merged into a so-called object.

In *functional* (*applicative*) *programming languages,* the relationships between inputs and outputs are described with the help of mathematical expressions by taking elementary expressions for simple functions as a basis and then representing more complex functions from them. A program consists of a set of expressions that define functions. The dominant language element is the application of a function. Essential parts of a program are definitions of functions and their application to terms. Thus, a calculation is the execution or application of a function to a list of values or expressions. In particular, recursive function applications play a major role, where a function is applied to itself. Examples of functional programming languages are the older languages Lisp or Logo as well as Haskell or, more recently, F# in the context of Microsoft's .NET Framework, whose importance is relatively small, but which, like Haskell, have significantly influenced the development of other programming languages.

Programming in *predicative programming languages* is understood as proving in a system of facts and inferences. A set of facts (valid predicates) and rules (how to get new facts from facts) are given by the user. The computer has the task to answer a posed question with true or false. Such languages are used to develop expert systems. An example of a predicative programming language is Prolog.

Object-oriented programming focuses on classes and objects. A class describes the union of (selected) functions or methods and constants or program variables (called attributes in object orientation). Attributes and the methods to be applied to them are grouped together in classes (called encapsulation in object orientation). Object-oriented programming languages have extensive class libraries by default. Programming here largely boils down to creating subclasses based on these classes, which inherit the functionality of the upper class and are also specially tailored to the respective task. The classes are then used to create objects at runtime that contain the data. An object corresponds to a data record that is referenced by the object name. The attributes are accessed (depending on their so-called visibility) using methods. The basic idea of object-oriented programming, which is currently the most important programming concept, is illustrated by an example in Sect. 3.1.4.3.

Object-oriented programming languages are based on a fundamentally different approach than imperative programming languages, which focus on mathematical calculation methods. Object-oriented programming languages correspond more to the way of thinking of the everyday life, in which objects and their properties are also (inseparably) linked (cf. Sect. 3.1.4.3). Directly for object-oriented programming, Smalltalk was already developed in the 1970s and Java in the early 1990s. The languages C++, Objectiv-C, Object COBOL, Object Pascal or also Oberon (to Modula-2) are extensions of conventional programming languages.

It should be noted that quite different dialects of a programming language can exist, which differ in terms of the scope of the language and the variety of functions. For example, an error-free source program in C from one software manufacturer can only be compiled by the C compiler of another software manufacturer after adjustments concerning manufacturer-specific extensions. To solve these problems, the Java programming language was developed, among others, which offers platform independence (cf. Sect. 3.1.1).

3.1.4 Programming: Concepts

The individual programming languages are based on different concepts of programming. This also shows a development process over time. In the meantime, the technique of structured programming has become a matter of course, but it was not (yet) possible with the languages available at the end of the 1960s. In the second half of the 1970s, modular programming emerged. Since the mid-1980s, object-oriented programming has steadily increased in importance, characterising the currently valid programming concept. However, the variable and the data type concept are still fundamental for all programming languages.

The *variable concept* abstracts from the actual underlying memory space. Addresses and operation codes are replaced by names. Thus, variables in a program identify a specific memory location in the computer, which is allocated during compilation and is usually unknown to the programmer. The assignment in Python notation “value = 5” means that the number 5 is entered as a bit sequence 00. . .0101 in the memory space of the variable “value” (e.g. 01001110). However, variables within a program are not identical to variables in mathematics. In a program, the instruction x = x + 1 is possible, which adds the value 1 to a variable and assigns the sum back to this variable, whereas in mathematics this equation has no solution.

Furthermore, the *data type* must usually be defined in advance (exception Python). Due to the task, often only integers or only real numbers should be used, text or a logical value can also be assigned to a variable. Different data types are available in the programming languages for the representation of such number or data areas, which also implement different storage accuracies or have different storage requirements. New data types such as arrays can be composed from simple data types (cf. in more detail Sect. 3.2 and especially Sect. 3.2.3).

3.1.4.1 Structured Programming

Structured programming aims to make a program easy to read and clearly arranged, and to structure it using simple rules and building blocks. This programming style basically avoids simple jumps (so-called goto statements), which characterised early programs and made them confusing (so-called spaghetti code). Above all, the consistent use of *structure blocks* serves to divide the program into individual parts and to structure it. These parts are enclosed with “{” and “}” in Java or C, for example. Python has a completely different structuring concept. Here, the indentation of lines or the use of spaces at the beginning of a line are used as structuring elements.

The following should apply to the decomposition of a program into structural blocks:

- A structure block has a unique function.
- A structure block consists of a single command (elementary command), of several commands or of several subordinate structure blocks.
- There are no overlaps between structure blocks: Either structure blocks are completely separated or a structure block is completely contained in a superior one.
- A structure block has exactly one input and one output (and therefore no jumps in the structure block).

In addition to structure blocks, *procedures* (*subroutines*) in particular are used for clear structuring. This breaks down the task into several subtasks, which are solved individually in several steps. The individual components can be easily separated, tested and verified individually, but also reused in other programs. So that groups of instructions can be used more often under different conditions (in other programs), it should be possible to adapt them as flexibly as possible to different individual cases. For this reason, procedures (subroutines) are provided with parameters to which different values can be assigned depending on the subroutine call.

Structured programming is also based on the use of clearly constructed data structures and constant agreements, which are declared separately (at the beginning) and not scattered throughout the program, as well as on the use of self-explanatory names and identifiers.

These rules for structured programming seem almost self-evident. It should be noted that the programming languages available at the beginning of computer development hardly supported such a programming style. For early programs, numerous program branches with forward and backward jumps were common, which made it difficult to keep an overview.

The program sequences can be reduced to a few basic forms, so-called *control structures*. The succession of individual steps of the algorithm is called a *sequence* of statements. Within a sequence, each statement is executed exactly once.

Selection (choice, case distinction) is a structure in which the execution of instructions depends on one or more conditions. This allows different solution paths to be followed depending on different application conditions. Several variants exist for selection, conditional branching (in two forms) and case discrimination. Adapted from Python, the code is as follows, with indenting lines or using spaces at the beginning of a line as structuring elements:

```
if condition:            If condition = true, then action 1 is executed,
          action 1       after that follows action z.
action z                 otherwise action z.
if condition:            If condition = true, then action 1 is executed,
          action 1       after that follows action z.
else                     If condition ≠ true, then action 2 is executed,
          action 2:      after that follows action z.
action z
```

```
if condition 1:          Depending on which of the parts condition 1 to
          action 1       condition3 is true, the corresponding action is
elif condition 2:        executed. A complex case distinction is made in a
          action 2       correspondingly large number of elif statements.
elif condition 3:        Some programming languages provide a so-called case
          action 3       statement for this purpose. In Python, however,
else:                    such an instruction does not exist.
          action 4
action z
```

Certain individual steps in an algorithm often must be repeated until a certain result is achieved (e.g., summing n values, sorting a list of n names). For such repetitions (iterations), the structural element of a *loop* is used, whereby several variants exist:

```
FOR index in RANGE (i,j,k)
     action 1            The body of the loop (i.e. here action 1, ...)
     action 2            is repeated until the counting index has passed
     action 3            through the values i to j. The value of i is
     ...                 increased by k at each step. With this loop type
     ...                 (simple counting loop) the number of repitions is
     action z            fixed beforehand.
```

```
WHILE condition
     action 1            First, the loop condition is always checked. As long
     action 2            as is true, the body of this loop is executed
     action 3            (loop with pre-test).
     ...
     action z
```

For the design of structured programs, so-called *Nassi-Shneiderman diagrams* are used, which are now standardised in Germany (DIN 66261) and are well suited for the documentation of structured programs. Program designs based on these diagrams, which are then inevitably composed of structural blocks, lead to a structured program (Fig. 3.4).

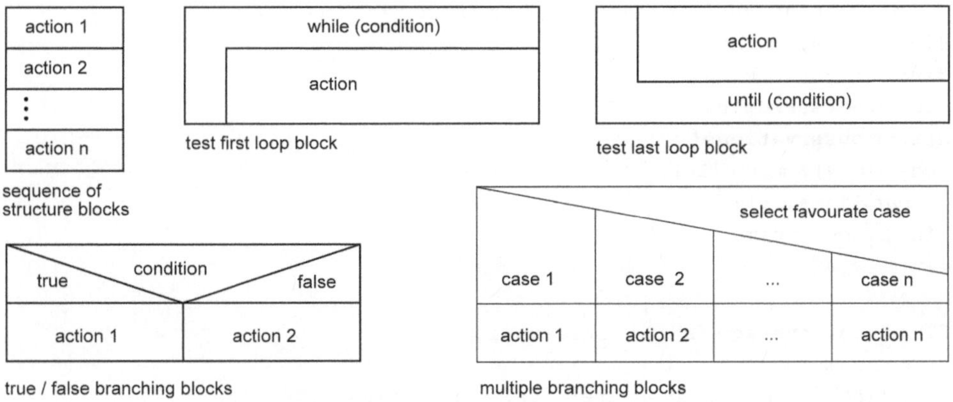

Fig. 3.4 Selection of Nassi-Shneiderman diagrams

Some principles of structured programming and object-oriented programming (cf. Sect. 3.1.4.3) are illustrated by a simple example. A file contains several data records, each with the name of a weather observation station, a further identifier (e.g. the month of data recording) and an unknown number of values or items (e.g. temperature). For each line, i.e. for each data set, the mean of these values must be calculated. The input data for this program can look like this (including any number of blanks):

```
A12 MAY       10.8 12.0 11.7 13.8   12.7
A01 APRIL           9.2
A23 MAY       12.1  11.7   12.8    13.2
A25 MAY       12.8 12.9 13.0 12.7 13.1
A14 MAY       12.4 11.9 14.0     12.5
A26 MAY       7.1 14.1 12.9 11.4
```

The program for calculating the arithmetic mean in C is easy to understand, although knowledge of the basic data types and data structures is required (cf. Sect. 3.2). Some explanations of the structured and concise program are helpful:

Each data record is processed as a so-called structure, i.e. as a user-defined data type (cf. the newly defined data type "observations"). The data records are read in and processed in a loop. Each input line is broken down based on the blanks it contains. Then the observations object is formed. The program is limited to a maximum of 15 values per line.

```
// Calculating the arithmetic mean of input data in C
#include <stdio.h>
#include <stdlib.h>
#include <string.h>
struct observations{
 char id[7], month[10];
 float items[15];
 int number_items;
};

float obsv_average (struct observations obsv){
 float sum = 0;
 for (int i = 0; i < obsv.number_items; i++) sum += obsv.items [i];
 return sum/obsv.number_items;
}

int main() {
//In addition, implementation of an error handling, if it cannot be opened.
 char xin[200];
 FILE *file = fopen("testdaten.txt", "r");
 for (int i= 0; fgets(xin, sizeof(xin), file) != NULL; i++){
  // Save the current line and split.
  char *copyLine = strdup(xin), *split = strtok(xin, " ");
  int tokens_pLine = 0;
   while (split != NULL){   //number of elements per row
    tokens_pLine++;
    split = strtok(NULL," ");
   }
  //Create storage structure and save data in it
  char *line[tokens_pLine], *split2 = strtok(copyLine, " ");
   for (int j = 0 ; j < tokens_pLine; j++) {
    line[j] = split2;
    split2 = strtok(NULL, " ");
   }
  //Create observations and fill it with data.
  struct observations x;
  strcpy (x.id, line[0]);
  strcpy (x.month, line[1]);
  x.number_items = tokens_pLine-2;
  float figure = 0;
   for (int k = 2; k < tokens_pLine; k++) {
     figure = strtof(line[k], NULL);   //Get value from input line.
     x.items[k-2] = figure;   //Save item value in data structure.
   }
  printf("| mean: %0.2f \n", obsv_average(x));
 }  //end of for
 fclose(file);
}  //end of main
```

The main program reads the data and prepares them in the data structure "x", which has the type "observations". The function "obsv_average" evaluates the data (i.e. "x"). Accordingly, further functions can be developed for other evaluations. This results in a logical separation according to different tasks.

Unfortunately, C has only meager operators for string handling. The instruction "fgets (xin), sizeof(xin), file)" reads one line of the input file and stores it in the array of characters xin. Of course, the file must still be specified exactly in the program with a path. Afterwards, the xin is divided into sections separated by blanks. The program uses *pointers* for this, which are marked by the character * in front of the name (cf. e.g. *split). A pointer does not directly contain a value, but "points" to the data set by merely storing the memory address of the set (for pointers cf. Sect. 3.2.2.6).

3.1.4.2 Modular Programming

Structured programming can be described as the first programming concept with regard to clarity and, above all, maintenance of a program. However, this concept does not solve all the problems that arise especially in the creation of large, complex software systems, in which a particularly large number of programmers are involved.

For larger programming tasks, it is necessary to break down individual parts into *modules,* each of which has a precisely defined interface and consists of type definitions, procedures and local variables. Services are offered to the user via this interface, which can be subroutines, but also constants, data types and variables. All other variables in the module are subject to the principle of confidentiality and are not visible from the outside. A larger task is broken down into function groups. These modules can be implemented independently of one another and their cooperation is manageable. For this, only the function and effect of these modules and their interfaces must be known. The implementation inside can be unknown.

The main advantages are great clarity and flexibility. When developing the program, only the interface needs to be defined or known, so that dummy modules can also be used initially, which are only later replaced by the specific modules. Programming and test runs can be carried out by different programmers. Maintenance and service are easier, since changes usually only affect a few modules.

The principle is to be illustrated by a sample program in a pseudo code (following Python), which calculates the distance of two points on the globe, if two location coordinates and the respective coordinate systems are known. The instructions associated with the programming language, such as PRINT, are written in upper case, while the variable definitions and specific program instructions are written in lower case.

```
# Main program in pseudo code showing modular programming
...
IMPORT LocationOnEarth

PRINT "Distance calculation between two locations".

# Comment:
# Input of a location by X- and Y-coordinate as well as an
# indicator P to identify the associated projection, approved
# Projections are 1 = Geographic coordinates, 2 = UTM coordinates
# each with respect to the WGS84 ellipsoid

PRINT "Enter the coordinates for the first location in the format: X Y P".
s1 = LocationOnEarth.readLocation()
PRINT "Enter the coordinates for the second location in the format: X Y P".
s2 = LocationOnEarth.readLocation()
PRINT "The distance is: ", LocationOnEarth.distance(s1, s2)
```

The LocationOnEarth module provides the services required for this subtask. The user only needs to know the interfaces that define the externally visible constants, variables, functions and methods (variable s1 and s2, methods readLocation and distance).

Contents of a LocationOnEarth file that defines the module of the same name:

```
...
DEF makeLocation (x, y, p):
  s = (x, y, p)
  RETURN s

DEF isLocation (s):
  # It must be checked whether the information is valid. For example
  # 6.30 52.40 is a correct geographical coordinate
  # 33.33 34.34 is an invalid UTM coordinate
  # If invalid: error message and return of false

DEF readLocation ():
  x = INPUT ("Enter X: ")
  y = INPUT ("Enter Y: ")
  p = INPUT ("Enter P: ")
  s = makeLocation(x, y, p)
  IF isLocation(s):
    RETURN s
```

```
DEF distance (s1, s2):
  distance = ...
  # If UTM coordinates are entered, they are converted into geographic
coordinates.
  # with respect to the WGS84 ellipsoid.
  # Using formulas of spherical trigonometry, the distance over great circle
is calculated
  # (cf. Section 4.2.3).
RETURN distance
```

The actual program solution, e.g. the implementation of an associated method, is unknown to the user. In the present case, the data structure must be specified: x and y coordinates as floating point numbers, an indicator for specifying the associated coordinate system as an integer, which is used internally, among other things, to control the distance calculation. Furthermore, the availability of a method "distance" as well as its call and performance must be known.

This example shows a program section for the definition of modules in modular programming. Here the implementation was disclosed to clarify the structure. However, the programmer does not need this knowledge. He only needs to know the interfaces and can be sure that the LocationOnEarth module is working correctly. In practice such a program will have further procedures for converting between different projections.

3.1.4.3 Object-Oriented Programming

Structured programming is used for clear structuring and distribution of tasks within a program, wherby data are subordinate but operations are in the foreground, which are primarily defined by functions and procedures. Modular programming aims to divide complex program systems into independent modules and (of course) takes into account the principles of structured programming. Object-oriented programming builds on these principles, but goes a decisive step further. Data types and program statements are no longer treated separately. The two are now combined into objects: Each object presents an abstract representation of a corresponding object in reality, combining the relevant data and the operations that can be performed on this data.

A central component of object-oriented programming are *classes* that describe special data structures with the associated methods. Object-oriented programming languages have extensive *class libraries* that can be extended user-specifically. Objects, which comprise data and properties as representations of the real world, are created from classes. For example, "r: = point.create (1.1, 2.3, 3.5)" creates an object of a class "point" to represent a point in a three-dimensional space, implicitly specifying the data structure (e.g., three floating-point numbers to represent the coordinates) and making all methods of the class "point" known to this object. If the class "point" has the method "move" for moving a point, "s: = r.move (3.0, 3.0, 3.0)" creates a new object from the old object "r", which is moved by the value 3 in x-, y- and z-direction. This approach brings all relevant information together and only the defined methods can process this information directly. This concept of *abstraction* provides greater software integrity and security. A concrete object of a specific class is only generated through *instantiation* during the runtime of a computer program. In many programming languages, a so-called constructor is executed, when an object is instantiated. An *instance of a class* is a concrete object with specific properties that can be used until it is destroyed. For example, a program that works with spatial objects could have a "River" class, in which the properties of a river are described as abstractly as possible. If a certain river is created in such a program, for example "Rhine", which is then available to the program for further use, then this is now referred to as an instance of the "River" class and the process of object creation is called instantiation.

Other central concepts of *object-oriented programming* are *inheritance, encapsulation*, and *polymorphism*:

All variables and methods that are common to all individual classes are defined in a superclass (e.g. superclass point with basic methods such as data read, move, rotate, mirror). Only derived classes that already have the basic properties (*inheritance*) are tailored to the actual task. Here, only the functionality that differs from the basic class or additional functionality is implemented (e.g. in the subclass pointnew the methods for determining the distance to another point). This results in class hierarchies.

By hiding the structural internals of a class (*encapsulation*), the consistency of data and associated methods is maintained. Data manipulations are performed exclusively by the defined and authorised methods. The above example already illustrates the encapsulation of data and methods. Furthermore, it illustrates the so-called secret principle (*information hiding*) of object-oriented programming. Thus, one does not need to know the concrete realisation of a class, but only its effect.

The term *polymorphism* describes the concept according to which a name can have different meanings depending on the context, which is useful, for example, for using the same name for several methods that perform an identical operation in the abstract sense but operate on very different data types. For example, the method "draw" can be defined for objects of the class "straight line" and inherited by the subclass "parcel boundary". If the method "draw" is then refined for this subclass (e.g. application of a specific signature), the execution of the call "x.draw" depends on the object type "x", i.e. whether the method "draw" of the class "straight line" or the method "draw" of the class "parcel boundary" is implemented. Late or *dynamic binding* during runtime and not already during compilation ensures that the "most specialised" version is bound to the object on which the operation is to be executed.

Some principles of object-oriented programming are to be illustrated with the same task as in Sect. 3.1.4.1 (implementation in C, structured programming) using the same data. A file contains several data records. For each data record that has the name of a weather observation station, a further identifier (e.g. the month of data recording) and an unknown number of values (e.g. temperature), the mean of these values (items) are to be calculated. The input data are separated by unknown numbers of blanks. The realisation takes place in a program in Java.

```
A12 MAY   10.8 12.0 11.7 13.8   12.7
A01 APRIL  9.2
A23 MAY   12.1  11.7  12.8  13.2
A25 MAY   12.8 12.9 13.0 12.7 13.1
A14 MAY   12.4 11.9 14.0  12.5
A26 MAY   7.1 14.1 12.9 11.4
```

First of all, it must be pointed out that object orientation has a special way of thinking. Although the initial data are available in the form of a table, it is not read as a table, e.g. into a two-dimensional array, which then has many empty spaces. Rather, each individual value (e.g. the temperature 11.7) is viewed as an individual object and is provided with a marker (e.g. A12) and the name of the month (e.g. MAY) as well as with a value identifier (id_value) for clear identification. These objects (here: "Value") are then combined to form the object "Measurements". This corresponds to the procedure of first placing each individual soil sample in individual bags (here class "Value") and then sorted into cardboard boxes according to the month of sampling (here class "Measurements").

First, two classes are defined for the two objects "Value" and "Measurements".

The class "Value" contains a so-called constructor for creating the object "Value", which consists of two strings for the marker and the name of the month, a consecutive number for value identification (id_value) and a variable for the numerical value ("xxxx"). It should be noted that this only defines the structure of the object. Initially, the object "Value" is still empty. In the illustration above, this procedure is comparable to the empty bags for the soil samples, whereby only the properties of the bags are defined first.

The class "Measurements" contains a so-called constructor for creating the object "Measurements" and three methods. A "Measurements" object consists of the class "LinkedList", which is available in Java and which brings any number of data values together in a linked list, as well as a variable for the number of values of this "Measurements" and a variable for the sum of the values. The first method ("add") adds exactly one new "Value" object to the "Measurements" object.

The second method ("addMultiple") first creates a new "Value" object from an array of individual strings, i.e. the fragmented input line, and adds this "Value" object to the Measurements object.

```
this.add(new Value(id, month, data));
```

This is the core of object-oriented programming. Finally, the method "mean" is used to calculate the mean value for all values of a Measurements object.

The method "main ()" represents the entry point into the execution of a Java application and must be "public static void main(String[] args)". The task here is particularly difficult, as the size of the values (e.g. two or four leading digits with two or three decimal places) and the number of values per input line are not known. The individual values are only separated by (one or more) blanks. In the "main" method, therefore, an entire input line is read in one after the other from a file as a sequence of text characters, and temporarily stored as a so-called string "inline" (cf. standard data types in Sect. 3.2.2). Subsequently, this string must be broken down piece by piece, whereby spaces are interpreted as separators.

In case of problems, the FileReader can throw a so-called exception. Therefore, the source code is included in a so-called "try-catch" block.

```
public class Value {
    //Global class variable
     static private int id_value = 0;
   //Object variables
   String month;
    String id_set;
    double xxxx;
   public Value (String id, String month, double xxxx) {
   Value. id_value = Value.id_ value +1;
       this.month = month;
      this.id_set = id;
       this.xxxx = xxxx;
   }
}

import java.util.LinkedList;
public class Measurements {
   //Define variables
    LinkedList measurements;
    int count;
    double sumxxxx;
   //Constructor for creating a Measurements object
   public Measurements () {
      measurements = new LinkedList ();
      count = 0;
     }
//Method for adding a single value object to a Measurements object
   public void add (Value value) {
      measurements.addLast (value);
     }
//Method for building a Measurements object from the fragmented input line
   public void addMultiple (String articles []) {
       String id = articles [0];
       String month = articles [1];
```

```
    for (int i=2; i<articles.length; i++){
     double data = Double.valueOf((String) articles [i]);
     this.add(new Value(id, month, data));
     count++;
     //Access to the numerical value of a value object and sums it up
     sumxxxx = sumxxxx + data;
    }
   }
//Method for calculating the mean of Measurements
  public double mean () {
   double mean = sumxxxx/count;
    return mean;
 }
}

//Computing
import java.io.BufferedReader;
import java.io.File;
import java.io.FileNotFoundException;
import java.io.FileReader;
import java.io.IOException;
import java.util.StringTokenizer;
import java.util.LinkedList;

public class Computing{

 public static void main(String[] args) {

  BufferedReader br = null;
    try {
     //BufferedReader for reading text lines
      br = new BufferedReader(new FileReader(new File("C:\crTEMPO
\crtestdaten.txt")));
      String inline = null;
      //Read line by line and store it in the string "inline".
      while((inline = br.readLine()) != null) {
      //decompose the input (i.e. inline) using a space separator
       StringTokenizer st = new StringTokenizer(inline, " ");
       int inlinelength = st.countTokens();
       //Save the decomposed strings into a string array
       String currentSet[] = new String[inlinelength];

       //output of current Measurements
       for (int i = 0; i<inlinelength; i++) {
        currentSet [i] = st.nextToken();
        System.out.print(currentSet[i]+" ");
       }
```

```
        //Instantiate and initialize the measurements object
        Measurements currentMeasurements = new Measurements();
        //Add the values from the string array
        //via the method addMultiple of the Measurements object
        currentMeasurements.addMultiple (currentSet);

        //Calculation and output of the mean of the current Measurements
        System.out.println("| Mean value of the current measurements: " +
         (Math.round(currentMeasurements.mean()*100.0)/100.0));
      }
    //catch and output of error messages
    } catch (Exception e) {
     e.printStackTrace();
     }

    //Close the input + catch errors
    try {
     br.close();
    } catch (Exception ex) {
     ex.printStackTrace();
    }
  }     /*von main*/
}     /*von class */
```

In geoinformatics, Java is of particular importance:

- For modeling and processing two-dimensional linear geometries the JTS Topology Suite, a free program library developed in Java, is available (cf. JTS 2023).
- Several open source products such as GeoTools 2023 (cf. Sect. 3.1.6) or the free geoinformation systems uDig (User-friendly Desktop Internet GIS) or OpenJUMP are based on JTS.

3.1.5 Programming with Python in Geoinformation Systems

Python is of great importance for several geoinformation systems. For example, the proprietary geoinformation system ArcGIS includes the Python API ArcPy, which provides access to all geoprocessing tools and makes script functions and special modules available that enable the automation of GIS tasks (for more information cf. ESRI 2023). Accordingly, the freely available geoinformation system QGIS offers the possibility to extend the range of functions with the help of plug-ins (cf. QGIS 2023a). The Python API PyQGIS is available for this purpose. With over 1000 plug-ins, a large number of extensions, some of them very special, are already available for free download (cf. QGIS

2023b). In addition to this constantly growing offer, efficient plug-ins tailored to one's own specific requirements can be programmed with Python.

```
def run(self):
#The Run method contains the logic behind the plug-in
#Start of the dialog event loop
  result = self.dlg.exec_()

#After pressing the OK button in the dialog box, the calculation starts:

  if result:
#Loading the locations of the power plants (point shape)
  nuclear_lyr = QgsVectorLayer("nuclear_stations.shp", "Nuclear", "ogr")
#Display the locations on a map
  QgsMapLayerRegistry.instance().addMapLayers([nuclear_lyr])
#Calculate and save the 200000 m buffer around the sites.
  buffer = QgsGeometryAnalyzer().buffer(nuclear_lyr, "buffer_out.shp",
200000, False, False, -1)
#loading the buffer
 buffer_layer = QgsVectorLayer("buffer_out.shp", "Buffer_Nuclear", "ogr")
#Displaying the buffer in a map
  QgsMapLayerRegistry.instance().addMapLayers([buffer_layer])
#loading the city polygons (polygon shape)
  city_layer = QgsVectorLayer("city.shp", "city", "ogr")
#Displaying the polygons on the map
  QgsMapLayerRegistry.instance().addMapLayers([city_layer])
#Calculate and save the spatial intersection of buffers and urban areas
  intersect = QgsOverlayAnalyzer().intersection(buffer_layer,
city_layer,"intersect_out.shp")
#loading the intersection
  intersect_layer = QgsVectorLayer("intersect_out.shp",
"intersect_Nuclear_city", "ogr")
#Display
  QgsMapLayerRegistry.instance().addMapLayers([intersect_layer])
```

Since the Qt-Designer software is installed directly with QGIS, it makes sense to also use it to develop the graphical user interface of a plug-in for QGIS (design of input masks with e.g. dialog boxes, drop-down fields and buttons). Qt itself is one of the most important frameworks for creating graphical user interfaces for applications on different operating systems. Qt-Designer provides a graphical user interface with code written in C++, so Qt-Designer can be integrated relatively easily into QGIS, which is also written in C++. With the PyQt library available in QGIS, functions exist to address the graphical user interface created with Qt with Python, enabling an efficient triggering of a plug-in created in Python.

If the user interface is implemented via a graphical user interface, a plug-in in QGIS initially contains quite a lot of code. Therefor, the sample program only shows the core of the plug-in with the so-called run method. A very simple task is to determine which cities or parts of cities are located within a radius of 200 km around a nuclear power plant. The locations of the power plants and the polygon of the cities are available for this purpose (cf. the data layer or the layer "nuclear_stations.shp", for the shapefile data format cf. Sect. 9.3.3). The buffer zones and the intersection of buffer zones and urban areas are calculated (for these standard functions of a geoinformation system cf. Sect. 9.4.4).

This task can be expanded as required (e.g. summing up the number of inhabitants in the catchment area) or adapted to similar issues (e.g. analysing the environments of volcanoes and evaluating the catchment areas). Overall, such plug-ins offer many options to automate the analysis of existing data in a user-specific way, e.g. input masks with database connection and linking to existing data, or to program new tools (cf. Garrard 2016 and Ulferts 2017).

3.1.6 Graphics Languages and Libraries

In the early days of *graphical information processing* interactive work, which is essential for the development of graphic software and the use of computer graphics, was hardly a matter of course. Only after the breakthrough of personal computers and the emergence of graphics-oriented user interfaces and games, which accelerated the further spread of graphically interactive personal computers in a self-reinforcing process that graphic information processing and the spread of graphic systems developed rapidly. Continuous advances in hardware led to the evolution of interactive graphical workstations, which have since become the standard human-machine interface. At the same time, the development and application process progressed from elementary, hardware-dependent graphics commands (so-called low-level graphic primitives), which were supplied by manufacturers for their respective devices, to device-independent software and complex graphic systems.

The sample program shows the use of the program library *OpenGL* (Open Graphic Library) in a Java program (for the result cf. Fig. 3.5).

```
//Import libraries
import com.jogamp.opengl.GL
public class JOGLExampleAWT implements GLEventListener {
static GLU  glu = new GLU();
static GLCanvas   canvas = new GLCanvas();
static Frame   frame = new Frame("Jogl Beispiel");
static FPSAnimator   animator = new FPSAnimator(canvas, 60); /60 frames per
sec..
```

```
public void display(GLAutoDrawable gLDrawable) {
  final GL2 gl = gLDrawable.getGL().getGL2();
  gl.glClear(GL.GL_COLOR_BUFFER_BIT);
  gl.glClear(GL.GL_DEPTH_BUFFER_BIT);
  gl.glClearColor(1.0f, 1.0f, 1.0f, 0.5f);    // set background color (RGBA)
  gl.glLoadIdentity();
  gl.glTranslatef(0.0f, 0.0f, -5.0f);
// Move the scene by - 5 units in the z-direction (move the scene into the field
of view)

// Draw polygon ("lake")
  gl.glBegin(GL2.GL_POLYGON);
  gl.glColor3f(0.0f, 0.0f, 1.0f);      // set color of polygon (RGB)
  gl.glVertex3f(-1.5f, 1.5f, 0.0f);    // 1st node (from top left, clockwise)
  gl.glVertex3f(-0.7f, 2.0f, 0.0f);    // 2nd node
  // ..
  gl.glVertex3f(-1.7f, 0.0f, 0.0f);    // 6th node
  gl.glEnd();

// Draw line ("road"):
  gl.glLineWidth(5.0f);  // set line thickness, must be set before glBegin!
gl.glBegin(GL2.GL_LINE_STRIP);
  gl.glColor3f(1.0f, 0.0f, 0.0f);        // set color of line (RGB)
  gl.glVertex3f(-0.5f, -1.5f, 0.0f);   // 1st node of the line
  // ..
  gl.glVertex3f( 1.5f, 1.5f, 0.0f);    // 5th node of the line   gl.glEnd();
   }

public void init(GLAutoDrawable gLDrawable) {
  GL2 gl = gLDrawable.getGL().getGL2();
  //...
}
public static void main(String[] args) {
  canvas.addGLEventListener(new JOGLExampleAWT());
  frame.add(cavas);
  //...
  animator.start();
}
```

The *OpenGL* program library represents the cross-platform standard (in contrast to DirectDraw for Windows) for graphic libraries. It is available for various computer platforms and several programming languages and provides a cross-platform and cross-programming language programming interface for the development of 2D and 3D computer graphics (cf. OpenGL 2023). With regard to the integration in Java programs, the

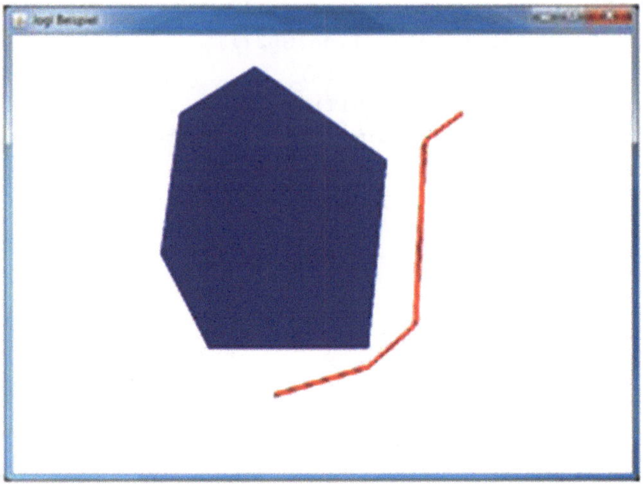

Fig. 3.5 OpenGL scene, generated with Java and JOGL

platform-independent open source program library JOGL (Java Bindings for OpenGL) is of great importance (cf. JOGL 2023). The Visualisation Toolkit (VTK) is an open source C++ program library for 3D computer graphics, image processing and visualisation, which has interfaces for integration into Python and Java programs, among others (cf. VTK 2023). In addition, with Java 2D and Java 3D, there are extensive class libraries for the generation, manipulation and display of two- and three-dimensional graphics (for an introduction to graphics programming cf. Klawonn 2010, Zhang and Liang 2007 and Zimmermann 2012).

Figure 3.5 with the associated code shows an example of the representation of an OpenGL scene in Java. The relevant program section illustrates how a polygon with a blue fill and a red line are displayed with the help of JOGL.

The so-called GeoTools, which comprise an open source Java code library that provides a wide range of methods for processing geodata (cf. GeoTools 2023), offer extensive options for more than just graphical presentation. With the help of the functions offered in this program library, geodata that are available in the proprietary shapefile data format of the company ESRI (cf. Sect. 9.3.3) can be processed directly. The sample program reads the street lines as UTM coordinates and visualizes the data (cf. Fig. 3.6). "JMapFrame" as a subclass of "JFrame" from the standard program library SWING for designing graphical user interfaces with Java provides a window under Windows. Even though the program consists of only a little Java code, it already has a considerable functionality that comes close to that of a geographic information system in terms of visualisation.

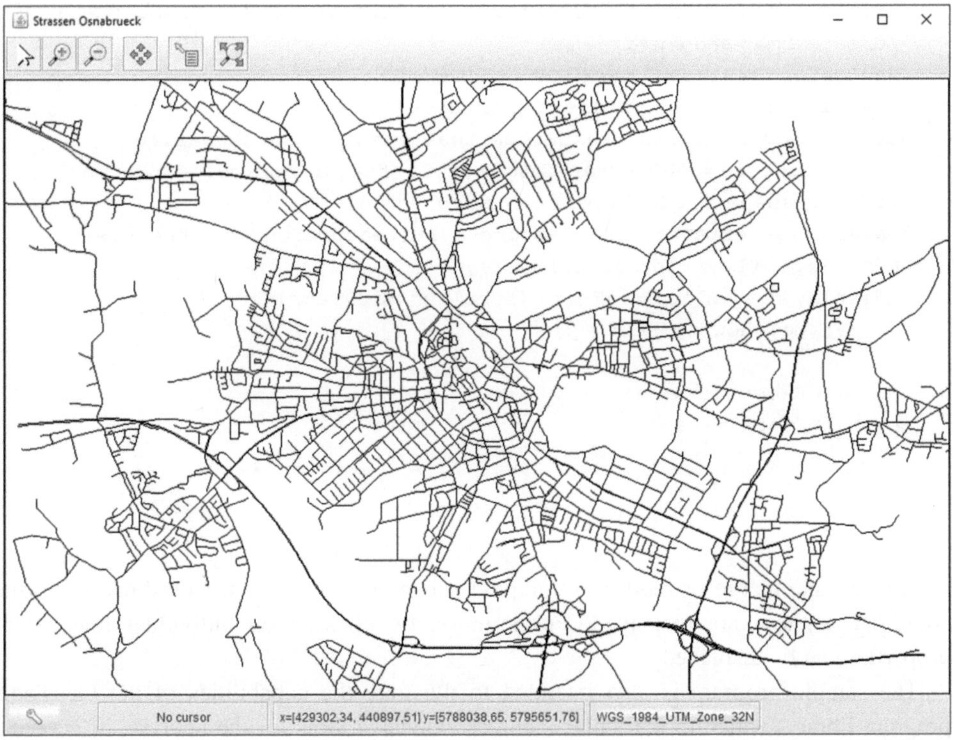

Fig. 3.6 Visualisation of linear geodata with GeoTools

```
import java.io.*;
import org.geotools.data.FileDataStore;
import org.geotools.data.FileDataStoreFinder;
import org.geotools.data.simple.SimpleFeatureSource;
import org.geotools.map.FeatureLayer;
import org.geotools.map.Layer;
import org.geotools.map.MapContent;
import org.geotools.styling.SLD;
import org.geotools.styling.Style;
import org.geotools.swing.JMapFrame;

public class Prog_OS {
 public static void main( String[] args ) {
  //path to shapefile
  File shape_osnabrueck = new File("path/roads_os_utm.shp");
  try {
   //add file to shapefile data store and get feature source
    FileDataStore fileStore = FileDataStoreFinder.getDataStore
(shape_osna-brueck);
```

```
      SimpleFeatureSource featureSource = fileStore.getFeatureSource();
     //create a map and set a title
      MapContent testmap = new MapContent();
      testmap.setTitle("Strassen Osnabrueck");
     define a simple style to render the features from the shapefile
      Style mapstyle = SLD.createSimpleStyle(featureSource.getSchema());
     //add the shapefile as layer to the map
      Layer strassenLayer = new FeatureLayer(featureSource, mapstyle);
      testmap.add Layer(strassenLayer);
     //display the map in JMapFrame (subclass of JFrame)
      JMapFrame.showMap(testmap);
    }
    catch (IOException e) {
      e.printStackTrace();
    }
  }
}
```

A user can enlarge or reduce the graphic and move the section. Coordinates and the coordinate system are displayed. Furthermore, by clicking on individual lines, their properties can be displayed.

The sample program is also intended to illustrate the capabilities of the GeoTools program library, which in particular includes extensive tools for the analysis of geodata. Furthermore, own algorithms can be implemented with the present approach.

3.1.7 Programming Applications for Browsers

The *Hypertext Markup Language* (*HTML*), which is a collection of commands and a simple language for designing web pages, is of central importance for applications in the World Wide Web (cf. Sect. 2.8.3). HTML5 is the core language of the Web. Compared to its predecessor HTML 4.01, the new standard offers a wide range of new functions such as video, audio, local storage, verifications of form entries, and dynamic 2D and 3D graphics, which were previously not directly supported by HTML4, i.e. not without additional plug-ins.

The integration of a Java program into a website is done via so-called *Java applets,* which are executed in a virtual machine (JVM cf. Sect. 3.1.1) isolated from the operating system. These applets are subject to security-critical restrictions which, for example, do not allow access to local resources such as the hard disk. These restrictions can be partially lifted by the user's consent. Java applets, which consist of hardware-independent Java bytecode (i.e. compiled Java source code), extend HTML pages with functions such as interactive animations. This is where the basic principle of Java comes into play. The Java byte code, which thanks to its enormous compactness is ideally suited for transmittance on the Internet, is translated and executed (only at runtime) by the so-called *Java Virtual*

Machine (*JavaVM, JVM*) into the corresponding machine instructions of the computer system. Many webbrowsers have a virtual machine as a fixed component or can be extended by corresponding plug-ins (for web technologies and application issues of Java applets cf. Sect. 2.8.3).

In contrast to Java applets, *JavaScript* is a so-called scripting language that now has many keywords and structures of Java, but was developed independently of the Java programming language (first under the name LiveScript and then as JavaScript for marketing reasons due to the great success of the Java language). JavaScript is integrated into an HTML page and only interpreted at runtime, if the browser has the appropriate functionality. Therefore, the execution usually takes more time compared to a (compiled) Java applet. The source code of JavaScript is open in an HTML page (or in a separate file with reference from the HTML page).

Similar to JavaScript, Microsoft developed the scripting language *JScript,* which directly competes with JavaScript, and the scripting language *VBScript*, which is based on Visual Basic. Microsoft also developed *ActiveX*, a collection of technologies for Internet applications, whereby *ActiveX controls* resemble Java applets. However, unlike the platform-independent Java applets, ActiveX controls are restricted to Windows operating systems. They grant access rights to the client, which is a significant security risk. For example, external programs for spying on data or computer viruses can be infiltrated.

With the freely available script language PHP, dynamic websites and (database-supported) applications can be created with quite little effort. PHP code is processed on the server side, but is not transferred to the webbrowser, but to an interpreter on the webserver (cf. Fig. 2.12) and therefore cannot be viewed. The PHP program can send an HTML page back to the client or, for example, generate an e-mail. PHP is characterised above all by its easy learnability, low server load, wide range of functions and broad support for various SQL databases.

With the explosive development of the World Wide Web, graphic applications have also increased by leaps and bounds almost at the same time. The *Scalable Vector Graphics* (*SVG*) format was introduced for vector graphics in order to provide a uniform language standard for graphics applications, which is essential for universal information transfer and display in every browser or platform. *SVG* is a language formulated in *XML (eXtensible Markup Language)* that was standardised by the World Wide Web Consortium (W3C) to describe two-dimensional vector graphics. As a completely open standard, the SVG format offers great flexibility due to its openness. The strict separation of content, structural and design information ensures clarity and good manageability. However, since this is a language formulated in ASCII code, such graphics can also be created with simple editors.

The following code can be executed in any modern browser without an additional SVG plug-in. This draws a blue polygon and a red line as in the above OpenGL program example (cf. Fig. 3.5).

```
<html>
 <body>
  <svg width="500" height="500">
   <polygon points="40,128
     90,128
     108,40
     60,10
     30,30
     15,80"
    style="fill:blue;
     stroke:black;
     stroke-width:0;" />
   <polyline points="60,140
     100,128
     120,50
     124,32
     140,20"
    style="fill:none;
     stroke:red;
     stroke-width:3;" />
  </svg>
 </body>
</html>
```

3.1.8 Programming Applications for Mobile Devices

The programming of applications for smartphones and tablet computers has become very important. These mobile devices have become an indispensable part of everyday life. They enable a wide range of applications that also address geoinformatics issues (e.g. visualisation of city maps, navigation and finding the shortest routes, GPS positioning). Two operating systems dominate the market for mobile devices. The IOS operating system for Apple devices has a market share of about 15%, while the Android operating system, which is used by all manufacturers, has a share of over 80%.

Android was created in the context of the reaction to the iPhone, which was released earlier in 2007 by Apple, when Google together with many technology companies such as Samsung or HTC of the Open Handset Alliance (OHA) developed a mobile phone operating system and made it available in 2008. The aim of the OHA was to develop a cheaper alternative to the iPhone. Accordingly, Android can be used on various mobile devices across manufacturers and is available as free software, which ultimately explains its high market share (cf. Android 2023a).

AndroidStudio is freely provided by Google as the official development environment for Android apps (cf. AndroidStudio 2023). AndroidStudio offers extensive tools for app development, such as the simulation of an app under Windows.

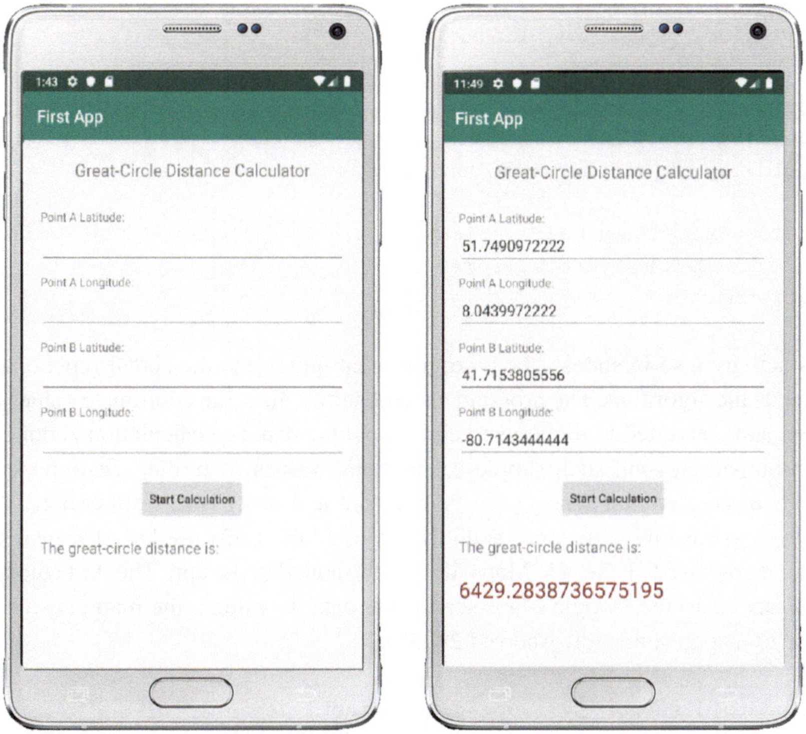

Fig. 3.7 App development with AndroidStudio

The basic principle of app development under the Android operating system is illustrated by an app that expects the user to enter the coordinates of two locations and then calculates the distance over the great circle between these two locations (for the calculation algorithm cf. Sect. 4.2.3). The user interface of the app on the smartphone is determined by so-called activities, each grouping the layout and the actual program into individual logical components. The layout of an *activity* is defined in an XML file. As an example, the definition of a text field and a button that should start the calculation is given (cf. Fig. 3.7):

```
<?xml version="1.0" encoding="utf-8"?>
<RelativeLayout
  xmlns:android="http://schemas.android.com/apk/res/android"
  android:layout_width="match_parent"
  android:layout_height="match_parent">

  <TextView
    android:id="@+id/heading"
    android:layout_width="wrap_content"
    android:layout_height="wrap_content"
```

```
...
    android:text="Great-Circle Distance Calculator" / >
...
<Button
  android:id="@+id/button_start"
  android:layout_width="wrap_content"
...
  android:text="Start Calculation"
  android:onClick="buttonClicked"/ >
</RelativeLayout>
```

This activity also includes a Java program that represents the actual functionality and implements the algorithm. The program excerpt shows, how the coordinate values are read in as text and converted to numerical values, how the distance calculation is done and the result is output. Beyond such simple calculations, values of further smartphone sensors (e.g. GPS or acceleration sensor) can be read out and processed. Furthermore, there are many ways to customize the app. With the Maps SDK (Software Development Kit), for example, maps based on Google Maps files can be added to the app. The API automatically manages access to the Google Maps server, the data download, the map display, and the response to map gestures (cf. Android 2023b).

```
import android.support.v7.app.AppCompatActivity;
import android.os.Bundle;
import android.view.View;
import android.widget.EditText;
import android.widget.TextView;
import android.widget.Toast;
public class MainActivity extends AppCompatActivity {
 private static final double EARTH_RADIUS = 6371000.0;
 @Override
  protected void onCreate(Bundle savedInstanceState) {
   super.onCreate(savedInstanceState);
   setContentView(R.layout.activity_main);
  }
  public void buttonClicked(View v) {
   EditText lat1  = findViewById(R.id.enter_lat_1);
   EditText lon1  = findViewById(R.id.enter_lon_1);
   EditText lat2  = findViewById(R.id.enter_lat_2);
   EditText lon2  = findViewById(R.id.enter_lon_2);
  if (lat1 != null && lon1 != null && lat2 != null && lon2 != null) {
   if (lat1.getText().length()  == 0 ||
    lon1.getText().length()  == 0 ||
    lat2.getText().length()  == 0 ||
    lon2.getText().length()  == 0) {
```

```
            Toast.makeText(this, "You have to enter 4 values!",Toast.
LENGTH_LONG).show();
    } else {
    // get the values from the EditText-elements and parse to double, convert
    // to radians.
        double double_lat1 = Math.toRadians(Double.parseDouble(lat1.getText
().toString()));
        double double_lon1 = Math.toRadians(Double.parseDouble(lon1.getText
().toString()));
        double double_lat2 = Math.toRadians(Double.parseDouble(lat2.getText
().toString()));
        double double_lon2 = Math.toRadians(Double.parseDouble(lon2.getText
().toString()));
    // calculate distance
    double distance = EARTH_RADIUS * Math.acos(Math.sin(double_lat1) *
    Math.sin(double_lat2) + Math.cos(double_lat1) * Math.cos(double_lat2) *
    Math.cos(double_lon2 - double_lon1));
    // now display the result
    TextView tv = findViewById(R.id.result);
    if (tv != null) {
     tv.setText(Double.toString(result));
    }
    }
   }
  }
}
```

3.2 Data and Data Types

3.2.1 Scale Levels

The treatment of task and issues with the help of computer systems takes place on the basis of input data that describe facts or system states. Features (attributes) and associated concrete feature values (attribute values) are used for this purpose. For example, a soil is characterised by the characteristics ph-value (characteristic value e.g. 5.5) or lime content, by the water content or the clay content, by the soil type (e.g. ranker) or also by the soil fertility, whereby, among other things, chemical and physical properties are defined. The feature soil type, instead, has no measured values and can only be determined qualitatively by looking at the soil horizons.

As this example shows, the features can be scaled very differently: *metric, ordinal* or *nominal scale level.* Therefore, different data types are available in computer systems to describe these measurement scales. In addition, very different accuracy requirements apply to individual data, depending on how it is used. Years, coordinates in the UTM coordinate system for a real estate cadastre or radiation data in remote sensing each require different levels of accuracy and different storage requirements. To solve these requirements, the various programming languages and also application programs (such as a geoinformation system) provide different data types.

The source data are often not disordered, but structured. So there are certain data structures with logical-content relationships. To model these data structures, user-specific data types can be derived from the standard data types. If the focus is on implementation in a programming language, one speaks of concrete data types. In contrast, abstract data types focus on the properties and operations that are defined or possible on a data structure (cf. Sect. 3.2.4).

3.2.2 Standard Data Types

The representation of information in computer systems by bit sequences was already discussed in Sect. 2.5. This section continues with data types in programming languages and in application programs.

3.2.2.1 Representation Data as Bit Sequences

The processing of information in a computer is based on bit sequences. Thus, all data must first be represented as bit sequences and, after internal processing by the computer or processor, must be transformed back into a generally readable form. The internal representation of data only as bit sequences makes this process of encoding and decoding necessary. In the internal representation of numbers and alphanumeric characters as bit sequences, the eight bits 01011010 can represent the integer 90 or the character Z (cf. Sect. 2.5). However, a distinction is only possible if the meaning of the bit sequence is known in addition to the content. By means of the corresponding data type it is recognised, for example, whether it is a number or a character. Thus, at least data types for numbers, characters (alphanumeric characters, special characters) and for logical values must be available.

In practice, it has proven to be useful to distinguish between integers and real numbers in the internal representation. In this way, integers can be represented in a more memory-efficient manner, which also means faster data access. Elementary data types or standard data types distinguish between integers (for example, int in Java), real numbers or floating point numbers (for example, double in Java), alphanumeric characters such as letters, and logical values. In addition, other standard data types are available in individual programming languages, such as date types or data types for handling complex numbers. In different programming languages or compilers as well as program systems, however, special features occur such as different names (instead of integer, for example, only int), different storage technologies or even value ranges.

The importance of the data types for the accuracy should not be underestimated in reality! When displaying UTM coordinates more than nine significant digits are necessary,

Table 3.2 Integer data types in Java

byte	−128... + 127	8 bit
short	−32768... + 32767	16 bit
int	−2147483648... + 2147483647	32 bit
long	−9223372036854775808... + 9223372036854775807	64 bit

the last digit is not rounded. This enables a centimetre-accurate calculation, which is required in the cadastral system. Here a number is not stored in 32 bits, but in 64 bits (cf. the data types "float" and "double" in Java, cf. Table 3.4):

East	434000.12 (in m)
North	5735256.15 (in m)

A data exchange of geodata often takes place in the UTM format with (at least) two decimal places. In the case of single precision storage, the values of the decimal places cannot be stored. For example, in Java with the data type float, a number can only be reproduced with an accuracy of 6–7 digits (cf. Table 3.4). Thus, the number 5,735,256.15 or $0.573525615 \cdot 10^6$ is only stored internally as $0.5735256 \cdot 10^6$. This can lead to the problem that in a geoinformation system polylines, which have to form a polygon without gaps, are not closed because their respective end coordinates do not match. In addition, a nasty trap is hidden here: The lack of precision is not displayed, rather a single precision number can even be printed with four decimal places. However, the digits behind the decimal point are determined randomly.

3.2.2.2 Default Data Type Integer

An integer is represented as a binary number by a bit sequence, for which 2 bytes (16 bits) are normally available. Since one bit is required to store the sign, the largest number to be represented is $2^{15} = 32{,}768$. Thus, there is a limited value range between –32,768 and +32,767 (cf. Table 3.2). In many programming languages at least a 4-byte representation (32-bit representation) is possible. In the present example, the highest bit indicates the sign, 0 stands for a positive and 1 for a negative sign. This results in the number 53 as an integer in 16-bit representation (now to be read from right to left): $53 = 1 \cdot 2^0 + 0 \cdot 2^1 + 1 \cdot 2^2 + 0 \cdot 2^3 + 1 \cdot 2^4 + 1 \cdot 2^5$ as an integer in 16-bit representation (now to be read from right to left):

15	14	13	12	11	10	9	8	7	6	5	4	3	2	1	0
0	2^{14}	2^{13}	2^{12}	2^{11}	2^{10}	2^9	2^8	2^7	2^6	2^5	2^4	2^3	2^2	2^1	2^0
0	0	0	0	0	0	0	0	0	0	1	1	0	1	0	1

The standard operations belonging to the standard data type integer are: addition, subtraction, multiplication, integer division with truncated remainder, modulo function (remainder formation for integer division), comparison operations.

3.2.2.3 Default Data Type Floating Point

A real number can only be converted as a finite decimal number, since only a finite number of "memory cells" are available (e.g. 23 for storing a mantissa cf. Table 3.3 and Table 3.4). Thus, even in the case of an infinite decimal fraction (cf. the number pi or 1/7), only a finite and therefore an "inaccurate" representation is achieved. Real numbers are represented in computer science as floating point numbers consisting of three parts: the sign V, the exponent E and the mantissa M. The decimal number 26625 is then written in the form $+0.26625 \cdot 10^2$ as a normalised floating point number to base 10. In this case, $E = 2$ and $M = 26{,}625$. The mantissa and exponent also have a sign.

To represent a number as a bit sequence, the number 2 is used as the base. Then a floating point number normalised to the base 2 is one in which the exponent is chosen so that the number can be represented in the form $\pm 1.m_1m_2m_3...m_n \cdot 2^E$. A normalised floating-point number is easily obtained in both the dual and decimal systems by successively shifting the decimal point one place to the left or right and simultaneously increasing or decreasing the exponent by 1. For the number 26.625 (= 11010.101, cf. Table 2.1) as a dual number, the following applies:

$$11010.101 = 11010.101 \cdot 2^0 = 11010.101 \cdot 2^{00000000}$$
$$11010.101 = 1101.0101 \cdot 2^1 = 1101.0101 \cdot 2^{00000001}$$
$$\ldots$$
$$11010.101 = 1.1010101 \cdot 2^4 = 1.1010101 \cdot 2^{00000100}$$

Further arrangements are made for storage:

- 0 is set for a positive and 1 is set for a negative sign 1.
- The normalisation is done by placing the first digit not equal to 0 directly in front of the point. This 1 is no longer stored, because it must always appear in the present definition of normalised floating point numbers.
- For a 32 bit representation, a so-called bias of 127, i.e. 01111111, is added to the exponent and the result, the so-called characteristic, is stored as an unsigned 8-bit

Table 3.3 Internal storage of 32-bit floating point numbers

Numeric value	sign (1 bit)	characteristics (8 bit)	mantissa (23 bit)
26.625	0	10000011	10101010000000000000000
1234711	0	10010011	00101101011100010111000
–293432165	1	10011011	00010111110101101011011
–0.00015	1	01110010	00111010100100101010001

Table 3.4 Floating point data types in Java

	float	double
Significant digits	6–7	15–16
Space requirement (bytes)	4	8
Bit for exponent	8	11
Bit for mantissa	23	52
Smallest positive number	$2^{-127} \approx 1{\cdot}10^{-38}$	$2^{-1023} \approx 1{\cdot}10^{-308}$
Largest positive number	$2^{+127} \approx 1{\cdot}10^{+38}$	$2^{+1023} \approx 1{\cdot}10^{+308}$

number. With this further method of representing positive and negative numbers, an exponent between -127 and +128 can be stored here. An exponent of -127 becomes the characteristic 00000000, 1 the characteristic 10000000, and $128 = 2^7 = 10000000$ the characteristic 11111111. For the characteristic in this example we get: $00000100 + 01111111 = 100000011$

The IEEE (Institute of Electrical and Electronics Engineers) has standardised two formats: 32-bit floating-point numbers (single precision) and 64-bit floating-point numbers (double precision) (cf. Gumm and Sommer 2013 pp. 28):

	sign	exponent	mantissa	bias
Single precision	1 bit	8 bit	23 bit	127
Double precision	1 bit	11 bit	52 bit	1023

The standard operations belonging to the standard floating point data type are: addition, subtraction, multiplication, division, comparison operations, various mathematical functions such as sqrt or trigonometric functions, although the scope can be vary different depending on the programming language.

3.2.2.4 Default Data Type Character

If text characters are to be displayed in a computer, the alphabet and punctuation marks must be encoded in bit sequences. For the representation of all characters, 7 bits are sufficient, which open up 128 different options (26 lower and upper case letters, punctuation marks, special characters such as & and non-printable characters or formattings marks such as the return key result in almost 100 characters).

The representation is normally done with 1 byte per position, i.e. 8 bits, which results in 256 different characters (coding according to the so-called extended ASCII code or according to the ANSI code, cf. Sect. 2.5.5). Then special characters such as characters from the Greek alphabet or simple graphic characters can be implemented. In addition, the UNIX computers only use the standardised ASCII characters from 0 to 127.

The standard operations belonging to the standard data type character (e.g. in Java char) are: conversion of a character into the corresponding decimal value of the ASCII or

ANSI character, comparison operations based on the ASCII or ANSI decimal values (e.g. "A" < "B" because 65 < 66, this allows sorting).

3.2.2.5 Default Data Type Logical Value

The range of values of this data type are the logical values "true" and "false". They are represented by means of exactly one bit. The standard operations belonging to the standard data type boolean (Java) are logical operations such as NOT, AND, OR.

3.2.2.6 Data Type String and Pointer

The data type string (Java) denotes a character string and is composed of a sequence of characters that are of the character type. Normally, a string variable offers space for a maximum of 255 characters.

Pointers are a special data type, but they are not implemented in all programming languages (e.g. not in Java and Python). In contrast to the other data types, a pointer does not contain a direct value, but the address of a value. In the C program mentioned (cf. Sect. 3.1.4.1), the function "strtok" splits up a string using separators (here spaces). At the first call "strtok" must be initialised with the input string, i.e. with xin (*split = strok (xin, " "). The pointer variable points to the first split. For subsequent calls, the value NULL is passed instead of xin, since "strtok" is already initialised. The pointer *split always points to the first character of the respective section, the end of the respective section is set with \0 in xin, the string is changed in this process. Therefore, when using strtok, only one copy of a string should be passed. In the while loop, the pointer "jumps" from section to section and counts the number. In the for-loop, the pointer to the respective section is stored in line[j], so that this section can be fetched in the second for-loop and included in the data structure "x" (cf. end of the program in C at the end of Sect. 3.1.4.1).

3.2.3 Structured Data Types

The standard data types enable the definition of structured data types, for which the term data structures is frequently used. Data structures are introduced to group together data elements that are related to each other. Based on the standard data types and the standard data structures, further user-specific or more complex data types can then be defined.

A very simple data structure is the *data set.* In the C programming language, this can be declared by the statement:

```
struct observations{
 char id[7], char month[10];
 float values[20];
 int number_values;
};
```

This represents a data set, i.e. a line of a file. The data type "struct" is useful when a record is composed of different standard data types. A single element of a record is accessed by its name (in the example, among others, by observations.month). In contrast, an *array* (field) is a series of similar elements (data of the same type). Access is possible with the help of an index, which identifies the position of the element within the array (field).

```
3, 9, 4, 7, 6, 1, 5, 0, 3, 2, 7, 2, 5, 1, 9, 1, 3, 4442, 4, 7
```

These data can be represented by a simple array: int vector[20] in the C programming language, where this array is of type integer, or as a list x = [3, 9, 4, 7, 6, 1, 5, 0, 3, 2, 7, 2, 5, 1, 9, 1, 3, 4442, 4, 7] in the programming language Python. The element vector(18) or x [2] has the value 4.

1	2	3	0	0	0	0	0	0	0	0	0	0	0	0	0	0	0	19	20
2	2	0	0	0	0	0	3	0	0	0	0	0	0	0	0	0	18	0	0
3	0	3	0	0	0	0	0	4	0	0	0	0	0	0	0	17	0	0	0
4	0	0	4	0	0	0	0	0	0	0	0	0	0	0	16	0	0	0	0
5	0	0	0	5	0	0	99	0	0	0	0	0	0	15	0	0	0	0	0
6	0	0	0	0	6	0	0	0	0	0	4	1	14	8	9	3	2	1	0
7	0	0	0	6	0	7	0	0	0	91	0	13	0	0	0	0	0	0	0

This amount of data can be represented by a two-dimensional array in the C programming language:

```
int matrix[7] [20];
```

or in the Java programming language:

```
int [ ] [ ] fffname= new int [7] [20];
```

The element j = matrix [4][7] or x = fffname [4][7] has the value 99, since in C, as in Java, the number is incremented starting from 0, in this case from 0 to 6 or from 0 to 19.

This two-dimensional array can also be represented as a one-dimensional array or list:

```
int [ ] vvv = new int [140]
```

The rows are stored one after the other, so that the element fffname[i, j] is assigned the element vvv[k] by the index calculation: k = i–20 + j.

The last mentioned structured data types seem to be obsolete in object-oriented programming languages. Java, for example, does not know a record type. An object emulates this concept (cf. class Measurements in the program example in Sect. 3.1.4.3) and extends it considerably (e.g. with associated methods). The above two-dimensional matrix can also be reproduced as an object. However, a two-dimensional array offers considerable advantages if numerical operations such as matrix multiplication are to be performed, i.e. if the indices of the matrix are used for calculation.

3.2.4 Abstract Data Types

A data structure together with operations defined on it is called an *abstract data type* (*ADT*). The basic forms of abstract data types include special types of linear lists (stacks, queues, chained lists) as well as trees that can efficiently map hierarchical relationships in terms of data technology. For a user, only the external view is interesting here, which describes the performance of the data types or the possible operations that are necessary to implement certain algorithms. The concrete implementation may be unknown or “hidden”.

Only basic fconcepts are outlined here. A detailed treatment of these data types must take place in connection with the explanation of algorithms (operations), which operate on these data types. In addition to the abstract data types listed here, to which special operations or standard procedures belong in each case (e.g. searching in trees), the term abstract data type is used in a general sense in object-oriented programming and an object is used as a data structure with the associated methods, i.e. understood as an abstract data type.

3.2.4.1 Stack

The abstract data type *stack* is a linear field in which only the topmost data element can be edited. Data elements can only ever be placed on top of the stack or removed. A stack is organised similar to a penny box for holding parking coins according to the lifo principle (last in – first out):

The basic functions associated with this data type are: place an element on the stack, remove the last element placed on the stack, get the element last placed on the stack, query whether the stack is empty or full. The implementation can be done in a variety of ways, e.g. using lists and pointers. A considerable degree of flexibility is achieved in the process. For example, the size of the memory does not have to be specified in advance, and a large amount of memory can be mobilised using pointer variables, since the stack is only full if the computer’s memory is completely occupied. With the help of a stack, any linear recursion can be formulated and calculated iteratively, i.e. without recursion (cf. Sect. 3.3.3).

3.2.4.2 Queues

A *queue* is a linear array that organises the fifo principle (first in – first out). An object can only be inserted at one end and removed at the other end. Thus, the element that has been in the queue for the longest time is accessed. Functions and implementations are similar to the stack.

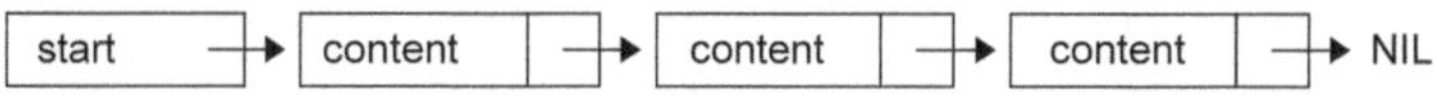

Fig. 3.8 Single linked list

3.2.4.3 Chained Lists

Chained lists are particularly efficient data structures that can be used to process data volumes of indefinite size without wasting memory. Single and double linked lists represent the most important dynamic data structures. For example, the type declaration does not have to specify a certain size of an array in advance, which in extreme cases can be too small and then lead to uncontrollable "results". If required, a linked list can be supplemented by further elements (Fig. 3.8).

A linked list can be implemented in C, for example, as a sequence of elements that are linked to one another by pointers. Each element of the list consists of the actual data element, which can be any data type, and a pointer to the next element in the list. In the case of the last list element, the pointer is equal to "NIL" (i.e., "nothing"). This chained list in C has a so-called head pointer which points to the first element of the chained list.

Basic operations on chained lists are: building a list from n elements, inserting, deleting and copying an element, traversing a list with searching for an element or determining the length of a list. In geoinformation systems, chained lists are used in many ways for storing lines, i.e. sequences of coordinate pairs. Typical applications occur here: At the beginning of digitising a line, the number of coordinates is unknown. The length of different lines and therefore the sequence and number of coordinates differ. Points must be inserted and deleted (to further, more complex data structures and associated algorithms for the storage of coordinates cf. Worboys and Duckham 2004 Section 6).

3.2.4.4 Trees

Trees represent the most important nonlinear data structures. Trees are mainly used for modeling hierarchical relationships (father-son relationships) and for recursive object structures. A classical field of application is graph theory including in particular the processing of issues on transport networks (for storage raster data in geoinformation systems with quadtrees or trees cf. Worboys and Duckham 2004 pp. 236). For example, the possible routes in a transport network between two points A and B can be represented as a sequence of edges between nodes A and B in a tree. The search for an optimal route then means determining the shortest path in a tree.

A *tree* is composed of nodes and edges. If the subtrees for each node are in a fixed order, the tree is ordered. A *binary tree* is empty or consists of one node and two binary trees (recursive definition, cf. Fig. 3.9). A tree is called balanced if the maximum path length is smaller than log n (n number of nodes). The path lengths from the root node to the leaves must not differ by more than 1. The depth of a tree is the maximum of the depths of its nodes, i.e. the lengths from the root to the nodes, where the length of a path is equal to the number of associated nodes minus 1. The depth of a tree with n nodes is between $\log_2$ n and

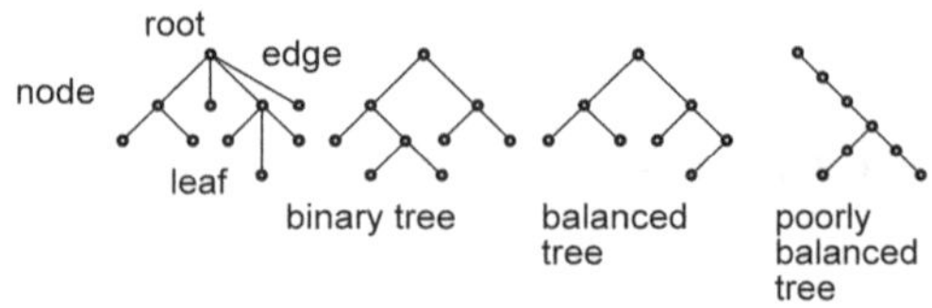

Fig. 3.9 Standard shapes of trees

(n – 1). In a weighted tree, values are assigned to the edges (e.g. path length between two nodes in a transport network).

Basic operations on trees include building a tree from n nodes, inserting, deleting and copying an element. However, so-called *traversals,* i.e. systematic options to visit each node, are of particular importance. This can be used above all for route planning, for which many applications in networks exist.

3.2.5 Files

A data record usually consists of several data fields (so-called items), each of which contains attribute values. A distinction is made between *logically* or *content-related defined data records* and *physical data records*, with regard to the technical storage form. A data record can contain several logical data records, which can therefore be distributed over several physical data records. Data records of the same type that belong together on the basis of content-relatedcriteria are collectively referred to as a file. Several files with logical dependencies or relationships form a database (for the distinction between file system and database system cf. Sect. 8.1 for more details). Thus, the logical data organisation is structured hierarchically:

- Database
- File
- Data record
- Data field (item)

From the point of view of computer science, the storage and access forms of data or files are interesting, whereby the mechanisms cannot be separated. According to the storage form, a distinction is made between how data or files are stored on external memories.

In the case of *sequential storage,* data records are stored one after the other without gaps, which are often initially sorted according to an ordering characteristic (i.e. according to attribute values of an item). A particular record can then only be accessed by reading all previously stored data. This can result in a very time-consuming search for specific data, so that this type of storage or access is suitable for data backup, but less suitable for ongoing operation with multiple reading or writing of data. This type of storage is typical, for example, for files on magnetic tapes which only allow sequential storage ("one-dimensional" storage in the movement direction of the tape).

Index tables are used for *indexed storage* work in which an order feature of a data record is stored with the associated absolute storage address.

In the case of *scattered storage,* the data are scattered on the data carrier, with a mathematical relationship between an order feature of the data record (key attribute) and the physical storage address. The physical address of the data record is then calculated using a special algorithm (so-called hash function).

3.3 Algorithms

3.3.1 Definition and Characteristics

The basis of a program that is executed on a computer system is always an *algorithm,* or in other words: A program is the realisation of an algorithm (cf. Sect. 2.4.1). In general, algorithms are (almost) inevitably of central importance in computer science. The efficiency of a program solution for a task is determined less by the programming language or by a sophisticated programming technique than by a suitable algorithm. It is precisely the careful selection of an optimal processing strategy and a suitable algorithm that can lead to a drastic reduction in the runtime of a program.

Both computer science as well as in numerical mathematics have a very large number of algorithms for different areas (cf. e.g. Ottmann and Widmeyer 2002, Sedgewick and Wayne 2011, Knuth 2011, Wirth 2013). Extensive standard procedures exist for many tasks such as sorting. In practice, extensive program libraries can be used. The NAG Library of the Numerical Algorithm Group (NAG) can be regarded as the most extensive, commercially available collection of numerical algorithms, with currently more than 1700 routines, which are now available for various programming languages such as FORTRAN, C, Java and Python (cf. NAG 2023). In contrast, the GNU Scientific Library (GSL) provides over 1000 functions for numerical computations in C and C++ under various UNIX derivatives (now also in a compiled version for Windows) (cf. GSL 2023). With regard to applications of Python in geoinformatics, reference should be made to NumPy, a program library for Python for easy handling of multidimensional arrays, to SciPy (cf. SciPy 2023), a Python library with numerical algorithms, as well as to Python Package Index (PyPI), a collection (so-called repository) of software for the Python programming language,

Of particular importance for geoinformatics is the JTS Topology Suite, a free program library developed in Java for the modeling and processing of two-dimensional, linear geometries (cf. JTS 2023), GEOS (Geometry Engine Open Source) is the transferring to C++ (cf. GEOS 2023). In addition, Shapely is a Python programming package for spatial geometries that is based on GEOS (cf. Shapely 2023).

Against the background of an almost unmanageable variety only general principles of algorithms are to be discussed here as well as examples of selected solutions are shown, which are also intended to illustrate the "ingenuity" of such procedures.

In general, an *algorithm* is a general calculation rule for solving a problem, which consists of several elementary steps that must be executed in a specified order. The implementation is independent of a particular programming language.

According to Levi and Rembold (2003 p. 136) a *procedural algorithm* has several general characteristics:

- An algorithm must be able to be executed by a machine. The information required for the algorithm to run must be available at the beginning.
- An algorithm must be universally valid. The size of the data set to which the algorithm is applied must not be restricted.
- The algorithm consists of a series of individual steps and instructions. Each step must be precisely defined in terms of its effect.
- An algorithm must end after a finite time (and after a finite number of steps). A termination condition must be formulated for the end of the algorithm.

A classic procedural algorithm therefore consists of several individual steps that are processed one after the other. A prime example of a procedural algorithm is the *bubble sort algorithm*, which belongs to the group of sorting algorithms. It is precisely these sorting procedures that are a central topic in computer science (cf. Gumm and Sommer 2013 pp. 315–345).

```
public static void sort (int [] x) {
 boolean sorted = false;   //sort completed?
 int temp;
 while (!sorted){   //as long as sorted false
   sorted = true;
   for (int i=0; i < x.length-1; i++) {
    if (x[i] > x[i+1]) {
     temp   =  x[i];    //if current number is greater than
     x[i]   =  x[i+1];    //subsequent, then swap
     x[i+1]   =  temp;
     sorted   =   false;    //to false, one more element to sort
    }
   }
 }
}
```

With the bubble sort algorithm, all neighboring elements are compared with each other in each run and swapped if necessary. In the k-th pass, the k-largest element moves to the back and the k-smallest element moves to the front – analog to the rising of a gas bubble in a liquid. The following ten numbers should illustrate the sorting principle:

```
7 4 1 8 5 2 9 6 3 0   1. pass (0 moves through):
0 7 4 1 8 5 2 9 6 3   2. pass (first 3, then 2, then 1 forward):
0 1 7 4 2 8 5 3 9 6   3. pass (first 3, then 2 forward):
0 1 2 7 4 3 8 5 6 9
0 1 2 3 7 4 5 8 6 9   ...
0 1 2 3 4 7 5 6 8 9   In the k-th pass only the elements need to be checked from the
0 1 2 3 4 5 7 6 8 9   back to the (k-1)-th position.
0 1 2 3 4 5 6 7 8 9   The elements from the beginning to the (k-1)-th position are
                      already sorted!
```

Procedural algorithms are mainly used in imperative programming. In contrast, problem solutions with the help of predictive programming languages, which are based on non-procedural algorithms or procedures, differ fundamentally. First of all, the existing knowledge is formulated by a set of true statements and rules. The actual problem is formulated as an assertion, which is attempted to be verified or falsified using a predictive programming language such as Prolog. Thus, facts and rules are the basic elements of such decision procedures, which are based on the principles of logic. These knowledge rules enable a decision based on logical conclusions.

3.3.2 Sequential and Parallel Algorithms

(Most) classical algorithms in computer science are based on the assumption that the central processing unit of a computer system is (only) capable of processing instructions one after the other. Accordingly, algorithms must be developed as a linear sequence of instructions, so that primarily sequential algorithms exist. A *sequential algorithm* for calculating the arithmetic mean of n values could have the following structure. Here, the loop is executed n times, so that the execution time is approximately proportional to n.

```
sum = 0.0
for index in range (1 ,n):
  sum = sum + value [index]
mean = sum/n;
```

Hardware development has led to the design of parallel computers, in which the central processing unit consists of several processors of the same type, so that several operations can be performed in parallel. However, special *parallel algorithms* (and parallel programming languages) must also be developed in order to use the advantages of parallel processing. Classic applications for parallel algorithms arise in matrix operations such as the multiplication of matrices (here matrices A, B and C with $A \times B = C$):

$$\begin{bmatrix} a_{11} & a_{12} & a_{13} \\ a_{21} & a_{22} & a_{23} \\ a_{31} & a_{32} & a_{33} \end{bmatrix} \cdot \begin{bmatrix} b_{11} & b_{12} & b_{13} \\ b_{21} & b_{22} & b_{23} \\ b_{31} & b_{32} & b_{33} \end{bmatrix} = \begin{bmatrix} c_{11} & c_{12} & c_{13} \\ c_{21} & c_{22} & c_{23} \\ c_{31} & c_{32} & c_{33} \end{bmatrix}$$

with:

$$\text{with } c_{ij} = a_{i1} \cdot b_{1j} + a_{i2} \cdot b_{2j} + a_{i3} \cdot b_{3j}$$

The computational cost for square $(n \times n)$-matrices is approximately proportional to n^3 (i.e. $O(n^3)$ cf. Sect. 3.3.4) for an intuitive algorithm that reproduce the calculation rule with three loops. Somewhat more favourable is the Strassen algorithm (with $O(n^{2,807})$, which e.g. decomposes a (4×4)-matrix into four (2×2)-sub- matrices, links them to seven auxiliary matrices, adds them to four (2×2)-sub result matrices, which are finally put together fo form the (4×4)-result matrix.

In addition to these sequential algorithms parallel algorithms exist that enable a simultaneous, parallel computation of individual sub-steps. Here, the parallel computation is divided into p different processes, which are then also executed on p different processors, where the rank of the matrices n is a multiple of the number p of processes or processors (in the simple case $n = \sqrt{p} \cdot \sqrt{p}$).

In the Cannon algorithm, similar to the Strassen algorithm, the two $(n \times n)$ matrices A and B are split into $(b \times b)$-blocks (with $b = \frac{n}{\sqrt{p}}$, i.e. in the simple case divided into $\sqrt{p} \cdot \sqrt{p}$ submatrices). After initialisation, the algorithm proceeds in two basic phases:

Calculation phase: Each process has two matching blocks. The matching blocks are multiplied in parallel in the p processes and added to the corresponding sub-block of C from the last iteration.

Alignment or communication phase: The blocks are moved cyclically and sent to the next p processes.

The iteration ends after $\sqrt{p}$ steps.

The algorithm is based on the idea of always processing different submatrices in one iteration step, which enables parallelisation. Besides the pure multiplication time, which is more favorable in the Strassen algorithm, the computation time can be improved by distributing the calculation over several processes ($O(n^3/p)$). However, the communication time between the processes must be taken into account. It can be shown that from a number of eight processors the Cannon algorithm is more effective (cf. Quinn 1993 p. 195 and pp. 281).

3.3.3 Iterations and Recursions

3.3.3.1 Iterative Strategies and Approximation Methods

In structured programming, the term *iteration* is used as a comprehensive term for repeated executions of instructions depending on conditions. Instructions, sequences of instructions or also functions are repeatedly run through or called. Normally, iterations are implemented by loops (cf. Sect. 3.1.4.1). In addition, iteration is also used to describe a process strategy for obtaining approximations for a solution by repeatedly using a calculation rule.

A classic iteration method in mathematics is *Newton's method* for calculating the zeros of a function $f(x)$ (cf. Schwarz and Köckler 2011 pp. 192). The iteration rule is generally (with $f'(x_i)$ as the first derivative of $f(x_i)$):

$$x_{i+1} = x_i - \frac{f(x_i)}{f'(x_i)}$$

A result value is repeatedly inserted into the formula and improved. The iteration is terminated if the last two values no longer differ significantly. Newton's iteration method is used in its special form to solve a square root as calculation the zero of the function $f(x) = x^2 - a$ (Heron method, cf. Sect. 2.4.1).

This iteration procedure can also be applied to functions of several variables $G(x) = y$ (with $x = (x_1, x_2, \ldots, x_n)$ and $y = (y_1, y_2, \ldots, y_n)$), which, however, are first brought to the form $F(x) = 0$, so that a zeropoint calculation can be carried out. In this way, nonlinear systems of equations can also be solved (see example below). The iteration rule is written in matrix notation, multiplied by the inverse Jacobi matrix J^{-1}.

$$\begin{pmatrix} x \\ y \end{pmatrix}_{i+1} = \begin{pmatrix} x \\ y \end{pmatrix}_i - J^{-1}\begin{pmatrix} x \\ y \end{pmatrix}_i \cdot F\begin{pmatrix} x \\ y \end{pmatrix}_i \quad \text{with } F\begin{pmatrix} x \\ y \end{pmatrix} = \begin{pmatrix} f_1\begin{pmatrix} x \\ y \end{pmatrix} \\ f_2\begin{pmatrix} x \\ y \end{pmatrix} \end{pmatrix}$$

The Jacoby matrix J contains the partial derivatives:

$$J\begin{pmatrix} x \\ y \end{pmatrix} = \begin{bmatrix} \frac{df_1}{d_x}\begin{pmatrix} x \\ y \end{pmatrix} & \frac{df_1}{d_y}\begin{pmatrix} x \\ y \end{pmatrix} \\ \frac{df_2}{d_x}\begin{pmatrix} x \\ y \end{pmatrix} & \frac{df_2}{d_y}\begin{pmatrix} x \\ y \end{pmatrix} \end{bmatrix}$$

The procedure is to be shown in the two-dimensional case for determining the coordinates of an unknown point, if the coordinates of two fixed points A (a, b) and B (c, d) as well as the distances from these fixed points to the unknown point P (u and v) are known (cf. Fig. 3.10). This task arises in a similar form when determining the position with the

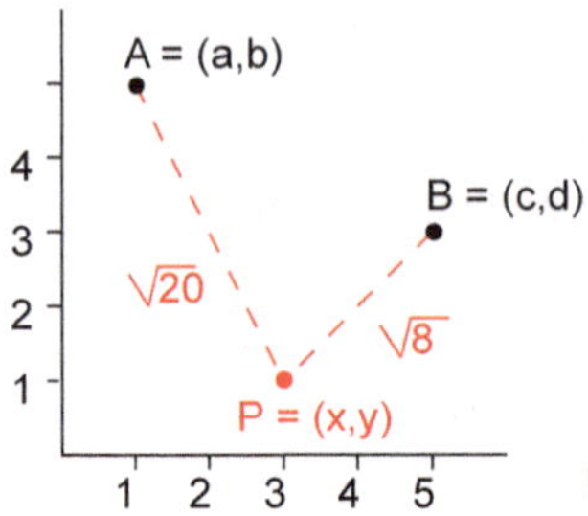

$$(a-x)^2 + (b-y)^2 = u^2 \quad \text{distance A to P}$$
$$(c-x)^2 + (d-y)^2 = v^2 \quad \text{distance B to P}$$

or

$$a^2 - 2ax + x^2 + b^2 - 2by + y^2 - u^2 = f1(x,y) = 0$$
$$c^2 - 2cx + x^2 + d^2 - 2dy + y^2 - v^2 = f2(x,y) = 0$$

or

$$F(x,y) = (f1(x,y), f2(x,y)) = (0,0)$$

Fig. 3.10 Determination of the coordinates of a point P from the known coordinates of two points A and B and the distances PA and PB

help of global navigation satellites (on GNSS cf. Sect. 5.3.3). In this three-dimensional case, the coordinates of the satellites, which are regarded as fixed points during the positioning, are known in a global coordinate system and the distances from the receiver to the satellites are known.

The iteration starts for the two known fixed points (a,b) = (1,5) and (c,d) = (5,3) as well as with their distances to the unknown point $u = \sqrt{20}$ and $v = \sqrt{8}$ as well as with the starting value $(x_0,y_0) = (3,3)$:

$$J\begin{pmatrix}3\\3\end{pmatrix} = \begin{bmatrix} -2a+2x & -2b+2y \\ -2c+2x & -2d+2y \end{bmatrix} = \begin{bmatrix} 4 & -4 \\ -4 & 0 \end{bmatrix}$$

$$J^{-1}\begin{pmatrix}3\\3\end{pmatrix} = \begin{bmatrix} 0 & -0.25 \\ -0.25 & -0.25 \end{bmatrix}$$

$$\begin{pmatrix}x\\y\end{pmatrix}_1 = \begin{pmatrix}3\\3\end{pmatrix}_0 - \begin{bmatrix} 0 & -0.25 \\ -0.25 & -0.25 \end{bmatrix} \cdot \begin{pmatrix}-12\\-4\end{pmatrix} = \begin{pmatrix}2\\-1\end{pmatrix}$$

Iteration	Start value	1	2	3	4	5
x-approx.	3	2	2.77272	2.98190	2.99987	3
y-approx.	3	-1	0.54545	0.96381	0.99973	1

3.3.3.2 Recursions

Recursive algorithms, in which a procedure calls itself within that procedure, are of great importance. A simple example of a recursive function is the recursive computation of the factorial of N, $N! = 1 \cdot 2 \cdot 3 \cdot \ldots \cdot N = N \cdot (N-1)!$ Thus, one can trace the computation of N! back to the computation of (N−1)!, resulting in a simple program design:

```
def fakult(n):
 if n == 0:
  return 1
 else:
  return n*fakult(n-1)
print("Enter the number from which the factorial is to be calculated: ")
xin=input()
if xin != "":
 try:
  xin_i = int(xin)
  print ("faculty of ", xin_i, "is ", fakult(xin_i))
 except:
  print("The input must be a single integer e.g. 4, 6, 10 or similar!")
```

An application of recursion results in a very effective sorting algorithm that operates on the frequently implemented principle of "divide and conquer". In this *quick-sort procedure,* an array is divided by a list separator into two sub-lists: one with values less than or equal to the list separator, and a second list with values greater than the list separator. This procedure is then recursively applied to both sublists.

This results in a binary partition tree with $\log_2$ (n) levels (cf. Fig. 3.11). The complexity required for the complete partitioning of each level is proportional to n. Overall, for the quick sort algorithm, the effort then averages n · log n and is proportional to n · n only for degenerate partitions where each partition results in a group with one object, (cf. Sect. 3.3.4).

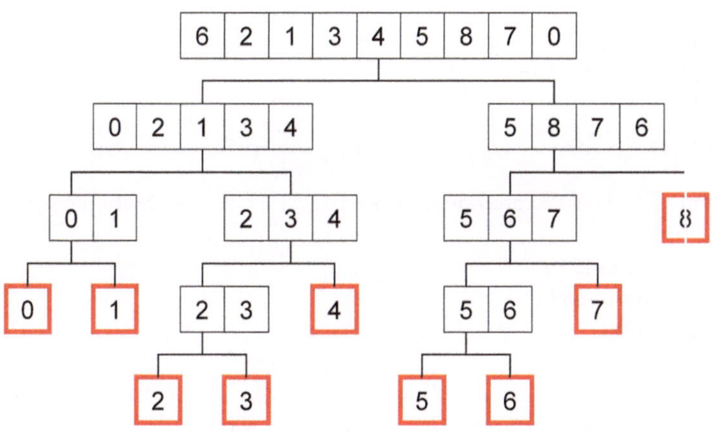

Fig. 3.11 Illustration of the Quick Sort procedure

```
public class QuickSort {
 private static int [] numbers;

 private static void quicksort(int low, int high) {
  int i = low, j = high;
  int pivot = numbers [low + (high-low)/2]; // find pivot element, middle of
array
  // division according to 'divide and conquer' principle
  while (i <= j) {
    // as long as the part has not yet been completely run through
    // search from the beginning of the array (value smaller than previous)
    while (numbers[i] < pivot) {
     i++;
    }
    //search from the end of the array (value greater than previous
    while (numbers[j] > pivot) {
     j--;
    }
    //When elements are found that need to be swapped - method call
    if(i <=j){
     exchange(i, j);
     i++;
     j--;
    }
  } /*from while*/
  recursion call of the own method - restart with new parts
  if (low<j)
   quicksort(low, j);    // sort left half
  if (i < high)// sort right half
   quicksort(i, high);
 } /*from quicksort*/

   private static void exchange(int i, int j) {    //method to exchange
2 elements
  int temp = numbers[i];
  numbers[i] = numbers[j];
  numbers[j] = temp;
 }

 public static void main(String[] args) {    // Main Method
  int[] numberlist = {0,9,4,6,2,8,5,1,7,3};
  numbers = number list;
  quicksort(0, numberlist.length - 1);
  for (int i=0; i<numbers.length; i++)    // Output of the sorted array
   System.out.print(numbers[i]+" ");
 }
}
```

3.3.4 Algorithm Complexity

The *efficiency of algorithms* is generally evaluated with respect to the computation time and the memory space. Here, it is not the absolute computation time or the absolute memory requirement that is decisive – the first quantity in particular is largely determined by the hardware – but the increase in computation time and the increase in memory space in each case as a function of the size of the input.

In the *linear search* of an element in an array of length n, the check starts with the first element of the array and continues with the other elements until the searched object is found. In the best case the element is found at the beginning, in the worst case at the end or not at all and in the average case in n/2 cases. However, the elements do not have to be in a specific sort order.

Binary search is applied to a sorted array divided at a point m: A[min] . . . A[m – 1] A [m] A[m + 1] . . . A[max]. For m, the middle between minimum and maximum is usually chosen, i.e. (min + max)/2. Then, for the element x to be searched for, it must be checked whether x = A[m], x < A[m] or x > A[m]. Then the search is terminated or continued accordingly in the left or right half. In the worst case, the parts must be halved so often until only one element remains. In a binary search (of an ordered array), $2^k - 1$ numbers cause at most k loop iterations, since after k bisections the interval length is 1. Thus, n numbers generate at most $\log_2 n$ loops.

The examples show (cf. also Quick-Sort-algorithm in Sect. 3.3.3.2) that the runtime of the algorithms and therefore their complexity is proportional to a parameter n, whereby the two algorithms listed have linear and logarithmic dependencies (cf. Table 3.5).

For other algorithms, this dependence can be expressed by certain functions $f_i(n)$. However, the exact dependence function is difficult to determine and of less interest than an estimate of the order of magnitude by a majorizing function, i.e., by a simple but known function with larger but approximate function values. In *O-notation,* the complexity of an algorithm is then expressed by the majorizing function. This shows the asymptotic behavior for a large N (cf. Table 3.6):

$$\text{f(n) is at most of order g(n)} \quad \text{if} : f(n) \leq c \cdot g(n), c = \text{const., for large n}$$
$$\text{written} : f(n) = O(g(n))$$

Table 3.5 Effort for linear and binary searches

	Best case	Average case	Worst case
Linear search	1	n/2	n
Binary search	1	$\log_2 (n) - 1$	$\log_2 (n)$

Table 3.6 Algorithm complexity

O		Examples
log n	Logarithmic	Search on a sorted set
$\log_2$ n	Logarithmic	Binary search
n	Linear	Linear search
n · log n	Log-linear	Efficient sorting
n^2	Square	Naive sorting
n^3	Cubic	
n^i	Polynomial	
2^n	Exponential	All subsets
n!	Faculty	All permutations

Table 3.7 Computing times as a function of complexity and number of cases for an assumed computing power of 1 million (= 10^6) floating-point operations per second (unless otherwise stated, computing time in seconds)

O	10	20	30	40	50	60
n	0.00001	0.00002	0.00003	0.00004	0.00005	0.00006
n^2	0.0001	0.0004	0.0009	0.0016	0.0025	0.0036
2n	0.001024	1.048576	17,896 min	12.7 days	35.7 years	36559 years
n!	3.62	77147 years	$8.4 \cdot 10^{18}$ Y	$2.6 \cdot 10^{34}$ Y	$9.6 \cdot 10^{50}$ Y	$2.6 \cdot 106^{8}$ Y

For comparison: The big bang happened about 12–15 billion ($1.2–1.5 \cdot 10^{10}$) years ago

If one assumes that a computer can execute one million floating point operations per second, then gigantic computing times result even with a relatively small numbers of cases for individual complexity functions (cf. Table 3.7).

Such complexity functions are not uncommon for tasks in geoinformatics. A frequent task in the analysis of networks is the search for an optimal route with n stations (here numbered from 1 to n). A complete route, where the starting point is reached again, can be viewed as a permutation of the numbers from 1 to n. Theoretically, this results in a total of n! routes that are not all implemented, but all of which would have to be checked in a simple search procedure (*brute force method* or *brute force algorithm*). This approach is hopeless even for a very small network. For 60 cities, computing the $60! = 8.321 \cdot 10^{81}$ (possible) paths would take more than $5.96 \cdot 10^{57}$ years, if the currently most powerful computer (Japan's Fugaku as of December 2021) with 442 petaflops or $4.42 \cdot 10^{17}$ floating point operations per second were used and, for the sake of simplistically, assumed that computing a path distance requires 10 floating point operations. If the computer power were to increase by a factor of 10^6, but not in twenty years (like recently), but perhaps in a shorter time, then the examination of all 60! distances would still take more than $5.96 \cdot 10^{51}$ years.

These calculations should show that ever faster computers often do not contribute to solving of a complex task. An efficient algorithm that finds a solution, if only approximately, but then in finite time, is of greater importance.

3.4 Basic Algorithms of Geoinformatics

3.4.1 Algorithms of Coordinate Geometry

In geoinformatics, especially in geoinformation systems, in which objects such as points, lines, and polygons are modeled with the help of coordinates, central tasks are solved with the help of graphic-geometric algorithms (cf. de Berg et al. 2008, Brinkhoff 2013 Section 7, Worboys and Duckham 2004 Chapter 5, Zimmermann 2012). In this context, tasks that seem simple at first sight often require more elaborate solutions. For example, the centroid of a closed polygon given by corner coordinates (cf. Fig. 3.12) is not computed as the arithmetic mean center from the mean of the x and y coordinates. The complex formula is with $(x_{n+1}, y_{n+1}) = (x_1, y_1)$ and F the area of the polygon (cf. similarly Worboys u. Duckham 2004 p. 197):

$$c_x = \frac{1}{6F} \sum\nolimits_{i=0}^{n} (x_i + x_{i+1}) \cdot (x_i \cdot y_{i+1} - x_{i+1} \cdot y_i)$$

$$c_y = \frac{1}{6F} \sum\nolimits_{i=0}^{n} (y_i + y_{i+1}) \cdot (x_i \cdot y_{i+1} - x_{i+1} \cdot y_i)$$

To calculate the area of a polygon, which is uniquely described by a sequence of the coordinates of the corner points $P_i = (x_i, y_i)$, the following calculation rule applies (cf. similarly Worboys and Duckham 2004 p. 196 and cf. Fig. 3.9):

$$F = 0.5 \bullet [P_1 \times P_2 + P_2 \times P_3 + \ldots + P_{n-1} \times P_n + P_n \times P_1]$$

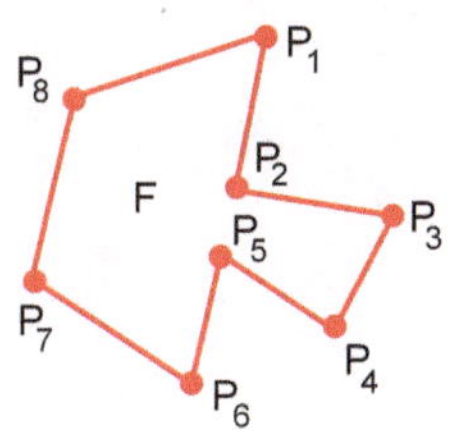

F = {(10,12), (8,8), (14,6), (12,4), (7,3), (6,2), (2,4), (4,10)}

= 60

U = 23,3

Fig. 3.12 Representation of a polygon bounded by line segments (edges)

where the symbol "x" denotes the vector product defined by:

$$P_i \times P_j = \begin{pmatrix} x_i \\ y_i \end{pmatrix} \times \begin{pmatrix} x_j \\ y_j \end{pmatrix} = \left(x_i \cdot y_j\right) - \left(y_i \cdot x_j\right) \qquad \text{(i.e. "crosswise")}$$

The following applies to the perimeter U of a simple polygon (without holes):

$$U = \sqrt{(x_1 - x_2)^2 + (y_1 - y_2)^2 + \sqrt{(x_2 - x_3)^2 + (y_2 - y_3)^2} + \ldots + \sqrt{(x_n - x_1)^2 + (y_n - y_1)^2}}$$

All points must be numbered clockwise. Negative y-coordinates are not allowed. Furthermore is $(x_1, y_1) = (x_{n+1}, y_{n+1})$.

The recording or digitising of lines by tracing a template (cf. Sect. 5.2.1) or the recording of a track with a GPS device provides coordinate sequences, whereby sometimes more points are determined than are of importance for the representation of the objects. Then the task arises to delete intermediate points and to reduce the density of the coordinates. The generalisation is mostly done according to the *Douglas-Peucker algorithm*, which is applied to the point sequence $\{P_a, P_i, \ldots, P_e\}$ representing the line after data collection (cf. Worboys u. Duckham 2004 p. 176). First, the start and end nodes are connected by a straight line P_aP_e (cf. Fig. 3.13). From this the perpendicular distance to the point P_c is formed, which is furthest from the straight line P_aP_e. This point is kept if the distance is greater than a given threshold. Then, the initial line is divided and the algorithm is recursively applied to the subsections P_aP_c and P_cP_e. The algorithm ends if for all segments P_iP_j the distance to the furthest segment point is below the threshold value. The smaller this threshold, the lower the generalisation.

In vector-based geoinformation systems, linear objects such as rivers or paths and boundaries of biotopes or parcels are modeled or approximated by sequences of straight lines (cf. Sect. 5.2.1, for modeling areas, i.e. polygons, in vector-based geoinformation systems cf. Sect. 9.3.2). Different thematic layers often must be superimposed and intersections have to be formed (cf. Sect. 9.4.4). If one thematic layer represents land use types and the second layer represent parcels, a typical question is to determine the land uses

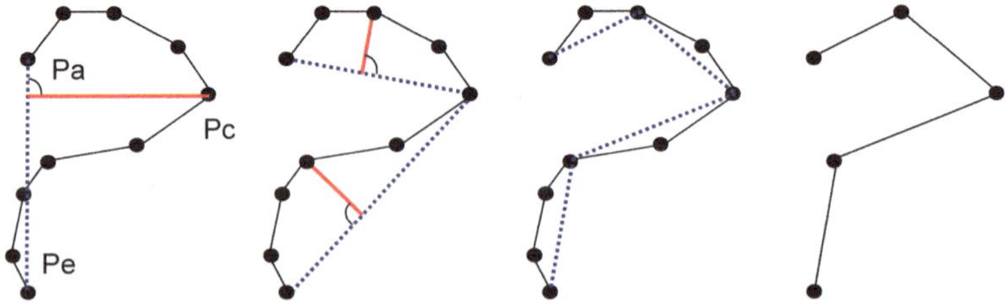

Fig. 3.13 Douglas-Peucker algorithm for the generalisation of lines

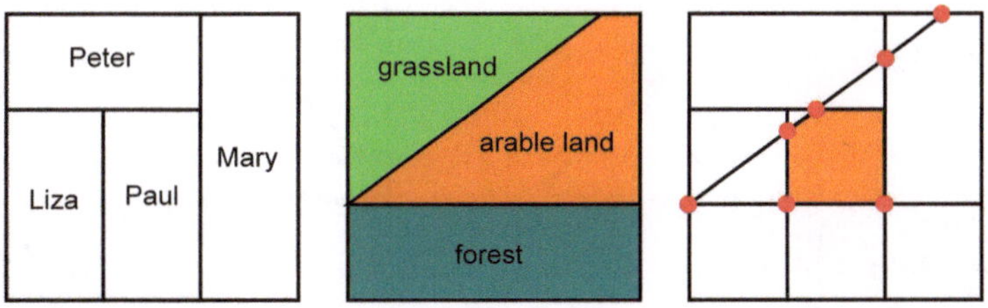

Fig. 3.14 Intersections of the two layers "land use" and "parcels" showing Paul's arable land

on parcels of a specific owner. Then the boundaries of the land use types, which are formed by single line segments, have to be intersected with the boundaries of the parcels, which are also formed by single line segments. This task finally leads to the determination of intersection points of straight lines (cf. Fig. 3.14). For example, in Fig. 3.14, Paul's property splits into three type of land use. The new polygon, which describes Paul's arable land, is made up of five new segments that have to be formed by crossing the lines.

In general, the following applies to the intersection (x_s, y_s) of two (straight) lines:

line 1 :	$y = a_1 + b_1 \cdot x$
line 2 :	$y = a_2 + b_2 \cdot x$
intersection :	(x_s, y_s) with $x_s = (a_1 - a_2)/(b_2 - b_1)$ and $y_s = a_1 + b_1 \cdot x_s$

where for the general equation of a straight line and points (x_i, y_i) on the straight line:

$y = a + b \cdot x$	with
$b = (y_2 - y_1)/(x_2 - x_1)$	(slope)
$a = y_1 - b \cdot x_1$	(intersection of the line with the $y-$axis)

In general, a few special cases must be considered when two line segments cross (cf. Fig. 3.15):

- The calculation rule is only to be used if an intersection can exist at all (cf. case 1).
- The existence of an intersection is always assured (apart from parallel lines) if the straight lines are of infinite length. However, it must be checked whether this intersection point can even lie on the segments in the intervals (x_1, x_2), (u_1, u_2) or (y_1, y_2) and (v_1, v_2) (cf. case 2).
- In the case of a vertical line (cf. case 3), the calculation of the slope according to the above formula leads to a division by zero. For the vertical line applies $x_1 = x_2$, but also $x_1 = x_2 = x_s$.
- Parallel lines have no point of intersection (cf. case 4).

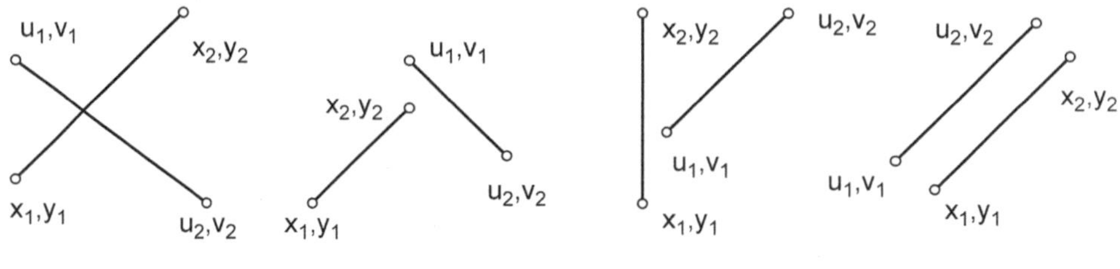

Fig. 3.15 Crossing of straight lines

A very simple approach is to try out all potential solutions (brute force method). In the example above, the straight line that marks the boundary between forest and arable land would have to be intersected with all property boundaries (cf. Fig. 3.14). Such an approach is usually easy to program, but mostly not very efficient either. Such a naive method has the complexity $O(n^2)$ for n segments and is unsuitable for larger applications. A *brute force algorithm* for solving the problem of finding intersections of line segments that represent a polyline has roughly the following form in a pseudo-programming language:

```
//LineSegments = list of line segments
//PointsIntersection = list of intersection points of segments
for i in range (1, LineSegments.length)
 for j in range (1, LineSegments.length)
   p = intersect (LineSegments[i], [j])
   if p ≠ null then
    PointsIntersection.insert(p)
```

However, very efficient techniques have been developed in the so-called *computational geometry*. This field of research, which emerged in the late 1970s, is dedicated to the analysis of graphical-geometric problems and the development of efficient geometric algorithms (including intersection determination, point localisation, visibility determination, triangulation tasks, cf. de Berg et al. 2008, Brinkhoff 2013 pp. 235, Klein 2005, Preparata and Shamos 1985, Schmitt et al. 1996, Zimmermann 2012). For the intersection problem of arbitrary lines, an (optimal) method is available in the form of the Bentley-Ottman algorithm, whose time complexity can be estimated as $O[(n + k) \cdot \log n]$, where k is the number of intersections found. However, in the extremely rare case of very many intersections, k estimated by $O(n^2)$, this method becomes even worse than the simple procedure with $O(n^2 \cdot \log n)$.

The *Bentley-Ottmann algorithm* belongs to the so-called *plane-sweep methods*, in which (in the two-dimensional case) an (imaginary) vertical line (so-called sweep line) is passed over the data in the plane (cf. Brinkhoff 2013 pp. 235, Klein 2005 pp. 64, Schmitt et al. 1996 pp. 33, Zimmermann 2012 pp. 171). In this process, operations are only carried out at selected points of contact between the sweep line and the objects.

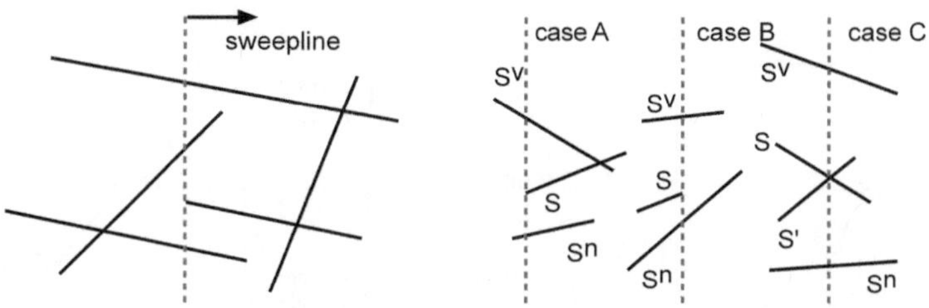

Fig. 3.16 Sweep-line algorithm for calculation all crossings of line segments

Those segments that intersect the sweep line are called active segments. Calculations are made at breakpoints of the sweep line, i.e. only at the start and end points of the segments and at the (found) intersections. Intersection checks are only carried out with the predecessor and successor segments S^v and S^n, respectively, where the predecessor and successor are ordered in the y-direction and S^v is "above" and S^n is "below" a breakpoint. This technique significantly reduces the computational cost, since predecessor and successor only have to be considered in one neighborhood of breakpoints. Two segments can only intersect if they are adjacent in the sequence of active segments (once before). Only one of three cases needs to be dealt with (cf. Fig. 3.16):

- If a breakpoint is the starting point of a segment S, it is checked whether the segments S^v, S and S^n intersect. Each intersection that is found is included in the set of breakpoints (cf. Fig. 3.16, case A).
- At an endpoint of a segment S, it must be tested whether S^v and S^n intersect. Again, any intersection that is found is included in the set of breakpoints (cf. Fig. 3.16, case B).
- If the segments S and S′ intersect, it must be checked whether S′ and S^v or S and S^n intersect. The intersections points found are included in the set of breakpoints (cf. Fig. 3.16, case C).

The Bentley-Ottmann algorithm requires efficient data structures in the form of queues and balanced trees (cf. Schmitt et al. 1996 pp. 32). Many applications exist for such sweep-line algorithms in computational geometry, e.g. for checking the position of a point in relation to a polygon or for triangulation of polygons.

In geoinformation systems, a rough *preliminary test* can considerably reduce the complexity, whereby additional information is often already stored during data recording or modeling. For example, when lines that consist of individual segments are intersected, a first check is made to determine whether the enveloping rectangles of the lines or the segments overlap (cf. Fig. 3.17, so-called *MER – Minimum Enclosing Rectangle* or *MBB – Minimum Bounding Box*). This alone excludes many cases. This procedure becomes very efficient in particular, if the extreme coordinates of a line are already stored and do not have to be recalculated each time. Furthermore, a complex line can be broken down into monotonous sections, for which the x- or y-coordinates either increase or decrease. Since the line in a segment always increases in one direction, the segment cannot "turn around" and intersect another line a second time. Such a segment, which may consist of several

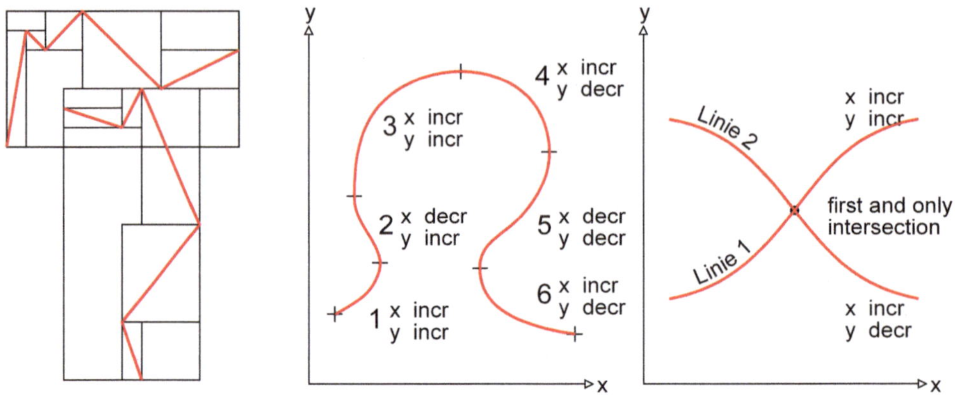

Fig. 3.17 Minimum bounding boxes of two polylines, monotone line segments and intersections of lines

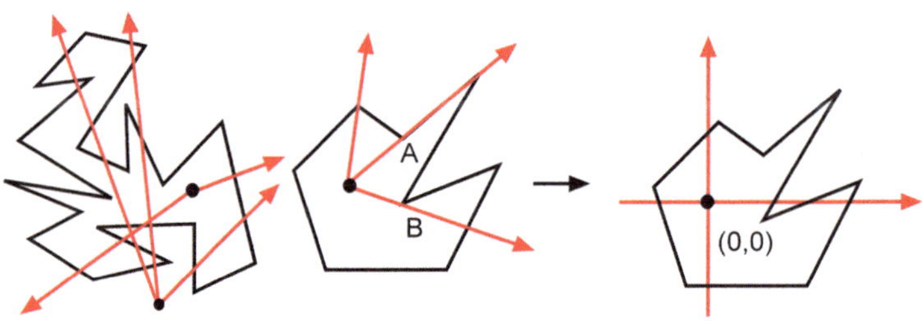

Fig. 3.18 Semi-line algorithm for the point-in-polygon test

smaller sections, can be treated as a single straight line segment as a whole. A pre-test based on monotonic segments can also lead to computational savings.

In geoinformation systems, it is very often necessary to determine whether a given point $P = (x_p, y_p)$ lies within a polygon defined by the corner coordinates of its edges: $F = \{(x_i, y_j)\}$. For this so-called *point-in-polygon test* a very simple algorithm can be given, which implements the theorem that every ray emanating from a point inside the polygon intersects the edges of the polygon in an odd number of intersections (application of Jordan's Curve Theorem, for further algorithms for point localisation cf. Schmitt et al. 1996 pp. 68, cf. also Worboys and Duckham 2003 pp. 197). However, several special cases must be taken into account here. Only the actual transition from the interior to the exterior is counted in each case (cf. Fig. 3.18, no transition at A, only one transition at B).

For the computational implementation, only the horizontal ray running to the right is considered in a simplified way. The polygon F with the corner coordinates $\{x_i, y_j\}$ is transformed or shifted in such a way that the point P coincides with the origin and the ray with the positive x-axis: $F' = \{(v_i, w_j) = (x_i\text{-}x_p, y_i\text{-}y_p)\}$. Then all the edges are traversed one after the other and it is checked wether they intersect the x-axis. No intersection calculations are necessary, only comparisons of the y-coordinates. For all i with $v_i > 0$ the cases are to be counted for which the following applies: ($w_i > 0$ and w_i+1 < 0) or ($w_i < 0$ and $w_{i+1} > 0$). This algorithm has the complexity O(n) with n number of corners of the polygon.

A frequent application of the point-in-polygon test occurs in geoinformation systems, if for a point P, which is given by its coordinates, the associated region and then the associated attributes are to be found in the database. The search effort with the above procedure is then O(m · n) for m polygons with n edges each. The effort can be reduced with a sweep-line algorithm (cf. Zimmermann 2012 pp. 173) or with an algorithm operating on so-called trapezoidal maps (cf. de Berg et al. 2008 pp. 122).

3.4.2 Graphs and Selected Path Algorithms

In geoinformatics, the so-called routing algorithms are a special group of methods that generally operate on so-called networks and for which a wide range of possible applications can be specified, especially in the modeling of transport networks and the determination of optimal routes. The mathematical basis for this is graph theory (cf. e.g. Jungnickel 2013). In general, *graphs* consist of a set of nodes and a set of edges connecting nodes (cf. Fig. 3.19). To describe relationships in a graph, the notions of adjacency and incidence are used. Two nodes A and E are adjacent if they have a connecting edge k(A, E). *Adjacency* describes the relationship between similar elements of a graph. Conversely, the two topologically adjacent nodes define an edge that is incident with the nodes. *Incidence* describes the relations between different elements of a graph. The nodes A and E are therefore incident with the edge k(A, E).

A path is a sequence of pairwise adjoint edges leading from one node to another node. A graph is called connected, if there is (at least) one path for any two nodes. A cycle consist of several distinct, connected edges, where the start and end nodes are the same. A complete graph is a simple undirected graph, in which each pair of nodes is connected by a unique

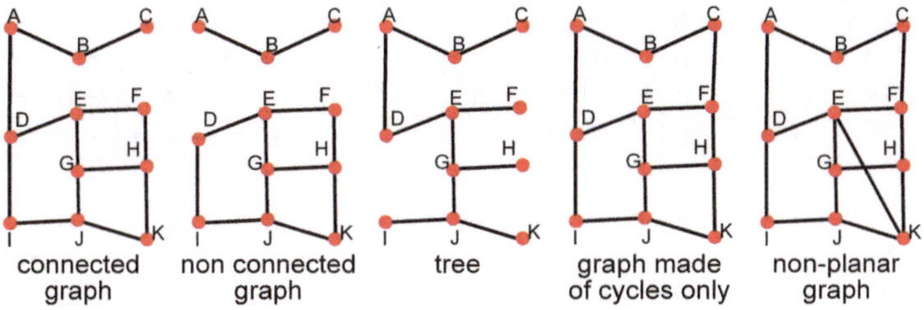

Fig. 3.19 Types of graphs

edge. If there are multiple paths between any two nodes, the graph has (at least) one cycle. A connected graph with no cycles is called a tree.

In addition to a wide range of applications in traffic and communication networks, graphs are also of great importance for the geometric-topological modeling of polygons in vector-based geoinformation systems (cf. Sect. 9.3.2). For example, edges of a connected graph define polygons, if a graph consists only of cycles. In particular, Euler's theorem applies to connected planar graphs (cf. Jungnickel 2013 p. 22):

$$k_n - k_a + p = 2 \quad (k_n = \text{number of nodes}, k_a = \text{number of edges}, p = \text{number of polygons})$$

An application results is the recording of lines and the subsequent modeling of polygons in a vector based a geoinformation system. The geometric-topological modeling of areas in the vector model is carried out on the one hand by a closed polygon (simple feature geometry object model) or on the other hand by connected line segments (node-edge-node modeling 9.3.2). In the second case, the line segments must meet certain conditions. However, digitising the lines often leads to results like those in Fig. 3.19, all of which are intended to represent individual areas. Only the fourth graph, for which $k_n = 11$, $k_a = 14$, $p = 5$, i.e. for which *Euler's theorem* applies, models all areas topologically correctly (cf. Sect. 9.3.3). The equation only becomes correct, if the "infinite" region outside the graph is counted as a polygon (here the fifth polygon). Euler's formula provides a consistency check of an exact modeling, which satisfies topological requirements (for the geometric-topological modeling of polygons in the vector model, which models areas from composite line segments and not as a closed polygon course cf. Sect. 9.3.2).

A graph such as a real traffic network can be described by a weighted *adjacency matrix*. The edges carry so-called cost attributes, which are to be regarded as *impedances*. Usually these impedances are defined by the length between two nodes in a length unit. Then the best route between two nodes is the shortest route. If the impedance is time, then the best route is the fastest route. Here, the matrix element a_{ij} expresses the impedance between nodes i and j (cf. Fig. 3.20). The impedance of a node to itself is given by 0. If two nodes are not connected by an edge, they are given the value ∞. In an unweighted adjacency matrix,

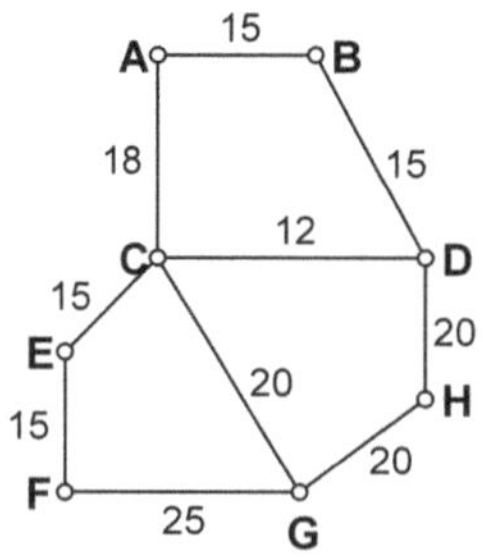

	A	B	C	D	E	F	G	H
A	0	15	18	∞	∞	∞	∞	∞
B	15	0	∞	15	∞	∞	∞	∞
C	18	∞	0	12	15	∞	20	∞
D	∞	15	12	0	∞	∞	∞	20
E	∞	∞	15	∞	0	15	∞	∞
F	∞	∞	∞	∞	15	0	25	∞
G	∞	∞	20	∞	∞	25	0	20
H	∞	∞	∞	20	∞	∞	20	0

A 15 → B 18 → C
B 15 → A 15 → D
C 18 → A 12 → D 15 → E 20 → G
D 15 → B 12 → C 20 → H
E 15 → C 15 → F
F 15 → E 25 → G
G 20 → F 20 → C 20 → H
H 20 → G 20 → D

Fig. 3.20 Network with associated adjacency matrix and list representation

the impedances of the edges are represented by 1. An adjacency matrix does not necessarily have to be symmetric (cf. e.g. directionally different travel times between two nodes). However, the presentation of a graph in form of an *adjacency matrix* is extremely memory consuming (memory complexity $O(n^2)$). In contrast, the presentation in the form of lists consumes less memory. Here, a list is defined for each node, which contains the immediate neighbors.

The so-called matrix methods for calculating shortest paths simultaneously determine all weights on paths between all nodes of a network. *Floyd's algorithm* (also called *Warshall's algorithm*) determines the minimum distance for all pairs of nodes (I,J) by calculating all paths via (exactly) one intermediate node for a pair of nodes (I,J) (cf. Jungnickel 2013 pp. 91). The calculation of the distance, i.e. in general the weight W (I,J), which can also be defined in terms of travel time or cost, between nodes I and J is done by:

```
W(I,J) = min { W(I,J); W(I,K) + W(K,J) }
```

From a computational point of view, the algorithm is programmed in such a way that an intermediate node is held and then the distances (weights) W(I,J) are calculated for all pairs of nodes:

```
For K = 1,..., N
   For I = 1, ..., N
   For J = 1, ..., N
   calculate W(I,J) = min { W(I,J); W(I,K) + W(K,J) }
```

In the end, the matrix W contains the shortest distances between any nodes. However, the individual routes cannot be read from the optimal distance matrix. Rather, they must be reconstructed subsequently with the help of the initial matrix. For this purpose, a so-called *predecessor matrix* is introduced in the calculation procedure, for which start values are set:

```
V (I,J) = I,   if:  I = J or I is directly adjacent to J
 0,             otherwise
```

The quotation of the respective predecessor must then be introduced into the calculation process of the Floyd procedure:

```
if (w[i,k] + w[k,j] < w[i,j]) then
 begin
   w[i,j]:=w[i,k] + w[k,j]
   v[i,j]:=v[k,j];
 end;
```

After completing the calculations, all desired routes can be determined from this predecessor matrix as a sequence of nodes. Since the predecessor is known for each node, one starts at the destination node and calculates backwards (destination → predecessor → predecessor → ... → beginning).

The so-called tree methods determine the shortest paths between a given node to all other nodes of the network. The best known method is certainly *Dijkstra's algorithm* (cf. Jungnickel 2013 pp. 83–87 and very descriptive Worboys and Duckham 2003 pp. 214). In the course of the procedure, starting from a starting node S, a sequence of successor nodes is recorded in a notelist, which, together with the successors of the successors, shows all possible routes from S to the destination Z. For these nodes, all routes are cleverly checked. For these nodes, all routes are skillfully tested, and nodes that have already been processed are not tested again. The algorithm in a pseudo notation (cf. Domschke 2007) needs an adjacency matrix A[i,j] as well as a notelist MERK, a list VORG[i] of the predecessors of node i (with VORG[i] is predecessor to node i) and a list WID[i] of the weights or distances of node i to the starting node S (weights WID[i] all nonzero).

```
Dijkstra's algorithm:
//Initialisation
repeat
  select node k from MERK with WID[k] : = min {WID[i], i ∈ MERK}
    delete node k from MERK
    for all successors j of k do //for these nodes j holds 0 < A[k,j] < ∞
begin
      ① if j ∉ MERK and WID[j] = ∞
         (*the successor j of S has not yet been reached*)
          WID [j] = WID [k] + A[k,j];
          VORG [j] := k;
          MERK:=MERK ∪ {j};
      ② if j ∉ MERK and DIST[j] < ∞
          choose the next successor j from node k, or if already
          all successors of node k have been checked,
          start next iteration
      ③ if j ∈ MERK
         if WID[k] + A[k,j] < WID[j] then
         begin
          WID[j]: = WID[k] + A[k,j];
          VORG[j]: = k;
         end;
    end;
until MERK = { }
```

First, the algorithm must be initialised. The start node is added to a watch list, the weight 0 is assigned to the path to the start node and the weight ∞ to the paths to the other nodes.

In the iteration, the foremost node is taken out of the notelist k (node k). Then, a case differentation is performed for all neighboring nodes j of the node k just taken out.

If node j has not been visited at all (i.e., WID[j] $= \infty$) and is not in the notelist, this node contains the distance WID[j], which is the sum of the cost to its predecessor (WID [k]) and the value A[k,j] of the weighted adjacency matrix of the edge from node k to node j that is currently under consideration.

If node j has already been visited (i.e. WID[j] $\neq \infty$) but is no longer in the notelist, another successor of k is considered. If all successors of node k have already been checked, the next iteration (with the next node) is continued.

If node j is present in the notelist, it is checked whether the weight WID[j] from the start node to node j is greater than the sum of the weight to the predecessor node k and the weight from node k to node j. If this is the case, a shorter path has been found via node j. The list of weights and the list of predecessors must then be updated.

The algorithm ends if there are no more nodes in the notelist.

These algorithms, like most other path algorithms, are memory and time critical. The (time) complexity is $O(n^3)$ for the Floyd algorithm (cf. loop organisation). For *Dijkstra's algorithm*, in the worst case, the complexity is proportional to n^2 (n number of nodes). In a modified form for a connected graph it is O(k * log(n)) (k number of edges, cf. Jungnickel 2013 pp. 87).

Improvements are provided by the A* algorithm, which, in contrast to uninformed search algorithms, uses an estimation function (heuristic). For example, the linear distance can be used as an estimation function, so that paths via nodes that are further away do not have to be considered further. A targeted search is characteristic, which reduces the runtime, since only a small part of all nodes must be examined. Like Dijkstra's algorithm, the A* *algorithm* also has the complexity $O(n^2)$ in the worst case, but in practice it offers time advantages on average (cf. Worboys and Duckham 2003 pp. 216–217).

In the *Travelling salesman problem* (TSP) a sequence of n nodes is sought such that each node (except the starting node) is visited exactly once, the trip ends back at the start, and the length of the round trip is minimal. All known methods so far amount to a complete analysis of all possibilities. A simple search procedure that checks all n! combinations is not applicable even for a small network (cf. Table 3.7). A basic method for solving such a problem is based on a so-called *branch-and-bound technique*. At each step, the set of remaining possible solutions is divided into two or more subsets, each of which can be represented by a branch in a decision tree. An obvious criterion for the Travelling Salesman Problem is to split all tours according to whether or not a certain route is included.

The Travelling Salesman Problem will be illustrated by a very simple example with only four stops. The example assumes four nodes with different distances. A decision tree with several branches is constructed. However, not all branches make sense (cf. Fig. 3.21). For example, the branch 1-2-1 may not be pursued further, since node 1 already exists in the tour. Only the start node may be contained twice in a tour, i.e. at the beginning and at the end, but only if a complete tour exists. For the example, the optimal route is: 1 – 3 – 2 – 4 – 1 with length 46.

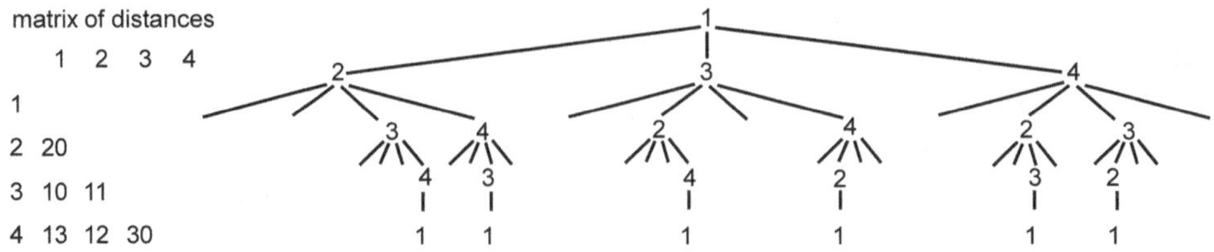

Fig. 3.21 Illustration of a branch-and-bound algorithm for the Travelling Salesman Problem

The processing of the various branches is shortened by calculating the length of the route covered so far up to the node k. The route length is then calculated. If this path length is already greater than the length of an already calculated complete route, all further paths via this node k are no longer followed. In the present example, such a bound does not come into action, since the tree is not yet strongly branched. However, if, for example, the distance between 1 and 3 is assumed to be 75, the middle part of the decision tree is no longer processed. In the left branch over edge 1-2, the value 74 (route 1 – 2 – 3 – 4 – 1) already resulted as the shortest distance.

For this procedure, the runtime increases exponentially. If in a graph exactly two edges lead away from each node, then 2^n paths have to be checked, each with n nodes, so that the estimation holds: $O(2^n)$. A realisation of the present branch-and-bound method is shown by Herrmann 1992 (cf. pp. 319, for further path selection algorithms or general algorithms on networks cf. Jungnickel 2013 pp. 65 and pp. 481 as well as Worboys and Duckham 2003 cf. pp. 211).

The Travelling Salesman Problem can be regarded as a classical basic problem of combinatorial optimisation. Besides the exact branch-and-bound method, many heuristic solution methods have been developed which come close to the exact solution after quite a short time, but do not guarantee its finding. Many program examples (i.e. Java applets) exist on the Internet that provide real-time solutions of the TSP for hundreds of cities. They mostly work with the help of so-called neural networks (cf. Kohonen 2001 and Scherer 1997, cf. Sect. 10.7.9.1). In addition, so-called natural algorithms such as the so-called ant colony optimisation algorithm provide a new class of algorithms. Here, the behavior of ant colonies during foraging, in which the colony as a whole finds the shortest path between nest and food source, is applied to optimisation problems.

3.4.3 Algorithms for Raster Data

Geoinformatics develops and uses many algorithms that evaluate raster data (cf. Sect. 9.5.3 for spatial analyses in a raster-bases geoinformation system). Here, an algorithm from the set of algorithms that are used outside of digital image processing (cf. Sect. 10.6) is presented as an example (cf. also algorithms e.g. for geomorphometry with height data on a raster basis in Hengl and Reuter 2008). With the help of *skeletonisation procedures* ("*thinning*"), areal binary objects such as shapes in black-and-white scans are to be reduced to linear skeleton lines, i.e. thinned out. Ideally, the algorithm returns a skeleton line that is only one pixel wide. The skeleton line must reflect the original shape, contiguous areas of an object must also be contiguous again in the skeleton.

To illustrate, assume a binary grid with 0 for a white pixel and 1 for a black pixel (background and foreground, respectively). For each pixel, the neighborhood is considered in a 3 × 3 mask.

P_8	P_1	P_2
P_7	P	P_3
P_6	P_5	P_4

A

0	0	0
0	P	1
1	1	0

B

1	1	1
0	P	1
1	1	1

P may be deleted in cases A, B.

C

0	0	1
0	P	0
1	1	0

D

0	0	1
0	P	1
0	0	0

E

0	1	0
1	P	1
0	1	0

P must not be deleted in cases C, D, E.

In this neighborhood the following must apply: A pixel P may only be deleted if this does not destroy the connection between the 1-pixels in the 3 × 3 neighborhood (maintaining of the connection, case A and C, respectively). A pixel may not be deleted if it is the endpoint of a skeleton line (case D). Only contour pixels that have at least one direct 0-neighbor may be deleted (case B and not E, respectively). The *Zhang-Suen algorithm* (cf. Worboys and Duckham 2004 pp. 209, cf. Fig. 3.22), which proceeds in two iterations, formally implements these conditions. For a final vectorisation, sequences of pixels are combined into chains with the help of so-called *chain coding algorithms* (cf. Sect. 9.3.5).

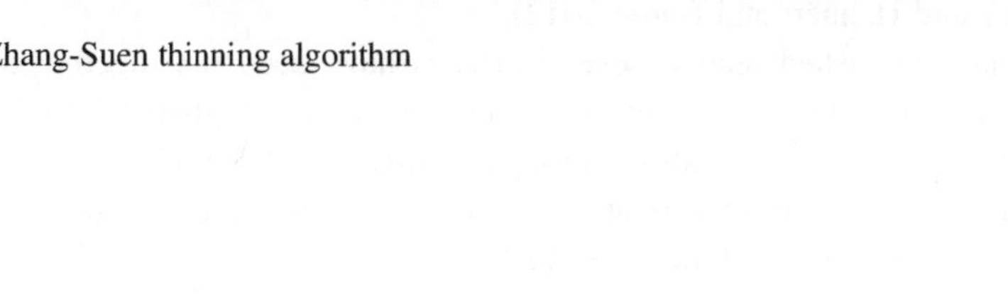

Fig. 3.22 Zhang-Suen thinning algorithm

```
Zhang-Suen algorithm
N(P) = P1+P2+...P8
S(P) = number of 0 to 1 connections from P to the direct neighbor P1, P2, ..., P8, P1
Repeat
 Mark all pixels P = 1 with
   2 ≤ N(P) ≤ 6) and (S(P) =1)and (P1 - P3 - P5 = 0) and (P3 - P5 - P7 = 0).
   Delete all marked points
 Mark all pixels P = 1 with
   (2 < N(P) < 6) and (S(P) = 1)and (P1 - P3 - P7 = 0) and (P1 - P5 - P7 = 0)
   Delete all marked points
until no more pixels can be marked
```

3.4.4 Advanced Algorithms

The almost unmanageable abundance of algorithms cannot be revealed (here). In addition, computational methods and solution strategies must always be seen in the context of the associated issues. Extensive collections of algorithms exist for very different tasks in geoinformatics (for an introduction cf. Worboys and Duckham 2004). In the context of graph theory, many algorithms exist for the analysis of networks, e.g. navigation and determination of shortest paths, traversals, postman problem (Chinese-Postman-Problem, for an introduction cf. Jungnickel 2013). Two types of classic problems should still be named:

Location-allocation problems are very old and have already been implemented in many cases by appropriate tools in geoinformation systems. A formulation of the task can be: Where should emergency service facilities be placed so that the largest number of people in a city can be reached within, say, four minutes? Thus, the optimal locations (and their number) are to be found and people assigned to them (for an introduction cf. Church and Medrano 2018 as well as the compilation of algorithms presented by the fathers of geoinformatics in Rushton, Goodchild and Ostresh 1973).

More recent tasks concern *map matching*, which assigns spatial objects exact positions in a digital map. Usually, the positions determined with (simple) GPS receivers do not intersect with the street lines of a digital map. Map matching algorithms are used to display the GPS positions at the "correct" positions on the lines of the digital maps (cf. Sultan and Haunert 2017 and Haunert and Budig 2012).

In a recently published reader, several articles have been put together that describe technologies and new methods for the automated processing of spatial data. In addition to fundamental methods of data analysis through combinatorial optimisation or data mining, these include the provision of geodata in geodata infrastructures through to visual data analysis and 3D visualisation (cf. Sester 2020).

3.5 Software Engineering

3.5.1 Software Engineering: Tasks and Goals

One of the main tasks of applied computer science is the development of application systems for subject-specific tasks. The general goal is to create high-quality software that basically has to meet functional requirements. In addition, software must be evaluated according to several criteria (cf. Gumm u. Sommer 2013 p. 828):

- correctness, reliability
- modularity, flexibility, elasticity, interoperability
- testability, changeability, reusability, maintainability
- portability, technical efficiency, economic efficiency
- transparency, intelligibility, integrity
- usability, validity, generality, documentation

Software engineering therefore also includes quality assurance and therefore requires appropriate measures during project implementation (including formulation of quality goals, controls and evaluation measures). Last but not least, documentation of the program and instructions that are easy to understand for the user are essential.

However, unsystematic programming fails completely when creating larger programs. Against this background, *software engineering* or (*software development*) is to be understood as the systematic development of software with the help of formal techniques, methods and tools (eg. systematic application of engineering approaches). The focus is on the development of application systems from an operational and economic point of view, whereby the development of an algorithm and the actual programming are important, but only partial services in the overall project process, which in the case of larger tasks only account for about 30% of the total software development.

The development of complex user software is often less a problem of computer science than a task of well-functioning project management. Software development is not the same as programming. This is about the development of a (software-related) solution for a technical problem, which is often characterised by a complex requirement profile and extensive data inputs as well as diverse users (cf. set up of an environmental information system with special applications such as contaminated site cadastre or land use planning). Before the actual programming, the requirements of the (sub-) components must be specified very precisely in order to then be able to commission a corresponding program work and subsequently be able to evaluate the service provided. Unfortunately, this almost self-evident principle is sometimes not followed in practice, so that deficiencies in the performance of the software are almost inevitable if the requirements profile is insufficiently elaborated.

In the development of software, the actual program development over time and the organisational implementation of the project in particular are closely linked. A superficially prepared process sketch makes communication between those involved more difficult and prevents efficient controlling (i.e. planning, monitoring and controlling development activities). Thus, on the one hand, it must be clarified what is to be done when and in what order, and on the other hand, it must be determined who has to do what and when. While the first view of software engineering can be formalised by process models, the second view focuses on project management.

3.5.2 Software Development Tools

The term *software engineering* is often used for the creation of software. This makes the reference to an engineering approach clear, in which general principles, methods, procedures, and tools are differentiated or used. These principles, methods, procedures, and tools can be used in any phase or stage of software development (cf.Sect. 3.5.3). However, the use of these terms is not always consistent; moreover, the concepts behind these terms often mesh. Here, the systematisation of Stahlknecht and Hasenkamp is followed (cf. Stahlknecht and Hasenkamp 2005 pp. 212).

Principles refer to general fundamentals of operations or strategies. In the context of software engineering, the most important are *stepwise refinement, top-down development*, and *bottom-up development*. With step-by-step refinement, a complex problem is gradually broken down into smaller and smaller individual problems that are clearer and usually easier to solve. This can be done from top to bottom and the overall functionality of the system can be broken down into smaller and smaller sub-functions or modules (top-down approach). Conversely, an overall system can be built up from subsystems or individual modules (bottom-up approach). With *modularisation*, completed tasks are defined as (software-technical) units. Then services and interfaces must be known or defined (function and data declaration). The internal processing remains hidden (so-called secret principle, *information hiding*). Especially general subtasks that occur several times in the same or similar form should be designed as modules. Furthermore, modules can also be developed independently of the specific problem and (re-)used in a toolbox for the design of very different algorithms or systems. In particular, software modules can be tested individually (cf. also *modular programming*, Sect. 3.1.4.2).

Methods describe the systematic approach based on certain principles. Important methods in software engineering (or development) are:

- the structured system design,
- the entity-relationship modeling of data structures (ER models),
- object-oriented system development.

While a vertically structured system design results from implementing of the step-by-ste refinement, ER modeling is explained in the design of databases (cf. Sect. 8.2). It was originally developed for the design of database structures. While object orientation was initially limited to programming, it has since evolved into a fundamental approach to software development (cf. Sect. 3.5.4).

Procedures represent instructions for the specific use of methods:

- the structured analysis (SA)
- the structure chart technique
- the object-oriented design method.

In some cases, certain representation techniques are characteristic of individual methods and procedures: among others, data flow diagrams and program flowcharts (cf. Sect. 2.4.2), hierarchy diagrams, ER diagrams (cf. Sect. 8.2), structure diagrams according to Nassi-Shneiderman (cf. Fig. 2.5), graphic representation forms for structured programs according to Jackson (JSP, Jackson Structured Programming). Structured analysis (SA), which has become the most widespread use in practice, focuses on data flow diagrams, with increasingly finer data flow diagrams beeing created according to the top-down principle. The structured design method uses the same representation forms as structured analysis and develops structure diagrams for larger software modules. However, the frequent separation between the data view and the functional view does not make sense and is somewhat unrealistic. Object-oriented system development removes this separation (cf. Sect. 3.5.4).

Tools, i.e. software development tools, are computer programs that support the development of new software. In the simplest case, these can include the development environments of a programming language (cf. Fig. 3.3). Here, a transition to the *CASE tools,* i.e. more precisely Lower CASE tools, is fluid. Comprehensive development tools are now available under the name *CASE* (*Computer Aided Software Engineering*), although (often) for historical reasons CASE is used to refer to transaction and graphic-oriented tools that are primarily used for structured system design and ER modeling. It is now common to use *upper CASE tools* for supporting the analysis and design phases (including presentation tools for creating hierarchy diagrams, data flow diagrams, and ER diagrams for data modeling) and *lower CASE tools* for supporting the implementation phase, i.e. programming and testing (including tools for editing, compiling, and testing).

3.5.3 Traditional Models of Software Development and Engineering

A variety of concepts have been presented to structure the process of software engineering. Of central importance here is the software life cyclemodel, which breaks down the entire project into smaller and mostly simpler sub-tasks that follow one another in time. Several typical questions have to be answered:

Which operations or tasks need to be provided for the software project?
What are the dependencies between these operations or tasks?
How is optimal time and cost planning carried out?

Table 3.8 Software life cycle activities

Preliminary stage	Project justification Develop project proposals Formulate project expectations Issue a project order
Analysis stage	As-is analysis Formulate the target concept
Design stage	System design Program specification Program draft
Realisation stage	Programming or purchase of standard software Program and system tests
Introduction stage	System introduction System operation

The stages or phases formulated in Table 3.8 can be further refined. For example, the system design can still be broken down into a rough and a detailed concept according to the level of detail. In particular, economic considerations and comparisons must be made in the analysis stage, and a decision must ultimately be made as to whether in-house developments should be carried out or whether standard software should be used. As a rule, such structuring tends to ensure that the project is complete, clear and free of contradictions. The following must be specified for each stage:

- tasks and principles, methods and procedures to be used,
- timetable including operational plan of all parties involved,
- cost plan.

Only through a subdivision into several stages that can be justified in terms of content and time the following can be achieved.

- to verify compliance with all requirements,
- to monitor the development effort,
- to initiation short-term control interventions .

The general software life cycle model has been extended by various modifications or supplemented by other process models. For example, the *waterfall* model (cf. Fig. 3.23) requires that a new stage can only be started when the previous stage has been completely finished. The stages are represented here as cascades of a waterfall. However, such a strict sequence can hardly be implemented in practice and is unrealistic. Thus, Royce, to whom the waterfall model is traced back, has already incorporated considerations of loopbacks and iterations (cf. Royce 1970). Starting from the waterfall model, subsequent changes to the requirements formulated by the users or organisational changes can lead (several times)

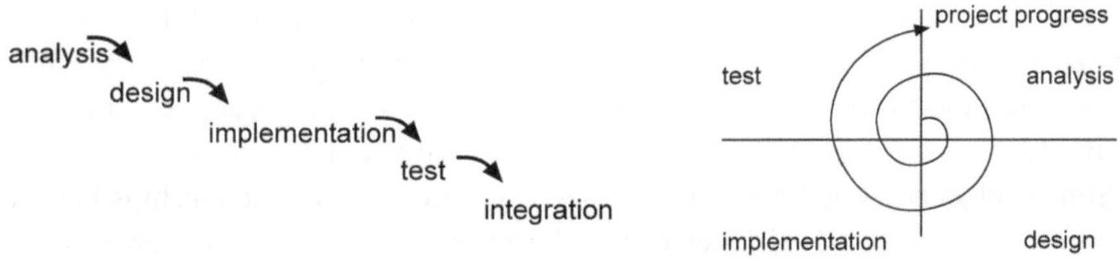

Fig. 3.23 Software development process models: waterfall and spiral model

to changes in the target concept. In this case, several (partial) stages must be repeated or run through cyclically (cf. other process models such as the *spiral model* according to Boehm 1988) (Fig. 3.23).

A frequent problem in practice is that the end user cannot assess all the options of the system to be developed at the beginning of the project. In most cases, only the current tasks can be named and a requirement profile can be derived from them. In most cases, however, new analysis and display options arise with a new system which up to now have not seem feasible due to a lack of tools or which were beyond the scope of imagination and therefore cannot be named by the users! Such extension requests arise when the new system is used for the first time and then (completely) new functionalities are demanded. In practice, this often results in the requirement to have a *prototype*, *i.e.* a testable version of the system, available at a relatively early stage, whereby not all functions have been implemented, but the functional principle can be recognised so that changes can be formulated. This has led to the process model of *prototyping,* which exists in different variants (rapid prototyping, evolutionary, explorative, experimental, vertical and horizontal prototyping). Prototyping can be used in individual stages as well as across stages and can usefully complement the life cycle model, but not replace it. In particular, there is a risk with prototyping that time and cost planning cannot be met.

3.5.4 Object-Oriented Software Development

Object-oriented programming characterises the most recent programming concept (cf. Sect. 3.1.4.3), whereby objectorientation can only be implemented meaningfully if it is applied at an early stage of software development. In particular, a method or structure break in the development process is then prevented because a single concept is consistently applied (use of the same abstraction mechanisms such as classes or inheritance). The stages are analog to the previous models of software development:

- *Object Oriented Analysis* (OOA),
- *Object Oriented Design* (OOD),
- Realisation (OOP for *Object Oriented Programming*).

In each of these stages, the same structures (including classes, objects, attributes, methods) are used. An attempt is made to model reality with regard to the present issue: During the analysis, the objects that are the objects of the task must be clearly specified and the associated features, i.e. attributes as well as functions operating on them, are to be worked out. Similar objects are grouped into classes. Furthermore, the relationships between the classes or objects are to be identified during the analysis (including inheritance or aggregation structures). Finally, an object or class structure model is created that combines the contents of the conventional data model and the function model. Instead of a top-down or bottom-up approach, this modeling is characterised by an incremental and iterative approach, i.e., a step-by-step development with increasingly finer detail, whereby several development steps are repeated several times.

A large number of methods and procedures have been developed for object-oriented system development, whereby standardisation efforts have meanwhile led to a generally accepted modeling language: the *Unified Modeling Language* (*UML*). This graphical description language for object-oriented modeling of complex systems and processes, which can be used universally for various problems, provides various diagram types for modeling.

3.5.5 V Models and Further Developments

The V-model of software development (cf. Fig. 3.24) is largely based on the waterfall model and also divides the development process into several phases that are only run through one after the other. In addition to the temporal component (horizontal alignment), the vertical characterises the detailing. The special feature is that the model contains phases for validation and integration in addition to the development phases. The left branch identifies the phases from specifying the software to implementation. The right branch contrasts the test phases with the specification.

The general V-model is also the basis of project management methods. Two other basic types exist:

In German the development standard V-Modell 97 had existed for a long time, which had been adopted by the German Federal Ministry of the Interior in 1992 and last updated in 1997, and which has since been a binding guideline for IT projects in federal

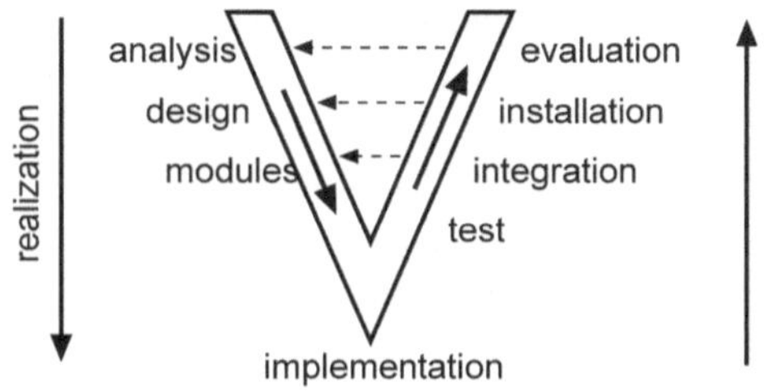

Fig. 3.24 Software development process models: V model

administrations. In 2005 the V-Modell XT replaced the V-Modell. The platform is suitable for companies as well as for public authorities and can also be adapted and used outside the public sector in the project economy under a common public license., The guideline VDI 2206, which includes a practice-oriented guideline for the development of mechatronic systems, proposes a process model that is based on the V-model. At the core of the development methodology is the V-model, which describes the transformation of requirements into a product. The essential feature of this model is the short-cycle, iterative property protection (cf. VDI 2206 2004 p. 26).

There is also a government V-model in the USA. The corresponding document is a resource and a learning tool on the topic of systems engineering. It is not a formal guidance from the Department of Transportation (cf. US Department of Transportation 2007 p. 1).

While the V-Modell was still very much based on the waterfall model, the V-Modell XT was designed according to the modular principle, which includes an iterative and incremental procedure in particular. The idea behind the model is the iterative approach for the step-by-step improvement of products – even over several phases. XT stands for "eXtreme Tailoring" and depending on the type of project should allow a tailor-made procedure, that is supported by ready-made document templates such as plan and proposal modules (cf. Rausch et al. 2008). The new XT model also considers the "roles" between client and contractor for the first time. This was the wish of the industrial representatives, since the formulation of the tasks is often not done with the conclusion of a solid contract, i.e. especially not in the case of new and innovative projects that enter unknown territory and that present a high risk. For example, it is not uncommon for a project specification to be created together with the contractor as part of the overall project. However, this poses the risk of making controlling more difficult or impossible, which exactly is to be prevented by the process model.

In software engineering, this type of approach, in which the specifications are developed jointly with the client only during the course of the project, is also referred to as *agile development* (cf. Wolf and Bleek 2011). These concepts place more emphasis on functional software than on comprehensive functional specifications and focus primarily on reacting flexibly to changing conditions as opposed to working off a plan (cf. buzzwords such as Adaptive Software Development, Dynamic System Development Methodology, Feature Driven Development, and Lean Development, cf. Agile Alliance 2023a and Agile Alliance 2023b). This also includes "eXtreme Programming" (XP), which is not uncontroversial because of its proximity to hacking, in which small teams of usually only two programmers dynamically take on subtasks in close and constant contact with the customer and feed their solution into the overall project almost on a daily basis (cf. Wolf and Bleek 2011 pp. 149–161).

The new V-Modell has opened up to these agile approaches and can support them. Thus, one can also turn around the so-called waterfall approach and start with the implementation and integration and only then create the documentation and specification (Fig. 3.25).

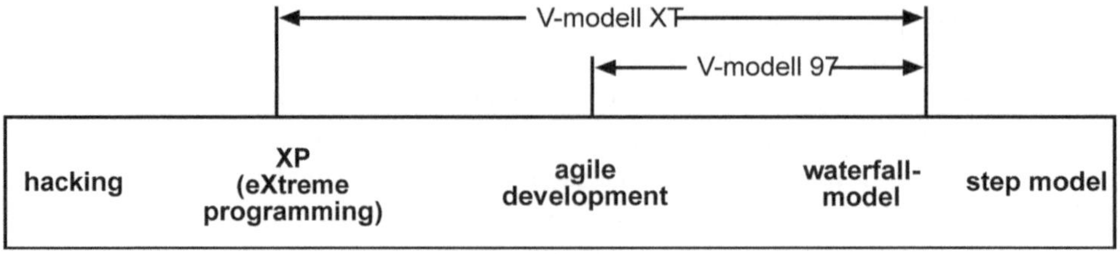

Fig. 3.25 Software development process models

References

Agile Alliance (2023a): What is Agile? https://www.agilealliance.org/agile101/ (14.04.2023).

Agile Alliance (2023b): The 12 Principles behind the Agile Manifesto. https://www.agilealliance.org/agile101/12-principles-behind-the-agile-manifesto/ (14.04.2023).

Android (2023a): Say hello to Android 13. https://www.android.com/intl/en_en/ (14.04.2023).

Android (2023b): Maps SDK for Android. https://developers.google.com/maps/documentation/android-sdk/overview (14.04.2023).

AndroidStudio (2023): Meet Android Studio. https://developer.android.com/studio/intro/ (14.04.2023).

Berg, de M., Cheong, O., Kreveld, van M. u. M. Overmars (2008). Computational Geometry: Algorithms and Applications. Berlin: Springer. 3. Ed.

Boehm, B. W. (1988): A Spiral Model of Software Development and Enhancement. Computer 21, S. 61–72.

Brinkhoff, T. (2013): Geodatenbanksysteme in Theorie und Praxis. Einführung in objektrelationale Geodatenbanken unter besonderer Berücksichtigung von Oracle Spatial. Heidelberg: Wichmann. 3. Ed.

Church, R. and Medrano, F.A. (2018). Location-allocation Modeling and GIS. In: Wilson, J.P. (Ed.): The Geographic Information Science & Technology Body of Knowledge (3rd Quarter 2018 Edition). DOI: 10.22224/gistbok/2018.3.4 (14.04.2023).

Domschke, W. (2007): Logistik: Transport. München: Oldenbourg. 5. Ed.

Eclipse Foundation (2023): About the Eclipse Foundation. https://www.eclipse.org/org/ (14.04.2023).

ESRI (2023): ArcGIS Pro. Python and Geoverarbeitung. https://pro.arcgis.com/en/pro-app/latest/help/analysis/geoprocessing/basics/python-and-geoprocessing.htm (14.04.2023).

Garrard, C. (2016): Geoprocessing with Python. Shelter Island, NY. Manning Publ.

GEOS (2023): Geometry Engine – Open Source. https://libgeos.org/ (14.04.2023).

GeoTools (2023): The Open Source Java GIS Toolkit. https://www.geotools.org/ (14.04.2023).

GSL (2023): GNU Scientific Library. http://www.gnu.org/software/gsl (14.04.2023).

Gumm, H.-P. u. M. Sommer (2013): Einführung in die Informatik. München: Oldenbourg. 10. Ed.

Haunert, J.-H. u. B. Budig (2012): An algorithm for map matching given incomplete road data. Pro-ceedings of the 20th Proceedings of the 20th International Conference on Advances in Geo-graphic Information Systems. Redondo Beach, California. S. 510–513.

Hengl, T. u. H. Reuter (2008, Hrsg.): Geomorphometry. Concepts, Software, Applications. Amsterdam: Elsevier. Developments in Soil Science 33.

Herrmann, D. (1992): Algorithmen Arbeitsbuch. Bonn: Addison-Wesley.

Kohonen, T. (2001): Self-Organizing Maps. Berlin: Springer. 3. Ed.

JOGL (2023): JavaBinding for the OpenGL API. https://jogamp.org/jogl/www/ (14.04.2023).
JTS (2023): JTS Topology Suite, API of 2D spatial predicates and functions. http://www.tsusiatsoftware.net/jts/main.html (14.04.2023).
Jungnickel, D. (2013): Graphs, Networks and Algorithms. Berlin: Springer. 4. Ed.
Klawonn, F. (2010): Grundkurs Computergrafik mit Java: Die Grundlagen verstehen und einfach umsetzen mit Java 3D. Wiesbaden: Springer. Vieweg, 3. Ed.
Klein, R. (2005): Algorithmische Geometrie. Grundlagen, Methoden, Anwendungen. Berlin: Springer. 2. Aufl.
Knuth, D. E. (2011): The Art of Computer Programming. Volume 1-4A. BoxedSet. Amsterdam: Addison-Wesley. https://www-cs-faculty.stanford.edu/~knuth/taocp.html (14.04.2023).
Levi, P. u. U. Rembold (2003): Einführung in die Informatik für Naturwissenschaftler und Ingenieure. München: Spektrum: Akademischer Verlag. 4. Ed.
NAG (2023): NAG Numerical Libraries. https://www.nag.com/ (14.04.2023).
OpenGL (2030): The Industry's Foundation for High Performance Graphics. https://www.opengl.org/ (14.04.2023).
Ottmann, Thomas u. Peter Widmeyer (2002): Algorithmen und Datenstrukturen. Heidelberg: Spektrum Akad. Verlag, 4. Ed.
QGIS (2023a): QGIS plugins web portal. https://plugins.qgis.org (14.04.2023).
QGIS (2023b): QGIS Python Plugins Repository https://plugins.qgis.org/plugins/ (14.04.2023).
Preparata, F. u. M. Shamos (1985): Computational Geometry. New York: Springer.
Quinn, M.J. (1993): Parallel computing: theory and practice. New York: McGraw-Hill. 2. Ed.
Rausch, A. u.a. (2008): Das V-Modell XT. Grundlagen, Methodik und Anwendungen. Heidelberg: Springer.
Royce, W. W. (1970): Managing the development of large software systems: concepts and techniques. In: Proceedings of IEEE WESCON 26, S. 1–9.
Rushton, G., Goodchild, M. u. L.M. Ostresh Jr. (1973): Computer Programs for Location-Allocation Problems. Dep. of Geography, Monograph Number 6, Iowa City.
Scherer, A. (1997): Neuronale Netze: Grundlagen und Anwendungen. Braunschweig: Vieweg.
Schmitt, A., Deussen, O. u. M. Kreeb (1996): Einführung in graphisch-geometrische Algorithmen. Stuttgart: Teubner.
Schwarz, H. u. N. Köckler (2011): Numerische Mathematik. Stuttgart: Teubner + Vieweg. 8. Ed.
SciPy (2023): Scientific Computing Tools for Python. https://scipy.org/ (14.04.2023).
Sedgewick, R. u. K. Wayne (2011): Algorithms. Amsterdam: Addison-Wesley Longman. 4. Ed.
Sester, M. (2020): Geoinformatik. In: Freeden, W. u. R. Rummel (Hrsg.): Handbuch der Geodäsie. Berlin/Heidelberg: Springer Spektrum.
Shapely (2023): Python package for manipulation and analysis of features in the Cartesian plane. https://pypi.org/project/Shapely/ (14.04.2023).
Stahlknecht, P. u. U. Hasenkamp (2005): Einführung in die Wirtschaftsinformatik. Berlin: Springer. 11. Ed.
Sultan, G. B. u. J.-H. Haunert (2017). Extracting spatial patterns in bicycle routes from crowdsourced data. In: Transactions in GIS, 21(6) S. 1321-1340.
Ulferts, L. (2017); Python mit ArcGIS. Einstieg in die Automatisierung der Geoverarbeitung in ArcGIS. Berlin: Wichmann.
US Department of Transportation, Federal Highway Administration (2007): Systems Engineering for Intelligent Transport. https://ops.fhwa.dot.gov/publications/seitsguide/seguide.pdf (14.04.2023).
VDI 2206 (2004): Entwicklungsmethodik für mechatronische Systeme. Beuth: Berlin.
VTK (2023): The Visualization Toolkit. https://vtk.org/ (14.04.2023).
Wirth, N. (2013): Algorithmen und Datenstrukturen: Pascal-Version. Stuttgart: Teubner. 5. Aufl.

Wolf, H. u. W.-G. Bleek (2011): Agile Softwareentwicklung. Werte, Konzepte und Methoden. Heidelberg: dpunkt. 2. Aufl.

Worboys, M. u. M. Duckham (2004): GIS. A computing perspective. Boca Raton: CRC Press, 2. Aufl.

Zhang, H. u. D. Liang (2007): Computer Graphics Using Java 2D and 3D. Upper Saddle River: Pearson Education.

Zimmermann, A. (2012): Basismodelle der Geoinformatik. Strukturen, Algorithmen und Programmierbeispiele in Java. München: Hanser.

4 Geoobjects and Reference Systems

4.1 Geoobjects

4.1.1 Geoobject: Concept

The spatial reference of information is a characteristic feature of geosciences and also of geoinformatics. The coupling of information to spatial reference units, spatial elements or objects with a spatial reference is typical of geoscientific issues. *Spatial reference units, spatial objects* or (more simply) *geoobjects* appear as:

points	e.g. boundary stone, counting or measuring point, source location of an emitter
lines	e.g. profile line, boundary line, row of trees, water pipe, connecting line
areas	e.g. parcel, biotope, municipal area, catchment area
3D solid	e.g. pollutant cloud, groundwater body, deposit, building

Geoobjects are spatial elements which have geometric and topological properties as well as thematic information and can be subject to change over time. Therefore, geometry, topology, topic and dynamic are characteristic of geoobjects.

The geometry of an object includes all information about the exact position and the shape or extent of a geoobject (e.g. position, size, perimeter). The geometry data are complemented by information about the relative position and neighborhood relationships (topological information, topology). Topological properties or concepts are neighborhoods (or neighborhood relations), subsets (or subset relations) or overlays (overlapping or intersection). Geoobjects can have different thematic information and also have temporal variability (temporal changes, dynamics), which is often used as a further distinguishing feature in addition to geometric, topological and thematic information (topic). In general,

N. de Lange, *Geoinformatics in Theory and Practice*, Springer Textbooks in Earth Sciences, Geography and Environment,
https://doi.org/10.1007/978-3-662-65758-4_4

Table 4.1 Object orientation of geoobjects

		Features	Methods
Object class	Street	Start, end, width, road surface other, special features:	Length calculation, slope, more special methods:
Subclass	Motorway	Width of the median strip, noise protection measures	Emissions calculation
Subclass	Road	Planting of the side strip	Estimation of planting costs

geoobjects can exhibit spatial and temporal variability, to which both the thematic information as well as the geometry and topology can be subject.

So far, the term geoobject has been used in the sense of a spatial element and has been detached from the object concept of computer science. However, the term geoobject was deliberately chosen in order to make the reference to the object orientation of computer science clear (cf. Sect. 3.1.4.3). In this way, object classes can be created that represent generalisations or types of objects with associated attributes and methods:

A special feature of this approach is the linking with methods which are specific for individual object classes. The other concepts of object orientation, such as inheritance (i.e. deriving a special subclass and inheriting the properties of the existing class) or class hierarchies (e.g. superclass, class, subclass, object, subobject) can also be seen in the example of the object class "street" (cf. Table 4.1 and Sect. 3.1.4.3). A specific object, i.e. in the terminology of object orientation, an instance of the object class, then also already has the attributes and methods of this class, but its own specific attribute values.

This approach or the use of the term geoobject in the sense of object orientation in computer science extends the definition mentioned at the beginning by further contents that can be derived from the object concept. However, in geoinformatics the term geoobject is mostly not understood or used in this broader sense. The existing processes in geoinformatics or the software systems that are currently mainly used still generally fall back on the simpler geoobject concept.

4.1.2 Geoobjects: Geometry

The *geometry* of a geoobject comprises the information on the position of the geoobject on the basis of a clear spatial reference system (coordinates). In geoinformatics, metric coordinate systems are used as a basis, which allow a quantifiable and objectifiable location determination. However, it must not be overlooked that human action ultimately takes place in a subjective reference system and perceptual space that is inadequately captured by computer science methods.

In geoinformatics different spatial reference systems are used. Thus, geoobjects can be represented in global coordinate systems by e.g. geographic coordinates (cf. Sects. 4.2.3 and 4.2.4). A map with a special projection that projects the (curved) earth's surface in a two-dimensional plane can also serve as a reference system for spatial orientation (cf. Sect. 4.3). Very often local coordinate systems and then (almost) exclusively Cartesian

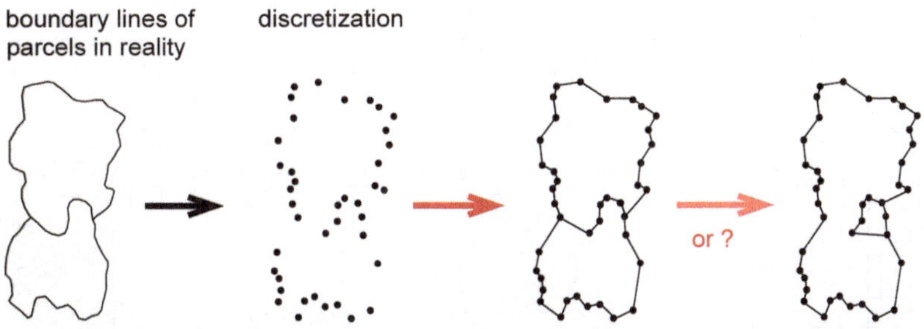

Fig. 4.1 Geoobjects in the vector model

coordinate systems are used (cf. Sects. 4.2.1 and 4.5). Reference systems are also required in a geoinformation system for displaying real geoobjects or for recording their geometry. The representation of the geoobjects takes place in a vector or in a raster model (cf. Fig. 2.6).

The *vector model* is based on directed lines, i.e. *vectors* in a coordinate system, which are clearly defined by specifying a starting point and an end point (in a two- or three-dimensional coordinate system by (x,y)- or (x,y,z)-coordinate information). A geoobject is described by points or vectors. Points are to be understood as vectors, which have their beginning in the origin of the coordinate system. This form of modeling requires that a linear geoobject is resolved, i.e. discretised, by a finite number of points. Accordingly, an area (polygon) is described by its bounding lines, which in turn are defined by individual points (modeling an area by coordinates of the delimiting "fence posts"). The data model on which the geoinformation system is based defines in more detail how areas are modeled in the vector model (cf. Sect. 9.3.2). After the discretisation has taken place, it is not always clearly recognizable from the coordinates alone how areas are composed (number of areas, clear definition of boundaries from coordinates Fig. 4.1). In addition to the coordinate information, topological information must be recorded and stored, which states, which vectors (points) define which line as well as which line define which area (for geometric-topological modeling in the vector model cf. Sect. 9.3.2). However, areas can also be modeled as closed polygons, i.e. as a closed sequence of vectors. In this case, the common boundaries of neighboring parcels are recorded twice, i.e. redundantly (cf. the Simple Feature Geometry Object Model, cf. Sect. 6.3.2 and Table 6.2).

Vector data, i.e. sequences of coordinates, are the standard data form of graphical information in cartography, surveying and cadastral surveying or in engineering ("line drawings"), from which a wide variety of applications of today's geoinformation systems have developed. Thematic data (cf. Sect. 4.1.4) must also be assigned to the vector information.

In the representation and discretisation of geoobjects in the *raster model,* the basic geometric element is of fixed shape and size, i.e. usually a square *mesh.* This rather simple *grid model* belongs to a much more general spatial model that can consist of any mosaics of different sizes and shapes that completely fill a flat surface or any surface without

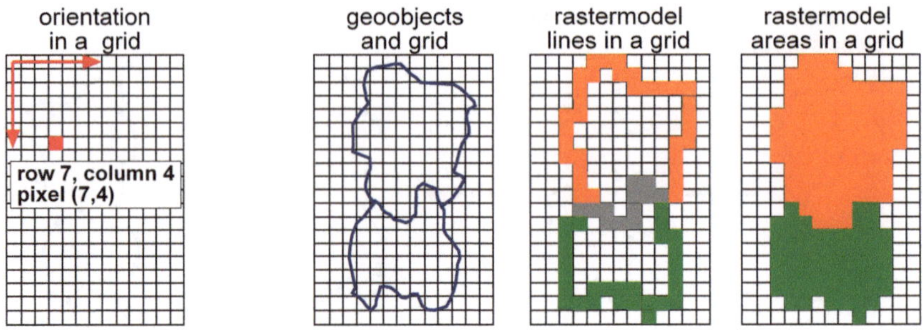

Fig. 4.2 Geoobjects in the raster model

intersections (tessellation). Although equilateral triangles or regular hexagons are also possible as mesh shapes that can tile a plane, squares or square pixels are used because they are easier to handle computationally. In particular, the raster model is then based on a Cartesian coordinate system (cf. Sect. 4.2.1).

The raster model plays a central role in digital image processing. Thus, the geometry in a raster image (e.g. in a digital satellite image or after recording an image with a scanner) is resolved by mostly square picture elements (*pixels,* cf. Sect. 2.5.6). The three-dimensional equivalent is a three-dimensional matrix of cubic cells, of *voxels.* The spatial reference system is formed by a regular grid covering arrangement of similar pixels, the size of which must first be fixed (mesh size). Each pixel is clearly described geometrically by specifying its row and column index. This is done according to the indexing of a matrix (cf. Figs. 2.6 and 4.2).

A point is approximated by a single pixel. A line is approximated by corresponding arrangements of connected pixels. Lines can then be described, for example, by sequences of index pairs (row, column) of the associated pixels. An area can also be represented by contiguous pixels. Thus, no further additional information is required to model areas as in the vector model (for geometric-topological modeling in the raster model cf. Sect. 9.3.4).

Several basic problems of the raster model are to be emphasised: The geometric shape of the geoobjects is changed by the rasterisation, curved lines are approximated by staircase-like raster structures, shape and size of the geoobjects are coarsened. In each case areas are considered. Points as well as lines are represented by two-dimensional pixels. In conclusion, this means that a single or even several adjacent pixels, which mark an area in the raster model, can also mean single points or lines in the real world. These difficulties can be reduced, but in principle not eliminated, by decreasing the mesh size. This increases the resolution, but at the same time the size of the raster matrix and the required storage space increase considerably. However, the accuracy of a coordinate specification in the vector model is not achieved.

4.1.3 Geoobjects: Topology

Topology characterises the spatial relationships between geoobjects. When considering the topology of geoobjects, the geometry is abstracted. The topological view can be illustrated very well by the example of a balloon, on which, for example, the outlines of a street map are drawn. If air is let out or pumped in, the geometry always changes. However, the topology, the relative position of the individual street lines to each other, does not change! These transformations as well as rotations, stretchings or compressions are topologically invariant.

Figure 4.3 shows the differences between geometry and topology. The bus lines are compared in a geometrically exact town map and in a topological map, which only shows the simplified route network and abstracts the geometrically exact routing. This representation can still be valid in the case of major changes to the routes such as diversions and bus stop relocations (e.g. when route 2 is diverted via the northern parallel road in Fig. 4.3). The topological representation contains all the information a passenger needs for orientation and route planning: information about transfer options or stopping options and generally about connection options. These everyday terms denote central topological concepts: *neighborhoods, overlaps* or *subset relationships*, which do not only apply to networks.

Figure 4.4 systematizes the six possible *types of spatial relationships* between the three basic geometric forms point, line and area in a two-dimensional space. The following are examples of relationships to be defined:

Two areas are adjacent if they have a common boundary (at least one common boundary point). Two lines are adjacent if the end point of one is identical to the start point of the other line. Two points are adjacent if they are connected by a line. However, it is also possible to define adjacency relations in terms of content. Thus, A and B are adjacent points if they are directly connected by a line with a certain property (e.g. connection of two cities by an ICE line).

Two objects *overlap* or *intersect* if they have (at least) one common point. This definition only makes sense for lines, areas and volumes. It is immediately obvious for the

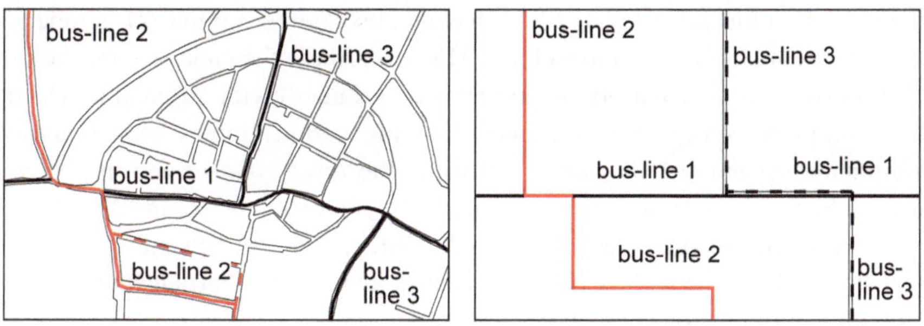

Fig. 4.3 Comparison of geometry and topology using the example of bus routes

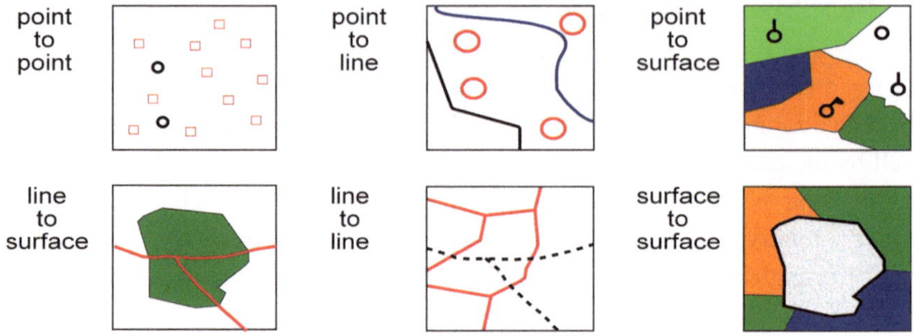

Fig. 4.4 Topological relationships between point, line and areas (polygons)

intersection of lines. The spatial average of overlapping areas e.g. of different types of land use (e.g. subdivision according to property parcels and according to crop types) can be regarded as *intersection of areas*.

A subset relationship exists, for example, when a point lies on a line or in an area, or when an area contains a subarea or a line.

It should be emphasised that topology and topological relationships between geoobjects among each other are explained here. The topological modeling of geoobjects with regard to their representation in geoinformation systems is to be seen independently of this (cf. Sects. 9.3.2 and 9.3.4).

4.1.4 Geoobjects: Thematic Information

A geoobject always has a *topic* or carries *thematic information*, which is generally characterised by several attributes (features, variables). The attributes can have different *measurements scales: nominal scale* (e.g. name, land use type: wet meadow), *ordinal or cardinal scale* (e.g. suitability rating: 13), *interval scale* (e.g. soil temperature: 2 degrees Celsius), *ratio scale* (e.g. depth of B horizon: 0.75 m). Table 4.2 names basic properties, whereby it should be emphasised that, compared to interval scaled data, ratios may only be calculated at the ratio scale level, which has an absolute zero point (cf. temperature in Celsius or Fahrenheit compared to Kelvin). Moreover, these thematic data can be stored in several databases (of different origin, currentness, accuracy) with different access rights.

The description, processing and storage of the various topics of geoobjects in a geoinformation system can be carried out using the *layer concept* and the *object class concept* (cf. Sect. 9.3.6). With the layer concept, the geometric data of the objects and their attributes are strictly separated according to the different thematic meanings and stored in different layers. The object class concept assumes a hierarchical arrangement of different themes with subset relationships of the themes.

Table 4.2 Types of measurement scales in statistics (cf. de Lange and Nipper 2018 pp. 44)

Scale	Properties and examples	Permitted operations
Nominal scale	Names, postal codes, numbers as codes	Present or absent and equal or unequal
Ordinal scale	Ranks, rankings, evaluation levels, school grades	Comparison in addition to the nominal scale
Interval scale	Metric data with fixed but not absolute zero point: degrees Celsius	Further to the ordinal scale addition and subtraction, consideration of intervals (30 °C is 10 °C warmer than 20 °C, **but not:** 30 °C is twice as warm as 15 °C)
Ratio scale	Metric data with absolute zero point: length in m, area in m^2, age in years	Further to the interval scale multiplication and division, consideration of ratios, now 200 °K is twice as warm as 100 °K

4.1.5 Geoobjects: Dynamics

Geoobjects can change over time with respect to their thematic information. For example, meteorological stations that record parameters for weather observation will measure different values of e.g. air temperature, air pressure or precipitation in the course of the year. The yield of vineyards in the Rheingau will change over the course of several harvest periods. The two examples cited have in common that their location and spatial reference do not change. The assumed geoobjects, i.e. the meteorological stations and the individual cultivated areas, have usually constant location coordinates or boundary lines over time. These geoobjects have a *temporal variability* only with respect to their thematic information. The features, i.e. the climatic parameters or the yield characteristics, can remain the same and their values can change. However, new variables can be recorded over time (e.g. additional wind direction and wind strength or, with regard to the cultivated areas, new economic characteristics). In contrast, geoobjects can also exhibit *spatial variability* over time and change their location or extent, so that variations in topology and neighborhood relationships are possible.

However, the *spatial-temporal variability* of geoobjects is difficult to record or to represent in a geoinformation system. Although measuring instruments exist that enable continuous data recording (e.g. seismographs or sunshine recorders), *discretisation* is always necessary for representation in a geoinformation system. In particular, spatio-temporal processes are only quantified in a discrete form. The geoobjects are discretised according to their geometry, topology and thematic information into several individual time slices (for the representation of spatio-temporal processes in geoinformation systems cf. Ott and Swiaczny 2001). Thus, a cloud of pollutants spreading from an emitter (e.g. an exhaust gasplume from a chimney) is recorded by values at different time at several individual meteorological stations, which are spatially distributed around the emitter. Thus, space and time are discretised. On the basis of such empirical values, a quantitative modeling of the spatio-temporal variability can then be attempted. In climatology and hydrology or also for the dispersion of noise or pollutants in environmental media, quite complex dynamic models and methods have been developed for spatio-temporal modeling (cf. e.g. the particle model AUSTAL 2000 for calculating the dispersion of dust and gaseous emissions as well as odors, cf. Umweltbundesamt 2021).

4.1.6 Geoobjects: Dimensions

The *geometric dimension of* a geoobject is identical to the number of coordinate axes in a Cartesian coordinate system that are necessary for a complete (geometric) description. To quantify the size of a geoobject, length, area size and volume can be calculated depending on its geometric dimension. Points have neither length nor area. Lines have only a (finite) length. Areas do not have a length, but rather a perimeter and an area size. A volume can be calculated for a 3D solid. In this case it is considered as a solid. Furthermore, the quantification of the surface is possible.

According to the geometric dimensions several *topological dimensions* can be distinguished (node, edge, mesh). Also according to the topic, different *thematic dimensions* arise, which denote the number of descriptive features of an object. Corresponding to the statistical methodology, n-dimensional feature spaces are differentiated.

4.2 Coordinate Systems

4.2.1 Metric Spaces and Cartesian Coordinates

Rectangular coordinate systems are of central importance in geoinformatics. Cartesian coordinate systems form the basis for the representation of geoobjects in vector or raster models and therefore for processing in geoinformation systems. The algorithms of geoinformatics, especially computational geometry (cf. Sect. 3.4), or the methods of graphical data processing (usually) require Cartesian coordinate systems.

In general, reference systems are defined as *metric spaces* consisting of a (non-empty) set M and a metric. A *metric* is a real-valued function, distance function or distance d(a,b) between two elements a and b of the set M, which fulfills three conditions:

1)	$\mathrm{d(a,b)} \geq 0$ for all a, b from M	$\mathrm{d(a,b)} = 0$ exactly when $a = b$
2)	$\mathrm{d(a,b)} = \mathrm{d(b,a)}$	(symmetry)
3)	$\mathrm{d(a,b)} \leq \mathrm{d(a,c)} + \mathrm{d(c,b)}$	(triangle inequality)

In geoinformatics, the *Euclidean metric* has the greatest importance (for other metrics or distance measures, especially in oblique coordinate systems cf. textbooks on cluster

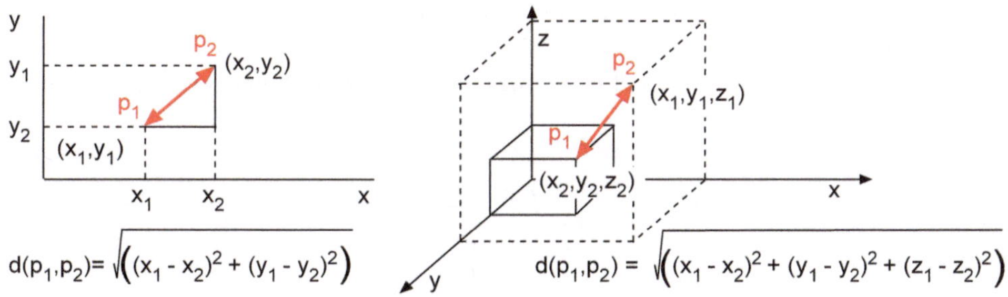

Fig. 4.5 Cartesian spaces with the Euclidean metric

analysis such as Bock 1974 or Bortz and Schuster 2010 p. 456 and Backhaus et al. 2016 pp. 457, for the use of metrics in classification methods cf. Sect. 10.7.2):

$$d_n\left(\overrightarrow{X_i}, \overrightarrow{X_J}\right) = \sqrt{\sum_{k=1}^{n} \left(x_{ik} - x_{jk}\right)^2}$$

$$\text{with } \overrightarrow{X_I} = (x_{i1}, x_{i2}, \ldots, x_{in}) \text{ and } \overrightarrow{X_j} = \left(x_{j1}, x_{j2}, \ldots, x_{jn}\right)$$

In the usual space of our everyday life, the Euclidean metric corresponds to the straight-line distance. Thus, Fig. 4.5 shows metric spaces with rectangular axes. In order to avoid double indexing in the above formula, here the coordinate axes are denoted by x, y, and z. Such n-dimensional *Cartesian coordinate systems* consist of n coordinate axes in pairs perpendicular (i.e. *orthogonal*) to one another with identical origin and same axis division. Then the position of an object can be clearly represented by specifying the sections on the coordinate axes, i.e. by the coordinates.

4.2.2 Homogeneous Coordinates

In computational geometry, especially in calculations and in transformations between (Cartesian) coordinate systems, *homogeneous coordinates* offer considerable advantages. Each point (x,y,z) can be represented by homogeneous coordinates (m,n,o,p), where: $m = p \cdot x$, $n = p \cdot y$, $o = p \cdot z$ with any value for p ($p \neq 0$). From Cartesian coordinates (x,y,z) it is very easy to go to homogeneous coordinates by (x,y,z,1) (vice versa: from (m,n, o,p) to (m/p,n/p,o/p)).

With the help of homogeneous coordinates the relative position of points and (directed) straight lines can be determined quickly. For example, if the straight line runs through the points from $P_1 = (x_1,y_1)$ to $P_2 = (x_2,y_2)$ (with direction from P_1 to P_2), is a third point $P_3 = (x_3,y_3)$ to the left or right or even on the straight line? To solve this problem, the matrix of homogeneous coordinates is formed from P_3, P_1 and P_2 (note the order!) and its determinant is calculated. For det M $(P_3,P_1,P_2) < 0$ the point is to the right of the straight line, for det M $(P_3,P_1,P_2) > 0$ to the left of the straight line, and for det M $(P_3,P_1,P_2) = 0$ on the straight line.

With $P_1 = (0,0)$, $P_2 = (5,3)$ and $P_3 = (4,2)$ the matrix M of homogeneous coordinates results:

$$M = \begin{pmatrix} 4 & 0 & 5 \\ 2 & 0 & 3 \\ 1 & 1 & 1 \end{pmatrix}$$

and then further using the expansion theorem for determinants (Laplace's expansion theorem, here expansion of the last row):

$$\det M = +1 \cdot \det\begin{pmatrix} 0 & 5 \\ 0 & 3 \end{pmatrix} + (-1) \cdot \det\begin{pmatrix} 4 & 5 \\ 2 & 3 \end{pmatrix} + 1 \cdot \det\begin{pmatrix} 4 & 0 \\ 2 & 0 \end{pmatrix}$$
$$= -1 \cdot (4 \cdot 3 - 2 \cdot 5) = -(12 - 10) = -2 < 0$$

Thus, point P_3 lies to the right of the straight line $\overline{P_1P_2}$.

More applications can be found in computational geometry and in particularly in transformations (cf. Sect. 4.2.5.2, cf. Pavlidis 1982 and Preparata u. Shamos 1985, Bartelme 2005 pp. 100, Zimmermann 2012 pp. 118 u. 120.).

4.2.3 Polar Coordinates and Geographic Coordinates on the Sphere

In addition to Cartesian coordinates, *polar coordinates* play a special role in geoinformatics. Figure 4.6 illustrates the representation in a two-dimensional and in a three-dimensional space. Here, a point is represented by the coordinates $P_a(r,\alpha)$ or $P_b(r,\alpha,\beta)$, i.e. by the distance to the coordinate origin and by angles.

For converting Cartesian and polar coordinates simple formulas exist, which are given here only for the three-dimensional case. The two-dimensional case results from this by $z = 0$, $\beta = 90°$ and then $\cos \beta = 0$ or $\sin \beta = 1$ (for orientation cf. Fig. 4.6):

Fig. 4.6 Polar coordinates

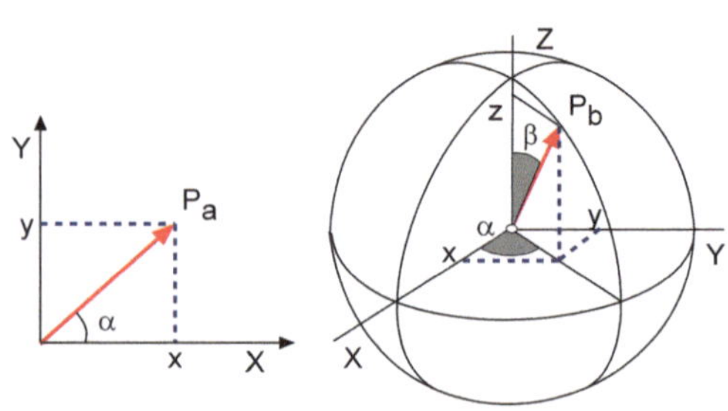

$$P(r,\alpha,\beta) \text{ then}: \quad x = r\cdot\sin\beta\cdot\cos\alpha \qquad P(x,y,z) \text{ then}: \quad r = \sqrt{(x^2+y^2+z^2)}$$
$$y = r\cdot\sin\beta\cdot\sin\alpha \qquad \alpha = \arctan\ y/x(\text{for } x \neq 0)$$
$$z = r\cdot\cos\beta \qquad \beta = \arctan\left(\sqrt{x^2+y^2}\right)/z$$
$$\text{for } x = 0 \text{ and } y = r \text{ is } \alpha = \pi/2$$
$$\text{for } x = 0 \text{ and } y = -r \text{ is } \alpha = -\pi/2$$

Analog to Cartesian coordinates, each point can be clearly identified and quantified by such polar coordinates. If r is constant, polar coordinates simply describe all points on a sphere.

Polar coordinates are of central importance for determing the position of a point on earth, which is represented in a simplified form by a sphere. The distance to the centre of the sphere is equated with the mean radius of the earth (R = 6,371 km), the position of a point then results only from two angles, which in the case of the globe are called latitude (φ) and longitude (λ) (but with $\varphi = 90° - \beta$, cf. Figs. 4.6 and 4.7).

The *latitude* (φ) and *longitude* (λ) build the illustrative geographic coordinate system. It should be emphasised that here the earth is approximated in a simplified way by a sphere (cf. Sect. 4.2.4). This simplification is quite acceptable for the representation of large parts of the earth's surface on a small scale (cf. Fig. 4.7).

- The 0° latitude is known as the *equator*, which divides the globe into the northern and southern hemispheres. The equator is the circle whose plane through the center of the earth is perpendicular to the earth's rotation axis.
- The circles running parallel to the equator are called *latitude circles* or *parallels. The latitude* is the angle between a point on the earth's surface and the equatorial plane along

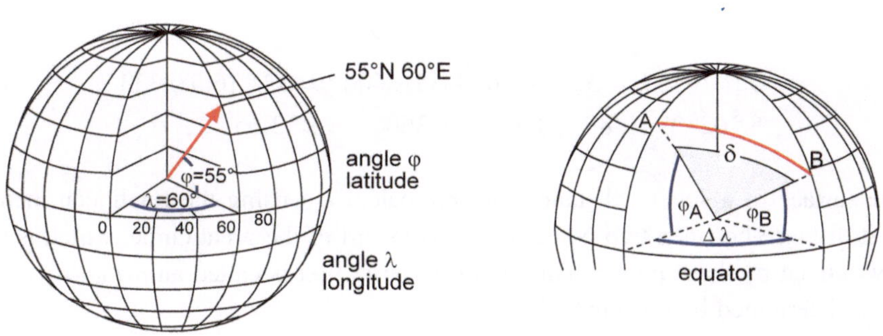

Fig. 4.7 Geographic coordinate system on the sphere and shortest distance between two points A and B on the surface of the sphere as distance on a great circle

the associated meridian (0° = equator, 90° = North Pole, −90° = South Pole). To avoid negative values, we speak of northern or southern latitude.

- The circles running vertically to the equator (and to the parallels) and through the two poles are called *meridians*. The meridian through the Greenwich Observatory near London was agreed to be the prime meridian. The *longitude* is the angle between the meridian plane of a point on the earth's surface and the prime meridian plane along the equatorial plane. Starting from the prime meridian longitudes are counted in an easterly direction from 0° to 180° and in a westerly direction from 0° to −180°
- The circles of latitude and longitude are used to construct the *geographic graticule.*
- Intersections of the globe and a plane through the center of the earth are called *great circles*. So they have all the same radius as the globe. All meridians and the equator are great circles.

To the north and south of the equator, the circumference of the circles of latitudes decreases (cf. Fig. 4.7). Due to the convergence of the meridians to the poles, the horizontal distance of one degree (longitude) at the equator corresponds to a distance of approx. 111 km, but at a latitude of 54° only approx. 65 km (cf. Table 4.3 also in Sect. 4.3.1).

The angles are mostly given in the 60 system: A circle has 360 degrees, where one degree consists of 60 minutes and one minute consists of 60 seconds. For the inner city of Osnabrück, for example, the latitude is 52° 16′ 35″ north and the longitude is 8° 02′ 39″ east. To convert to the decimal system, divide the minutes by 60 and the seconds by 3600 and add both results to the degrees: 52.276388 north and 8.044167 east.

$$52^\circ 16' 35'' = 52^\circ + 16/60^\circ + 35/3600^\circ = 52.276388^\circ$$
$$8^\circ 02' 39'' = 8^\circ + 02/60^\circ + 39/3600^\circ = 8.044167^\circ$$

as well:

$$52^\circ 16' 35'' = 52^\circ + 16' + 35/3600^\circ = 52^\circ 16.583333'$$
$$8^\circ 02' 39'' = 8^\circ + 02' + 39/3600^\circ = 8^\circ 2.65'$$

On the surface of a sphere, distances are not calculated using the Euclidean metric.The shortest distance between two points A and B is part of the great circle, which is already clearly defined by these points. This shortest connection is called an *orthodrome,* whose length is determined by (cf. Fig. 4.7):

$$\text{Distance}(A, B) = R \cdot \delta \quad \text{with}$$
$$R = \text{radius of the sphere,} \quad \delta \text{ angle in radians between A a.B}$$

If latitude and longitude are used to determine A (λ_A,φ_A) and B (λ_B, φ_B), the distance between A and B is calculated using the cosine theorem of spherical trigonometry. Here simplified, it is assumed that the earth is a sphere:

$$\text{Distance}(A, B) = R \cdot \cos^{-1} (\sin \varphi_A \cdot \sin \varphi_B + \cos \varphi_A \cdot \cos \varphi_B \cdot \cos (\lambda_A - \lambda_B))$$

4.2.4 Geodetic Coordinates on an Ellipsoid

In contrast to simplifications, in which the earth is considered as a sphere, large-scale tasks in national surveys are based on a more precise model of the earth, which is, however, still mathematically manageable. Thus, surveying around the world is generally based on ellipsoids, which best approximate the globe for the area of the respective national survey (i.e. regionally, cf. Fig. 4.17). This has several consequences:

- Since different ellipsoids are used by national surveys worldwide, different geographic coordinates are available for a single point on the earth's surface depending on the modeling by different ellipsoids.
- The national surveys use elliptical coordinates to calculate plane coordinates (i.e. UTM or, formerly in Germany, Gauss-Krüger coordinates, cf. Sects. 4.5.2 and 4.5.5).
- Satellite-based navigation (cf. Sect. 5.3) requires a globally standardised reference ellipsoid and unambiguous coordinate data (WGS84 reference ellipsoid). This global ellipsoid approximates the earth as best as possible worldwide, but no longer regionally for the area of a national survey (cf. Sect. 4.4.2).
- The ellipsoids do not only differ with respect to the size of their axes (cf. Table 4.4), they are also differently positioned with respect to a global coordinate system (e.g. WGS84). Thus, they have a mutually shifted origin and a different orientation of their axes (cf. Fig. 4.20). The conversion of the geographic coordinates of a point with respect to an older reference system of a national survey into the global reference system WGS84 requires a so-called *datum transformation* (cf. Sect. 4.4).

A point is then defined by specifying the *geodetic latitude* (φ), the *geodetic longitude* (λ) and *geodetic height* h (cf. Becker u. Hehl 2012 pp. 21 u. p. 26). The geodetic latitude of a point P is the angle between the equatorial plane and the ellipsoid normal at point P, i.e. the line that is perpendicular to the tangential surface at point P of the ellipsoid. The geodetic longitude is the angle between the prime meridian plane and the meridian plane at the point P. The length of the ellipsoid normal is equal to the ellipsoidal height of the point P

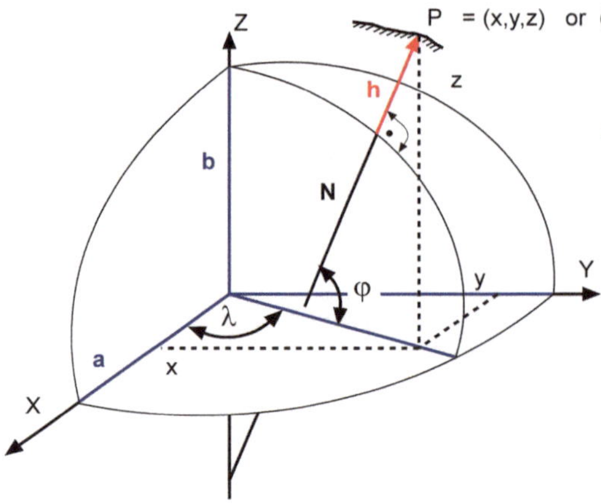

Fig. 4.8 Cartesian coordinates (x,y,z) and ellipsoidal coordinates (λ,φ,h) (cf. Hofmann-Wellenhof et al. 2008 p. 280)

(cf. Fig. 4.8). The radius of curvature N at a given point on a curve (2D-case) or ellipsoid (3D-case) is the radius of a circle or sphere that best approximates the curve or ellipsoid at that point. Because at the North Pole, the minor axis of an ellipsoid is smaller than its sphere, the radius of curvature is larger than the radius of the sphere. A larger sphere is required to cling against the ellipsoid (cf. N in Fig. 4.8).

$$X_a = (N + h) \cdot \cos\varphi \cdot \cos\lambda$$

$$Y_a = (N + h) \cdot \cos\varphi \cdot \sin\lambda$$

$$Z_a = \left[N \cdot \left(1 - e^2\right) + h\right] \cdot \sin\varphi \quad a = \text{major}, \ b = \text{minor semiaxis}$$

$$N = \frac{a}{w} \qquad w = \sqrt{1 - e^2 \cdot \sin^2\varphi} \qquad e^2 = \frac{\left(a^2 - b^2\right)}{a^2}$$

4.2.5 Plane Coordinate Transformations

4.2.5.1 Georeferencing

In geoinformatics, geoobjects are recorded in many coordinate systems. Often data recording is not based on a well-defined reference system with e.g. coordinates of the national survey, but on device coordinates. This is the rule if geometries are recorded from a template during on-screen digitisation, which are then initially available in any rectangular coordinate system (cf. Sect. 4.2.1). Since, however, it is not these coordinates that are of interest, but rather the coordinates in a certain coordinate system, transformations become necessary (changes in scale, rotations of the coordinate system, as the template is almost always not aligned exactly on the digitising tablet or on the scanner). A geoinformation system, i.e. the software, usually has tools to carry out georeferencing (cf. Sect. 4.6.1). In order to understand the options and to use them correctly (e.g. selecting the option "1st order polynomial (affine)"), however, knowledge of coordinate transformations is necessary.

First of all, it should be noted that this section assumes that the georeferencing templates are not distorted. Rectification of the template and resampling are not covered here. However, older paper templates are often disorted due to aging processes of the paper. Airborne scanners usually provide more distorted images due to the relatively unstable flight attitude (e.g. yaw, roll and pitch, cf. Figs. 5.16 and 10.17). Therefore, a correction of the geometric distortion, i.e. a rectification of the original data, is necessary. The image rectification is often a hybrid approach. The pixels of the input image are simultaneously rectified and assigned to a cartographic or geodetic coordinate system. In remote sensing, it is common to speak of registration instead of georeferencing. In addition to the transformation into a new reference system, the brightness values of the pixels in the original image must be converted into brightness values of the new, not distorted pixels in the result image (resampling). Section 10.6.1.2 discusses rectification, registration and resampling in digital image processing in more detail.

Here, the central task is to transform the source data into a standard coordinate system such as the usual coordinate system of the national survey. This process, which is part of everyday geoinformatics, is called *georeferencing* or *geocoding*, in which the device coordinates are referenced to a real geographic reference system (cf. Fig. 4.9). After that it is also possible to change the coordinate system or projection. Geoinformation systems offer a wide range of map projections and transformations. At the end of Chap. 4, after the

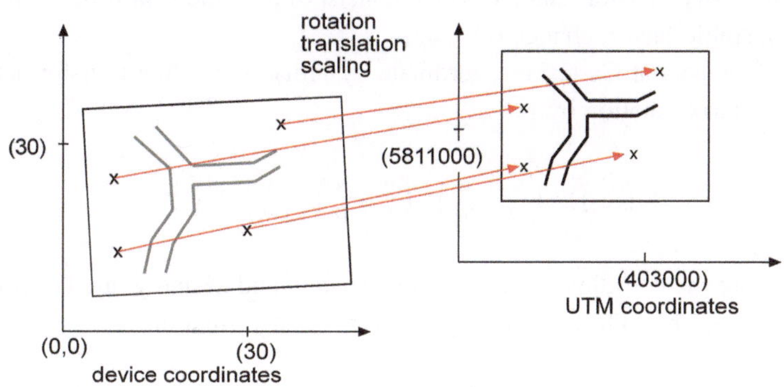

Fig. 4.9 Georeferencing or geocoding

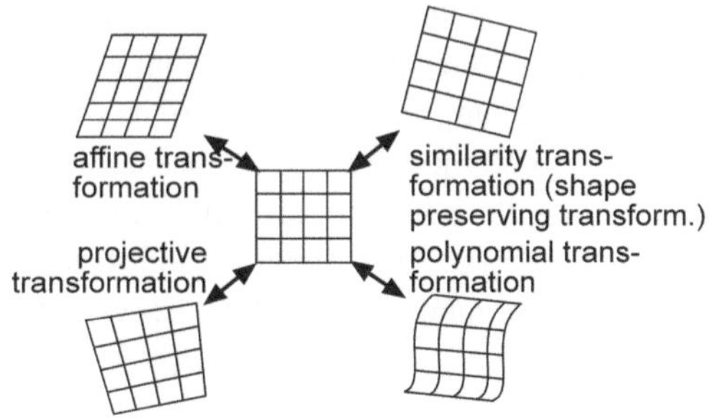

Fig. 4.10 Coordinate transformations (cf. Resnik and Bill 2018 p. 182)

discussion of transformations, map projections and coordinate reference systems, an example of georeferencing with a geoinformation system is given (cf. Sect. 4.6).

Typically, a similarity transformation will be carried out, which transforms a rectangular coordinate system into a rectangular coordinate system (cf. Fig. 4.10). This is the case, for example, when a scanned map is available that can be assumed to have no distortion. Similarity transformations belong to the larger group of affine coordinate transformations, which in turn belong to the group of linear transformations, which can be represented by (simple) matrix multiplications. Projective coordinate transformations are used to rectify aerial images that represent a perspective image. Nonlinear coordinate transformations can only be represented by polynomial equations of higher order (for rectification and resampling of (satellite) images cf. Sects. 10.6.1.2 and 4.2.5.3).

4.2.5.2 Affine Coordinate Transformation

Here, only *affine transformations* A are considered, which are defined between two Cartesian coordinate systems ($A: R_n \rightarrow R_n$), which preserve lines, parallelism and partial ratios on each straight line, but lengths, angles, areas and orientations can change. The transformation is reversible unless the determinant of the transformation matrix is 0 (cf. matrix notation in homogeneous coordinates). Affine transformations include, above all, similarity transformations, which consist of a rotation, shift or scaling (multiplication with a single factor, change of scale).

For two-dimensional Cartesian coordinate systems, any affine transformation can be described in matrix notation by:

$$\begin{pmatrix} x' \\ y' \end{pmatrix} = \begin{pmatrix} S_x & 0 \\ 0 & S_y \end{pmatrix} \cdot \begin{pmatrix} A & B \\ C & D \end{pmatrix} \cdot \begin{pmatrix} x \\ y \end{pmatrix} + \begin{pmatrix} T_x \\ T_y \end{pmatrix}$$

Here x and y are the coordinates in the source system and x′ and y′ are the coordinates in the target system. The following also apply to this transformation:

– The simple *rotation* around the origin of the coordinate system, which preserves the perpendicularity and the proportions (no additional stretching or compression and displacement), is described by a matrix, where α is the angle of rotation (counterclockwise):

$$\begin{pmatrix} A & B \\ C & D \end{pmatrix} = \begin{pmatrix} \cos\alpha & -\sin\alpha \\ \sin\alpha & \cos\alpha \end{pmatrix}$$

If a rotation with a given angle is to be carried out around any fixed point, the fixed point must first be moved to the origin, the rotation must be executed and finally the movement must be reversed.

- The *translation* (shift) of the coordinate origin in the x-direction by the value T_x and in the y-direction by the value T_y is expressed by the shift or translation vector:

$$\begin{pmatrix} T_x \\ T_y \end{pmatrix}$$

- The *change in scale* is represented by two different scale factors (S_x,S_y) that cause the affine distortion. A square becomes a parallelogram, a circle becomes an ellipse. A similarity transformation results, if S_x and S_y are identical.

The three-dimensional case can be described by a corresponding equation, where S_i means changes in scale and T_i means translations:

$$\begin{pmatrix} x' \\ y' \\ z' \end{pmatrix} = \begin{pmatrix} S_x & 0 & 0 \\ 0 & S_y & 0 \\ 0 & 0 & S_z \end{pmatrix} \cdot \begin{pmatrix} A & B & C \\ D & E & F \\ G & H & I \end{pmatrix} \cdot \begin{pmatrix} x \\ y \\ z \end{pmatrix} + \begin{pmatrix} T_x \\ T_y \\ T_z \end{pmatrix}$$

The *rotation* around the origin can be decomposed into three rotations, i.e. three rotation angles around the three coordinate axes, which are to be executed one after the other. For the rotation (only) around the z-axis the rotation matrix is:

$$\begin{pmatrix} \cos\alpha & -\sin\alpha & 0 \\ \sin\alpha & \cos\alpha & 0 \\ 0 & 0 & 1 \end{pmatrix}$$

The entire transformation can be represented as a single matrix multiplication using homogeneous coordinates:

$$\begin{pmatrix} x' \\ y' \\ z' \\ 1 \end{pmatrix} = \begin{pmatrix} S_x & 0 & 0 & 0 \\ 0 & S_y & 0 & 0 \\ 0 & 0 & S_z & 0 \\ 0 & 0 & 0 & 1 \end{pmatrix} \cdot \begin{pmatrix} 1 & 0 & 0 & T_x \\ 0 & 1 & 0 & T_y \\ 0 & 0 & 1 & T_y \\ 0 & 0 & 0 & 1 \end{pmatrix} \cdot \begin{pmatrix} A & B & C & 0 \\ D & E & F & 0 \\ G & H & I & 0 \\ 0 & 0 & 0 & 1 \end{pmatrix} \cdot \begin{pmatrix} x \\ y \\ z \\ 1 \end{pmatrix}$$

This simplifies the calculations and justifies (also) the use of homogeneous coordinates. The three-dimensional case (multiplication of 4 × 4 matrices) can of course also be restricted to the two-dimensional case (multiplication of 3 × 3 matrices, omission of the third row or column).

A two-dimensional affine projection ($R^2 \rightarrow R^2$) is often also described as a first-order polynomial (cf. Sect. 4.2.5.4):

$$\begin{pmatrix} x' \\ y' \end{pmatrix} = \begin{pmatrix} S_x & 0 \\ 0 & S_y \end{pmatrix} \cdot \begin{pmatrix} A & B \\ C & D \end{pmatrix} \cdot \begin{pmatrix} x \\ y \end{pmatrix} + \begin{pmatrix} T_x \\ T_y \end{pmatrix}$$

Following from this:

$$x' = S_x \cdot (A \cdot x + B \cdot y) + T_x$$

$$y' = S_y \cdot (C \cdot x + D \cdot y) + T_y$$

and finally the representation as polynomials:

$$x' = a_0 + a_1 \cdot x + a_2 \cdot y \quad \text{with} \quad a_0 = T_x \text{ and } a_1 = S_x \cdot A \text{ and } a_2 = S_x \cdot B$$

$$y' = b_0 + b_1 \cdot x + b_2 \cdot y \quad \text{with} \quad b_0 = T_y \text{ and } b_1 = S_y \cdot C \text{ and } b_2 = S_y \cdot D$$

An example is intended to explain an affine transformation starting from three points $P_1(1,1)$, $P_2(3,2)$ and $P_3(2,3)$. By executing:

a rotation of 30 °

$$\begin{pmatrix} \cos 30° & -\sin 30° \\ \sin 30° & \cos 30° \end{pmatrix} \cdot \begin{pmatrix} 1 & 3 & 2 \\ 1 & 2 & 3 \end{pmatrix}$$

$$\begin{pmatrix} \frac{1}{2} \cdot \sqrt{3} & -0.5 \\ 0.5 & \frac{1}{2} \cdot \sqrt{3} \end{pmatrix} \cdot \begin{pmatrix} 1 & 3 & 2 \\ 1 & 2 & 3 \end{pmatrix} = \begin{pmatrix} 0.366 & 1.598 & 0.232 \\ 1.366 & 3.232 & 3.598 \end{pmatrix}$$

and then scaling by the factor 10 in x- and in y-direction

$$= \begin{pmatrix} 0.366 & 1.598 & 0.232 \\ 1.366 & 3.232 & 3.598 \end{pmatrix} \cdot \begin{pmatrix} 10 & 0 \\ 0 & 10 \end{pmatrix} \begin{pmatrix} 3.66 & 15.98 & 2.32 \\ 13.66 & 32.32 & 35.98 \end{pmatrix}$$

and translation (shift) by a factor of 10 in x- and in y-direction

$$\begin{pmatrix} 3.66 & 15.98 & 2.32 \\ 13.66 & 32.32 & 35.98 \end{pmatrix} + \begin{pmatrix} 10 \\ 10 \end{pmatrix} = \begin{pmatrix} 13.66 & 25.98 & 12.32 \\ 23.66 & 42.32 & 45.98 \end{pmatrix}$$

the points $P_1(1,1)$, $P_2(3,2)$ and $P_3(2,3)$ are transformed to the points $Q_1(13.66,23.66)$, $Q_2(25.98,42.32)$ and $Q_3(12.32,45.98)$.

A two-dimensional affine projection is also used in georeferencing a template in Sect. 4.6. The options for selecting a suitable transformation in a geoinformation system are "affine" or "first-order polynomial".

4.2.5.3 Projective Transformation

Projective coordinate transformations extend the definition of affine transformations, whereby projective coordinate transformations also include non-reversibly unique projections. The so-called *central projections* (perspective projections as e.g. photographs) are of special importance especially in photogrammetry. Here, every point of a plane is clearly mapped as a point of another plane and every straight line is mapped again as a straight line. However, in contrast to an affine transformation, the projection rays originate from a fixed center, so that formerly parallel straight lines in the original plane now intersect in the projection plane at the infinite point. Homogeneous coordinates are particularly suitable for the representation of projections (cf. Bartelme 2005 pp. 111–114 and Zimmermann 2012 pp. 126).

$$x = \frac{a_1 \cdot x' + a_2 \cdot y' + a_3}{c_1 \cdot x' + c_2 \cdot y' + 1} \qquad y = \frac{b_1 \cdot x' + b_2 \cdot y' + b_3}{c_1 \cdot x' + c_2 \cdot y' + 1}$$

In the case of terrain with little relief, which does not require the inclusion of an elevation model, the rectification of aerial images can be based on simple projective relationships between two planes (i.e. aerial image and terrain surface). Here, x' and y' are the coordinates of control points in the (digital, scanned) aerial image and x and y are the coordinates in the rectified image (or map). The eight unknowns a_i, b_i and c_i can be calculated using four control points. The entire image can then be transformed. For the determination of the parameters a_i, b_i and c_i linear equations are set up, which are similar to the equations for determining the parameters for an affine transformation (for affine transformations and for the procedure in case of overdetermination, i.e. here with more than four control points, cf. Sect. 4.2.5.5).

4.2.5.4 Polynomial Transformation

As shown an affine transformation equation can be described by a *first-order polynomial* instead of in matrix notation (cf. Sect. 4.2.5.2):

$$x' = a_0 + a_1 \cdot x + a_2 \cdot y$$

$$y' = b_0 + b_1 \cdot x + b_2 \cdot y$$

Such simple equations often do not provide satisfactory results in remote sensing in connection with rectification. In digital image processing mostly polynomials up to a maximum of third order are used to transform the pixel coordinates of the original image to real coordinates (for the rectification of raster data in remote sensing cf. Sect. 10.6.1.2).

The representation as a *second order polynomial* is:

$$x' = a_0 + a_1 \cdot x + a_2 \cdot y + a_3 \cdot x \cdot y + a_4 \cdot x^2 + a_5 \cdot y^2$$

$$y' = b_0 + b_1 \cdot x + b_2 \cdot y + b_3 \cdot x \cdot y + b_4 \cdot x^2 + b_5 \cdot y^2$$

4.2.5.5 Determining the Transformation Equations

In order to perform a transformation, the associated polynomial transformation (cf. Sect. 4.2.5.4) must first be set up and its coefficients a_i and b_i must be determined. For this purpose selected control points (so-called reference points) with coordinates in both systems are required. Since the control points with coordinates (x_i,y_i) in the source coordinate system correspond to the actual control points with coordinates (x'_i,y'_i) in the target coordinate system, the system of equations can be solved, which yields the coefficients a_i and b_i. Then, this transformation equation is used to transform all the initial coordinates or all the pixel coordinates of the initial image into the new coordinate system. For the case of first-order polynomials (affine transformation) and second-order polynomials, respectively, we get (x_i,y_i) and (x'_i,y'_i) in matrix notation for m reference points:

First order Second order

$$\begin{pmatrix} x'_1 \\ x'_2 \\ \cdots \\ x'_m \end{pmatrix} = \begin{pmatrix} 1 & x_1 & y_1 \\ 1 & x_2 & y_2 \\ \cdots & \cdots & \cdots \\ 1 & x_m & y_m \end{pmatrix} \cdot \begin{pmatrix} a_0 \\ a_1 \\ a_2 \end{pmatrix} \text{ or } \begin{pmatrix} x'_1 \\ x'_2 \\ \cdots \\ x'_m \end{pmatrix} = \begin{pmatrix} 1 & x_1 & y_1 & x_1y_1 & x_1^2 & y_1^2 \\ 1 & x_2 & y_2 & x_2y_2 & x_2^2 & y_2^2 \\ \cdots & \cdots & \cdots & \cdots & \cdots & \cdots \\ 1 & x_m & y_m & x_my_m & x_m^2 & y_m^2 \end{pmatrix} \cdot \begin{pmatrix} a_0 \\ a_1 \\ \cdots \\ a_5 \end{pmatrix}$$

or $X' = W \cdot A$. Similarly for the y-coordinates $Y' = W \cdot B$.

It should be noted that this describes linear equations with the unknown variables a_i and b_i. Since the control points are known, the equations contain specific numbers for x_i, x_iy_i, y_i, x_i^2 and y_i^2.

If m is equal to the number of coefficients k of the polynomial (with $k = [(n + 1) \cdot (n + 2)]/2$, n order of the polynomial), the matrix W can be inverted (for first order polynomials: $n = 1$ and $k = 3$). The inversion of the $m \times m$ matrix W then leads to exact solutions for the coefficients a_i and b_i and for the matrices A and B, respectively:

$$W^{-1} \cdot X' = A \text{ and } W^{-1} \cdot Y' = B$$

The following applies to the example from Sect. 4.2.5.2 (transforming $P_1(1,1)$, $P_2(3,2)$, $P_3(2,3)$ to $Q_1(13.66,23.66)$, $Q_2(25.98,42.32)$ and $Q_3(12.32,45.98)$):

$$\begin{pmatrix} x'_1 \\ x'_2 \\ x'_3 \end{pmatrix} = \begin{pmatrix} 1 & x_1 & y_1 \\ 1 & x_2 & y_2 \\ 1 & x_3 & y_3 \end{pmatrix} \cdot \begin{pmatrix} a_0 \\ a_1 \\ a_2 \end{pmatrix} \text{ or } \begin{pmatrix} y'_1 \\ y'_2 \\ y'_3 \end{pmatrix} = \begin{pmatrix} 1 & x_1 & y_1 \\ 1 & x_2 & y_2 \\ 1 & x_3 & y_3 \end{pmatrix} \cdot \begin{pmatrix} b_0 \\ b_1 \\ b_2 \end{pmatrix}$$

Inserting concrete numbers results in:

$$\begin{pmatrix} 13.66 \\ 25.98 \\ 12.32 \end{pmatrix} = \begin{pmatrix} 1 & 1 & 1 \\ 1 & 3 & 2 \\ 1 & 2 & 3 \end{pmatrix} \cdot \begin{pmatrix} a_0 \\ a_1 \\ a_2 \end{pmatrix} \text{ or } \begin{pmatrix} 23.66 \\ 42.32 \\ 45.98 \end{pmatrix} = \begin{pmatrix} 1 & x_1 & y_1 \\ 1 & x_2 & y_2 \\ 1 & x_3 & y_3 \end{pmatrix} \cdot \begin{pmatrix} b_0 \\ b_1 \\ b_2 \end{pmatrix}$$

Multiplication with the inverse matrix results:

$$\frac{1}{3} \cdot \begin{pmatrix} 5 & -1 & -1 \\ -1 & 2 & -1 \\ -1 & -1 & 2 \end{pmatrix} \cdot \begin{pmatrix} 13.66 \\ 25.98 \\ 12.32 \end{pmatrix} = \begin{pmatrix} 10 \\ 8.66 \\ -5 \end{pmatrix} = \begin{pmatrix} a_0 \\ a_1 \\ a_2 \end{pmatrix} = A$$

$$\frac{1}{3} \cdot \begin{pmatrix} 5 & -1 & -1 \\ -1 & 2 & -1 \\ -1 & -1 & 2 \end{pmatrix} \cdot \begin{pmatrix} 23.66 \\ 42.32 \\ 45.98 \end{pmatrix} = \begin{pmatrix} 10 \\ 5 \\ 8.66 \end{pmatrix} = \begin{pmatrix} b_0 \\ b_1 \\ b_2 \end{pmatrix} = B$$

The polynomial equations are calculated from these matrix equations:

$$x' = a_0 + a_1 \cdot x + a_2 \cdot y$$

$$y' = b_0 + b_1 \cdot x + b_2 \cdot y$$

$$x' = 10 + 8.66 \cdot x - 5 \cdot y$$

$$y' = 10 + 5 \cdot x + 8.66 \cdot y$$

Inserting the coordinates e.g. for $P_1(1,1)$ results in Q_1 (control calculation):

$$x' = 10 + 8.66 \cdot 1 - 5 \cdot 1 = 13.66$$

$$y' = 10 + 5 \cdot 1 + 8.66 \cdot 1 = 23.66$$

However, in most cases (significantly) more reference points are used beyond the minimum required control points. Then the system of equations is overdetermined. Then for $m \geq k$:

$$X' = W \cdot A + E_X \quad \text{or} \quad X' - W \cdot A = E_X \rightarrow E_X = X' - U$$

$$Y' = W \cdot B + E_y \quad \text{or} \quad Y' - W \cdot B = E_Y \rightarrow E_Y = Y' - V$$

Here $E_x = \{(x'_i - u_i)\}$ and $E_y = \{(y'_i - v_i)\}$ are the deviations between the actual coordinates of the control points in the target system (x'_i,y'_i) and the coordinates (u_i,v_i), which are predicted by the model (transformation polynomial) using the initial coordinates (x_i,y_i) (cf. Fig. 4.11). Due to inaccuracies in determining the control points, which can also be very different for distinct control points, it is not possible in practice to find a transformation which transforms all reference points of the initial system exactly to the corresponding points in the target system so that $(x'_i - u_i) = 0$ and $(y'_i - v_i) = 0$.

Expressed mathematically, this means that with $m > k$ the $m \times k$ -matrix W cannot be inverted, which leads to the *minimum mean-square-error equalisation* that minimizes the deviations (cf. Niemeier 2008 pp. 129):

$$\|X' - W \cdot A\|^2 = (X' - W \cdot A)^T \cdot (X' - W \cdot A) = \sum (x'_i - u_i)^2 \text{ minimal}$$

$$\|Y' - W \cdot B\|^2 = (Y' - W \cdot B)^T \cdot (Y' - W \cdot B) = \sum (y'_i - v_i)^2 \text{ minimal}$$

Here $\|\|$ stands for the norm of a matrix. One can show (cf. Freund u. Hoppe 2007 pp. 259 u. 262) that the minimal conditions are fulfilled with A^0 and B^0

$$A^0 = (W^T \cdot W)^{-1} \cdot W^T \cdot X \text{ and } B^0 = (W^T \cdot W)^{-1} \cdot W^T \cdot Y$$

with $(W^T \cdot W)^{-1} \cdot W^T$ the so-called pseudoinverse of W.

The coefficients of A^0 and B^0 provide the coefficients a_i and b_i of the transformation polynomials, as in the example given. This calculation minimizes the so-called *root mean square error (RMS error).*

$$\text{RMS error} = \sqrt{\frac{1}{n} \cdot \sum \left((x'_i - u_i)^2 + (y'_i - v_i)^2 \right)}$$

This error measures the (mean squared) differencies that exists between the actual coordinates of the control points in the target system (x'_i,y'_i) and the coordinates (u_i,v_i), which are predicted by the model (transformation polynomial) using the initial coordinates (x_i,y_i). The positions of the control points after the transformation are compared with the given positions (cf. Fig. 4.11). A two-dimensional affine projection is also used in georeferencing a template in Sect. 4.6. The RMS error is also given when georeferencing a template in Sect. 4.6 (cf. last column in Table 4.10).

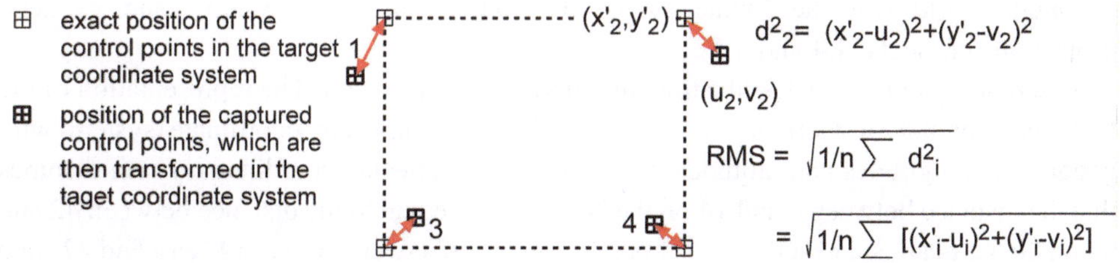

Fig. 4.11 Illustration of the RMS error

Determining the control points is usually an iterative process. Removing critical reference points with large deviations and adding new control points can try to reduce the transformation error. However, the aim should not be soleley to reduce the RMS error. This error is only meaningful for the control points! Rather, attention should also be paid to an optimal distribution of the control points (e.g. broad spatial distribution over the entire investigation area) and to an exact determination of the position in the initial system.

4.3 Map Projections

4.3.1 Presenting Geoobjects: General Issues

In ordert to locate geoobjects in a geoinformation system, which are based on a two-dimensional Cartesian reference system, the three-dimensional, curved surface of the earth has to be projected in a flat plane. However, the curved earth's surface or the associated graticule cannot be mapped onto a flat map or a flat rectangular grid. This is a consequence that the earth's surface is not isometric to a plane (due to Gauss's Theorema Egregium). A transformation which preserves all properties of geometries on the spherical surface also in the plane (cf. Sect. 4.3.2) is not possible in principle. However, many projections have been developed which preserve parts of these properties respectively (conformal, equal-area or equidistant projections). These projections map the graticule or the network of longitude and latitude lines, which spans the earth's surface as a mesh, into the plane (for a brief overview cf. Dana 2023a).

Cartographic projections are usually based on maps at scales of 1: 500000 and smaller, whereby, in addition to the topography, the graticule is usually also presented for selected parallels and meridians (cf. e.g. small-scale atlas maps for continents). Only locally small sections of the earth's surface can be projected into a Cartesian coordinate system, whereby projecting errors and distortions have to be accepted. Thus, a distinction is made between *cartographic* and *geodetic projections*. Large-scale maps as well as geodesy and land surveying are based on geodetic projections and always transform geoobjects into a rectangular grid (for the fundamentals of geodetic reference systems and geodetic projections cf. Sects. 4.4 and 4.5).

Figure 4.12 and Table 4.3 illustrate the fundamental problem. The representation of the Federal Republic of Germany is not possible in a Cartesian coordinate system with geographic longitudes and latitudes. For example, a Cartesian coordinate system assumes that the distance between 6 and 14 on the horizontal is equal to the distance between 42 and 50 on the vertical. But exactly this is not the case (e.g.) between 6° and 14° east and 42° and 50° north. Thus, it is mandatory to specify the corresponding projection when displaying geoobjects that are determined by geographic coordinates. If this step is omitted and rectangular pixel coordinates are simply assigned to geographic coordinates, considerable distortions can occur, especially in the case of a large-scale view (cf. Fig. 4.12). In particular, no distances may be calculated with the help of the Pythagorean theorem, e.g. between (53°, 7°) and (47°, 12°) (for the calculation of the length of the *orthodromes* cf. Sect. 4.2.3 and Fig. 4.7).

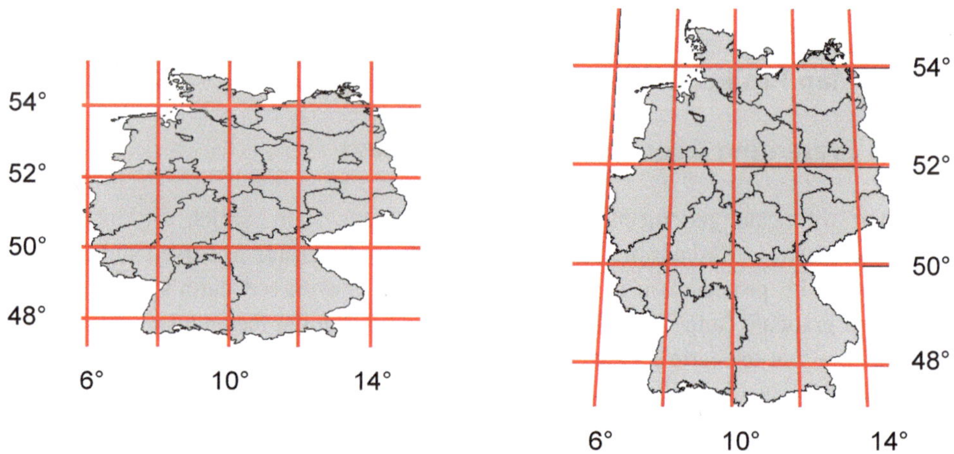

Fig. 4.12 Outlines of the Federal Republic of Germany in a Cartesian coordinate system and in the Lambert conformal conic projection (with two parallels of latitude 48° 40′ and 52° 40′, central meridian 10° 30′)

Table 4.3 Horizontal distance between two locations A and B on different latitude circles between Munich and Lübeck (distance in each case between 8° and 10° east)

Location A		Location B		Distance in km	Simplified assumption: earth as a sphere with radius 6371 km
Longitude	Latitude	Longitude	Latitude		
8	54	10	54	130,713	
8	52	10	52	136,913	
8	50	10	50	142,945	
8	48	10	48	148,804	

Geodata are provided by the national surveys in different spatial reference systems. The common representation in a geoinformation system then requires conversions into a single reference system. For these conversions, which are usually performed by default by the functionality of a geoinformation system, the parameters of the map projection of the source coordinate system as well as of the target coordinate system must be known (for map projections and in particular for the mathematical formulation of the projections see detailed textbooks from cartography, among others, such as Bugayevskiy and Snyder 1995, Grafarend and Krumm 2006 and Snyder 1987).

In general, knowledge of map projections is required when geoobjects have to be recorded and displayed in a geoinformation system. The following task certainly makes unusual requirements, but basic issues can be exemplified. Therefore this task is placed at the beginning, before discussing map projections and coordinate reference systems:

Based on water samples (spatial sampling) in Prince William Sound (Alaska), the distribution of pollutants is to be modeled. For this purpose, water samples are taken with a boat. The coordinates of the samples are only recorded by mobile data recording via GNSS (cf. Sect. 5.3), i.e. without the option of marking the location on a map due to the lack of reference points on the water. This makes it difficult to subsequently check the locations which can be identified solely on the basis of the received GPS coordinates. Thus, it must be known at the time of data recording, which coordinate system is set on the GPS device. This is by no means unambiguous, since geographic coordinates as well as UTM coordinates can exist for different reference ellipsoids. A GPS device generally offers a wide range of options for specifying a location in very different coordinate systems. For example, the previous user of the GPS device may still have set the Tokyo datum for data recording in Japan. The coordinates determined in Alaska (and then ultimately also the water samples) are then completely useless.

It is further assumed that the project has two small-scale maps for the area north of the 60th parallel, one of which is intended to serve as a background map in a geoinformation system. The associated information is printed at the bottom of the maps: The first map is based on the 1866 Clarke ellipsoid (NAD27/Alaska Albert, EPSG 2964, conic equal area Albers projection), the second map is based on the GRS80 ellipsoid (NAD83/Alaska Albert, EPSG 3338, conic equal area Albers projection). As it can be seen on the map each has a different graticule and different coordinates. The georeferencing and representation of this map in the geoinformation system must take the respective projection into account in order to ensure correct reproduction of the map. A geoinformation system as software has usually implemented many map projections from which the "correct" one has to be selected. This can be a special challenge.

It must be achieved that the locations of the water samples and the map "fit together" when displayed in the geoinformation system. The easiest way to do this is to transform both the sample coordinates and the map into a new coordinate system, i.e. preferably into the UTM system (UTM with respect to WGS84, alternative NAD 1927 or NAD 1983). Only now, after Cartesian coordinates are available, distances can be calculated according to the Pythagorean theorem, which is not possible with geographic coordinates. This is a mandatory prerequisite for modeling the spatial dispersion of pollutants using methods of spatial interpolation such as IDW or Kriging (cf. Sect. 9.7.3).

The easiest way would be to record the water samples directly as UTM coordinates (with respect to WGS84) and to use a map that is also based on a UTM projection (with respect to WGS84).

4.3.2 Projection Properties

In principle, distance, angle and area distortions can occur with every projection from the curved surface of the earth into the plane. Depending on how a map is used, only such projections can be used which avoid or reduce one or the other distortion:

In case of a *conformal projection*, the angle between intersecting lines (i.e., more precisely, between the tangents to these lines at their intersection) is preserved. However, conformality can only exist on a small scale. Conformal projections preserve the shape of objects only locally. No map projection can maintain conformality and therefore exact shape for a larger area. A conformal projections cannot be equal-area, individual areas are distorted (cf. polar regions in a Mercator map in normal orientation, cf. Fig. 4.16).

A projection is called an *equal-area projection*, if all areas are represented by the projection in the correct relative size (equal-area or equivalent projection). An equal-area map projection cannot be conformal, so most angles and shape are distorted.

In the case of an *equidistant projection* the distance between two points in the original matches the distance between the projected points in a map except for a scale factor. However, the earth's surface can only be projected equidistantly in the plane to a limited extent (partial equidistance only along specific lines, e.g. equidistant projection of the equator in a Mercator projection in normal orientation).

4.3.3 Types of Map Projections

4.3.3.1 Projection Surfaces

Projections are based on poorly illustrative, mathematical equations that transform points on the curved surface of the earth into the plane. But a simple way to illustrate the projections and their basic concepts is to place a light source to a selected point relative to the globe (e.g. in the center of the globe). Thus, the properties of these projections can easily be shown by considering the grid on a *developable surface* that is a surface that can be unfolded or unrolled into a plane without stretching, tearing or shrinking anywhere. Here, a flat, a cylindrical or a conical surface are used as mapping surfaces, so that *azimuthal, cylindrical* and *conical projections* can be distinguished (Fig. 4.13):

It is also possible to distinguish how the developable surface is arranged relative to the globe (cf. Fig. 4.14). In the case of a normal projection, the axis of the earth and the rotation axis of the cylinder or cone coincide. Transverse projections occur when the axis of the

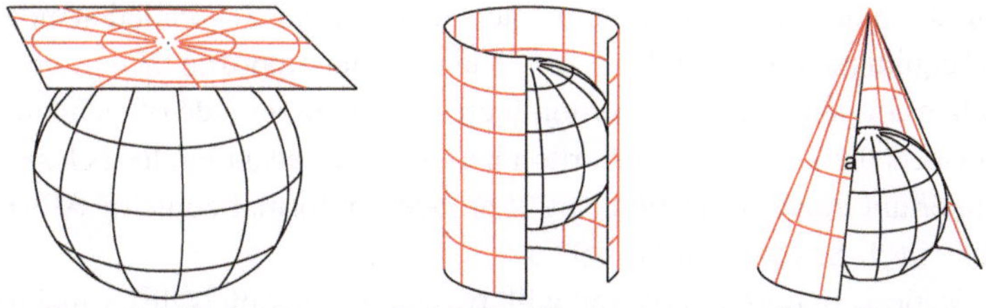

Fig. 4.13 Classification of map projections according to the type of developable surfaces

Fig. 4.14 Position of projection surfaces

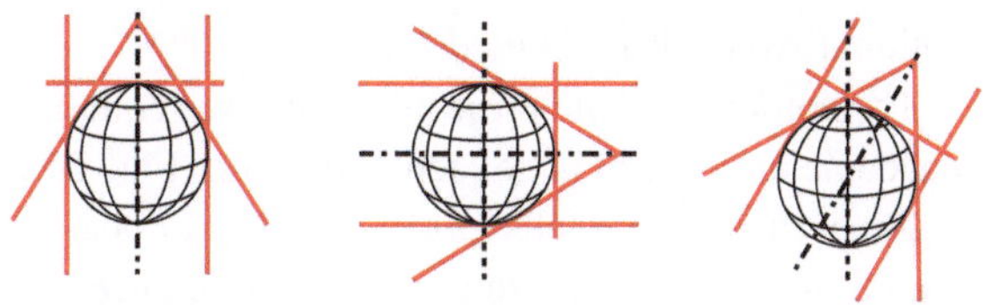

earth and the axis of rotation of the cylinder or cone are perpendicular to each other. If the axis of the earth and the axis of rotation of the cylinder or the cone enclose any angle, one speaks of oblique-axis projections. Similar designations apply to azimuthal projections.

4.3.3.2 Azimuthal Projection

This type of projection results from a tangential plane that touches the globe at a point. The point of contact (e.g. the North Pole) determines the position, whereby the polar position (i.e. normal position), the equatorial position (i.e. transverse position) and the oblique position are distinguished. *Azimuthal projections* are characterised by the fact that the direction – or the azimuth – is correctly reproduced from the point of contact to any other point. The polar position represents the simplest form of this projection, in which the latitude circles are shown as concentric circles around the point of contact (here a pole) and the meridians as straight lines, which intersect at the point of contact with their real angles.

Azimuthal projections can be further differentiated according to the location of their projection center. In the case of the gnomonic projection, this center lies in the center of the earth, in the case of the stereographic projection generally at the opposite pole of the point of contact, and in the case of the orthographic projection it is at infinity.

Azimuthal projections are mostly used for the cartographic representation of polar regions. Of greater importance, especially for the representation of the western and eastern hemisphere or of continents in atlases, is Lambert's azimuthal equal-area projection.

4.3.3.3 Conic Projection

This type of projection can be illustrated by unwinding a cone into the plane. The cone has been wrapped around the globe and touches it along a single latitude (one tangent) or intersects the cone in two circles (two secants). The cone is cut along a meridian and unfolded into the plane. The touching or intersecting circles (so-called standard parallels) are projected equidistant into the flat plane. The meridian opposite the cut line is called the central meridian. In general, the distortion increases on either side of standard parallels. If the cone intersects the sphere, the distortion between the secant circles is less than outside. The most important conic projections are: Lambert conformal conic or Albers equal-area conic projection (cf. Snyder 1987 pp. 97).

Conic projections in normal position with two cutting parallels differ mainly in the way the parallels between the intersection circles are projected. In the Albers equal-area conic projection the distortion between the standard parallels is minimal, although scale and shape are not preserved, This design is used for numerous single and atlas maps, especially in the USA, e.g. by the US Geological Survey or the US Census Bureau or within the National Atlas oft he United States.

The conformal cone projection with two equidistant projected parallels (Lambert conformal conic projection) is frequently used (cf. Fig. 4.12). Distortions can be minimised within a region of interest that falls mostly between or near the two standard parallels. This projection is used, for example, on the basis of a reference ellipsoid for maps at 1:500000 (e.g. in the Federal Republic of Germany or in Austria) or also for new editions of the International World Map at 1:1000000. The Bundesamt für Kartographie und Geodäsie (BKG, Federal Agency for Cartography and Geodesy of the Federal Republic of Germany) offers free of charge digital maps in the Lambert conformal conic projection with two equidistant parallels (48° 40′ and 52° 40′) and the central meridian 10° 30′ with the WGS84 as well as the WGS84 reference ellipsoid (cf. DTK 1000, BKG 2023a and BKG 2023b).

4.3.3.4 Cylinder Projection

Cylinder projections can be illustrated by transferring the geometries from the earth's surface to a cylinder that has been wrapped around the globe and touches it in one circle or intersects it in two circles. Figure 4.15 illustrates the construction principle. Any point P on the globe can be mapped onto the cylinder by extending the location vector. The cylinder, which is a developable surface, is then unrolled into a plane. In contrast to this illustration, the cylinder projections are based on mathematical functions: In the east-west direction, the angle λ is plotted on the cylinder, but *in* the north-south direction, a function value of the latitude φ is plotted.

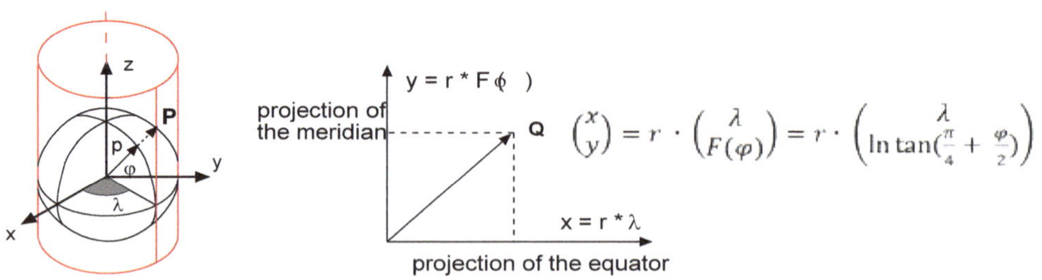

Fig. 4.15 Concept of a cylinder projection in normal orientation and the Mercator projection

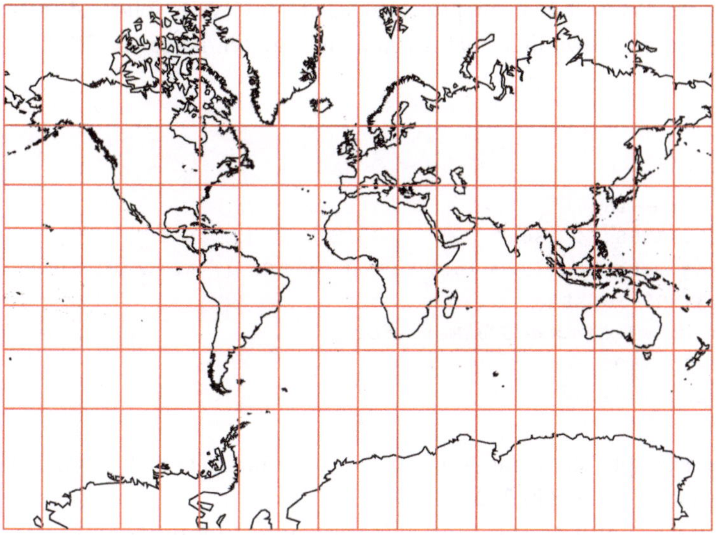

Fig. 4.16 Conformal cylinderprojection with equidistant projected equator (Mercator map)

Conformal cylinder projections are commonly called *Mercator projections*, which can be further differentiated according to the position of the developable cylinder (cf. Fig. 4.16). In the representation with normal orientation, where the earth's axis and the cylinder axis coincide (cf. Fig. 4.13 or 4.15), the enveloping cylinder touches the equator. In this so called normal Mercator projection, parallels and meridians are straight and perpendicular to each other. Since the cylinder is tangential to the globe at the equator, the equator is projected equidistantly. The Mercator projection inflates the size of areas away from the equator. At the pole the distortions become infinitely large. For example, Greenland appears much larger than South America or India, contrary to the actual area situation. Thus, the use of this projection for atlas maps or small-scale maps is severely limited. However, Mercator maps are of great importance in navigation and aviation (therefore also called nautical charts): The course of a ship is usually determined (with a compass) as a fixed course angle with respect to the north direction. Thus, the course line (the *rhumb line)* intersects each meridian at a constant angle. This course line is shown as a straight line on the Mercator map.

4.3.3.5 Web Mercator Projection

The *Web Mercator* or *Pseudo-Mercator projection* has become the standard for web mapping applications (cf. Battersby et al. 2014). The projection is used by all major online map providers (including Google Maps, Bing Maps, Open Street Map). The formulas of the Web Mercator and the standard Mercator projection are basically the same (cf. Fig. 4.15). But the Web Mercator projection (for all scale levels) only assumes a simplified shape of the earth as a sphere and not as a rotational ellipsoid. However, this does not yet fully describe the Web Mercator projection, in which the underlying coordinates are WGS84 but are projected as if they were spherical coordinates. The conversions can be shown in a small Python program (cf. Aitchison 2011):

```
import math
lon=8.0
lat=55.0585
#WGS84toGoogleBing
x = lon * 20037508.34/180
y = math.log(math.tan((90 + lat) * math.pi/360))/(math.pi/180)
y = y * 20037508.34/180
#Inverted: GoogleBingtoWGS84Mercator
#lon = (x/20037508.34) * 180
#lat = (y/20037508.34) * 180
```

Thus, for example, the main portal of the castle in Osnabrück has the Web Mercator coordinate (895,478.4; 6,849,342.8).

Overall, the use of these coordinates is very critical for distance or area calculations. While UTM coordinates are designed for small-scale applications and thereby keep the projection errors from the ellipsoid to the plane low (cf. Sect. 4.5.5), the Web Mercator Projection does not aim at reducing projection errors, but rather at the high-performance, i.e. high-performance transmission of 256 × 256 pixel images (so-called tiles), which show the respective section of the earth's surface in a zoom level from 0 to 18 (cf. Open Street Map 2023).

Thus, considerable projection errors can occur as the following example is intended to show. With tools of the geoinformation system QGIS, a point in Bavaria (on Lake Starnberg) and a point about 800 km to the north in Schleswig-Holstein are buffered based on an OpenStreetMap (EPSG 3857), so that a square with a diagonal of 200 m is created that stands at the tip of a corner. The area of both squares is 20,000 m^2 each, which is also displayed exactly below the Web Mercator Projection in QGIS. An export of these two areas into the shapefile data format (c.f. Sect. 9.3.3) and a subsequent import with transformation of these two areas into the UTM coordinate system (zone 32) lead to considerable differences. Thus, the area in Bavaria is only 9.003,46 m^2 and that in Schleswig-Holstein is only 6.634,03 m^2.

4.4 Coordinate Reference Systems

4.4.1 Overview

The task explained in Sect. 4.3.1 aims to ensure that the different coordinates of the water samples and the map "fit together" when presented in a geoinformation system. This implies that the different coordinate systems can be referenced to one another or can be converted into one new consistent coordinate system. It must be defined how the coordinate

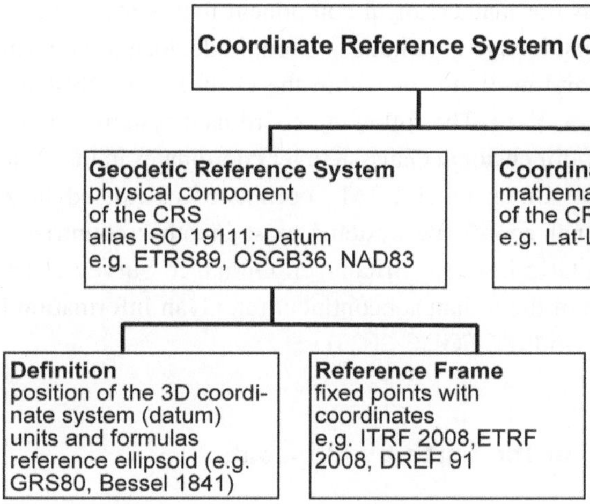

Fig. 4.17 Coordinate reference system (cf. sachsen.de 2023)

systems are related to each other. For the purposes of international standards the ISO 19111 (Geographic Information – Spatial referencing by coordinates) provides a conceptual scheme for the relationship between coordinate reference system, geodetic datum and coordinate system. Furthermore the basic conditions for coordinate transformations are presented (for further reading cf. Rummel 2017, Seitz 2017 and Torge 2017).

A coordinate reference system (CRS) is a framework which is used to uniquely describe and represent locations on the earth's surface using coordinates (cf. Fig. 4.17). In addition to the coordinate values, further information is required for the correct handling of coordinates. The coordinate reference system always consists of a geodetic reference system and a coordinate system.

The geodetic reference system – in ISO 19111 referred to as *geodetic datum* – is the physical component (e.g. ETRS89 (Europe), OSGB36 (Great Britain), NAD83 (USA)). A datum describes the position of a coordinate system in relation to the earth. A geodetic reference system is initially defined theoretically by a number of specifications, e.g. the reference ellipsoid and certain physical quantities and formulas to be used. This makes it possible to implement the concept of coordinate system for practical applications and tasks of land surveying. A reference system is then defined using fixed points and their coordinates, the so-called reference frame. This makes the geodetic reference system usable for measurements. For historical reasons there are different coordinate systems in individual national surveys around the world, so that a certain point can have significantly different coordinates.

The coordinate system is the mathematical component that is the code used to assign numerical values to point positions (e.g. latitude and longitude, location information within the UTM System (easting and northing) or within the Gauß-Krüger-System in Germany (Hoch- und Rechtswert) or X-Y-Z). The following coordinate systems are mainly used in surveying in Germany: two-dimensional Gauss-Krüger coordinates in the 3° stripes system (3GK, cf. Sect. 4.5.2), two-dimensional UTM coordinates calculated in zone 32 and 33 (cf. Sect. 4.5.3). National coordinate systems exist in other countries just like the Ordnance Survey National Grid in Great Britain (cf. Ordnance Survey 2018), the Indian Grid System for large areas of the Indian subcontinent (cf. Gyan Information Pedia 2023), the United States National Grid (cf. FGDC 2023).

4.4.2 Approximation of the Earth by Ellipsoids

The earth is often viewed simply as a sphere. For exact position determinations and map projections, however, the actual shape of the earth must be approximated by an ellipsoid. Among other things, the flattening of the earth at the poles and the bulge at the equator are taken into account. An ellipse is generally defined by two radii: The longer axis is called the major axis, and the shorter axis is called the minor axis. A (rotational) ellipsoid is then created by rotating an ellipse about one of its axes. An ellipsoid that approximates the shape of the earth is formed by rotating about the minor axis, i.e., the so-called polar axis. However, the earth is actually not an ellipsoid either. It has (besides the bulge at the equator and the flattening at the poles) further smaller dents and bulges. Thus, different ellipsoids are used in different regions of the earth to achieve the best approximation locally. The national surveys of different countries therefore use different reference ellipsoids as a basis for setting up their surveying network and then for determining the geographic coordinates of geoobjects (cf. Fig. 4.18 and Table 4.4).

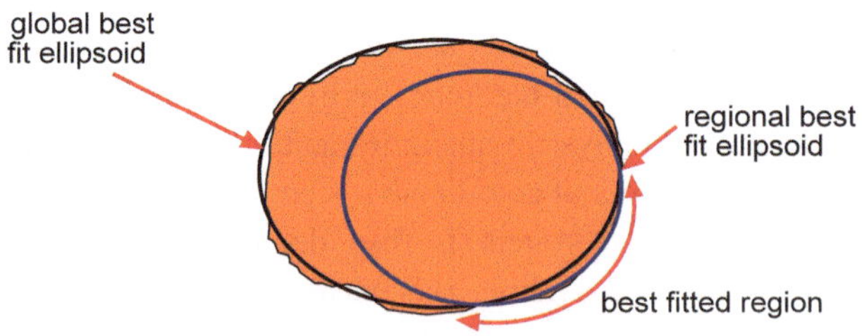

Fig. 4.18 Approximation of the earth by best-fit ellipsoids

Table 4.4 Parameters of important reference ellipsoids (cf. EPSG 2023a)

Name	EPSG	a (in m)	b (in m)	1/f	Distribution
Airy, 1830	7001	6,377,563.396	6,356,256.909	299.3249646	Great Britain, Ireland
Bessel, 1841	7004	6,377,397.155	6,356,078.963	299.1528128	Europe, Asia
Clarke, 1866	7008	6,378,206.4	6,356,583.8	294.9786982	North America and Central America
Clarke, 1880	7011	6,378,249.2	6,356,515	293.466021294	a.o. Africa, Israel, Jordan, Iran
GRS80, 1980	7019	6,378,137	6,356,752.314	298.257222101	Worldwide, intern. accepted
Hayford, 1909 International, 1924	7022	6,378,388.0	6,356,911.946	297.0	Europe, Asia, South America, Antarctica
Krassowski, 1940	7024	6,378,245.0	6,356,863.019	298.3	USSR and other Eastern European states
WGS72, 1972	7043	6,378,135.0	6,356,750.520	298.26	Worldwide
WGS84, 1984	7030	6,378,137.0	6,356,752.314	298.257223563	worldwide

a = major semi-axis, b = minor semi-axis, f = (a − b)/a = geometric flattening

Although various ellipsoids are still in use worldwide, the World Geodetic System 84 (WGS84), which was defined in 1984 and is valid worldwide, has become of central importance since it is the basis for satellite-based positioning (GPS, cf. Sect. 5.3.3).

4.4.3 Traditional Coordinate Reference Systems in Germany and Geodetic Datum

In the last few centuries, the national surveys in Germany have built up a geodetic reference network with geodetic control points (in German: Lagefestpunktfeld), which together with control points of fixed altitude (in German: Höhenfestpunktfeld) form the basis for the topographical survey and for other surveys such as cadastral surveys. Similar to the terrestrial triangulation in Germany national surveyings in other countries were carried out. For this purpose, a system of control points was developed, which are identified by specifying their position, altitude and gravity in the respective reference system. These set of control points were defined by triangulation and measured precisely by classical techniques of terrestrial surveying (measurements of directions, distances and angles between adjacent points) and are based on a non-geocentric coordinate system. They are mostly built up in four levels:

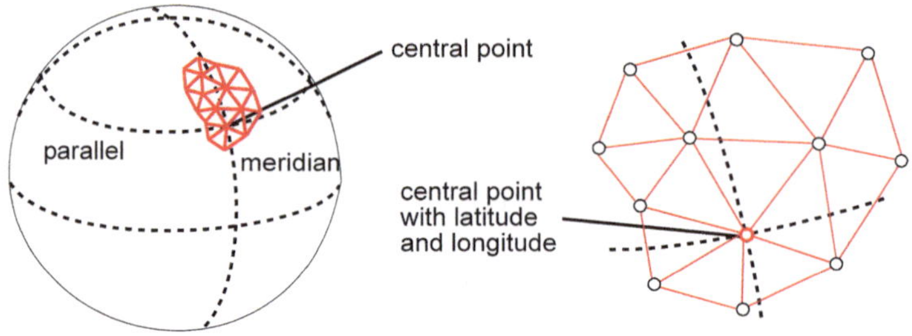

Fig. 4.19 Local reference system and geodetic reference network (cf. Resnik and Bill 2018 p. 38)

Table 4.5 Selected datums: location of ellipsoids with respect to WGS84 (e.g., δ x = [(X in WGS84) – (X of specified datum)], in m cf. Dana 2023b a. Dana 2023c)

Datum	Ellipsoid	δ x	δ y	δ z
European 1950 (concerning Germany)	Intern. 1927	−87	−98	−121
North America 83	GRS80	0	0	0
Ordinance Survey GB 36 (concerning England)	Airy	+375	−111	+431
Pulkovo 1942 (concerning Russia)	Krassowski	+28	−130	−95
WGS72	WGS72	0	0	0

In Germany the triangles of the network of first order triangles having a side length of 30 to 50 km with the first order trigonometric points as vertices. By mesh refinement, the fourth order trigonometric points reach distances of 1 to 3 km. As a last step of mesh refinement, polygons are inserted into the third and fourth order trigonometric points (cf. Kahmen 1997 p. 592 and Kahmen 2006 pp. 492). The *Deutsches Hauptdreiecksnetz* DHDN (German Main Triangle Network) was created around 1870 from the union of several networks.

To clearly define the geodetic control points (two-dimensional, non-geocentric position system), the exact determination of a so-called *fundamental point* or *central point* in a global, absolute coordinate system as well as the determination of the azimuth to another point is necessary (cf. Fig. 4.19). Furthermore, information about the selected ellipsoid as well as about the position of its center in relation to the earth's center of gravity must be given. These factors are summarised in the *geodetic datum,* which includes a set of parameters that indicate the origin, orientation and scale of a reference system and describe the ellipsoid with respect to a base frame. Different position systems are made comparable with each other by these specifications (for the transition between two reference systems cf. Sect. 4.4.4). Of particular importance are the geodetic datums, which describe the relationship between a local ellipsoid of a national survey and the global reference system WGS84 (cf. Table 4.5).

The *Potsdam datum* with the Bessel ellipsoid applied to the *German Hauptdreiecksnetz* (main triangulation network), which essentially goes back to the Prussian land survey in the nineteenth century. The fundamental point was the point Rauenberg (in Berlin), which had not existed since 1910 and which was formally replaced by the trigonometric point Potsdam. This received the coordinates in the system Rauenberg, so that a recalculation of all trigonometric points to the new datum was avoided. After the Second World War, this led, initially in the military field, to the not entirely correct, but meanwhile also common name Potsdam datum. The coordinates of Rauenberg obtained from astronomical observations were introduced as ellipsoidal coordinates. The orientation in the geodetic reference networkwas done by the azimuth Rauenberg – Berlin/Marienkirche.

The geodetic reference networks of the former German Democratic Republic and all former East Block countries were based on the Krassovsky ellipsoid (1940) and the geographic coordinates of the Pulkovo observatory near St. Petersburg *(Pulkovo-St. Petersburg datum), which was* calculated in 1942 (1942 system). This system were made up and refined several times, so that after the so-called reunification in Germany at the beginning of the 1990s, several networks existed. The *Deutsches Hauptdreiecksnetz 90* (*DHDN90*) can be viewed as a compound coordinate reference system (CCRS). It consists of the following components (AdV-online 2023a):

- DHDN in the western Bundesländer including Berlin,
- S42/83 in Mecklenburg-Western Pomerania, Saxony-Anhalt and Brandenburg,
- RD/83 in Saxony,
- PD/83 in Thuringia

This situation required a new regulation with regard to harmonisation and standardisation in Germany and the EU.

4.4.4 New Reference Frames: ITRF, ETRF and ETRS89

The International Astronomical Union (IAU) and the International Union of Geodesy and Geophysics (IUGG) are promoting the development of high-precision terrestrial reference frames (cf. Kahmen 1997 p. 599, Kahmen 2006 pp. 3 and pp. 499). An *International Terrestrial Reference Frame* (*ITRF*) is calculated annually on the basis of highly precise position determinations for about 180 stations distributed worldwide. The configuration of the terrestrial grid points is not rigid and constant over time due to, among other things, the tectonic behaviour of the earth's crustal plates. The coordinate sets are therefore given a year (e.g. ITRF89). In 1990 it was decided that the ITRF coordinates calculated for 35 European stations in early 1989 define the *European Terrestrial Reference Frame (ETRF-89*). It is assumed here that the relative position of the European ITRF stations on the rigid continental plate remains unchanged. These coordinates form the framework for the Europe-wide uniform reference system *ETRS89*, the *European Terrestrial Reference System. The* reference ellipsoid for ETRS89 is the ellipsoid of the Geodetic Reference System 1980 (GRS80) defined by the International Union for Geodesy and Geophysics, which is almost identical to WGS84 (for an overview cf. GEObasis.nrw 2020 u. Sachsen.de 2023).

The wide-meshed framework created for Europe was further condensed at the European level (EUREF89, European Reference Frame, approx. 200 km point spacing), at the national level (e.g. German Reference Frame DREF91) and at the level of the Bundesländer (federal states) in order to use it for practical work. In Germany, the network consists of 110 points, most of which coincide with first order trigonometric points (cf. Kahmen 2006 p. 502).

In 1991, the Arbeitsgemeinschaft der Vermessungsverwaltungen der Länder der Bundesrepublik Deutschland (AdV, Working Group of the Surveying Authorities of the Federal States of the Federal Republic of Germany) decided to introduce the ETRS89 as the reference system for land surveying and real estate cadastre, whereby the Universal Transversal Mercator projection was also established as the associated coordinate system (cf. GEObasis. nrw 2020). Thus, for the national survey in Germany, a new reference frame applies to the geodetic control point network, which is now connected to international networks. The basis is no longer the Potsdam datum with the Bessel ellipsoid, but the GRS80 ellipsoid.

4.4.5 Datum Transformation

Satellite-based positioning refer to a globally defined reference ellipsoid (WGS84 or GRS80). This results in the need to establish a relationship between the systems of a national survey and the WGS84 as well as a coordinate transformation. The national surveying authorities offer suitable transformation and conversion programs for this purpose. The so called *datum transformation* takes place in several steps, which are illustrated using the example of the transformation from DHDN to WGS84 (cf. Hofmann-Wellenhof et al. 2008 pp. 277) (Fig. 4.20). This makes it clear once again that the associated geodetic datum should always be given for coordinate data:

In the first step, the coordinates in the DHDN datum are converted into Cartesian coordinates (calculations based on the parameters of the Bessel ellipsoid as the coordinates

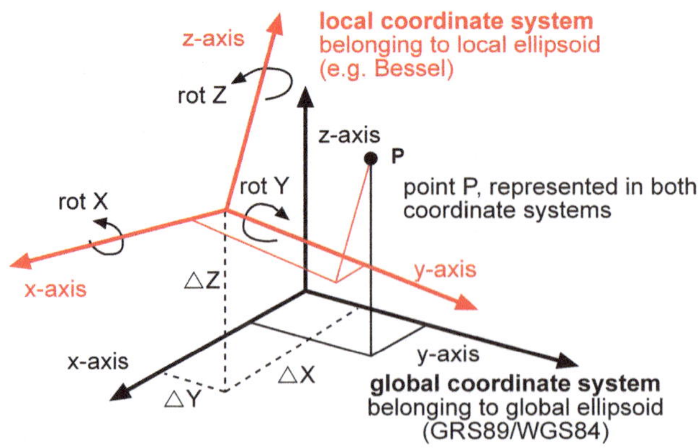

Fig. 4.20 Datum transformation (transition from a global to a local coordinate system or ellipsoid)

Table 4.6 Transformation parameters in relation to WGS84 (cf. BKG 2023c)

Parameter	Transition DHDN to ETRS89 Total Germany 2001	Transition S42/83 to WGS84
delta x	+598.1 m	+24.9 m
delta y	+73.7 m	−126.4 m
delta z	+418.2 m	−93.2 m
rot x	0.202″	−0.063″
rot y	+0.045″	−0.247″
red z	−2.455	−0.041″
scale factor	+6.7 * 10^{-6}	+1.01 * 10^{-6}

of DHDN are based on Bessel). The result are xyz-coordinates in an earth-centered coordinate system.

The second step transforms these coordinates into xyz-coordinates in the WGS84 reference system. This actual datum transformation is done with the help of the so-called 7-parameter *Helmert transformation*, which consists of three translation factors, three rotation factors and one scale factor (cf. Table 4.6). The translation factors indicate how many meters the origin of the new xyz-system lies away from the center of the Bessel ellipsoid. The rotation factors describe by how many arc seconds the axes are rotated during the transformation. The scale factor expresses changes in size.

In the third step, the xyz coordinates in the WGS84 reference system are converted into geodetic coordinates. Subsequently, UTM-coordinates can be determined from the geodetic coordinates (UTM coordinates in WGS84).

1. Conversion of geographic coordinates (φ, λ, h) into a cartesian system. Here the calculation is still done with respect to the Bessel ellipsoid:

$$X_a = (N + h) \cdot \cos\varphi \cdot \cos\lambda$$

$$Y_a = (N + h) \cdot \cos\varphi \cdot \sin\lambda$$

$$Z_a = \left[N \cdot \left(1 - e^2\right) + h\right] \cdot \sin\varphi$$

with: φ, λ geodetic latitude and longitude and h height above the Bessel ellipsoid, a and b major and minor semimajor axis of the Bessel ellipsoid, and

$$N = a/w \qquad w = \sqrt{1 - e^2 \cdot \sin^2\varphi} \quad \text{with e the first numerical eccentricity}$$

2. The so-called Helmert transformation is used for the datum transition:

$$\begin{pmatrix} X_b \\ Y_b \\ Z_b \end{pmatrix} = (+scale\ factor) \cdot \begin{pmatrix} 1 & rot_z & -rot_y \\ -rot_z & 1 & rot_x \\ rot_y & -rot_x & 1 \end{pmatrix} \cdot \begin{pmatrix} X_a \\ Y_a \\ Z_a \end{pmatrix} + \begin{pmatrix} \delta_x \\ \delta_y \\ \delta_z \end{pmatrix}$$

The transformation parameters are defined using identical points in the German Reference System. The coordinates of these points are available in the WGS84 and in the Potsdam datum or in the Pulkovo-St. Petersburg datum for the areas of the former GDR.

3. Conversion of the Cartesian coordinates (X_b, Y_b, Z_b) into geodetic (ellipsoidal) coordinates (φ, λ, h). Now the calculation is done with respect to the WGS84 (ellipsoid). While the equations for the calculation of the Cartesian coordinates from the ellipsoidal ones are closed and strict, only an iterative solution exists for the inversion (cf. Landesvermessungsamt NRW 1999 p. 33, cf. Seeber 2003 p. 24). To avoid iterative processes the following formulas are given, which calculate ellipsoidal from Cartesian coordinates with a sufficient approximation (cf. Hofmann-Wellenhof et al. 2008 p. 280):

$$\lambda = arctan(Y_b/X_b)$$

$$\varphi = arctan\left(Z_b + e\prime^2 \cdot sin^3\theta\right)/\left(p - e^2 \cdot a \cdot cos^3\theta\right)$$

$$h = p/cos\varphi - N$$

with: φ, λ, h geodetic latitude, longitude and altitude above the WGS84 ellipsoid, X_b, Y_b, Z_b geocentric Cartesian coordinates, a and b major and minor semimajor axis respectively, and other parameters now with respect to the WGS84 ellipsoid:

$$p = \sqrt{X_b^2 + Y_b^2} \qquad \theta = arctan(Z_b \cdot a)/(p \cdot b)$$

$$e\prime^2 = \left(a^2 - b^2\right)/b^2 \qquad e^2 = \left(a^2 - b^2\right)/a^2$$

$$\mathrm{N} = \mathrm{a}/\mathrm{w} \qquad \mathrm{w} = \sqrt{1 - \mathrm{e}^2 \cdot \sin^2\varphi}$$

With change of the geodetic reference system within the area of the Federal Republic of Germany and the neighboring foreign countries deviations of a maximum of three meters are to be expected with coordinate transformations. The inaccuracy follows from the large-scale 7-parameter transformation sets. By the use of local 7-parameter transformation sets higher accuracies can be achieved. The local transformation sets can be obtained from regional authorities (Landesvermessungsämter). The Bundesamt für Kartographie und Geodäsie (BKG, Federal Agency for Cartography and Geodesy) offers extensive services for coordinate transformation (e.g. Helmert transformation with an accuracy in the range of 3 metres, cf. BKG 2023d).

Datum transformations are particularly relevant for working with geometric data in a geoinformation system. If, for example, more recent ATKIS data based on the UTM system with the ETRS89 ellipsoid and geoobjects of an older biotope cadastre, which are still based on Gauss-Krüger data and the Bessel ellipsoid, are to be displayed together, a datum transformation is required.

4.4.6 Height Reference Surfaces in Germany Until 2016

To determine positions, the earth or regions of the earth can be approximated by an ellipsoid or by different ellipsoids (cf. Sect. 4.4.1). A spheroid can be used as a reference surface with the advantage of being able to calculate on a mathematically manageable surface. In contrast, such simplifications are not suitable for height measurements where gravity influences the measurement results, e.g. in levelling. For example, the vertical axes of surveying instruments are aligned in the direction of gravity. However, gravity is dependent on latitude and altitude and even varies locally. Thus, theoretically, the most suitable surface for determining height is a surface that is perpendicularly intersected in all its points by the respective direction of gravity. This surface is called a *geoid*, which is difficult to determine metrologically (cf. Fig. 4.21). Thus, mainly for historical reasons, different height systems exist, which operationalize the earth's gravitational field in different ways. A distinction should be made (cf. Fig. 4.22, for levelling-based and geoid-based height systems cf. in detail Gerlach et al. 2017 pp. 362 and BKG 2023e):

- the ellipsoidal height H_E,
- the orthometric height H_o,
- the normal height N_N,
- the normal orthometric height H_{NO}.

The *ellipsoidal height H_E* is the metric distance from the ellipsoidal surface along the surface normal to the point P. This height determination does not take into account gravity or local deviations from the earth's normal gravity field due to density anomalies of the earth's crust. The satellite-based positioning (e.g. a satellite positioning receiver) detects the height of a point in relation to this reference surface, which can be clearly described mathematically (e.g. in relation to WGS84 or GRS80).

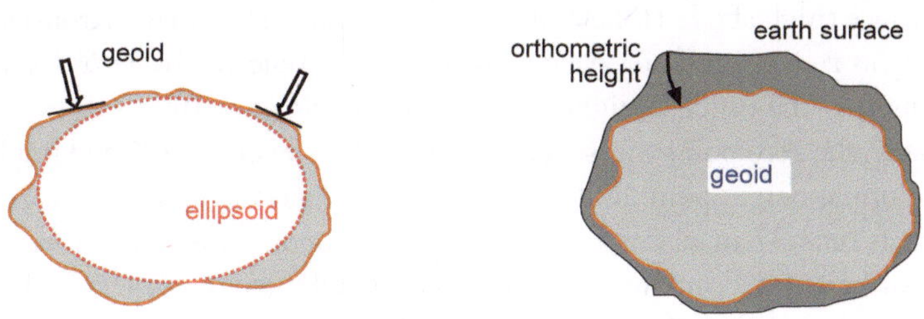

Fig. 4.21 Geoid, ellipsoid and physical earth surface

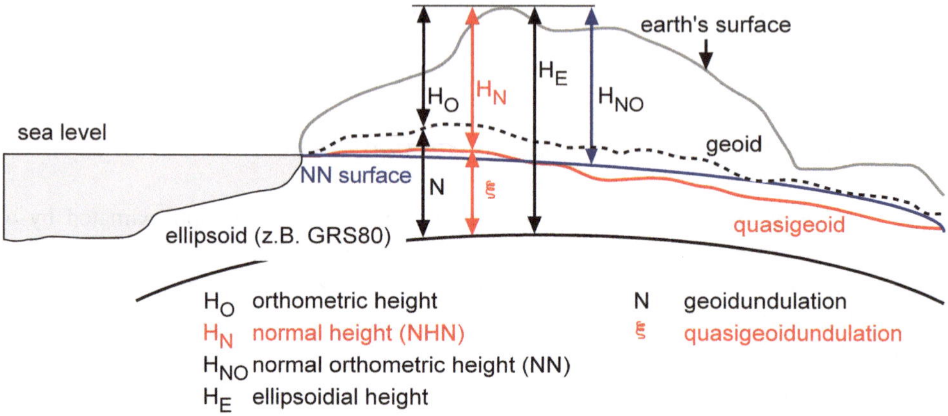

Fig. 4.22 Height systems and height reference surfaces

The *orthometric height* H_O, which is the scientifically best definition of a height or an altitude, is defined as the length of the plumb line from the geoid to the earth's surface. To determine this (curved) line, however, knowledge of the gravitational acceleration values along this line is necessary, which cannot be determined by measurement technologies. In some European countries, these values are therefore derived using mathematical models that operationalize the density of the earth's structure.

The *geoid* is (at least in the German national survey) of almost only theoretical interest, it does not meet the requirements of high-precision height measurements and cannot be precisely defined at every point of the earth. Instead, a quasigeoid is used as a height reference surface. The *quasigeoid* is a computational model that corresponds to the smoothed geoid. The *normal height* H_N (or NHN), defined as the height above the quasigeoid, characterises the distance from a clearly reproducible reference surface. The *quasigeoidundulation*, i.e. the distance from the quasigeoid to the ellipsoid, can be calculated in order to establish relationships with ellipsoidal heights. In particular, normal heights can be calculated from ellipsoidal heights and values of the quasigeoidundulation: $\mathrm{H_N} = \mathrm{H_E} - \xi$ (cf. Fig. 4.22). Thus, normal heights and height determination with GPS devices, which provide ellipsoidal heights, have a high practical benefit.

Normal heights were already calculated in the former GDR with a quasigeoid as a reference surface (high above Höhennull HN). The heights refer to the Kronstadt level near St. Petersburg as zero point. The system was completely renewed by 1976 (system HN 76), which was used as the official height system of the GDR until 1990.

In Germany, the zero point was set at the Amsterdam level in 1875 and implemented by a so-called normal height point at the former Royal Observatory in Berlin. The levelling surface, which runs 37 m below this normal height point, was defined as the reference surface for all height measurements in Germany (height above sea level). This reference point was moved to Berlin-Hoppegarten in 1912 after the demolition of the old observatory. The entire levelling network was recalculated taking into account normal gravity values. These so-called *normal orthometric heights* H_{NO} (height above NN) are

Table 4.7 Former height reference systems in Mecklenburg-Western Pomerania until 2016 (following State Office for Internal Administration Mecklenburg-Western Pomerania 2008, cf. update Ministry of the Interior and Europe Mecklenburg-Western Pomerania 2017)

Height reference system	Reference point	Comments
DHHN12 (NN elevation system 1912)	Normal height Hoppegarten (near Berlin) (level Amsterdam)	with normal orthometric reduction
SNN 56 (HN 56) SNN 76 (HN 76)	Normal elevation point Hoppegarten (level Kronstadt)	part of the uniform levelling network of Eastern Europe, calculation of normal heights, 1976 repetition levelling of the GDR
DHHN92 (NHN)	REUN/UELN point church Wallenhorst (near Osnabrück) (level Amsterdam)	joint adjustment of SNN 76, DHHN85 (old federal states) and connection measurements in geopotential nodes, calculation of normal heights

REUN/UELN United European Levelling Network
SNN State level network (of the GDR)

approximate values for orthometric heights. They are the result of the geometric levelling and the normal orthometric reduction along the levelling line from a known height point. These heights depend therefore on the levelling route. For these normal orthometric heights, the designation Deutsches Haupthöhennetz or DHHN12 was common in the western Bundesländer after 1945 (cf. Table 4.7).

After the German reunification the surveying administrations in Germany decided in 1993 to introduce uniform normal heights. The Deutsche Haupthöhennetz 1992 (DHHN92, German Main Elevation Network 92) was the nationwide uniform height reference system, which uses the quasigeoid calculated according to Molodenski's theory. This clearly reproducible reference surface allows to calculate distances to the reference ellipsoid: the so-called *quasigeoidundulation* ξ (cf. AdV-Online 2023b u. GEObasis.nrw 2018). In this way, relationships to ellipsoidal heights can be established which have a very high practical relevance with the spread of satellite-based location determinations. The DHHN92 has since been replaced by the DHHN2016 (cf. Sect. 4.4.7).

When doing field work with a drone (unmanned aerial vehicle, UAV), such as recording the height of buildings or the growth of vegetation, it is important to set the "correct" altitude reference to calibrate the flight. The default GPS height above the WGS84 may cause a crash (cf. Fig. 4.22).

4.4.7 The Integrated (Geodetic) Spatial Reference 2016

As early as 2005, the Arbeitsgemeinschaft der Vermessungsverwaltungen der Länder (AdV, Working Group of the Surveying Authorities of the Federal States) decided to update the previous official geodetic spatial reference in order to create a new, nationwide uniform spatial reference (for the guideline cf. AdV 2017 and AdV-Online 2023c). In 2007, the process was completed with the introduction of the integrated geodetic spatial reference, which provides uniform and highly precise coordinates for position and height as well as gravity values. These data are based on a complete re-measurement of Germany, which was carried out between 2006 and 2012 (cf. BKG 2023f). The integrated (geodetic) spatial reference 2016 links the reference systems ETRS89/DREF91 (realisation 2016), DHHN2016 and DHSN2016 via the quasigeoid model GCG2016.

Components of the integrated spatial reference are:

- *ETRS89/DREF91* (implementation 2016) for ellipsoidal coordinates and heights
  The first realisation was based on the spatial coordinates of the points of the Deutsches Referenznetzes 1991 (German Reference Network 1991, DREF91) fixed in 1994. In the meantime, a third realisation is available, which is based on a new GNSS campaign, which links height and gravity fixed points as well as points of the basic geodetic network. This new reference frame was introduced on 1.12.2016.
- *DHHN2016* (Deutsches Haupthöhennetz 2016, German Main Elevation Network 2016) for physical heights from precision levellings.
  This nationwide official height reference frame was introduced in Germany on 1.12.2016 after extensive levellings.
- *DHSN2016* (Deutsches Hauptschwerenetz, German Main Gravity Network 2016)
  This network of gravity stations replaces the older DHSN96 and implements the international gravity standard through modern measurement methods and equipment.
- *GCG2016* (German Combined Quasigeoid 2016).
  This new quasigeoid of the Federal Republic of Germany serves as a height reference surface for the transition between geometric, satellite determined heights in ETRS89/DREF91 and physical heights determined by levelling in DHHN2016 (cf. Fig. 4.21).

With the German Combined Quasigeoid 2016, the physical heights in DHHN2016 that are common in practice can be calculated from the geometric heights in ETRS89. As a result, the height determination using satellite technology, which for example is carried out with SAPOS, can be executed more economically and with significantly higher accuracy and partially replace existing measurement methods (for the satellite positioning service SAPOS cf. Sect. 5.3.5 and AdV-Online 2023d).

4.5 Geodetic Projection

4.5.1 Geodetic Projection: Use

Transformations into rectangular, plane coordinates (*geodetic coordinates*) are referred to as geodetic projections, which project latitude and longitude coordinates into a square grid. A flat, geodetic coordinate system is formed by a Cartesian coordinate system. Due to the mathematical derivation of the transformation equations by C.F. Gauss, the positive x-axis points to the north (in the UTM system: northing) and the positive y-axis (in the UTM system: easting) points to the east. Such projections, which in particular are the basis of the national survey, have to meet high demands for accuracy. They are therefore usually defined only locally, i.e. in a regionally narrowly limited area around a reference point. In addition, and this is particularly important, they approximate the earth by a rotational ellipsoid and are based on ellipsoidal calculations (cf. Bugayevskiy and Snyder 1995 pp. 159, Kuntz 1990 pp. 62).

In Germany currently several reference systems exist in the national survey (cf. Table 4.8 and Sect. 4.4.3). In the old Federal Republic of Germany, the Gauss-Krüger coordinate system with the Bessel ellipsoid and meridian stripes with a width of three degrees of longitude exists (cf. Sect. 4.5.2). In the area of the former GDR, a similar Gauss-Krüger coordinate system existed, but with the Krassowski ellipsoid and six degree wide meridian stripes. In 1991 and 1995, respectively, the Arbeitsgemeinschaft der Vermessungsverwaltungen der Länder der Bundesrepublik Deutschland (AdV, Working

Table 4.8 Coordinates for each identical points in Osnabrück and Clausthal in different reference systems (data request: State Office for Geoinformation and Surveying of Lower Saxony, State Surveying and Geobasis Information, update 15.08.2019)

				Heights	
	Gauss-Krüger coordinates			HS 160	
Osnabrück	3 435 038.438	5 791 675.323	LS 100	105.754	
Clausthal	3 592 935.536	5 741 403.369	LS 100	605.282	
	Geographic coordinates				
Osnabrück	52° 15′ 17.08907″	8° 2′ 51.46011″	LS 889		
Clausthal	51° 47′ 56.58302″	10° 20′46.14807″	LS 889		
	UTM coordinates			DHHN2016	H.ETRS89
Osnabrück	32 434 991.651	5 789 799.754	LS 489	105.763	149.656
Clausthal	32 592 825.304	5 739 545.459	LS 489	605.289	650.519
	x,y,z 3D coordinates				
Osnabrück	3 874 144.805	547 759.966	5 020 323.303	LS 389	
Clausthal	3 888 685.134	709 930.154	4989 518.309	LS 389	

LS, Lagestatus 100: Gauss-Krüger coordinates, Potsdam datum (Bessel ellipsoid)
LS, Lagestatus 389: 3D coordinates, ETRS89 (GRS80 ellipsoid)
LS, Lagestatus 489: levelled UTM coordinates, ETRS89 (GRS80 ellipsoid)
LS, Lagestatus 889: Geograph. Coordinates in ETRS89
HS, Height status 160: normal altitude (altitude in DHHN12 system), altitude above sea level
DHHN2016: normal altitude (altitude in the DHHN2016 system), altitude above NHN, also altitude status 170 in other Bundesländer
H.ETRS: height above the ETRS ellipsoid
The heights in status 160 are derived from a levelling and a height transfer. The heights in DHHN2016 are derived from a transformation of the heights in height status 160 with the nationwide uniform model HOETRA2016

Group of the Surveying Authorities of the States of the Federal Republic of Germany) decided to introduce a new reference system for the national survey and to introduce the Universal Transverse Mercator Projection (cf. Sect. 4.5.5, cf. also Sect. 5.5.4.2). This means that for a single location point, several coordinate specifications can exist, so that the reference systems must always be specified with the coordinates in order to achieve clarity (cf. Table 4.8).

4.5.2 The Gauss-Krüger Coordinate System in Germany

The *Gauss-Krüger system*, which has been fundamental in Germany for land surveying and for cadastral maps as well as for topographic maps, corresponds to a transversal, conformal cylindrical projection (Mercator projection). The construction principle can be illustrated by a transverse cylinder, which is rotated horizontally around the globe and which touches it in several meridians at intervals of three degrees of longitude (Fig. 4.23):

Thus, the area is covered by several meridian stripes with the central meridians 6°, 9°, 12° and 15° eastern longitude, which are projected equidistant. On each of these meridian stripes – i.e. locally – a rectangular coordinate system is created with the central meridian as the vertical axis. In contrast to this graphic description, the Gauss-Krüger projection is based on a complex calculation rule of so called *Rechtswert* and *Hochwert* (i.e. easting and northing, for formulas cf. Bugayevskiy and Snyder 1995 pp. 159 and Snyder 1987 pp. 60).

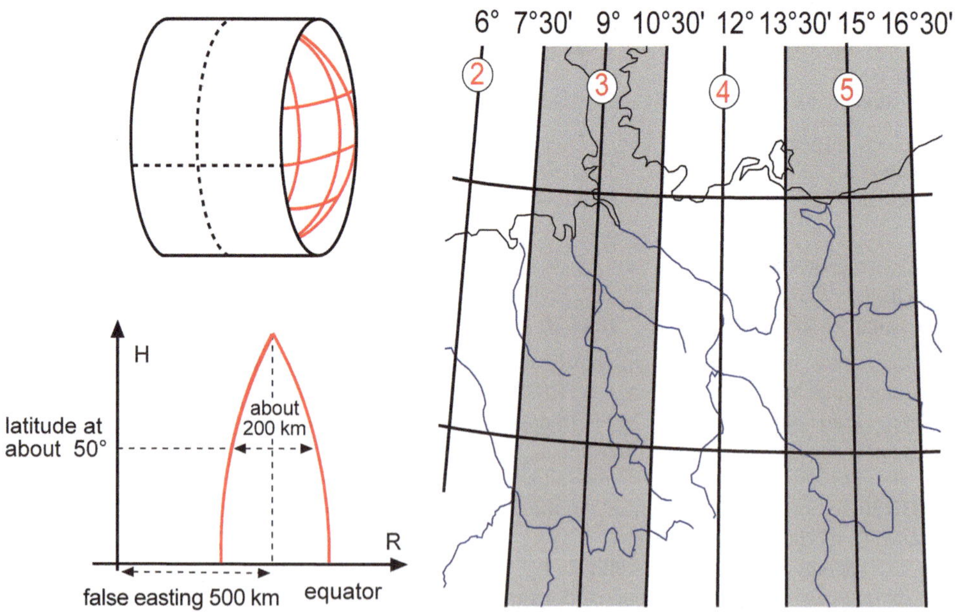

Fig. 4.23 Germany in the Gauss-Krüger system

This formulation is quite correct in the practical application, but mathematically strictly speaking incorrect, since the exact value of an elliptic integral required for the representation can only be approximated by a finite sum of terms. The mathematical formulation of the conformal projection goes back to C. F. Gauss, who developed it for the Hanoverian national survey (1822–1847), which he directed. It was further developed by Schreiber (1966) and especially by Krüger (1912/1919). Sect. 4.5.4 gives an impression of the formulas and the mathematical idea, which are basically identical for UTM and Gauss-Krüger coordinates (but with different ellipsoids and widths of the stripes or zones).

With increasing distance from the central meridian, distortions occur (as expected) between the position on the surface of the sphere and the projection on the developabel cylinder. In order to reduce these distortions, a meridian strip is limited to an extension of 1° 40′ in longitude (i.e. about 100 km) on both sides of the central meridian according to a decision of the Arbeitsgemeinschaft der Vermessungsverwaltungen der Länder der Bundesrepublik Deutschland (AdV, Working Group of the Surveying Authorities of the Federal States of the Federal Republic of Germany) of 1966 (cf. Kahmen 2006 p. 238). Accordingly, adjacent systems overlap in a strip 20 minutes wide in longitude (on average about 23 km wide), where points are calculated as needed in both systems. This keeps the length distortion so small that it can be neglected in many practical cases (maximum 12 cm per 1 km, cf. Kahmen 2006 p. 239).

The use of Gauß-Krüger coordinates is shown using the example of the location of the town hall tower Berlin-Mitte (so-called "Rotes Rathaus"), which has the geographic coordinates 13° 24′ 36.01″ east and 52° 31′ 11.65″ north (Potsdam Datum, Bessel ellipsoid). Thus, the tower is located in the overlapping area of the fourth and fifth meridian stripes. Its Gauss-Krüger coordinates are (according to Hake and Heissler 1970 p. 136):

	Within the fourth stripe central meridian 12°	Within the fifth stripe central meridian 15°
Rechtswert:	4 595 696.00 m	5 392 088.39 m
Hochwert:	5 821 529.20 m	5 821 783.04 m

The rectangular coordinate system with the central meridian as the vertical axis defines the grid of eastings and northings. In order to avoid negative Rechtswerte (eastings), the value 500000 (meters) is assigned to each main meridian. Furthermore, each Rechtswert (easting) is preceded by the identification number of the meridian stripe, i.e. the number of longitude of the central meridian divided by 3. In the present example, the first number of the Rechtswert (easting) indicates the corresponding central meridian (here fourth main meridian, 12°). Taking into account the addition of 500000 m for the central meridian, the ordinate base is exactly 95696.00 m east of the central meridian of 12°. In this stripe the point is 5821529.20 m from the equator. In the fifth meridian stripe, the tower is 107911.61 m west of the central meridian of 15°. In this stripe, the ordinate base is 5821783.04 m from the equator.

This example illustrates the problem of projecting a spherical surface into a plane, which cannot be solved exactly. Since Cartesian coordinate systems, which are the basis of many applications in geoinformatics and geoinformation systems, can only be set up locally, leaps are unavoidable when identifying a position of a geoobject in two adjacent meridian stripes. Although the deviation of the Hochwerte (northing) appears large at first sight, it is clearly below the given maximum value of 12.3 cm per 1 km.

4.5.3 The UTM Coordinate System

The *UTM coordinate system is* based on a conformal transverse cylindrical projection (Universal Transverse Mercator projection), which was introduced in 1947 by the U.S. Army (and then later by NATO, among others) to mark rectangular coordinates in military maps of the (entire) world. It is now used worldwide by various national surveys and mapping agencies, although the underlying ellipsoids must be taken into account (mostly the Hayford International Ellipsoid, most recently GRS80).

The UTM system can be illustrated in a similar way to the Gauss-Krüger system by a transversal enveloping cylinder, which is systematically rotated around the earth depending on the zone (cf. Fig. 4.24). The earth is divided into 60 zones (meridian stripes) each with an extension of six degrees of longitude, whereby the UTM system covers the earth

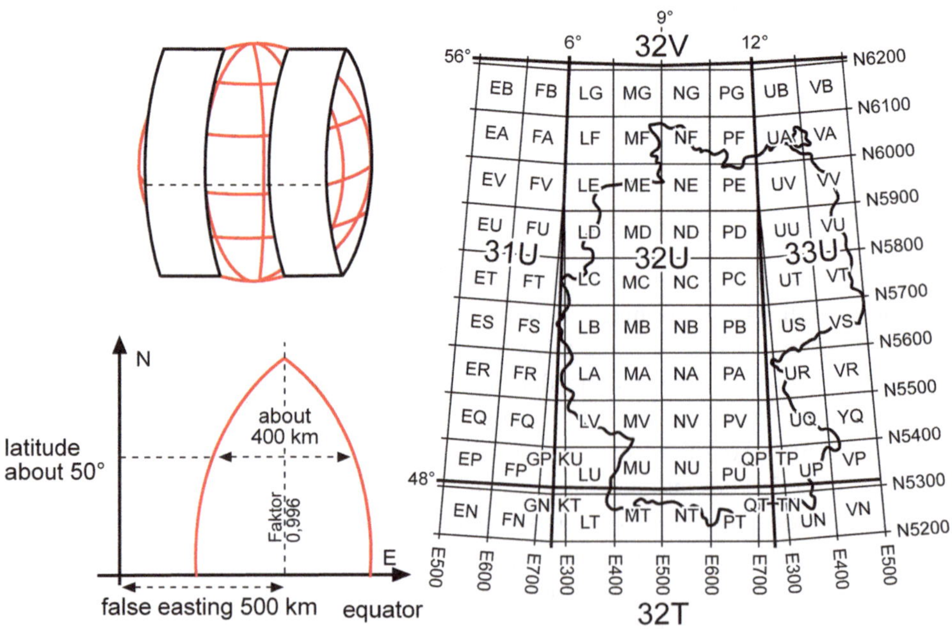

Fig. 4.24 Germany in the UTM system

between 84° north and 80° south latitude. Thus, a meridian strip in the UTM system is twice as large as in the Gauss-Krüger system. Each zone then has a central meridian e.g. at 3°, 9° or 15° eastern and western longitude. However, in the UTM system, this cylinder does not touch the globe, but intersects it in two parallel circles, so that the central meridian is not equidistant projected, but with the factor 0.9996. As a result the projection is only equidistant at about 180 km on both sides of the central meridian. At the border meridian of a zone, a length distortion of about 15 cm per 1 km can be assumed at a latitude of 50° (cf. Hake et al. 2002 p. 77). Overall, the good mapping properties lead to a (world) wide application for maps in a medium scale.

The UTM system is based on a universal reporting system (UTM Reference System, UTMREF), which was used for military purposes. Thus, the zones are numbered eastwards from 1 to 60 starting with the 180th meridian with respect to Greenwich. The first zone extends between 180° and 174° west longitude and has the central meridian 177° west longitude. The zones each extend from 80° south to 84° north. They are divided into bands of 8° latitude, marked alphabetically with capital letters beginning with C (at 80° south). A large part of the Federal Republic of Germany is located in zone field 32U (between 6° and 12° east, central meridian 9° east) (cf. Fig. 4.24). Starting from the central meridian, the zones are further divided into a square grid with a mesh size of 100 km, the squares are marked by double letters. Within a square, points can then be defined by coordinates. For example, the church of List (northeast tip of Sylt) is located in the UTM grid 32UMF6397, i.e. within the field 32 U, within the 100 km square MF and then within the 1 km square with the southwest corner 63 (right) and 97 (high).

Similar to the Gauss-Krüger notation, the counting of the x-coordinates starts at the equator (indicated with N, North), the counting of the y-coordinates starts at the central meridian (indicated with E, East). The coordinates are given in meters. To avoid negative coordinates, additions are added to the respective coordinate origin. Thus, a UTM coordinate specification constist of northing and easting:

The central meridians in each zone is defined as 500000 meters east. This value is called *"false easting"*.

For the southern hemisphere, the number 10000000 is added to the negative x-values (in meters). This value is called "false northing". Thus, for the southern hemisphere, northing decreases from 10000000 at the equator to the south.

The eastings are sometimes preceded at the beginning by the two-digit designation of the meridian strip (e.g. for the western half of Germany 32, for the eastern half 33).

The coordinate (434777 east, 5791572 north) marks a point in the garden of the castle of Osnabrück. It is located about 65 km west of the central meridian of 9° and about 5,792 km north of the equator (in the WGS84 or GRS80 reference system, cf. Table 4.4). The identical point has UTM coordinates (434859 easting, 5791776 northing) in the reference system ED 50 (European Datum 1950, International Ellipsoid according to Hayford). This shows that UTM coordinates can refer to different geodetic datums and that the associated geodetic datum must therefore generally be specified for an exact coordinate specification. It should be noted that the point in the UTM system is located in zone 32 and 5791572 m north of the equator (i.e. along the WGS84 ellipsoid), but in the Gauss-Krüger system it is located in the third stripe and 5793449 m north of the equator (here along the Bessel ellipsoid). This shows in turn that the associated reference ellipsod must be specified for an exact specification of a localisation.

Overall, it can be stated that the worldwide introduction of the UTM system (with the WGS 84) leads to a standardisation and to a strong simplification of the country-specific coordinate systems, so that the exchange of data is made easier. The use of data sets becomes simpler, since the sometimes time-consuming specification of geodetic reference systems in a geoinformation system is no longer necessary.

4.5.4 UTM Coordinates: Computation

National surveys and the calculation of UTM coordinates are based on ellipsoidal coordinates. However, the conversion of ellipsoidal coordinates to UTM coordinates is a mathematically complex process. The notes follow the derivation of a conformal mapping of an ellipsoid to the plane according to Krüger 1912, which has formed the basis of the Gauss-Krüger coordinate system in Germany (cf. Krüger 1912). The projection is mathematically identical to UTM mapping (except for the ellipsoid and width of the stripes) (Fig. 4.25):

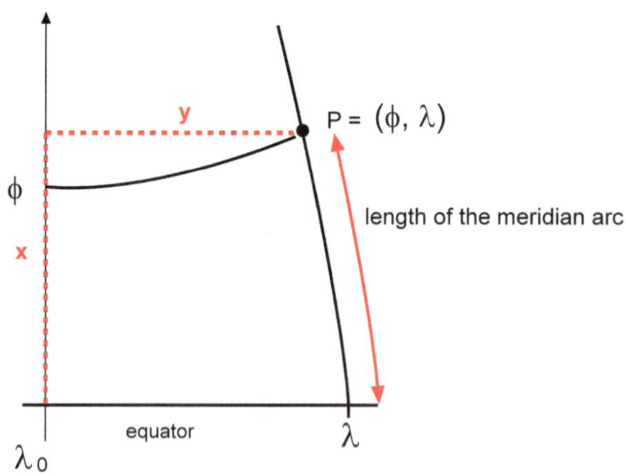

Fig. 4.25 Conformal projection from an ellipsoid to the plane – basic principle (cf. Krüger 1912 p. 3)

- The UTM coordinates are understood as complex numbers with $z_k = u_k + i \cdot v_k$ (u_k is called the real part, v_k is called the imaginary part), whereby in this derivation the coordinate axes are interchanged. In the standard case of the complex or Gaussian plane y just does not indicate the horizontal axis.
- The starting point is the basic equation $F(\varphi) + i \cdot \lambda = f(x + i \cdot y)$, which describes a relation between a point $w = (\varphi, \lambda)$ on the ellipsoid and a z-coordinate (y,x) in the plane, where $F(\varphi)$ is known (theoretically). The function $f(x + i \cdot y)$ is expanded as a Taylor series of $i \cdot y$.
- The Taylor series is broken down into a real part (which yields the northing) and an imaginary part (easting). This is shown by the following formulas for easting and northing.
- Another Taylor series is created to solve the elliptic integral to determine the length of the meridian arc (first element of the real part).

The information given above ist to indicate that the coefficients a_i appearing in the formulas result from a Taylor series expansion (cf. Krüger 1912 pp. 3 u. 37). By convention, y is given first and then x, since the point (φ, λ) on the ellipse is mapped to the point (y,x) in the plane (cf. Hofmann-Wellenhof et al. 2008 pp. 286, calculation of N simplified):

$$\text{easting} = a_1 \cdot l^1 + a_3 \cdot l^3 + a_5 \cdot l^5 + a_7 \cdot l^7 + \ldots$$

$$\text{northing} = B(\varphi) + a_2 \cdot l^2 + a_4 \cdot l^4 + a_6 \cdot l^6 + a_8 \cdot l^8 + \ldots$$

with:

$$
\begin{aligned}
a_1 &= N \cdot \cos\varphi \\
a_2 &= \frac{t}{2} \cdot N \cdot \cos^2\varphi \\
a_3 &= \frac{t}{6} \cdot N \cdot \cos^3\varphi \cdot \left(1 - t^2 + \eta^2\right) \\
a_4 &= \frac{t}{24} \cdot N \cdot \cos^4\varphi \cdot \left(5 - t^2 + 9 \cdot \eta^2 + 4 \cdot \eta^4\right) \\
a_5 &= \frac{t}{120} \cdot N \cdot \cos^5\varphi \cdot \left(5 - 18 \cdot t^2 + t^4 + 14 \cdot \eta^2 - 58 \cdot t^2 \cdot \eta^2\right) \\
a_6 &= \frac{t}{720} \cdot N \cdot \cos^6\varphi \cdot \left(61 - 58 \cdot t^2 + t^4 + 270 \cdot \eta^2 - 330 \cdot t^2 \cdot \eta^2\right) \\
a_7 &= \frac{t}{5040} \cdot N \cdot \cos^7\varphi \cdot \left(61 - 479 \cdot t^2 + 179 \cdot t^4 - t^6\right) \\
a_8 &= \frac{t}{40320} \cdot N \cdot \cos^8\varphi \cdot \left(1385 - 3111 \cdot t^2 + 543 \cdot t^4 - t^6\right)
\end{aligned}
$$

$$B(\varphi) = \textit{length of the meridian arc}$$

a and b are the major and minor semiaxes of the reference ellipsoid, respectively.

$N = \frac{a^2}{b \cdot \sqrt{1+\eta^2}}$	radius of curvature
$\eta^2 = e'^2 \cdot cos^2\varphi$	auxiliary variable
$e^2 = \frac{a^2 - b^2}{a^2}$	first numerical eccentricity
$t = \tan\varphi$	auxiliary variable
$l = (\lambda - \lambda_0)$	distance to central meridian
λ_0	longitude of the central meridian (e.g. 9 °, 15 ° east)

The length of the meridian arc, i.e. the distance from the equator to the point w on the ellipsoid (elliptical distance), is calculated by:

$$B(\varphi) = a \cdot (1 - e^2) \cdot \int_0^{\varphi} (1 - e^2 \cdot \sin^2 t)^{\frac{-3}{2}} dt$$

The elliptic integral in the formula to determine the length of the meridian arc cannot be solved directly, i.e. analytically. An expansion of the integrand as a Taylor series and integration of the single summands provides (cf. Hofmann-Wellenhof et al. 2008 p. 287):

$$B(\varphi) = \alpha \cdot (\varphi + \beta \cdot \sin(2\varphi) + \gamma \cdot \sin(4\varphi) + \delta \cdot \sin(6\varphi) + \varepsilon \cdot \sin(8\varphi)) \text{ with :}$$

$$\alpha = \frac{a+b}{2} \cdot \left(1 + \frac{n^2}{4} + \frac{n^4}{64} + \ldots\right)$$

$$\beta = -\frac{3}{2} \cdot n + \frac{9 \cdot n^3}{16} - \frac{3 \cdot n^5}{32} + \ldots$$

$$\gamma = \frac{15}{16} \cdot n^2 - \frac{15 \cdot n^4}{32} +$$

$$\delta = -\frac{35}{48} \cdot n^3 + \frac{105 \cdot n^5}{256} + \ldots$$

$$\varepsilon = \frac{315}{512} \cdot n^4 +$$

$$n = \frac{(a-b)}{(a+b)}$$

These formulas go back to Helmert (cf. Helmert 1880 pp. 46). Other authors implement the numerical eccentricity e in their formulas (cf. BEV-Bundesamt für Eich- und Vermessungswesen 2023).

If the parameters of the Bessel ellipsoid are used as a basis, (y,x) gives the Hochwert (northing) and Rechtswert (easting) in the Gauss-Krüger system. In the UTM projection, the reference meridian is not mapped equidistant, but multiplied by the factor 0.996 (cf. Sect. 4.5.5). Thus, the (provisional) UTM coordinates result to:

$$\text{northing} = 0.996 \cdot y$$

$$\text{easting}^{\#} = 0.996 \cdot x$$

In the formula for calculating x, the odd powers of $l = (\lambda - \lambda_0)$ mean that x always have the same sign as l. Thus, points east of the central meridian have a positive sign and points west of it have a negative sign. To avoid negative signs, the value 500000 m is added to the (provisional) easting$^{\#}$. Finally:

$$\text{northing} = 0.996 \cdot y$$

$$\text{easting} = 0.996 \cdot x + 500000$$

For $\varphi = 52°$ and for the Bessel ellipsoid, $B(\varphi) = 5762750.674$ m and for the WGS84 ellipsoid, $B(\varphi) = 5763343.550$ m, respectively. These values are obtained using the formulas given and, for verification, also by direct calculation of $B(\varphi)$ using numerical integration. For a point exactly on the central meridian, $B(\varphi)$ multiplied by the factor 0.9996 equals the northing, since $l = (\lambda - \lambda_0) = 0$. Therefore the remaining members in the above formula receive the value 0.

The representation of a UTM coordinate without zone indication is ambiguous. For example, a point northwest of Rostock, which is 194.4 km east of 9° E and 6005.9 km north of the equator with respect to WGS84, has the UTM coordinate (694400 and 6005900). This specification also applies, among other locations, to a point in the municipality of Skorzewo, south-west of Gdansk. Therefore zone 32 or zone 33 must be added, when defining the coordinate system in a geoinformation system.

4.5.5 EPSG Codes

The so-called *EPSG codes*, which were compiled by the original European Petroleum Survey Group Geodesy (EPSG), are used to identify the various coordinate systems. This working group was founded in 1986 and replaced in 2005 by the Surveying and

Table 4.9 Selected EPSG codes (cf. EPSG 2023b)

Coordinate system	EPSG
Gauss-Krüger zone 2 (Bessel)	31466
Gauss-Krüger zone 3 (Bessel)	31467
Gauss-Krüger zone 4 (Bessel)	31468
Gauss-Krüger zone 5 (Bessel)	31469
Gauß-Krüger fourth strip (Krassowski 3 degrees)	2398
Gauß-Krüger fifth strip (Krassowski 3 degrees)	2399
UTM with respect to ETRS89 32N	25832
UTM with respect to ETRS89 33N	25833
WGS84	4326

Positioning Committee of the International Association of Oil and Gas Producers, which proceeds the key system.

The data sets comprise 4- to 5-digit, globally unique key numbers for coordinate reference systems and descriptions of coordinate transformations (cf. Table 4.9). The key system is very detailed: For example, EPSG-31466 identifies the second Gauss-Krüger zone with the coordinate sequence Hochwert (northing) and Rechtswert (easting), while EPSG-5676 numbers the identical values in the reverse coordinate sequence.

4.6 Georeferencing in a Geoinformation System: Example

4.6.1 Georeferencing Based on Geodetic Coordinates

A common practical task is to display a template such as an analog aerial photo or a paper map in a geoinformation system with real-world coordinates. To do this, the template must first be digitised (i.e.scanned) and then georeferenced. The older approach involved georeferencing of an analog map using a digitising tablet (cf. Sect. 2.5.6 and Fig. 2.7). In contrast, almost only digital templates are used today, i.e. digital aerial photos or satellite images or scans, which can be georeferenced directly on the monitor using tools of a geoinformation system. For reasons of clarity an example is presented, which uses a section of a topographical map 1:25000 near Berlin (sheet 3548 Rüdersdorf, cf. Fig. 4.26). On the one hand, there are clear control points in the template, for which exact coordinates of the national survey are available. On the other hand, several coordinate systems can be recognised, so that this map section is also well suited to illustrate different coordinate systems and the datum transformation (cf. Sects. 4.6.2 and 4.6.3).

After scanning, the TIF image is loaded into the graphic editor of a geoinformation system. For georeferencing, the control points within the template are “clicked” with the selection tool of the geoinformation system (i.e. usually with the mouse), then coordinates are assigned to them. Table 4.10 documents the process of georeferencing (strongly rounded values):

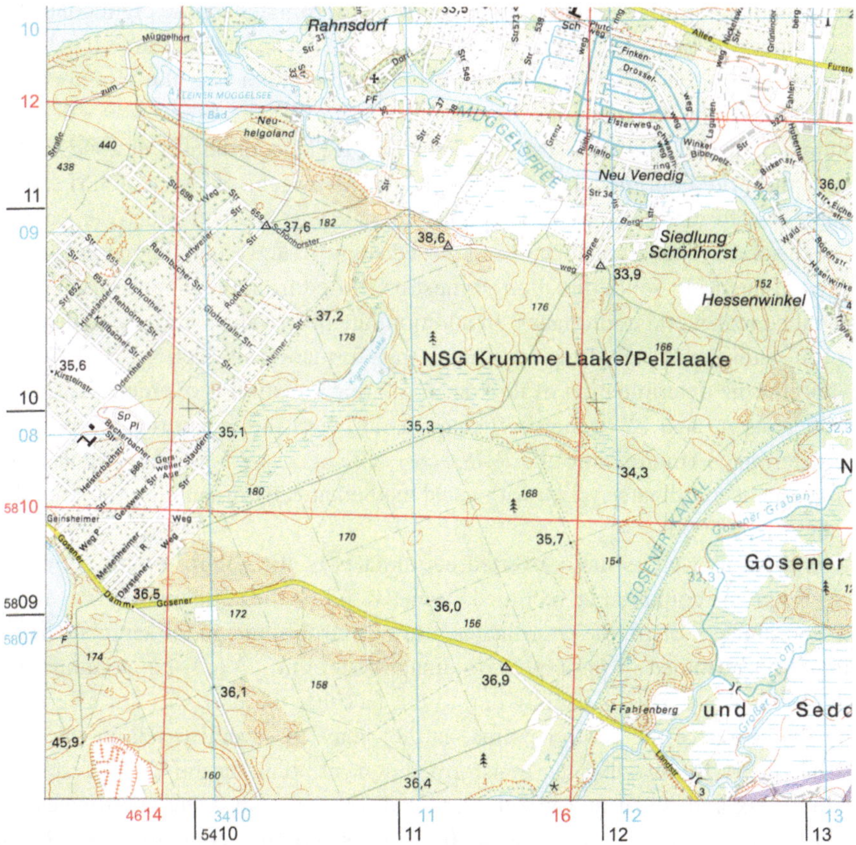

Fig. 4.26 Georeferencing (Geobasis data: ©GeoBasis-DE/LGB,GB 13/19)

Table 4.10 Georeferencing and calculation of the RMS error (cf. the blue grid in Fig. 4.26)

	Screen coordinates of the control points		Given UTM	Given UTM	Transformed screen coordinates of control points		
i	X-coord.	Y-coord.	easting	northing	easting	northing	$qdiff_i$
1	1260.463	−2583.096	410000	5807000	409999.724	5806999.287	0.585
2	4049.034	−642.088	413000	5809000	413000.035	5808999.323	0.460
3	3144.887	−2549.304	412000	5807000	412000.365	5807000.880	0.908
4	1223.673	−690.846	410000	5809000	410000.467	5808998.948	1.324
5	3125.603	−1601.980	412000	5808000	411999.818	5808001.994	4.010
6	4086.498	−2535.332	413000	5807000	412999.986	5806998.599	1.963
7	2164.727	−673.269	411000	5809000	410999.573	5809000.486	0.418
8	1241.983	−1635.680	410000	5808000	410000.033	5808000.483	0.234
						RMS =	1.113

The second and third columns contain the pixel coordinates for the eight control points of the template. Comparing with Fig. 4.26, it can be seen that the origin of the Y coordinates lies in the upper left corner of the graphic editor (i.e. the monitor) and that the orientation of the Y axis points downwards.

Columns four and five show the corresponding UTM coordinates of zone 33. This is based on the grid imprinted in blue (for the UTM system cf. Sect. 4.5.5). An affine coordinate transformation is used since a flat template was scanned which can be assumed not to be distorted. The template has been placed slightly skewed on the scanner, so it must be rotated in the north-south direction. It must be shifted from the pixel coordinate system into so-called real-world coordinates. In addition, the pixel values must be scaled to meter values (cf. definition of an affine coordinate transformation in Sect. 4.2.5.2). With eight control points the determination of the transformation equation is overdetermined. Before the coefficients A^0 and B^0 of the coefficients a_i and b_i of the transformation polynomials can be determined (minimizing the *RMS error*, c.f. Sect. 4.2.5.5), the y-values must be converted into a coordinate system with usual mathematically orientation (i.e. from bottom to top, here $y_i = 5000 - y$).

The original as well as the converted coordinates of the control points show that the regular grid points could not be recorded correctly. While the target coordinates reflect the regular coordinate system, the coordinates of the control points in the template have been inaccurately determined in the pixel coordinate system (x_i,y_i) (c.f. i.e. columns two and three in Table 4.10). This is usually due to the fact that the user was not able to clearly identify the center of the control points, as they may be several pixels in size. Thus, it almost inevitably follows that the transformed control points in the UTM-System (u_i,v_i), i.e. columns six and seven in Table 4.10, also differ from the target coordinates (x'_i,y'_i), i.e. columns four and five in Table 4.10. The last column contains these deviations, i.e. more precisely the squared deviation distances, i.e. $d_i^2 = (x'_i - u_i)^2 + (y'_i - v_i)^2$. For the first row of Table 4.10, this partial sum is calculated by:

$$\text{qdiff}_1 = (410000 - 409999.724)^2 + (5807000 - 5806999.287)^2 = 0.585$$

The values d_i are illustrated by the small arrows in Fig. 4.11. The *RMS* error results by summing the values in the last column, then dividing the sum by the number of control points, and finally taking the root. Since the target system is in meters, the RMS error is also given in meters. This value gives a good illustration of this measure of accuracy. For the last column of Table 4.10, the RMS error is calculated by:

$$\text{RMS} = \sqrt{\frac{1}{8} \cdot \sum\nolimits_{i=1}^{8} \text{qdiff}_i} = 1.113$$

In practice, determining the control points is an iterative process. Since the fifth control point makes the largest contributor to the RMS error, it should be deleted and set again.

However, it is important to ensure that not only a small RMS error should be aimed for. Rather, care must be taken to ensure that the entire study area is optimally covered by the distribution of the control points.

In the present example, which describes a typical application, it is assumed that control points can be clearly identified in the template. In contrast, the georeferencing of a template with points which cannot be clearly defined as control points, such as the georeferencing of an aerial photograph, is more complex and probably less precise. Thus, geodetic coordinates of prominent points such as building corners have to be determined in advance.

In addition to the transformation into a new reference system, the pixel values of the template (here mostly RGB-values) must be converted into pixel values of the result image. This process is called resampling. Several pixel values of the template can be involved in a value for a single pixel in the georeferenced image. Digital image processing provides several methods for resampling (cf. Sect. 10.6.1.2). The nearest neighbor method is usually used for georeferencing a map template.

4.6.2 Definition of the Spatial Reference After Georeferencing

After georeferencing, a coordinate system must be assigned to the georeferenced template, since the new coordinate values as pure numbers alone say nothing. Only then can the geometries of the geoobjects (e.g. country borders, river courses) recorded in further steps be converted into other coordinate systems. If no coordinate system is assigned to the georeferenced template, only a graphical reproduction as a background image in a geoinformation system is possible. Thus, knowledge of the underlying projection is necessary, as the geoinformation system expects detailed information on the spatial reference.

Geoinformation systems often distinguish between "geographic coordinate systems" and "projected coordinate systems". Coordinate specifications of the first group are specified by coordinate systems such as "Deutsches Hauptdreiecksnetz" (EPSG 4314), "ETRS 1989" (EPSG 4258) for Germany or "Amersfort" (EPSG 4289) for the Netherlands. The second group includes the geodetic coordinate systems such as Gauss-Krüger and UTM or the systems of the US states as well as other national grids such as the coordinate system "Lisboa_Hayford_Gauss_IPCC" (Transversal Mercator Projection based on the Hayford ellipsoid defined by Instituto Português da Cartografia e Cadastro) or the "British National Grid" (Transversal Mercator Projection based on the Airy ellipsoid defined by the British Ordnance Survey). It can be assumed that a geoinformation system (i.e. the software) has information of almost all coordinate systems. In most cases, geodetic coordinates from a national survey can be used as the reference system of the template (cf. e.g. National Grid in Ordinance Survey). Very often the procedure is simplified if worldwide available UTM coordinates are used. However, it should be noted that UTM coordinates may be based on different ellipsoids.

Only in the very few cases in which the template provides geographic coordinates, whose specification is not covered in the software, a reference must be defined independently. Knowledge of projections is helpful for this task (i.e. specification of an ellipsoid with associated parameters). As a rule this task arises only for templates for larger regions, which are usually not based on a Cartesian coordinate system. Georeferencing follows the usual procedure, but geographic coordinates are assigned to the control points. It is important to note how the geographic coordinates are specified. Does the declaration 3.50 mean a geographic coordinate of 3° 50′ or of 3° 30′, i.e. half of a degree?

4.6.3 Multiple Reference Systems and Datum Transformation

The use of a datum transformation can also be shown on the basis of Fig. 4.26. On the one hand, two grids are drawn which represent UTM coordinate systems based on the WGS84 or ETRS89 ellipsoids, respectively. The red grid indicates UTM coordinates in zone 32, the blue grid UTM coordinates in zone 33. In addition faint black crosses can be seen corresponding to the black coordinates on the edge of the map. This is the grid that denotes the Gauss-Krüger coordinate system, which is based on the Bessel ellipsoid. In order to transform the Gauss-Krüger coordinates into UTM coordinates, a datum transformation is necessary. When converting a UTM coordinate from zone 32 to zone 33, however, no datum transformation is necessary since the same reference ellipsoid is used in each case.

References

AdV, Arbeitsgemeinschaft der Vermessungsverwaltungen der Bundesrepublik Deutschland (2017): Richtlinie für den einheitlichen integrierten geodätischen Raumbezug des amtlichen Vermessungswesens in der Bundesrepublik Deutschland. https://www.adv-online.de/icc/extdeu/nav/dab/binarywriterservlet?imgUid=59770a88-3739-b261-4b34-98951fa2e0c9&uBasVariant=11111111-1111-1111-1111-111111111111 (14.04.2023).

AdV-Online (2023a): Länderspezifische Lagebezugssysteme. https://www.adv-online.de/icc/extdeu/nav/238/broker.jsp?uMen=238707b7-f12f-9d01-3bbe-251ec0023010&SP_fontsize=0 (14.04.2023).

AdV-Online (2023b): Deutsches Haupthöhennetz 1992 (DHHN92). https://www.adv-online.de/icc/extdeu/broker.jsp?uMen=a49707b7-f12f-9d01-3bbe-251ec0023010 (14.04.2023)

AdV-Online (2023c): Integrierter geodätischer Raumbezug. http://www.adv-online.de/icc/extdeu/nav/dab/dab4dc09-a662-261b-5f8d-14201fa2e0c9&sel_uCon=90d0dc09-a662-261b-5f8d-14201fa2e0c9&uTem=73d607d6-b048-65f1-80fa-29f08a07b51a.htm (14.04.2023).

AdV-Online (2023d): SAPOS. https://www.adv-online.de/AdV-Produkte/Integrierter-geodaetischer-Raumbezug/SAPOS/ (14.04.2023).

Aitchison, A. (2011): The Google Maps / Bing Maps Sperical Mercator Projection. https://alastaira.wordpress.com/2011/01/23/the-google-maps-bing-maps-spherical-mercator-projection/ (14.04.2023).

Backhaus, K., Erichson, B., Plinke, W. u. R. Weiber (2016): Multivariate Analysemethoden: Eine anwendungsorientierte Einführung. Springer: Berlin 14. Ed.

Bartelme, N. (2005): Geoinformatik: Modelle, Strukturen, Funktionen. Berlin: Springer. 4. Ed.

Battersby, S. et al (2014): Implications of Web Mercator and Its Use in Online Mapping. In: Cartographica. The International Journal for Geographic Information and Geovisualization 49(2). P. 85-101. https://doi.org/10.3138/carto.49.2.2313 (18.01.2022).

Becker, M. u. K. Hehl (2012): Geodäsie. Darmstadt: Wiss. Buchgesellsch.

BEV-Bundesamt für Eich- und Vermessungswesen (2023): Transformation von Gauß-Krüger(GK)-Koordinaten des Systems MGI in Universal Transversal Mercator(UTM)-Koordinaten des Systems ETRS89. https://www.bev.gv.at/pls/portal/docs/PAGE/BEV_PORTAL_CONTENT_ALLGEMEIN/0200_PRODUKTE/PDF/TRANSFORMATION_GK_MGI_UTM_ETRS89.PDF (14.04.2023).

BKG, Bundesamt für Kartographie und Geodäsie (2023a): Dokumentation, DTK1000. https://sg.geodatenzentrum.de/web_public/gdz/dokumentation/deu/dtk1000.pdf (14.04.2023).

BKG, Bundesamt für Kartographie und Geodäsie (2023b): Open Data. https://gdz.bkg.bund.de/index.php/default/open-data.html (16.06.2022).

BKG, Bundesamt für Kartographie und Geodäsie (2023c): Coordinate Reference Systems in Europe. http://www.crs-geo.eu/crs-national.htm (14.04.2023)

BKG, Bundesamt für Kartographie und Geodäsie (2023d): Webanwendung zur Koordinatentransformation. https://gdz.bkg.bund.de/index.php/default/koordinatentransformation.html (14.04.2023).

BKG, Bundesamt für Kartographie und Geodäsie (2023e): Die Höhenbezugsfläche von Deutschland. https://www.bkg.bund.de/DE/Das-BKG/Wir-ueber-uns/Geodaesie/Integrierter-Raumbezug/Hoehenbezugsflaeche/hoehenbezug.html (14.04.2023).

BKG, Bundesamt für Kartographie und Geodäsie (2023f): Integrierter Raumbezug. https://www.bkg.bund.de/DE/Das-BKG/Wir-ueber-uns/Geodaesie/Integrierter-Raumbezug/integrierter-raumbezug.html (14.04.2023).

Bock, H.H. (1974): Automatische Klassifikation. Göttingen: Vandenhoeck u. Rupprecht.

Bortz, J. u. C. Schuster (2010): Statistik für Human- und Sozialwissenschaftler. Lehrbuch mit On-line-Materialien. Berlin: Springer. 7. Ed.

Bugayevskiy, L.M. u. J.P. Snyder (1995): Map Projections. A Reference Manual. London: Taylor u. Francis.

Dana, P.H. (2023a): Map Projection Overview. https://foote.geography.uconn.edu/gcraft/notes/mapproj/mapproj.html (14.04.2023).

Dana, P.H. (2023b): Geodetic Datum Overview. https://foote.geography.uconn.edu/gcraft/notes/datum/datum_f.html (14.04.2023)

Dana, P.H. (2023c): Reference Ellipsoids and Geodetic Datum Transformation Parameters (Local to WGS-84). https://foote.geography.uconn.edu/gcraft/notes/datum/edlist.html (14.04.2023).

de Lange, N. u. J. Nipper (2018): Quantitatice Methodik in der Geographie. Grundriss Allgemeine Geographie. Paderborn: Schöningh.

EPSG (2023a): EPSG specifications e.g. 7004. https://epsg.io/7004-ellipsoid (14.04.2023).

EPSG (2023b): EPSG Geodetic Parameter Dataset. https://epsg.org/home.html (14.04.2023).

FGDC, Federal Geographic Data Committee (2023): United States National Grid. https://www.fgdc.gov/usng (14.04.2023).

Freund, R.W. u. R.H.W. Hoppe (2007): Stoer/Bulirsch: Numerische Mathematik 1. Berlin: Springer. 10. Ed.

GEObasis.nrw (2020): ETRS89/UTM. Das amtliche Lagebezugssystem in Nordrhein-Westfalen. https://www.bezreg-koeln.nrw.de/brk_internet/publikationen/abteilung07/pub_geobasis_etrs89.pdf (14.04.2023).

GEObasis.nrw (2018): Normalhöhen und Höhenbezugsflächen in Nordrhein-Westfalen. Bonn: Bez. Reg. Köln. https://www.bezreg-koeln.nrw.de/brk_internet/publikationen/abteilung07/pub_geobasis_normalhoehen.pdf (14.04.2023).

Gerlach, C. et al. (2017): Höhensysteme der nächsten Generation. In: Rummel, R. (2017, Hrsg.). Erdmessung und Satellitengeodäsie. Handbuch der Geodäsie. S. 349-400. Berlin/Heidelberg: Springer Spektrum.

Grafarend, E.W. u. F.W. Krumm (2006): Map projections: cartographic information systems. Heidelberg: Springer.

Gyan Information Pedia (2023): Indian Grid System. http://gyaninformationpedia.blogspot.com/2015/10/indian-grid-system.html (14.04.2023).

Hake, G. u. V. Heissler (1970): Kartographie I. Kartenaufnahme, Netzentwürfe, Gestaltungsmerkmale, Topographische Karte. Berlin: Walter de Gruyter, Sammlung Göschen Bd. 30/30a/30b.

Hake, G. et al. (2002): Kartographie. Visualisierung raum-zeitlicher Informationen. Berlin: de Gruyter, 8. Ed.

Helmert, F.R. (1880): Die mathematischen und physikalischen Theorien der höheren Geodäsie. Einleitung und 1 Teil. Leipzig: Teubner, p. 46–48. https://archive.org/stream/diemathematisch01helmgoog#page/n67/mode/2up (14.04.2023).

Hofmann-Wellenhof, B., Lichtenegger, H., Wasle, E. (2008): GNSS – Global Navigation Satellite Systems. GPS, GLONASS, Galileo and more. Springer: Wien.

Kahmen, H. (1997): Vermessungskunde. Berlin: de Gruyter. 19. Ed.

Kahmen, H. (2006): Angewandte Geodäsie. Vermessungskunde. Berlin: de Gruyter. 20. Ed.

Krüger, L. (1912): Konforme Abbildung des Erdellipsoids in der Ebene. Veröffentl. Königl. Preuss. Geodät. Inst. NF 52. Leipzig. Teubner. https://gfzpublic.gfz-potsdam.de/rest/items/item_8827_2/component/file_130038/content (14.04.2023).

Kuntz, E. (1990): Kartenentwurfslehre. Grundlagen und Anwendungen. Heidelberg: Wichmann. 2. Ed.

Landesvermessungsamt Nordrhein-Westfalen (1999): Transformation von Koordinaten und Höhen in der Landesvermessung. Teil I: Theoretische Grundlagen. Bonn-Bad Godesberg: Selbstverlag.

Ministerium für Inneres und Europa Mecklenburg-Vorpommern (2017): Landesbezugssystemerlass. http://www.landesrecht-mv.de/jportal/portal/page/bsmvprod.psml?doc.id=VVMV-VVMV000008736&st=vv&doctvp=vvmv&showdoccase=1&paramfromHL=true#focuspoint (14.04.2023).

Niemeier, W. (2008): Ausgleichsrechnung. Statistische Auswertemethoden. Berlin: de Gruyter. 2. Ed.

Open Street Map (2023): Slippy map tilenames. https://wiki.openstreetmap.org/wiki/Slippy_map_tilenames (14.04.2023).

Ordnance Survey (2018): A Guide to Coordinate Systems in Great Britain. https://www.ordnancesurvey.co.uk/documents/resources/guide-coordinate-systems-great-britain.pdf (14.04.2023).

Ott, Th. u. F. Swiaczny (2001): Time-integrative Geographic Information Systems. Management and Analysis of Spatio-Temporal Data. Berlin: Springer.

Pavlidis, T. (1982): Algorithms for graphics and image processing. Berlin: Springer. 2. Ed.

Preparata, F. u. M. Shamos (1985): Computational Geometry. New York: Springer.

Resnik, B. u. R. Bill (2018): Vermessungskunde für den Planungs-, Bau- und Umweltbereich. Berlin: Wichmann. 4. Ed.

Rummel, R. (2017, Hrsg.). Erdmessung und Satellitengeodäsie. Handbuch der Geodäsie (Hrsg. Freeden u. R. Rummel). Berlin/Heidelberg: Springer Spektrum.

Sachsen.de (2023): Geobasisinformation und Vermessung. Grundlagen und Begriffe. Koordinatenreferenzsystem. https://www.landesvermessung.sachsen.de/grundlagen-und-begriffe-5585.html (14.04.2023).

Seeber, G. (2003): Satellite Geodesy. Berlin: de Gruyter 2. Ed.

Seitz, M. u.a. (2017): Geometrische Referenzsysteme. In: Rummel, R. (2017, Hrsg.). Erdmessung und Satellitengeodäsie. Handbuch der Geodäsie. S. 324-348. Berlin/Heidelberg: Springer Spektrum.

Snyder, J.P. (1987): Map Projections – A Working Manual. Washington, D.C.: United States Gov. Printing Office. = US Geological Survey Professional Paper 1395.

Umweltbundesamt (2021): Ausbreitungsmodelle für anlagenbezogene Immissionsprognosen. https://www.umweltbundesamt.de/themen/luft/regelungen-strategien/ausbreitungsmodelle-fuer-anlagenbezogene/uebersicht-kontakt#textpart-1 (14.04.2023).

Torge, W. (2017): Geschichte der Erdmessung. In: Rummel, R. (2017, Hrsg.). Erdmessung und Satellitengeodäsie. Handbuch der Geodäsie. S. 1–71. Berlin/Heidelberg: Springer Spektrum.

Zimmermann, A. (2012): Basismodelle der Geoinformatik. Strukturen, Algorithmen und Programmierbeispiele in Java. München: Hanser.

Digital Geodata: Data Mining, Official Basic Geodata and VGI

5

5.1 Basic Terms

5.1.1 Primary and Secondary Data Recording, Primary and Secondary Data

Geoobjects are characterised by geometric, topological, thematic and temporal information. Together, these data form the *geodata* of a geoobject (for the definition of geoobjects cf. Sect. 4.1). A distinction is made between *primary data*, which are based on surveys or measurements and have not yet been (significantly) processed by users, and *secondary data*, which represent data derived from the primary data and processed. In addition, there are *metadata,* which can contain descriptive information on, among other things, the measurement or survey method, the reason for collection or the data quality (cf. Sect. 6.5). The collection of geometric data is to be systematised into:

Primary recording methods

field exploration
    survey as well as telephone interview, census, description, mapping, observation, measurement in the field, collection of e.g. soil samples
continuous recording of measurements
    recording of data such as air pressure or precipitation at weather observation stations
terrestrial-topographic survey
    traditional geodetic surveying with tape measure, opto-electronic measuring methods with theodolite and tachymeter, terrestrial laser scanning (TLS), Global Positioning System (GPS)

N. de Lange, *Geoinformatics in Theory and Practice*, Springer Textbooks in Earth Sciences, Geography and Environment,
https://doi.org/10.1007/978-3-662-65758-4_5

topographic survey by remote sensing
- aerophotogrammetry, aero- or airborne laser scanning (ALS, airborne or by means of unmanned aerial vehicles (UAV)), radar methods, satellite imaging methods

Secondary recording methods

recording and derivation of data from existing sources
- geometric construction methods
- digitising with a digitising tablet
- scanning

The recording of three-dimensional position coordinates with the help of satellite-based positioning systems is assigned to the terrestrial recording methods here (cf. Sect. 5.3), as this technique has become established in the practice of surveying. Furthermore, the basic principle of laser scanning methods for large-scale recording of 3D data is explained, from which e.g. building information can be derived (cf. Sect. 5.4). Apart from these two approaches surveying methods of geodesy and photogrammetry are not discussed here (cf. Resnik and Bill 2018, Kahmen 2006 and Kohlstock 2011).

Primary recording methods mentioned above especially of geometric data provide original or primary data from individual surveys or measurement campaigns. In addition to this approaches geometric information can be obtained from the official surveying authorities and cadastral authorities. These data have the status of primary data, for which the term *official basic geodata* (in German: Amtliche Geobasisdaten) is now used (cf. Sect. 5.5).

With regard to thematic information or *thematic data*, primary recording methods mostly provide analog data. Only a few sensors, such as the group that is important in the geosciences, which measures climate parameters such as temperature, solar radiation or wind speed, can directly record primary, digital thematic data. In contrast, terrestrial surveying methods now almost exclusively generate digital geometric primary data, i.e. two- or three-dimensional position coordinates in a spatial reference system in vector format. Remote sensing methods only provide digital raster data initially without a reference system, which requires georeferencing (cf. Sect. 10.6.1.2).

In the frequent case of thematic information, the secondary recording methods create new analog data from existing analog data (e.g. description, text evaluation), which are manually converted into digital form via the keyboard or via a scanner with character recognition. In addition, there are also digital secondary methods of recording thematic data. For example, statistical methods on a computer system derive new digital data from existing digital data. This data processing provides a wealth of thematic information, which in the simplest case of geo-related data includes density values such as population density, but also, for example, vegetation indices from remote sensing data.

In contrast to this thematic information or thematic data, geometric data can also be obtained, i.e. derived, using digital, secondary recording techniques. Geometric construction methods calculate new coordinates from existing digital ones. This includes moving a line in a geoinformation system or designing a building in a CAD system. In geoinformatics, the group of secondary recording methods primarily includes the digitisation of existing geometric data from primary or secondary data, i.e. from already recorded data and especially from maps in paper form. In the simplest case, this digitisation can be carried out manually by entering coordinates via the keyboard, semi-automatically or fully automatically with the help of analog-to-digital converters. Semi-automatic *analog-to-digital conversion* includes the recording of geometric information from drawings or paper maps using a digitising tablet or directly on the screen (cf. Sect. 5.2.1). It provides two-dimensional device coordinates in vector format. In contrast, scanners only generate, e.g. from analog line drawings, exclusively digital raster data without reference system. In both cases, georeferencing to a known and well-defined coordinate system becomes necessary (cf. example in Sect. 4.6).

5.1.2 Discretisation

The discretisation associated with analog-to-digital conversion, i.e. a *temporal* and *spatial discretisation*, is of great importance. Thus, continuous (measurement) data such as air temperature, precipitation, water levels or traffic flows are only collected at certain time intervals or are related to time periods and stored as individual values. In particular, the spatial discretisation of geoobjects is an essential prerequisite for recording and modeling in geoinformation systems:

- Point geoobjects are already discrete data.
- Linear geoobjects such as a stream or a path are broken down into individual sections, whereby only their start and end points are digitally recorded and the course in between is assumed to be straight (cf. Fig. 5.3). This principle corresponds to the definition of boundaries of properties, where a boundary stone is set at each corner or at each change of direction of the boundary. Sometimes the start and end points are recorded and the intermediate section is modeled and discretised by specifying a function (e.g. as a curved section by specifying a radius).
- Areas are recorded by boundary lines that are modeled according to the principle just described. Thus polygons discretize irregularly shaped areas.
- With raster data, a line (or area) is discretised by individual or adjacent pixels.

Three-dimensional continuous surfaces or bodies require special effort for (digital) recording and discretisation. Several approaches can be named for this, and specific methods are provided for their implementation in geoinformation systems (cf. Sect. 9.7 and especially 9.7.5):

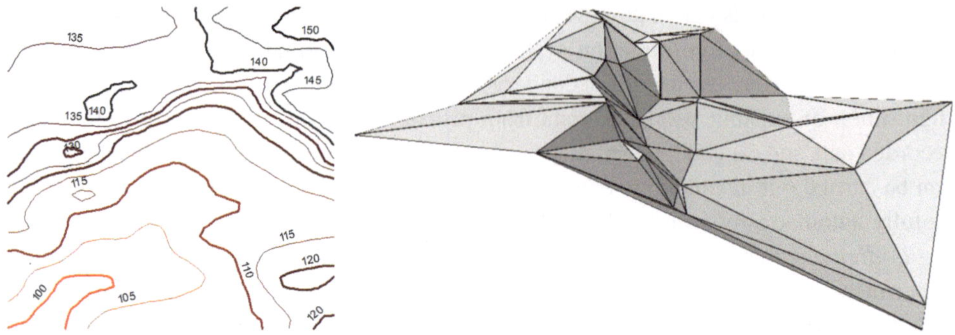

Fig. 5.1 Modeling of surfaces by discretisations (Höxberg near Beckum in Germany)

- decomposition into regularly distributed grid points or grid cells and visualisation of the height information by different column heights (cf. Fig. 9.34),
- calculation and visualisation of isolines (cf. Fig. 5.1 left),
- calculation of a triangular mesh and visualisation as a three-dimensional surface (cf. Fig. 5.1 right and Fig. 9.34).

5.2 Digital Recording of Geometry Data Using Secondary Techniques

Figure 5.2 systematises the methods of recording geometry data using digital secondary techniques:

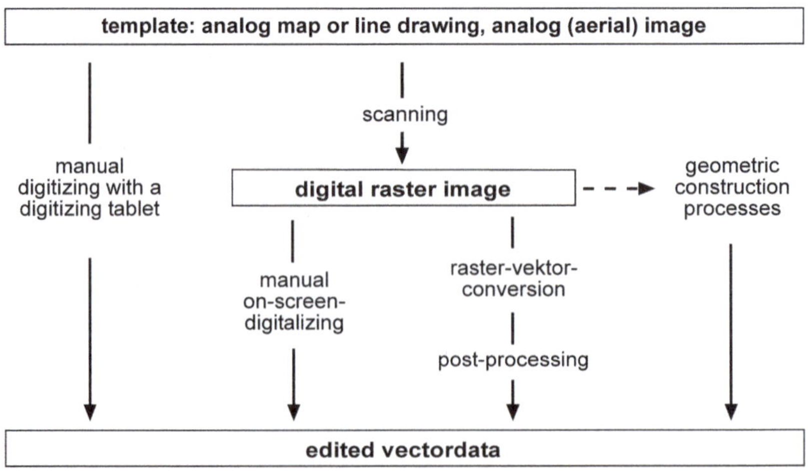

Fig. 5.2 Secondary recording methods of geodata in vector format

5.2.1 Digital Recording of Geometry Data in Vector Format

The recording of points, which are to be understood as vectors in a coordinate system, with the help of a graphics tablet (e.g. digitiser) and a pointing device such as a stylus or a so called puck is now outdated (cf. Fig. 2.7). However, this technique is suitable to illustrate the basic principle of data recording. The puck is a mouse-like pointing device with a magnifying glass that can detect its absolute position on the tablet. The puck is used to pick up selected points of a graphic on the tablet. A pulse is transmitted from the digitising magnifier to a grid system integrated in the tablet and two electrical conduction paths are activated, which provide the corresponding coordinates. The grid lines closest to the position of the selected point are selected. The resolution and the detection accuracy are determined by the spacing of the grid. Of course, there are other design principles of a graphics tablet, all of which make it possible to determine the exact position of the pointing device on the tablet.

With the help of a graphics tablet, all geometries are recorded by individual points, i.e. discretised. A line is approximated by a sequence of coordinates. It is initially assumed that there is a straight line connection between the coordinates. A curved line is therefore broken down into a sequence of straight lines, whereby only their end and start points are recorded (cf. Fig. 5.3).

The digital recording of geometric data with a graphics tablet has the advantage that large-format analog maps and drawings can also be processed. This was particularly important in the 1990s, when extensive analog map stocks had to be digitised in order to build up digital information systems. This phase of (initial) recording of geometries, especially in spatial planning and land surveying (recording of old inventories), has been completed. The maps are available as georeferenced raster maps, so that the recording of digital vector data directly from analog originals with a graphics tablet has disappeared. In contrast, digital data recording through *on-screen digitisation* has become the standard.

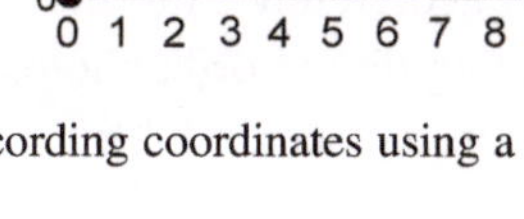

L1: (0,0), (1,1), (2,2)
L2: (2,2), (3,1), (4,1), (5,1), (6,1), (7,2), (8,3), (8,4), (8,5), (7,6)
L3: (7,6), (4,6), (3,6), (2,6)
L4: (2,6), (1,5), (1,4), (2,2)
L5: (7,6), (7,7)
L6: (7,7), (5,8), (4,8), (2,8), (2,7), (2,6)
L7: (7,7), (7,8), (8,8)

Fig. 5.3 Recording coordinates using a digitising tablet

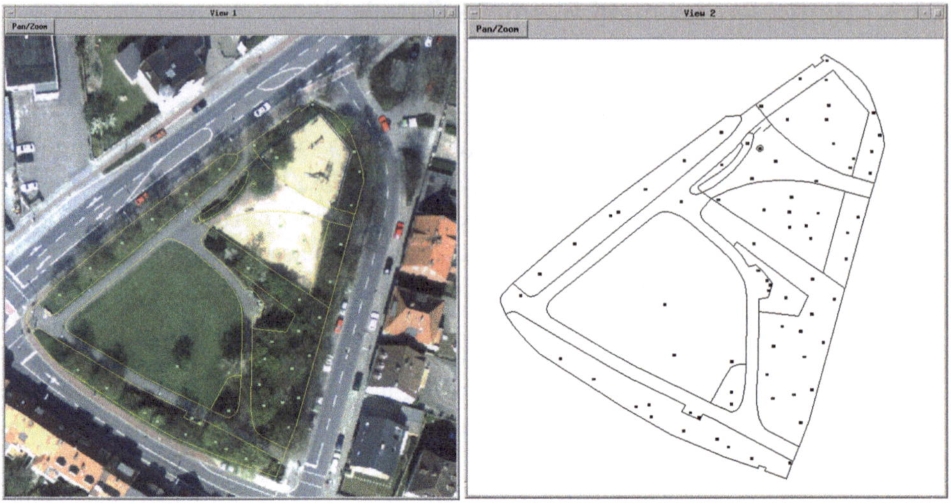

Fig. 5.4 On-screen recording (digitising) of geometries

This refers to the digital data recording of geometries directly on the screen (i.e. "on screen"). With this technique, graphics (e.g. ground plans, pictures) in the screen background serve as templates from which geometries can be traced on the screen with the computer mouse. On the one hand, raster maps can be used, e.g. digital aerial photographs or scanned maps and drawings, to create new geometries. On the other hand, digital databases can be updated, above all building plans can be updated or new point objects such as individual trees can be added.

Figure 5.4 shows the on-screen data recording for the green area cadastre of the city of Osnabrück. This represents a classic approach that merged the digital official basic geodata (here: information from the ALK, the automated real estate map, in German Automatisierte Liegenschaftskarte, cf. in detail Sect. 5.5.2) with the existing special data in the form of analog aerial photos and maps in a digital information system (cf. de Lange and Wessels 2000 and Nicolaus 2000). On the basis of several digital layers, green areas and associated areas such as certain shrub plantings or lawns were identified on the screen (cf. in Fig. 5.4 the lines in the aerial photograph, which are only faintly reproduced, and on the right the result of the data recording without an aerial photograph). The main source for the actual inventory were true-color aerial photographs at an average scale of 1:3000. The analog orthophotos were scanned and had to be rectified and georeferenced to the Gauss-Krüger coordinate system (cf. Sects. 4.2.5 and 10.6.1.2, creating digital orthophotos). This work was a necessary prerequisite to link the information on the screen with the other digital information sources.

A geoinformation system has various functions to model the course of the line between two digitised coordinates, whereby the basic problem is the same for data recording with a digitising tablet or on-screen. Since Fig. 5.3 is based on a very coarse grid and wide spacing

Fig. 5.5 Smoothing effects with different grain values

of the traces, the data recording is coarse and angular. A smoothing effect is possible by specifying a so-called *grain value*. If the interpolation points are further apart than the set grain value, further interpolation points are added algorithmically. If the distance is smaller, points are deleted and therefore the course of a line is simplified (cf. Figs. 5.5 and 3.12). However, whether such a generalisation is valid depends on the application. This is forbidden, for example, when recording legally binding parcel boundaries for a real estate cadastre.

It should be noted that on-screen digitising is always based on the orthogonal screen coordinates and not on an existing cartographic projection. If the template does not have an orthogonal projection, the recorded orthogonal device coordinates must be projected into the associated map projection. In general, a coordinate system in so-called real-world coordinates (e.g. UTM coordinates) must be assigned to the digital raster image, which only has screen coordinates, directly after georeferencing (cf. example in Sect. 4.6). A geoinformation system offers many tools for this.

The digitising system also supports the recording of geometric data in many ways and helps to eliminate errors or prevent them from occurring in the first place. Incomplete recordings of linear data structures cause most of the errors. This often occurs when lines must meet exactly, such as at a street crossing when the end point of one line does not exactly coincide with the starting point of another line. When recording, a so-called coordinate snap should be set, so that the coordinate to which a line is to be connected only has to be selected approximately. The start of the new line jumps or "snaps" to an existing coordinate.

A strict *node-edge-node recording* is advantageous, in which line segments or edges are digitised, which always start at a node and go exactly to the next node. This is the classic "arc-node-recording" as it is called within the geoinformation system Arc/Info of the software company ESRI (cf. Sect. 9.3.2). Never digitise beyond an intersection with other lines or edges. A line always ends at a node, whereby a node is defined as the meeting point of at least three lines. This digitising variant is very stringent, but also laboriously. In contrast, the so-called *spaghetti digitising* does not have any specific requirements. The lines are recorded arbitrarily, they do not have to end at intersections with other lines and can overlap. The recording is much more simply. However, such an unsystematic analog-to-digital conversion should always be avoided. Errors, especially missing points or lines and so-called dangles (cf. Sect. 9.3.2), are more likely to occur with spaghetti digitising. A geoinformation system provides some functions to convert a spaghetti digitisation into a node-edge-node data structure. Then more attempts are made to eliminate these problems.

The node-edge-node recording leads to a strict geometric-topological modeling of geoobjects in the vector model, which forms the basis of some geoinformation systems (cf. Sect. 9.3.2). An area such as a lake is modeled as a polygon made up of individual edges. Since polygons are stored topologically, the polygon F1 in Fig. 5.3 is defined by its segments L2, L3 and L4. This topological information is stored in the data model in addition to the geometric information. The adjacent polygon F2 is defined by only storing its boundaries L3, L5, L6). A boundary (here L3) is therefore only recorded once (cf. Sect. 9.3.2).

In contrast to this strict modeling of polygons, simplified data models exist. The standardised *Simple Feature Geometry Object Model* for describing two-dimensional vector geometries defines an area as a closed polyline (cf. Sects. 6.3.2 and Table 6.2). Likewise, in the proprietary shapefile data format of the software company ESRI, a polygon is formed by a closed polyline (cf. Sect. 9.3.3). It should be noted that with this approach of recording, boundaries of adjacent areas must be recorded twice, which can lead to inconsistencies and errors (cf. Sects. 9.3.2 and Fig. 9.4).

5.2.2 Digital Recording of Geometry Data in Raster Format

The automatic analog-to-digital converters include scanners that transform templates such as pictures, but also ground plans and maps into digital data, thereby breaking down the original template into a fine grid of pixels (Fig. 2.6). After being recorded with a scanner, the data are available in a Cartesian coordinate system with pixel coordinates. Thus, georeferencing to a well-defined reference system, e.g. of the national survey, becomes necessary (cf. Sects. 4.2.5 and 10.6.1.2). Frequently, post-processing is also necessary, such as removing the legend and deleting superfluous information.

It should be emphasised that the pixels are carriers of geometric and topological information as well as of thematic information, which is recorded as brightness or color values! The focus here is initially only on the geometric data. Digital image processing has methods for evaluation the thematic information, with which pattern recognition or classification of the pixels is possible (cf. Sect. 10.7).

5.2.3 Conversion Between Raster and Vector Data

In environmental information systems, data are available in both vector and raster format. Thus, an emission inventory can initially show the various emitters in their original spatial reference (e.g. power plants as point objects, traffic on road segments as line objects or the emitter group domestic heating on building blocks as area objects). Likewise, all calculations of emissions, e.g. based on emission factors from heat equivalents, are carried out on this original spatial reference level. If, however, all emissions of a single air

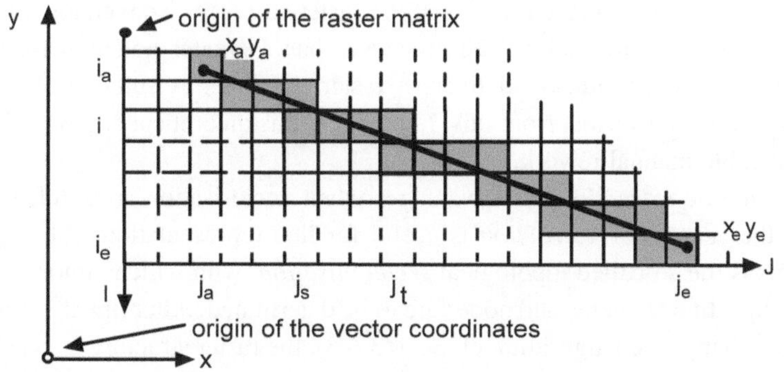

Fig. 5.6 Principle of vector -raster data conversion

pollutant are to be represented together for all emitter groups, a uniform spatial reference basis must be selected, for which a square grid then makes sense. This results in the need to convert the vector data into raster data (so-called *rasterisation*) as well as converting the attribute data (cf. Sect. 9.5.2).

The reverse direction, the conversion of raster data into vector data (so-called *vectorisation),* becomes necessary if further processing as vector data is to be carried out after data recording in raster form (e.g. analog-to-digital conversion with a scanner). Suitable algorithms and functions usually exist in geoinformation systems for both conversion directions. While the vector-to-raster conversion leads to coarser output information and is relatively unproblematic, vectorisation can sometimes lead to ambiguous results and will hardly do without manual post-processing. For example, a pixel may represent a point object as well as a small line segment or an area. The decision must ultimately be made in manual post-processing depending on the context.

With regard to the representation of lines and areas using a raster only the important geometrical conversion of lines should be treated here (for the conversion of attribute data cf. Sect. 9.5.2). Here, the row and column indices of the pixels, which are intersected by a line segment, are determined. Since the start (x_a,y_a) and end coordinates (x_e, y_e) of the segment as well as the pixel size are known, the row and column indices of the start (i_a,j_a) and end pixel (i_e, j_e) can be calculated (cf. Fig. 5.6). For all rows i between i_a and i_e, the column indices j_s and j_t are then to be determined (setting up the linear equation, intersection with the gridlines), so that the pixels between j_s and j_t receive the value 1 and appear “blackened”. In contrast to this simplified representation, it should be noted that the two coordinate systems are oriented differently. Mostly the so-called Bresenham algorithm is used, which is a standard procedure of computer graphics, which is easy to implement, which requires the addition of integers as the most complex operation and which therefore does not require multiplication, division and floating point numbers (cf. Foley et al. 1996 pp. 72).

Compared to vector-to-raster conversion, raster-to-vector conversion plays a more important role in geoinformatics. The procedure has a greater potential for automation. For example, a scanner quickly delivers a raster template. A suitable algorithm could generate data in vector format from this. In practice, this theoretical approach is associated with considerable manual rework.

While *edge line extraction* has advantages when vectorisation areas (cf. Fig. 5.7), the method of the *center line extraction* is useful for line representations (cf. Fig. 5.8). The basis for this is the so-called topological *skeletonisation,* with which, among other things, line beginnings, line elements and nodes are to be determined. After line thinning (e.g. with the common Zhang-Suen algorithm, cf. Sect. 3.4.3), the raster structure is transformed into a set of pixel chains, each representing a line segment (so-called chain coding, cf. Sect. 9. 3.5). Subsequently, each pixel chain is transformed into a sequence of vectors (to the course of the procedure cf. Worboys and Duckham 2004 pp. 208). However, the result of the vectorisation may still have deficiencies, which can be partially eliminated automatically (cf. line smoothing), but which usually require manual correction (cf. Figs. 5.8 and 5.9).

Practical work shows that automated raster-vector conversion requires a clear template that is as simple and unambiguous as possible, i.e. high-contrast line drawings without superimposed text. Complex line structures, as they already exist in a simple development plan with texts, are usually not automatically converted satisfactorily.

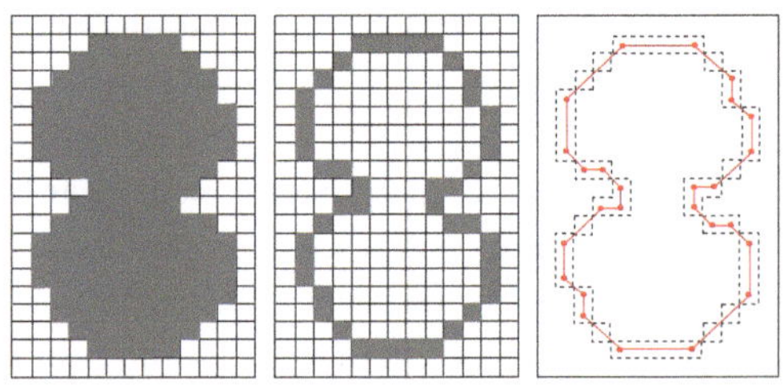

Fig. 5.7 Principle of edge line extraction in raster-to-vector conversion

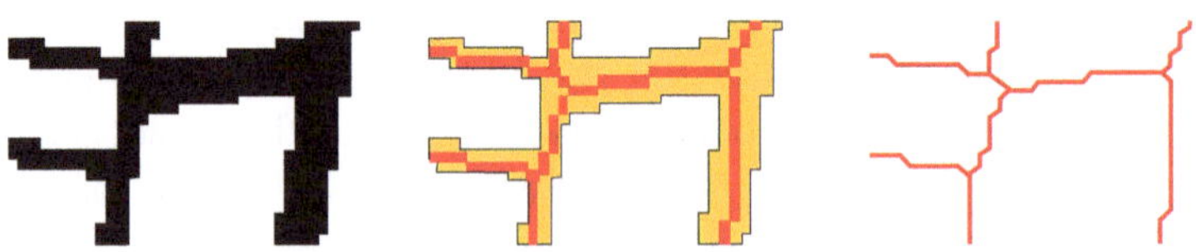

Fig. 5.8 Principle of center line (medial axis) extraction in raster-vector conversion (cf. Hake et al. 2002 p. 258)

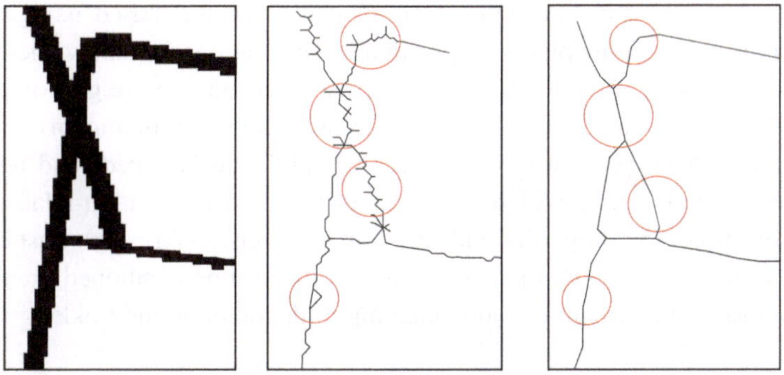

Fig. 5.9 Problems with a raster-vector conversion of lines (cf. Hake et al. 2002 p. 259)

5.3 Satellite-Based Positioning and Recording of 3D Position Coordinates

5.3.1 GPS and GNSS

Satellite-based positioning has become the standard method for determining coordinates and positions in surveying and for navigation (cf. the comprehensive standard work by Bauer 2018). While initially referred to only as the *Global Positioning System (GPS)*, the terminology, content and applications have shifted towards *Global Navigation Satellite Systems (GNSS)*. Satellite-based positioning was almost always equated with the US satellite system NAVSTAR/GPS or just GPS. In the meantime, this equating is no longer appropriate due to other systems such as GLONASS, BeiDou and Galileo. The new term GNSS is intended to make it clear that it is no longer just about determining the locations, but also on the use of this information by navigation systems for persons, vehicles, ships and aircraft. It is precisely these applications that place further demands on the positioning systems, which go beyond the geometric accuracy of the position determination to be achieved (cf. Sect. 5.3.4).

Currently, NAVSTAR/GPS, GLONASS, BeiDou and Galileo are the only fully global operational navigation satellite systems. In the meantime, many receivers simultaneously use GPS, GLONASS, Galileo and already recently also BeiDou signals. On the evening of June 21st 2022, 11 GPS, 7 GLONASS, 4 Galileo and 8 BeiDou satellites were available over Beckum, a small town in North Rhine-Westphalia in the middle of German, whose signals could even already be evaluated by a smartphone. With NAVIC (Navigation with Indian Constellation), a regional satellite-based positioning system covering the Indian subcontinent was established in 2016.

The possible uses of satellite-based navigation and positioning are enormously diverse. The classic area of application is surveying. This also includes the exact determination of control points, i.e. the determination of reference points for rectification and geocoding for aerial photo evaluations during flight. Commercially, satellite-based navigation has the greatest significance in route planning, positioning of locations for land vehicles, ships and aircraft. Mobile, location-based data recording can also be cited (e.g. biotope mapping, taking water samples in a lake without any reference points for orientation). In agriculture, new applications have become possible, such as application of seeds and fertilizers to a specific small-scale location, which in turn requires the exact location-related determination of, for example, soil properties and crop yields (precision farming). Last but not least, application examples in the tourism and leisure sector can be mentioned (general orientation and compass replacement or route planning in the outdoor and trekking sector).

5.3.2 NAVSTAR/GPS: Basic Principles Before Modernisation

NAVSTAR/GPS (Navigation Satellite Timing and Ranging Global Positioning System), commonly used mostly under its simplified name GPS, is a satellite-based navigation system developed, operated and controlled by the U.S. military to provide immediate positioning of any object anywhere in the world . Its development began in 1978 with the launch of the first GPS satellite and can be considered complete since 1993, when 24 GPS satellites were in orbit and available for navigation, while the US Air Force only declared in 1995 that GPS was full (militarily) operational (for history cf. NASA 2012).

The US Global Positioning System consists of several segments (cf. NASA 2014):

The *space segment* comprises at least 24 satellites orbiting the earth at an altitude of about 20,200 km in six orbits with an orbital period of about 12 hours. However, there are always more satellites in orbit, since in addition to reserve satellites, new satellites are launched to replace old ones. The orbits are inclined at 55° to the equatorial plane. This constellation means that between four and eight satellites are visible high enough (more than 15°) above the horizon for any point on earth. The system has now been modernised. The older generation includes the so-called Block IIA and Block IIR satellites, of which none and 7 respectively were still operationally in orbit in February 2022 (cf. GPS.gov 2023a).

The *control segment* consists of a global network of ground facilities with a master control station in Colorado, an alternate master control station in California, and 11 command and control antennas and 16 monitoring sites (cf. GPS.gov 2023b). The monitor stations observe the satellites and calculate their orbits. The master station uses the data from the monitor stations to compile the so-called navigation message with precise flight path data and data on the satellite clocks, which are transmitted to the satellites (for sending to the users' receiver stations).

The *user segment* consists of the GPS receivers and the users. The GPS receivers determine the exact location on earth from the signals transmitted by the satellites. At least four satellites are needed simultaneously for three-dimensional positioning in real time (cf. Sect. 5.3.3). It should be emphasised that GPS is a passive system which only receives data.

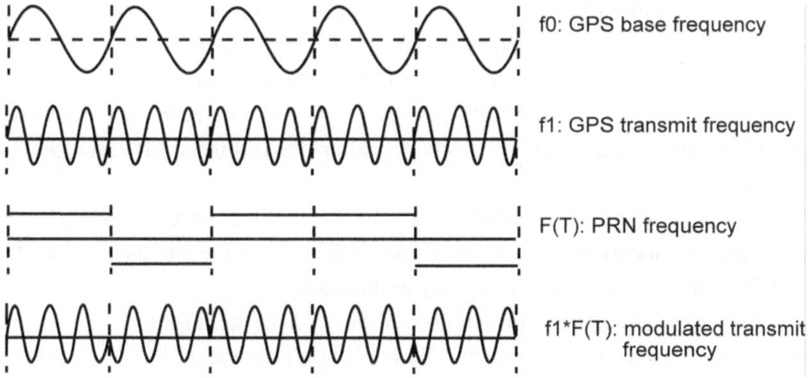

Fig. 5.10 Modulation of signals

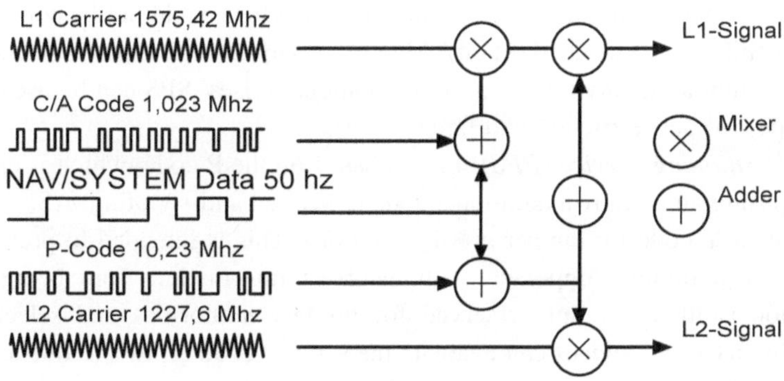

Fig. 5.11 Basic principle of GPS satellite signals (cf. Dana 2023)

In accordance with the principle commonly used in radio technology, the satellite information is transmitted to the GPS receiver via suitable carrier waves. The so-called L1 carrier wave (1575.42 MHz) is available for civil, free use. For military use, the L2 carrier wave (1227.60 MHz) is also available. Within the scope of GPS modernisation, further frequency bands will be added for civilian use (cf. Sect. 5.3.6). A regular signal (code) is superimposed on the carrier wave by phase modulation (for the basic principles cf. Figs. 5.10 and 5.11; for further details cf. the very detailed description in Bauer 2018 pp. 142). The objectives of modulation are:

- The signals are structured so that the problem of measurement ambiguity is solved.
- The signals carry the navigation message (orbit data of the satellites and additional information).
- The codes used for modulation (so-called pseudo-random noise codes, PRN codes) are kept secret so that only desired users can benefit from GPS.

The carrier waves are modulated by three different binary codes (for the signal structures of NAVSTAR/GPS before modernisation cf. Bauer 2018 pp. 310):

The modulation with the *C/A code* (C/A = Coarse Acquisition) is a generally accessible coding, so that anyone can use the GPS signals – with limited accuracy – ("coarse code"). The C/A code repeats itself every 1023 bits (i.e. after one millisecond) and modulates only the L1 carrier wave.

The *P code* (P = Precise) modulates both the L1 and L2 carrier waves and enables a particularly precise location. It is a very long PRN code (seven days). Its decryption requires authorisation from the US military authorities.

The *D-code* (D = Data) transmits the actual navigation messages.

Global Positioning System accuracy falls into two categories:

- *Standard Position Service* (SPS): SPS is based on the C/A code. The accuracy was intentionally reduced by the US Department of Defense (so-called *Selective Availability, SA).* Thus, the navigation message of the satellites was garbled by fluctuations in the signal. These manipulations were irregular and therefore could not be corrected. Selective Availability was switched off on 1 May 2000. However, it can be switched on again by the US military at any time, e.g. in international crises. SPS can be used by anyone worldwide without restriction and free of charge.
- *Precise Positioning Service (PPS):* PPS is based on the P code, which is encrypted by superimposing an unknown additional code (so-called anti-spoofing, AS, generation of the so-called Y code for tamper-proof operation). This is intended to prevent a signal sent out by a military opponent with incorrect information from being evaluated. Decryption is therefore only reserved for military authorities and selected civilian users. Military GPS devices can evaluate the same P-code on two frequencies (L1 and L2) by dual-frequency measurement and therefore achieve a navigation accuracy of 0.8 to 1 m. However, the signals transmitted with the encrypted codes can now also be used for civilian applications and for dual-frequency GPS devices, but without being able to assess whether incorrect data are present.

The carrier frequencies or the C/A and P codes are used to determine the propagation time of the signals between the receiver and the visible satellites in order to calculate the distance (cf. Sect. 5.3.3). At the same time, which is ultimately the genius of the approach, the D-code is used to transmit the actual navigation message, which primarily contains the position of the satellites. From these two information, distance to fixed points (in this case satellites) and position of the fixed points, the unknown position of the receiver can be determined. In the two-dimensional case, where the distances r_1 and r_2 to two known fixed points are given, this principle can be understood very easily (cf. Fig. 5.12). The circles with radii r_1 and r_2 around the fixed points intersect in the initially unknown location.

Each satellite transmits a so-called *data frame (*the navigation message*)*, which consists of five *subframes*, each of which requires 6 seconds for transmission. Each subframe begins with a so-called telemetry word and a transfer word, which are the same in all subframes.

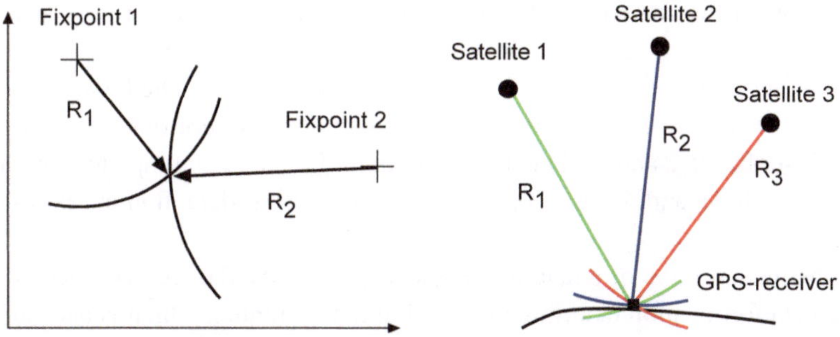

Fig. 5.12 Principle of the location determination in the two- and three-dimensional case with known fixed points and known distances to the fixed points

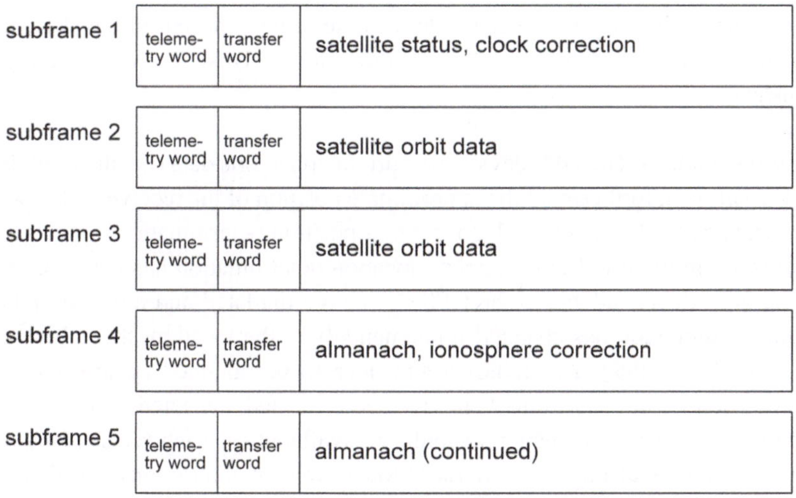

Fig. 5.13 Basic structure of the navigation message (dataframe)

These are generated individually in each satellite and are used, among other things, for identification (cf. Fig. 5.13). The remaining information (e.g. describing the satellite orbit) is sent from the main control station to the satellite, where it is implemented in the five subframes, modulated onto the carrier frequency and transmitted. In the GPS receiver, the navigation message, which is modulated onto the carrier, is extracted again (cf. Bauer 2018 pp. 308 and Dodel and Häuptler 2009 pp. 186).

The first three subframes are always the same for a single satellite (cf. Fig. 5.13). The first subframe contains, among other things, information on the status and accuracy of the satellite (status of signal transmission, status of the orbit), actuality of the ephemeris (i.e. the mathematical description of the orbit) or clock corrections. The second and third subframes contain parameters of the ephemeris. Part of this data is reserved for military users.

Only the fourth and fifth subframes contain the data of the so-called *almanac,* which has information about the orbital parameters of all satellites, i.e. not only of the transmitting satellite (cf. subframe 2 and 3). Due to the amount of data, the (complete) almanac cannot be sent as a whole and is split up into 25 consecutive data frames, partitioned into subframes 4/5.

Since the transmission of a data frame takes 30 seconds, the receiver receives the most important data for location determination within half a minute, which is also sufficient for real-time navigation of vehicles. However, the complete transmission of the entire almanac (master frame) requires $25 \bullet 30\ sec = 12.5$ minutes due to the partition over 25 data frames.

TTFF (time to first fix) describes the time required for a receiver to determine its position for the first time after it has been switched on. This time essentially depends on how up-to-date the information stored in the receiver about the visible satellites is. On the one hand the GPS device needs information, which satellites are to be expected in the sky, and on the other hand the orbital data of these satellites themselves. Three situations can be distinguished:

- *Hot start* (standby): The GPS device has current ephemeris data (i.e. the orbit data of the visible satellites), furthermore the approximate position of the receiver is known, so that a few seconds (< 15 s) are needed for a first position determination.
- *Warm start:* In the last 4 to 6 hours no location determination has been carried out, so that the orbit data available in the GPS device are outdated and must be updated. The satellite constellation has changed fundamentally at the location, the ephemeris of the (new) visible satellites are unknown and have to be obtained again. Since the most important data are transmitted in 18 seconds and repeated every 30 seconds (cf. Fig. 5.13), current ephemeris should be available for each received satellite after about 48 seconds at the latest, if reception is free of interference. In this case it is assumed that the almanac is still up-to-date. The GPS receiver can use the almanac data (together with the local time and the rough country identification as basic settings of the receiver) to determine the satellites, which are expected to be approximately overhead. The range determination can then be restricted to these satellites, which ultimately speeds up the location determination. The more new satellites are visible, the longer this warm start will take.
- *Cold start:* If neither current ephemeris nor almanac data are available, the complete reception of the entire almanac must be awaited. This cold start then takes 12.5 minutes.

However, this information is almost only of theoretical interest. Thus, methods exist to shorten the TTFF considerably. The techniques designated A-GNSS for Assisted GNSS use additional data that are provided via the Internet, WLAN or mobile communication. After switching on a smartphone with an integrated GNSS receiver and transmitting information from the mobile network, an approximate determination of the receiver position becomes possible in a few seconds. Furthermore, there are manufacturer-specific, partly patented algorithms for the pre-calculation of the orbit data (in sleep mode when the device is switched off).

5.3.3 Distance Determination: Principle

The distance measurement is carried out using a special time determination. The carrier frequency, including the modulated P or C/A codes, which are generated in the satellite and are different for each satellite, are also generated synchronously in the same way in the receiver on earth. The satellite signals picked up by the receiver (and corrected for Doppler shift) show a time shift compared to the signals generated in the receiver (cf. Fig. 5.14). The GPS receiver compares the two signals and can determine the distance to the satellite by measuring the transit time of the signal.

The time ΔT is proportional to the distance between satellite and receiver. However, the product $\Delta T \bullet c$ (c velocity of light in vacuum) is not the actual distance. The product $\Delta T \bullet c$ is therefore called the *pseudo-distance*. But the receivers with their quartz clocks make errors in timekeeping compared to the atomic clocks of the satellites.

$$R = (\Delta T + \Delta t) \cdot c$$

with:

ΔT: the time difference measured in the receiver
c: the velocity of light in vacuum
R: the distance (range) receiver to satellite
Δt: the error of the receiver clock with respect to the satellite clock

In total, there are four unknowns (X_e, Y_e, Z_e, Δt), so that a system of equations with four equations must be set up. Thus, at least four pseudo-ranges to different satellites are needed to determine the location:

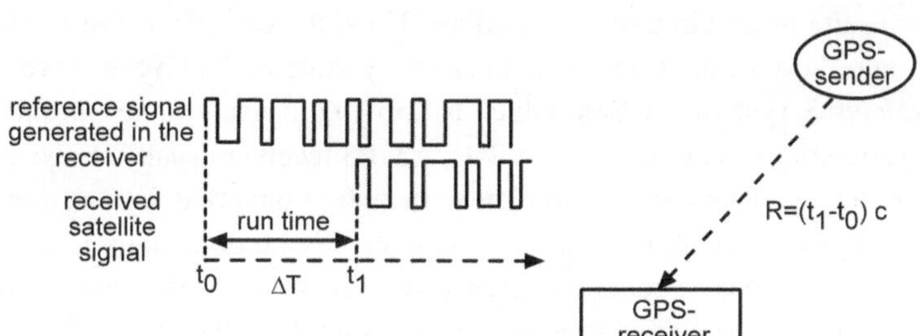

Fig. 5.14 Principle of distance determination via signal transit times

$$[(\Delta T_i + \Delta t) \cdot c]^2 = (X_i - X_e)^2 + (Y_i - Y_e)^2 + (Z_i - Z_e)^2$$

with:

ΔT_i:	the measured transit times of the satellite signals
c:	the velocity of the signal in vacuum
X_i, Y_i, Z_i:	the known coordinates of the satellites
X_e, Y_e, Z_e:	the unknown coordinates of the receiver
Δt:	the unknown time error of the receiver

This non-linear system of equations provides the basis for calculating the receiver coordinates (cf. Fig. 3.10 and Sect. 3.3.3.1 for the calculation principle), although in practice the distance determination is much more extensive. Usually distances to more than four satellites are present, so that it is more complex to approximate the solution of this overdetermined system. Furthermore, the distance determination shown is not unambiguous. For example, the C/A code repeats after just one millisecond, which at the velocity of light of 300000 km/s corresponds to a distance of 300 km, so that basically the distance to the satellite can only be determined up to a maximum of 300 km. However, the satellites are more than 20000 km away. Therefore, in the initialisation phase, approximate location coordinates within several hundred kilometers of the receiver are assumed (basic setting of the receiver, for further evaluation methods and for setting up further distance equations, including consideration of the earth's rotation during the transit time of the signals, cf. Hofmann-Wellenhof et al. 2008 pp. 238, Bauer 2018 pp. 213).

The Global Positioning System is based on a Cartesian, spatially fixed coordinate system, whose origin is the earth's center of mass *(Conventional Inertial System, CIS).* The satellites move precisely around this centre. The calculations are therefore initially carried out in this CIS. The z-axis points in the direction of the earth's mean axis of angular momentum, the x-axis points in the direction of the vernal equinox. However, this spatially-fixed CIS is not suitable for location determinations on earth. This requires a so-called *Convential Terrestrial System (CTS),* i.e. a reference system that is fixed to the earth, i.e. one that rotates with the earth (also commonly used name: *ECEF, Earth Centred Earth Fixed System*). The z-axis is identical to the z-axis of the CIS, but the x-axis is associated with the mean Greenwich meridian. Thus, the coordinates of the CIS must be transformed into those of the CTS, which is already done in the GPS receiver. The global geocentric reference system WGS84, which is firmly connected to the earth, is used as the Convential Terrestrial System (cf. Sect. 4.2.4). A GPS receiver usually shows exactly these WGS84 coordinates. This information must then be converted into a user coordinate system (e.g. UTM coordinate system). This can already be done automatically in simple GPS receivers, although the geodetic reference system as well as the geodetic datum of the national survey must be specified in advance (cf. example at the end of Sect. 4.3.1).

5.3.4 Error Influences and Accuracies of a GPS Location Determination

The possible error influences can be differentiated according to path, propagation and receiver errors. Selective availability, which led to an artificial degradation of the navigation message and therefore inevitably to inaccuracies, has been switched off in the meantime. However, in addition to unspecific error sources such as general hardware and software errors, there are still system-related error influences (cf. Dodel and Häupler 2009 pp. 187–188):

- ionospheric delays (about up to 10 m) due to attenuation/refraction of waves in the ionosphere
- error of the satellite clocks that are not corrected by the control stations (about 1 m). Thus, the satellite clocks go slower due to their faster movement compared to the earth, but faster than on earth due to the lower gravity
- ephemeris data errors orbits (about 1 m)
- tropospheric delays (about 1 m) through attenuation of wave propagation due to weather phenomena in the troposphere
- delays due to multipath effects (about 0.5 m) due to reflection of satellite signals on surfaces or buildings in the vicinity of the receiver

The user can assume that these problems are partly solved by the satellite receiver (cf. corrections of the propagation of the waves in the ionosphere and troposphere). The accuracy of the position determination depends not only on the visibility of the satellites but also on the position of the satellites relative to one another and to the unknown location. This deviation, which is lowest for a two-dimensional positioning at a satellite separation angle of 90°, can be illustrated relatively easily for the two-dimensional case using Fig. 5.15. The receiver is located within the displayed error areas (cf. in more detail and for the three-dimensional case Mansfeld 2010 pp. 193).

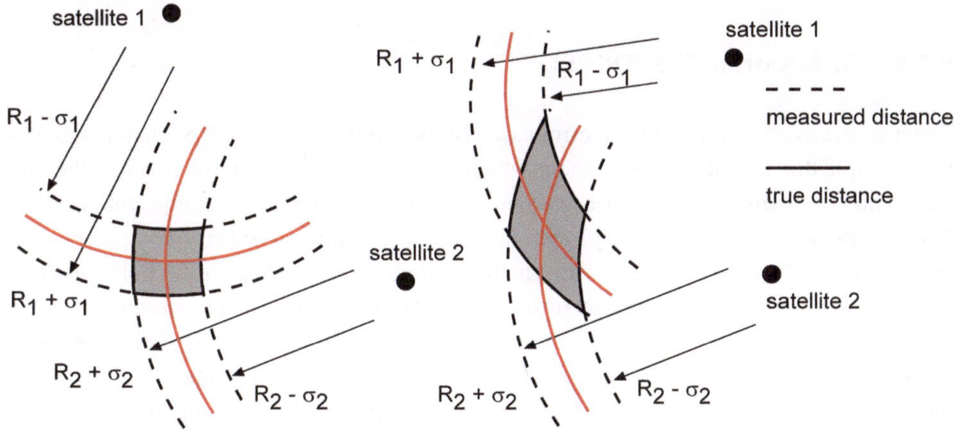

Fig. 5.15 Position errors and error areas in location determination based on two distance measurements with uncertainty for two different satellite constellations (cf. Flühr 2010 p. 103)

The constant distance measurement to a stationary receiver fluctuates due to the constantly changing satellite constellations and the listed error influences. The distance measurement error can be described by the standard deviation σ_R of the distance measurements. For the same reasons, the calculated position of the receiver varies and so does the associated position error, expressed by the standard deviation σ_P of the position determinations.

The term "Dilution of Precision" (DOP) has been introduced for the factor of the relative increase or decrease of the position error.

The DOP factor indicates the factor, by which the position error increases compared to the error of the measured distance. DOP is dimensionless and in no way describes the average deviation from the receiver's true position.

$$\text{DOP} = \frac{\text{standard deviation of the position error } \sigma_P}{\text{standard deviation of the distance measurement error } \sigma_R}$$

Further factors are:

- PDOP (Position Dilution of Precision) for three-dimensional position determination
- HDOP (Horizontal Dilution of Precision) for two-dimensional position determination in the horizontal plane
- VDOP (Vertical Dilution of Precision) for position determination in the vertical plane
- TDOP (Time Dilution of Precision) for the time deviation
- GDOP (Geometrical Dilution of Precision) denotes the error specification of the entire system. A value of 1 indicates the best possible arrangement, a value of 6 is still good, while values above 10 do not allow any evaluation:

$$GDOP = \sqrt{PDOP^2 + TDOP^2}$$

5.3.5 Differential GPS (DGPS)

The increase in accuracy of the determination of the Global Positioning System is based on the principle that the GPS measurements at adjacent receiver points have errors of the same size. Thus, two simultaneous measurements are performed with two adjacent receivers. The coordinates of the stationary reference station are known exactly compared to the measurement of the second receiver, which is usually used mobile in the field (so-called rover). Thus, the difference between the (currently) measured pseudo distances and the exact calculated, geometric distances to the satellites can be determined. These difference values (correction values) are transmitted to the receiver (rover station), which calculates the corrections of the pseudo distances. Due to this error correction in the so-called

Differential GPS (DGPS) a very high level of accuracy can be achieved for single point measurements.

On the one hand, reference stations can be stationary and immovable, e.g. mounted on roofs, i.e. very often on the roofs of the local surveying and cadastral offices, and on the other hand they can be stationary in the field for a single measuring campaign, but can be generally mobile (cf. also mobile reference stations in precision farming).

There are generally two different methods for differential GPS:

- real-time differential adjustment by transmitting the correction signals directly to the field receiver via a radio data link or via the Internet,
- post-processing differential adjustment after field measurements.

The surveying authorities of the Bundesländer of the Federal Republic of Germany offer the *satellite positioning service SAPOS* (cf. SAPOS 2023a, in English SAPOS 2023b). SAPOS provides correction data that can be used to increase positional accuracy down to the millimetre range, depending on the effort involved. This service is based on a network of permanently measuring reference stations, which are set up and operated by the respective surveying authorities. The data can be used uniformly throughout Germany. The SAPOS services include three categories, each with different characteristics, accuracies and costs (cf. for North Rhine-Westfalia SAPOS 2023c):

- EPS (real-time positioning service) enables an accuracy of 0.3 to 0.8 m (horizontal) and 0.5 to 1.5 m (ellipsoidal height). The DGPS correction data are transmitted every 1 second in the standardised RTCM format via the Internet (so-called NTRIP procedure).
- HEPS (High Accuracy Real-Time Positioning Service) provides an accuracy of 1 to 2 cm (horizontal) and 2 to 3 cm (ellipsoidal height). In addition to the correction data of the EPS, further data are made available to the user in real time. The correction data are transmitted every 1 second via the Internet (NTRIP procedure) and mobile phone (GSM standard, Global System for Mobile Communication) in the standardised RTCM format.
- GPPS (Geodetic Postprocessing Positioning Service) enables accuracies in the centimeter or millimeter range in position and height. These data are made available to the user in the standardised RINEX format by e-mail, via the Internet (webserver) or on data carriers.

It can be assumed that high-quality geodetic receivers can receive parallel satellite signals from several satellite systems (GPS, GLONASS, Galileo) as well as correction signals. The state of Lower Saxony will provide all SAPOS services free of charge from October 2019.

5.3.6 NAVSTAR/GPS: Modernisation

As part of the GPS modernisation, the space and ground segments are being renewed. In addition to an M signal reserved exclusively for military applications, three new civilian signals have been added (L1C, L2C, L5). In Table 5.1; the additional signals on the modernised satellites are highlighted in grey. The new signals require new hardware on the satellites, so that a parallel renewal of the space segment is also necessary (modernised satellites Block IIR-M, Block III satellites, cf. Table 5.1). The selection of suitable frequencies was determined, among other things, by avoiding interfering influences. For example, ionospheric delays in signal transmission are to be reduced. Furthermore, the signal propagation should not be disturbed by weather influences. Above all, the propagation speed of the (electromagnetic) signals should deviate as little as possible from the velocity of light (cf. GPS.gov 2023a).

L2C is the second civilian GPS signal designed specifically for commercial purposes. In combination with the older L1 signal (i.e. L1 with C/A code), civilian users can make dual-frequency measurements without restrictions by the U.S. military, provided that the receiving device is also capable of doing so. Then the ionospheric propagation delays can be determined and taken into account, so that the same accuracy can be achieved for civilian as for military applications. L5 is the third civilian GPS signal to meet the high demands of traffic safety and other high performance applications. L1C is the fourth civil GPS signal that enables interoperability between GPS and international satellite navigation systems, while preserving U.S. national security interests.

Table 5.1 Overview of the evolution of GPS signal types and signals (cf. GPS.gov 2023a)

		Satellite type			
			Modernised		
Signal		II R	II R-M	II F	III
L1	C/A	•	•	•	•
	P	•	•	•	•
	M		•	•	•
	C				•
L2	P	•	•	•	•
	C		•	•	•
	M		•	•	•
L5	C/A			•	•
		1997–2004	2005–2009	2010–2016	Since 2018
Launched/Operational:		7	7	12	4

5.3.7 GLONASS

Parallel to the US-GPS, the Russian *Globalnaya Navigatsionnaya Sputnikovaya Sistema* (*GLONASS*), which is very similar in structure and function, was developed and continues to be operated by the Ministry of Defence of the Russian Federation. Construction began in 1972, launch of the first satellites was in 1982, and full deployment was achieved in 1996 with 24 satellites (including three reserve satellites). As of summer 2022, 25 satellites were in constellation (22 in operation, 3 satellite in maintenance (cf. in English IAC 2023).

While in NAVSTAR/GPS all satellites use the same frequencies but with different codes, in GLONASS all satellites transmit with the same code but on different frequencies. Both a precise and a coarse code can be used by civilian receivers, whereby the precise one can be changed to another (secret) code at any time for military reasons. Thus, dual-frequency measurements are also possible for civilian applications.

5.3.8 Galileo

The global satellite system Galileo, which is the European Union's Global Satellite Navigation System (GNSS), is interoperable with GPS and GLONASS, but is under civilian control. The full system will consist of 24 satellites and an additional 6 reserve satellites orbiting the Earth at an altitude of 23,000 km in three orbital planes at an inclination of 56° to the equator (cf. EUSPA 2023a). As of spring 2022, the constellation consisted of 21 usable satellites plus two satellites in the test phase (cf. EUSPA 2023b). The ground segment consists of two control centres (Galileo Control Centres (GCC) in Oberpfaffenhofen/Germany and in Fucino/Italy) as well as a worldwide network of Ground Control Segments (GCS) and Ground Mission Segments (GMS) and other service facilities (cf. EUSPA 2023c).

The operating principle of GPS and Galileo is basically identical. Galileo uses three frequency bands E1, E5 and E6, whereby E1 or L1 and E5 or L5 are (partly) shared by GPS and Galileo (cf. Bauer 2018 pp. 138). The Galileo system, once fully operational, will offer six high-performance services worldwide (cf. EUSPA 2023d):

- Open service (OS): Galileo offers a free and unencrypted service. The signals are to be transmitted in two frequency ranges as standard, so that positioning accuracy in real time down to the metre range can be achieved with appropriate dual-frequency receivers.
- Open Service Navigation Authentication (OSNMA): Galileo offers a free access service complementing the OS by providing authenticated data, assuring users that the Galileo navigation message they receive is from the system itself and has not been modified.

- High accuracy service (HAS): In addition to the OS, other navigation signals and services (e.g. weather warnings) are to be provided. The HAS signal is to be encrypted in order to control access to the chargeable HAS services. HAS results from the redesign of the former Commercial Service
- Public regulated service (PRS): This encrypted and chargeable service is intended to be restricted to users authorised by government agencies with regard to sensitive applications that require a high level of quality and integrity as well as continuity even in crisis situations.
- Commercial Authentification Service (CAS): This service complements the operating system and provides users with controlled access and authentication functions.
- Search and rescue service (SAR): This service will be Europe's contribution to COSPAS-SARSAT, an international satellite-based search and rescue alert detection system.

5.3.9 BeiDou

China's BeiDou positioning system (BDS, BeiDou Navigation Satellite System) has expanded tremendously in recent years. In March 2020, the 54th satellite was launched (BeiDou Navigation Satellite System 2020). The (preliminary) system was already globally operational with 19 satellites by the end of December 2018. A three-stage development strategy was pursued (cf. China Satellite Navigation Office 2018):

- Completion of BDS-1 to provide a regional positioning service for the country of China by the end of 2000.
- Completion of BDS-2 to cover the Asia-Pacific region by the end of 2012
- Complete BDS-3 to provide a global service by the end of 2020. The constellation will then consist of 27 MEOs (satellites in medium earth orbit, such as GPS and GLONASS satellites), 5 GEOs (satellites in geosynchronous orbit) and the 3 IGSOs (satellites in geostationary orbit) that have existed since phase 1 (cf. BeiDou Navigation Satellite System 2020).

The BeiDou positioning system will have several special features (cf. BeiDou Navigation Satellite System 2023):

- The space segment will consist of satellites in three types of orbits. Compared to other navigation satellite systems, more satellites will be in high orbits to provide better coverage especially in low latitude areas.
- BeiDou will provide navigation signals in multiple frequencies to improve positioning accuracy by using combined multi-frequency signals.
- BeiDou will integrate navigation and communication functions and will have multiple service capabilities: positioning, navigation and timing, short message communication,

international search and rescue, satellite-based augmentation, ground augmentation and precise point positioning.

As of August 2020 the actual measured average global positioning accuracy of BDS is 2.34 m (cf. BeiDou Navigation Satellite System 2020). Data transmission is to take place over three frequencies. More detailed information is available on signal B1 for the open service (cf. China Satellite Navigation Office 2019 u. Bauer 2018 pp. 360). Two global positioning services are planned, an open service (OS) and an authorised service (AS), in addition to several regional services (cf. China Satellite Navigation Office 2018 p. 7).

5.3.10 GNSS Data

The National Marine Electronics Association (NMEA) has developed an exchange format for communication between navigation devices on ships. The data records are in ASCII format. A record starts with a "$" or "!" followed by a device ID (usually two characters), a record ID (usually three characters) and comma-separated data fields according to the record definition. In the context of GNSS, several device IDs such as BD (Beidou), GA (Galileo), GL (Glonass), GP (GPS) as well as several record IDs are relevant. Here only the Global Positioning System Fix Data (GGA) will be discussed in more detail (cf. NMEA 2023 in more details). The line:

$GPGGA,110727.00,5217.054002,N,00801.403383,E,1,29,0.3,72.9,M,46.5,M,,

describes a point on the campus of the University of Osnabrück, Germany, taken at 11:07:27 UTC with longitude 52° 17. 054002' N and latitude 8° 1.403383′ E. The subsequent "1″ in the data set indicates the reception quality (fix quality with e.g. 0 = invalid, 1 = GPS fix, 2 = DGPS fix, 3 = PPS fix, 4 = Real Time Kinematic, 5 = Float RTK). 29 satellites were recorded, HDOP was 0.3. The high of the antenna was 72.9 m above mean sea level (i.e. above the geoid). The difference between the mean sea level and the WGS84 ellipsoid was 46.5 m. An NMEA file contains further information on the satellites involved, such as the number of a satellite, its altitude and azimuth (each in degrees) and the indication of the system (e.g. GPS or GLONASS).

The open and license-free GPS Exchange Format (GPX) was originally developed to store GPS data and is now used for the general exchange of geodata. The structure of the much simpler *XML* file can be read almost intuitively:

```
< trk>
 < trkseg>
  < trkpt lat="52.284218" lon="8.02304">
   <ele>75.0</ele>
   <time>2019-10-01T11:29:51.999Z</time>
  </trkpt>
```

```
    < trkpt lat="52.28429" long="8.023027">
      <ele>74.2</ele>
      <time>2019-10-01T11:29:52.999Z</time>
    </trkpt>
    < trkpt lat="52.28429" long="8.023027">
      <ele>67.9</ele>
      <time>2019-10-01T11:30:00.999Z</time>
    </trkpt>
...
  <trkseg>
<trk>
```

5.3.11 Evaluation of Positioning Systems and Further Development of GNSS

The quality of the location determination essentially decides the applications, whereby the geometric position accuracy is an important, but only a first feature. Thus, several criteria describe the overall reliability (i.e. system performance, cf. Dodel u. Häuptler 2009 pp. 302):

Liability (legal basis) aims above all at political independence (i.e. also independence from military decisions) and unrestricted insight into the system.

Integrity comprises two sides: Technical integrity describes the ability to verify the integrity and credibility of data within a very short period of time. Almost more important is the institutional integrity, how quickly a system operator reports satellite malfunctions and makes corresponding commitments. Inevitably, military operators fail to meet these obligations.

Availability describes, how often a positioning system is available worldwide.

Continuity indicates, how stable the data supply is and whether a service is available continuously (e.g. during the entire landing approach of an aircraft).

The achievable positioning *accuracy* essentially depends on the frequency of the emitted signals (influence of ionospheric errors) and the consistency of the on-board clock systems.

Furthermore, there must be *robustness* against interference and jamming.

Finally, *affordability (usage cost) must* also be considered in the overall evaluation.

The accuracy and reliability associated with NAVSTAR/GPS and also with GLONASS are not sufficient for safety-critical applications. For example, the exact location is not secured during an entire flight and especially in the landing phase. However, it should not be forgotten that GPS originally had purely military objectives. The improvement (so-called augmentation) of the satellite systems is to be understood as a civilian response to the restrictions imposed by the military. Basically, a distinction can be made between Ground Based Augmentation Systems (GBASs), which are the classic DGPS applications, and Space Based Augmentation Systems (SBASs).

The U.S. Federal Aviation Administration was an early developer of the *Wide Area Augmentation System (WAAS),* which, among other things, increases the accuracy and availability of GPS signals. In addition to the basic GPS, GPS services are strengthened and improved in a large area (cf. FAA 2023a). WAAS is designed to meet the highest safety standards. Within 6 seconds users are informed about disturbances that can lead to an incorrect location determination. Thus, there is a very high reliability for the calculated position of the GPS/WAAS receiver. In the spring of 2022, almost 2000 U.S. airfields could be approached by aircraft with a WAAS-enabled GPS receiver (cf. FAA 2023b).

The Wide Area Augmentation System (WAAS) uses a network of 38 reference stations in the USA that receive GPS satellite signals. These stations, for which precise position determinations have been carried out and which are interconnected, calculate, among other things, the existing position error. WAAS corrects the GPS signal errors resulting from tropospheric and ionospheric disturbances, timing errors, and satellite orbit inaccuracies, and also provides information on the satellite integrity. Three master stations collect data from the reference stations and compute a GPS correction message. The correction signals are transmitted to three geostationary satellites. From there, the signal is transmitted to the GPS receivers (cf. FAA 2023a).

The European Geostationary Navigation Overlay Service (EGNOS) is the regional satellite-based augmentation system (SBAS) in Europe (cf. EGNOS 2023a). The functional principle of EGNOS corresponds to that of WAAS, but here the signals from GPS and GLONASS as well as Galileo satellites are processed. The ground segment consists of a network of 40 monitoring stations (RIMS, Ranging Integrity Monitoring Stations), two control centres (MCC Control and Processing Centres) and two ground stations for each of the three geostationary satellites. The dense network of 40 precisely surveyed ground stations receives the (raw) GPS data, which is forwarded to the control stations. These compute the correction messages, which are then finally transmitted to the geostationary satellites, from where the information is routed to the receiver. EGNOS offers three services. The Open Service, which is available free of charge, has been available since October 2009 (cf. EGNOS 2023b). The Safety of Life Service, which has the same accuracy as the Open Service, but additionally provides integrity information, has been in existence since March 2011 (cf. EGNOS 2023c). Since July 2012, the EGNOS Data Access Service has been providing correction data via terrestrial networks (Internet) for registered users, who cannot always see the EGNOS satellites, such as in urban street canyons, and who wish to use additional services (cf. EGNOS 2023d).

In addition to WAAS for North America and EGNOS for Europe, an Indian and a Japanese satellite-based augmentation system are in operation (GAGAN and MSAS respectively), which correspond to the standard developed for civil aviation. Further systems are under development (including the Russian SDCM, System for Differential Corrections and Monitoring, cf. Bauer 2018 pp. 389).

5.4 Airborne Laser Scanning

LiDAR (Light Detection and Ranging) is an active remote sensing technology that basically works like an echo sounder. The distance measurement is based on measuring the time that elapses between the emission of the light pulse and the returning, reflected pulse. Other acronyms are used which have slightly different meanings: LaDAR (Laser Detection and Ranging) implies that laser light is used to measure distance. *ALS (Airborne Laser Scanning)* concretizes that a laser beam scans the surface from an aircraft. This technique in particular, which is used to create digital surface models, has become very important in geoinformatics (for an introduction cf. Heritage and Large 2019, Vosselman and Maas 2010, Shan and Toth 2008 and in detail Pfeifer et al. 2017).

The measuring instruments mounted vertically under the carrier aircraft emit a pulsed or continuous laser pulse which, at a flight altitude of e.g. 1000 m and the velocity of light, requires 6.671 microseconds for the entire measurement distance (cf. Fig. 5.16). The time elapsed between sending and receiving the signals is used to determine the distance between the sensor and the surface.

However, this only determines the distance between the objects on earth and the measuring device. Thus, the exact position of the aircraft, i.e. its position in a three-dimensional reference system, must also be known. This location determination is made possible with the help of the Differential Global Positioning System (DGPS). Since the aircraft continues to fly during a measurement and is also not fixed located vertically above the terrain surface, deviations due to the movement and attitude of the carrier aircraft must also be taken into account. For this purpose, the third component is a so-called *Inertial Navigation System* (INS), which determines the three flight parameters based on laser technology (therefore also called LINS): rotation around the x-axis running in the direction of movement (roll), rotation around the y-axis running perpendicular to the x-axis (pitch) and rotation around the z-axis perpendicular to the x-y plane (yaw). From the exact position of the aircraft and the distance between the aircraft and a point on the surface the position of this point can be determined. The ALS method finally provides a 3D surface model that

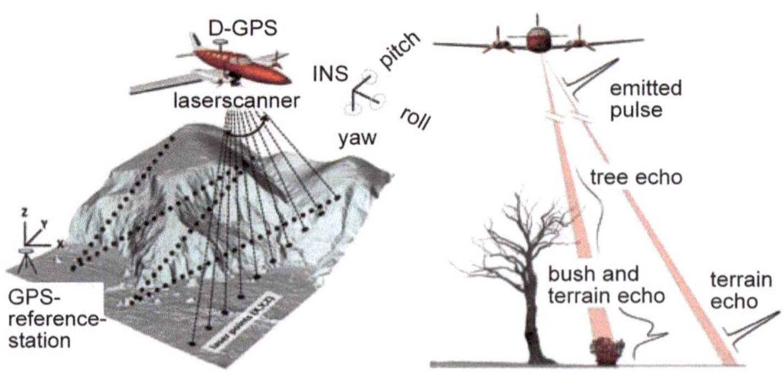

Fig. 5.16 Airborne laser scanning: principle and system components (cf. Höfle and Rutzinger 2011 p. 4)

consists of a set of measuring points. Each measuring point only has information about its position and height (x-, y-, z-coordinates). It is not recorded whether the measuring point is located on a tree or a roof, for example. The positional accuracy of the 3D coordinates depends on the accuracy of the individual system components and is less than 40 cm horizontally and less than 10 cm vertically.

Individual objects can generate more than one echo. For example, a tree canopy can generate a first echo (so-called first pulse) and then the underlying surface a last echo (so-called last pulse), provided that the tree canopy is at least partially transparent. By matching first and last pulses, vegetation can be detected relatively easily. However, not every recording system is capable of receiving multiple echoes.

In summary, airborne laser scanning offers several advantages:

- The 3D recording of objects occurs directly, i.e. without the usual stereo image analysis of photogrammetry.
- There are no requirements for lighting conditions or texture of the surface.
- Data recording and data processing take place in a completely digital process chain.

Compared to these systems, which record discrete echoes, the latest sensors can record the full time-dependent variation of the received signal strength, i.e. the complete waveform of the returned signal. This opens up new possibilities for object recognition (cf. Harding 2008).

Meanwhile, the first commercially available multispectral laser scanner, the Optech Titan Sensor, also allows simultaneous recording in three wavelength ranges (1550 nm, mid-infrared, 1064 nm near-infrared, 532 nm visible green) with up to 4 returns per laser pulse (cf. Teledyne Optech 2023, cf. also Matikainen et al. 2017 and Wichmann et al. 2015).

ALS has become established. Airborne laser scanning has been used to record the digital terrain model e.g. in Bavaria since 1996 or in North Rhine-Westfalia since 2002 or in Hesse (cf. Bayerische Vermessungsverwaltung 2023, GeobasisNRW 2023 or Hessische Verwaltung für Bodenmanagement und Geoinformation 2023). Airborne laser scanning data also provide the basis for the construction of solar roof cadastres, which show the suitability of roofs for photovoltaic systems. In the process, roof areas are extracted from ALS data (cf. Fig. 5.17) and analysed according to size, inclination and exposure of the areas, including potential suitability and usage expectations (cf. Hilling and de Lange 2010a and 2010, cf. e.g. the state-wide solar roof cadastre of Thuringia based on ALS data, cf. Solarrechner Thüringen 2023).

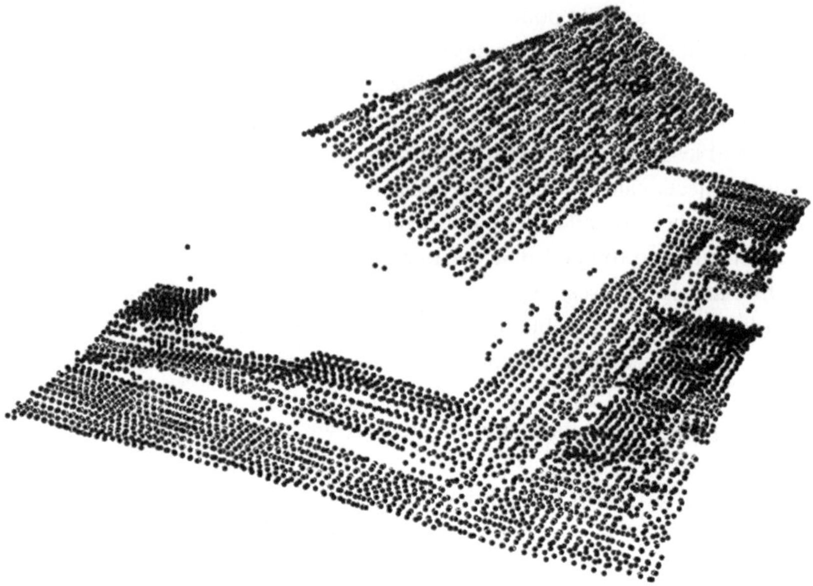

Fig. 5.17 3D surface model of a building by ALS elevation points (for better print reproduction without the roof lying in the background)

5.5 Official Basic Geodata

5.5.1 Official Basic Geodata: Tasks and Challenges for the Surveying Authorities

In general, such spatial data can be described as official basic geodata that are available over a wide area, are updated continuously or periodically, correspond to a general standard and are not compiled and maintained for a specific purpose but serve as a basis for a wide range of applications. Here the term *official basic geodata* is restricted to the geodata of the official surveying system. It is true that market data from the field of geomarketing are of great commercial importance, e.g. in the location planning of retail companies. However, these geodata are (usually) assigned to the thematic data outside of surveying.

In Germany, as in other countries worldwide, the surveying authorities have the legal mandate to manage and provide the national topographic map series and the real estate cadastre. In Germany, this cadastre formerly consisted of the manually maintained real estate map and the real estate register (in German: Liegenschaftskarte and Liegenschaftsbuch). Both together documented the ownership of a parcel of land.

For this purpose, the territory has to be surveyed and presented cartographically. The real estate map is the basis for parcel-specific planning and inventory verification in business and administration. It contains, among other things, information on the geometry (parcel boundaries, boundary markers), significant data (parcel number, cadastral district, place name) as well as descriptive data (actual use, results of soil assessment). Figure 5.18, which is a print of the already digital version, gives an impression of the content that is displayed in the real estate map (cf. Sec. 5.5.2). The real estate register (as a database without graphics) contains, among other things, a description of the parcels according to location, use, land register number, parcel size and, above all, owner information. All of this information is required in particular for the sale of land, the planning of construction projects, financing or funding applications.

A reference can be made between both via the parcel number and the district number which are unique identifiers of a parcel in Germany. Extensive (analog) databases have emerged from the above-mentioned areas of responsibility of the official surveying authorities in Germany, which exactly meet the requirements for official basic data. The *official basic geodata* (in German: Amtliche Geobasisdaten) of the surveying authorities in Germany.

- are created by official surveying authorities or equivalent institutions on the basis of official documents, so that accuracy and liability are guaranteed,
- have a high quality due to the integration of all cadastre-relevant data,
- are regularly updated, so that a high degree of actuality is guaranteed,
- have a (largely) uniform structure throughout Germany,
- are based on a uniform reference system (now UTM coordinates).

Since the 1970s, the surveying authorities in Germany, which are organised in the Arbeitsgemeinschaft der Vermessungsverwaltungen der Länder der Bundesrepublik Deutschland (*AdV,* Working Group of the Surveying Authorities of the Federal States of the Federal Republic of Germany), have been working on the development of digital, spatial information systems in order to store these official basic geodata and also to offer them to (private) users to set up their own geoinformation systems (for obtaining the data cf. Sect. 5.5.4.6).

A digital real estate cadastre, as used in Germany, consists of a digital information system including a digital real estate map and a digital real estate register. It certainly shows a unique development. The development of digital bases for the automated creation of topographic maps can also be regarded as unique. The German surveying authorities have mastered major challenges before ALKIS, the official real estate cadastral information system (in German: Amtliches Liegenschaftskatasterinformationssystem, cf. Sect. 5.5.4.4) and ATKIS, the official topographic cartographic information system (in German: Amtliches Topographisch-Kartographisches Informationssystem, cf. Sect. 5.5.4.5) were established. These automated systems have the potential to be transferred to other countries, where they can be used as the basis for official and cadastral surveying. For this reason, they are explained in more detail here.

5.5.2 Former ALK

In a joint effort of the surveying authorities in Germany, a uniform concept for the structure and digital management of the real estate cadastre was developed. The digital real estate cadastre consisted of the *ALK*, the *automated real estate map* (in German: Automatisierte Liegenschaftskarte) and the *ALB*, the *automated real estate register* (in German: Automatisiertes Liegenschaftsbuch), which were built from their analog counterparts. The Objektartenkatalog OBAK (object mapping catalog) was developed to record the content from the analog real estate map and to define the objects in the ALK. In Lower Saxony, the ALK has been set up nationwide since 1984 (cf. Sellge 1998 p. 91).

The ALK contained the data in three primary files:

- The *ground plan file* included all geometric and semantic information for the representations of the contents of the real estate map in Gauss-Krüger coordinates (e.g. parcels, buildings).
- The *point file* consisted of information on point data such as position coordinates and elevation data as well as further information for managing the points of the geodetic network.
- The *file of measurement elements* contained data for the coordinate calculation such as point identifiers, measurement units or remarks.

An interface enabled the communication between the processing part, database and data users (EDBS, in German: Einheitliche Datenbankschnittstelle). This interface, which was designed to be system-neutral, but significantly hindered the exchange of data, because it did not correspond to any industrial standard.

The ALK objects represented a functional unit of the real estate map (e.g. parcels, buildings). The functionally related objects were assigned to a so-called foil (in German: Folie). An ALK object was described by specifying the foil (e.g. foil 001 "parcels") and the object type according to the object key catalog OSKA (in German: Objektschlüsselkatalog). OSKA was a central encryption catalog and was used to identify all foils and object types. It has been replaced by the ALKIS object type catalog (cf. Sect. 5.5.4.4).

Figure 5.18 shows the layout of the property plots on the left and the building outlines on the right. It should be pointed out that the streets are not reproduced and that a building can stand on several parcels. In contrast, the ALB as a database without graphics provides, among other things, owner information.

The concept of ALB and ALK, which dated back to the 1970s, was already outdated in the 1990s. The parallel management of ALB and ALK meant that data had to be recorded

Fig. 5.18 Excerpt from foil 1 (only parcel boundaries) and foil 11 (only building outlines) for a part of Osnabrück city centre

and updated twice. This required a considerable effort of synchronisation with the risk of inconsistent information. In addition, ALK and ATKIS had parallel developments, the data models of both systems were not compatible. Therefore a new technical concept was developed by the AdV. ALK and ALB have been integrated and are managed together in ALKIS (cf. Sect. 5.5.4).

5.5.3 Former ATKIS

In the 1980s, the surveying authorities in Germany decided to develop ATKIS, the official topographic cartographic information system (in German: Amtliches Topographisch-Kartographisches Informationssystem). The surveying authorities wanted to set up a nationwide digital, topographic-cartographic database, to administer it in an interest-neutral manner, and to provide it as part of the state's public services and to offer it as a state service. In particular, the immanent objective was always to be able to produce the official topographic maps more efficiently using digital procedures (cf. Harbeck 1995 pp. 19).

The ATKIS system design of 1989 planned a stepwise development, among other things for reasons of cost, but also in order to push the project at all and to obtain data as quickly as possible. The first stage of implementation, *DLM25/1* (in German: *Digitales Landschaftsmodell, digital landscape model*) should be available in 1995, at least in the old federal states. The abbreviation should make clear that the contents were roughly oriented to the contents of the topographic map 1:25000 (TK25). The basis for the recording work was primarily the DGK5, 1:5000, which had the positional accuracy of ±3 m, but also aerial maps at a scale of 1:5000 and other sources of information such as city plans. The necessary development work to derive the DKM25 (in German: *Digitales Kartographisches Modell, digital cartographic model*) from the DLM25 should be carried out parallel to the second expansion stage.

The ingenious approach of the original ATKIS concept was to represent the earth's surface using object-structured landscape models and then further using object-structured, symbol-oriented cartographic models. A real geoobject should be replaced by a model object, which is described in form of point-, line- or polygon-shaped components that have different meanings, and which is finally represented by a specific symbol on a map. The contents of the DLM and the regulations for modeling the landscape objects were defined by an ATKIS object type catalog, which was based on cartographic principles, since a topographical map should come out at the end.

The special feature of ATKIS was the symbol-independent storage of topographic objects in the DLM. Abstract geometric information is stored. The "translation" of vector information into a line symbol, for example a double line, takes place in a later work step. The graphic design depends only on the choice of the symbol key, which transforms a single point of the DLM into a distinct pictorial symbol, for example for a chapel, a church or a cathedral. This basic idea also applies to the new AAA model (cf. Sect. 5.5.4.5).

The original ATKIS concept of 1989 still provided a DKM, which was to be set up with a fixed symbol key according to the standards of the national topographic maps. The ATKIS symbol catalog is used for this. The derivation of analog or digital topographic maps should take place in a multi-stage process (cf. Jäger 1995 pd. 233). In the first step, a so-called raw DKM should be automatically generated from the DLM according to the specifications of the ATKIS symbol catalog. In a second step, the raw DKM should be developed interactively into the final DKM25. Since the geometries were recorded at the (larger) scale 1:5000, cartographic generalisation (simplification, displacement) and the optimal placement of text must be carried out for the DKM25 at a scale of 1:25000. The DKM produced in this way should be stored in the ATKIS database. In the last step, the digitally stored map information should be presented, e.g. on the graphics screen, using a presentation program, which no longer has any special requirements, since the cartographically difficult tasks have been solved in the previous steps.

At the beginning of the entire development of ATKIS, the creation of the DLM had priority, the development of the DKM was not far advanced and was still a conceptual model. First, a so called presentation graphic was developed in order to be able to visualize the database at all in advance of the DKM. However, the presentation graphics should not be confused with the planned DKM and the topographic maps! Generalisation and displacement as design processes are missing.

The concept of the DKM from 1989 was developed as a theoretically brilliant system. In the future the German topographic map series should be automatically derived with the help of computer algorithms from the DLM, which is continuously updated. The true benefit of the ATKIS project would have been particularly evident, if the smaller-scale subsequent maps 1:50000 or 1:100000 could have been automatically derived from the DLM25.

However, these ambitious objectives have not been achieved. There have been three main arguments against the DKM.

1. The old concept provided two data models for the DLM and the DKM within ATKIS. This is not optimal in terms of conception.

2. No usable automated solutions have been found for deriving the DKM25 from the raw DLM25 (based on the modeling of the real geoobjects in a scale of 1:5000). The non-trivial problems of cartographic generalisation prevent the derivation of a DKM from the DLM taking place automatically (generalisation, among other things, by selecting, combining, omitting, simplifying or moving objects).
3. Despite the convincing conception, the construction of independent cartographic models was questioned by the majority of AdV member administrations as an impractical solution. It was difficult to make potential users understand that on the one hand the location-accurate DLM was used for the recording and evaluation of the thematic data, and on the other hand a cartographically generalised DKM for their presentation. To top it, the presentation graphics of the DLM achieved a cartographic quality sufficient for many specialist applications, so that it seems questionable, whether the DKM would be in demand at all.

As a consequence, the AdV presented a new ATKIS concept within the new AFIS-ALKIS-ATKIS project (cf. Sect. 5.5.4.5). Furthermore, a new conceptual and material product was introduced, the *Digital Topographic Map (DTK)*, which acts as the sole end product and ultimately replaces the DKM.

It has to be pointed out that a convincing conception was developed:

- recording and modelling of the topographic objects with great accuracy at a scale of 1:5000 or larger,
- automatic derivation of the subsequent maps 1:25000, 1:50000 or 1:100000.

Perhaps the methods of artificial intelligence will soon solve the task of automatic generalization.

5.5.4 AFIS – ALKIS – ATKIS

5.5.4.1 AFIS-ALKIS-ATKIS Concept: Main Features

The ALK and the ALB as well as ATKIS each have their own conceptual problems (cf. Sect. 5.5.3). The ALK and ATKIS, which could perfectly complement each other in terms of the holistic use of official basic geodata, are not compatible due to different data models and object type catalogs. Thus, the AdV has decided on a new concept for an automated, integrated management of all official basic geodata: the AAA concept, which is documented in the so-called GeoInfoDok (still valid GeoInfoDok 6.0, GeoInfoDok NEW will be valid from 31 December 2023, cf., GeoInfoDok 2023a, GeoInfoDok 2023b and GeoInfoDok 2023c):

- The fixed points of the national survey, which are not original components of ALK and ATKIS, are managed in *AFIS*, the *official fixed point information system* (in German: Amtliches Festpunkt-Informationssystem).

– The previous information systems ALK and ALB are integrated in the information system *ALKIS*, the *official real estate cadastral information system* (in German: Amtliches Liegenschaftskatasterinformationssystem). Furthermore, a formal, content-related and semantic harmonisation with ATKIS will be carried out.
– *ATKIS*, the *official topographic cartographic information system* (in German: Amtliches Topographisch-Kartographisches Informationssystem) continues to describe the topography of the Federal Republic of Germany with digital landscape and terrain models.

The relationships between the individual systems are documented in the AFIS-ALKIS-ATKIS reference model, which is structured into a control layer, a production layer and a communication layer (cf. GeoInfoDok 2019 main document V.6.0.1 p. 10). It is fundamental that all legal and thematic features of the real world, which are important as information for the official surveying, are structured from a technical point of view and mapped as technical objects in a common AAA data model. The AFIS-ALKIS-ATKIS application scheme describes the formal data structures and data contents of the AAA data model. It reflects the conceptual view.

The application schema contains the complete formal description of a dataset. The real world is abstracted through the introduction of (technical) objects and rules on how to record and continue them. Compared to the previous individual approaches it is essential that the common AFIS-ALKIS-ATKIS application schema offers a uniform and object-oriented model approach for AFIS, ALKIS and ATKIS. The modeling of the AAA data model corresponds to international standards (using UML and XML, cf. Sects. 3.5.4 and 6.3.3).

The AAA basic scheme, as part of the application scheme, is the basis for the technical modeling of AFIS, ALKIS and ATKIS objects and for the data exchange as well as for further technical applications with spatial reference. It describes the basic properties for one or more applications and contains all base classes that are required for modeling geoinformation (cf. GeoInfoDok 2009 main document p. 18). The AAA technical objects are described in the associated object type catalogs of the GeoInfoDok. The catalogs are very extensive and comprise more than 500 pages for ALKIS alone (cf. GeoInfoDok 2008a).

A new data exchange interface NAS, the standards-based exchange interface (in German: Normbasierte Austauschschnittstelle) is defined, which replaces the previous interface (EDBS). By using the NAS interface based on XML structures, which is therefore based on international norms and standards, the ALKIS data can be displayed in any browser. The new interface is primarily used for communication between the surveying authorities and externally for users of AAA data. The interface is very extensively defined and offers a large data depth. Frequently, however, users do not require this density, but rather simpler data structures that can be processed directly by common geoinformation systems. This problem already existed with EDBS, which was not accepted by users. As with EDBS, special converters are also required for NAS in order to convert the data into the data formats of the widely used geoinformation systems, i.e. into the so-called industry standards. In order to enable as many users as possible to use the AAA data and to market the gigantic mountain of data, it is also possible to deliver the data, for example, in the proprietary shapefile data format of the computer company ESRI (for the shapefile data format cf. Sect. 9.3.3). In this case, however, the data depth is lost (cf. Fig. 5.25).

The AFIS-ALKIS-ATKIS concept provides versioning and a history of objects. Versions are current and historical information about an object. In addition, a new procedure was introduced to update the databases of the individual users in a user-related and individual manner (in German: Nutzerbezogene Bestandsdatenaktualisierung, NBA). This significantly expands and replaces the procedure of updating changes that was previously implemented. However, the newly introduced data model does not allow any further updates of old datasets of the ALK and the ALB. For the previous users, especially utility companies, this has considerable consequences for their pipeline cadastres. In any case, a new initial supply of ALKIS data must take place.

5.5.4.2 Location Reference System in the AAA Concept

The paradigm shift through the introduction of the AAA concept to uniform official basic geodata is completed by the fact that, at the same time as the migration of the data to the new AAA model, the new reference system ETRS89 is introduced with the UTM-projection (cf. Sect. 4.4.4). Consequently, the change of the reference system led to a shift in the coordinate values of the property boundaries and building outlines. This is due to the change from the Bessel ellipsoid to the ellipsoid of the Geodetic Reference System 1980 (GRS80), the transition from the German Main Triangulation Network (central point Rauenberg) to the ETRS89 and the change from the Gauss-Krüger to the UTM projection (cf. Sects. 4.4.7 and 4.5.3). This has the far-reaching consequence that the data in special information systems, e.g. of the utility companies, which are based on old geodata, must also be transformed into the new reference system.

Overall, the new reference system has considerable advantages. In this way, a nationwide and EU-wide uniform spatial reference system (not only) for the provision of official geodata is created, which forms the basis for a sustainable, EU-wide spatial data infrastructure (cf. Sect. 6.7.2). Satellite-based data recording on the basis of WGS84, which is almost identical to ETRS89, now represents a de facto standard (cf. Sect. 5.3.1). Thus, coherent coordinates are recorded as by using the ETRS89 system no change of datum and no transformations are necessary for further processing of the coordinates. Locations can therefore be measured directly with GNSS in the official reference system.

5.5.4.3 AFIS in the AAA Model

Within AFIS the fixed points of the national survey (trigonometric points, levelling points and gravity fixed points) are modeled in digital and object-structured form according to the specifications of the AAA model (cf. GeoInfoDok 2008b). The position fixed points (in German: Lagefestpunkt), the height fixed point (in German: Höhenfestpunkt) and the gravity fixed point (in German: Schwerefestpunkt) as well as the relevant information for the SAPOS reference stations are verified.

5.5.4.4 ALKIS in the AAA Model

The old data model of the ALK has changed completely due to the migration to ALKIS. This is immediately apparent to the user because the structure in foils has disappeared and the structure of the dataset has fundamentally changed. There is a new thematic structure (cf. Fig. 5.19).

The AdV has defined a nationwide set of core data that is to be maintained by all state surveying administrations, among others: parcel, special parcel boundary, limiting point, building, site designation, person, address. In addition, each federal state has the option of defining its own state profiles beyond the basic dataset.

Figure 5.20 shows a section of an ALKIS dataset that was imported into a geoinformation system. In contrast to the very simple and plain presentation of the geometries in this geoinformation system, the ALKIS symbol catalog defines the more complex and meanwhile also colored presentation of the real estate map, which is reproduced for the same section in Fig. 5.21. The illustration shows a screenshot of the geodata viewer TIM-online, which is used as part of the "OpenGeodata.NRW" initiative within the SDI-NRW for the free visualisation and provision of geodata from the NRW surveying and cadastral administration (cf. Sect. 6.7.5).

10000-parcel, location, point
  11000-information of the parcel
  11001-AX_parcel
  ...
20000-owner
  21000-personaldata
  21001-AX_person
  ...
30000-buildings
  31000-information of the building
  31001-AX_building

40000-actual use
  41000-built up area
  ...
  42000-traffic
  42001-AX_road
  42006-AX_path
  ...
  43000-vegetation
  ...
  44000-waters
  ...
50000-constructions, miscellaneous
60000-relief
70000-legal-restrictions
90000-migration

Fig. 5.19 Hierarchical structure of the object type catalog of ALKIS (cf. Lower Saxony Surveying and Cadastral Administration 2010)

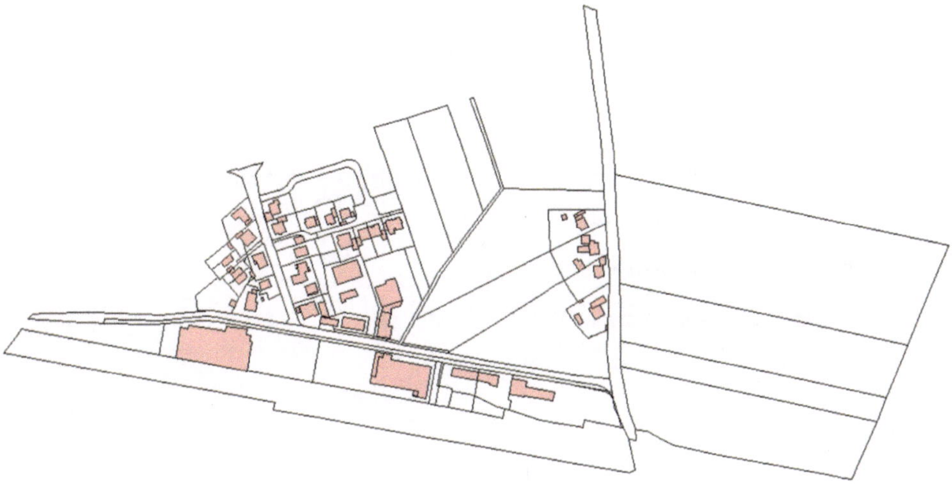

Fig. 5.20 Data extract from ALKIS object type AX_Gebaeude_polygon and AX_Flurstueck_polygon, municipality of Welver / district of Soest. (Data download via TIM-online. Land NRW (2019) Daten Lizenz Deutschland – Version 2.0 https://www.govdata.de/dl-de/by-2-0). cf. Fig. 5.21)

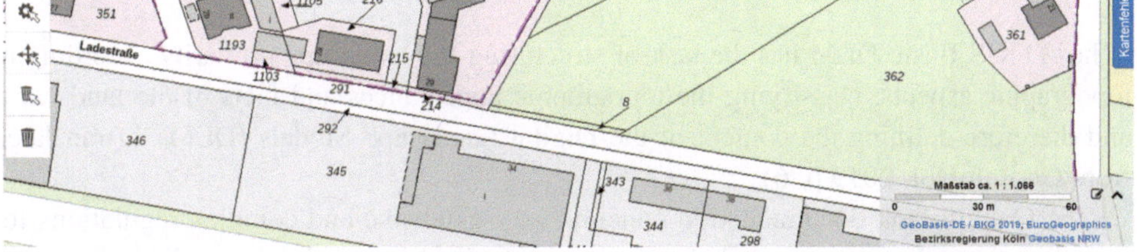

Fig. 5.21 Presentation of properties via TIM-online, municipality of Welver / district of Soest. (Data download via TIM-online. Land NRW (2019) Daten Lizenz Deutschland – Version 2.0 https://www.govdata.de/dl-de/by-2-0, cf. Fig. 5.20)

5.5.4.5 ATKIS in the AAA Model

With the introduction of the AAA model, the ATKIS data were remodeled, adapted to the new standards and standardised with ALKIS. The overall ATKIS product now comprises four components:

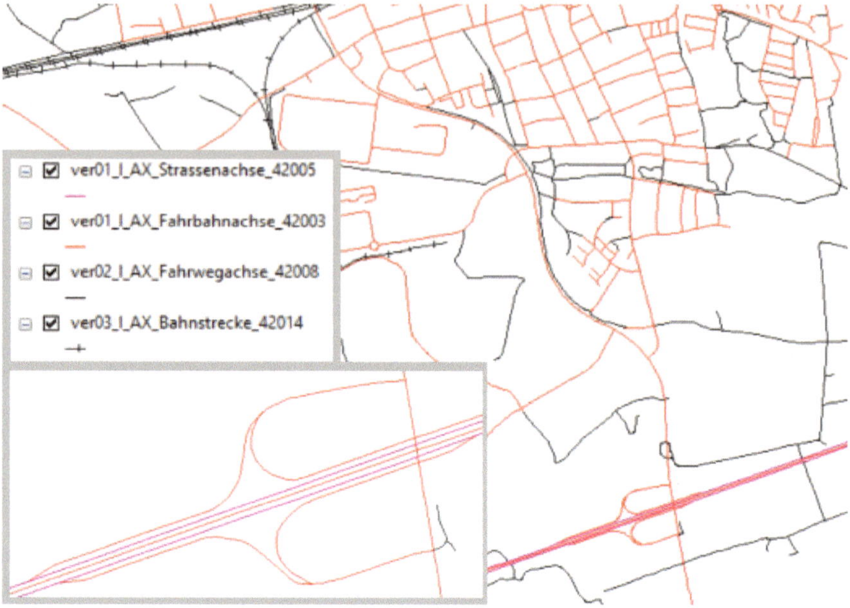

Fig. 5.22 ATKIS-Basic-DLM, traffic; line objects, Neubeckum / Kreis Warendorf. (Data source: https://www.opengeodata.nrw.de/produkte/geobasis/dlm/basis-dlm/ Land NRW (2019) Daten Lizenz Deutschland – Version 2.0 https://www.govdata.de/dl-de/by-2-0, cf. Fig. 5.24)

- DLM, digital landscape model (in German: Digitales Landschaftsmodell)
- DGM, digital terrain model (in German: Digitales Geländemodell)
- DTK, digital topographic map (in German: Digitale Topographische Karte)
- DOP, digital orthophoto (in German; Digitales Orthophoto)

"The ATKIS Basic DLM has the task of structuring the landscape primarily according to topographic aspects, classifying the topographic appearances and facts of the landscape, and therefore defining the content of the Digital Landscape Models (DLM)." (translated from GeoInfoDok 2019 p. 6).

The GeoInfoDok documentation contains very extensive and complex regulations for modeling the real geoobjects and also, for example, for the non-trivial modeling of over- or underpasses or pedestrian bridges (cf. GeoInfoDok 2019). Compared to ALKIS, a different level of accuracy is emphasised. The model accuracy of ±3 m refers to the geometry of essential linear objects of the basic DLM. This refers to the roads to be modeled in linear form, the rail-bound traffic routes and the water bodies as well as the nodes (e.g. intersections) in the network of roads and rail-bound traffic routes. All other objects of the basic DLM on the surface have a positional accuracy of ±15 m (cf. GeoInfoDok 2019 p. 29).

The result of the modeling is shown in Fig. 5.22, which is based on a data retrieval of the entire ATKIS-DLM of North Rhine-Westfalia. The data are made available free of charge as part of the "OpenGeodata.NRW" initiative within the SDI-NRW (cf. Sect. 6.7.5). Selected line objects are shown in the ATKIS-DLM, which were transferred as simple

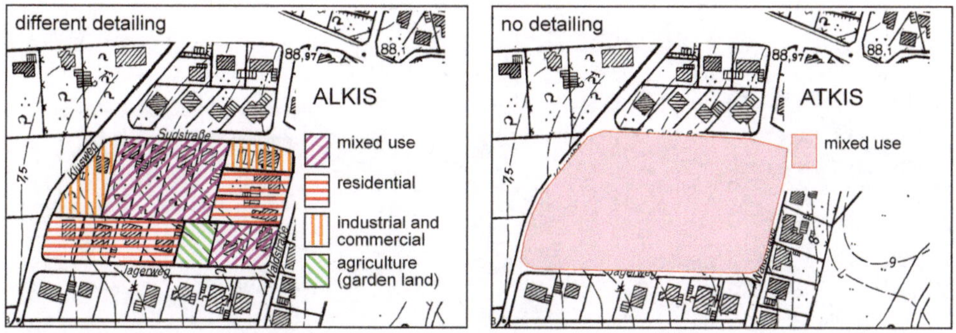

Fig. 5.23 Different detailing in ALKIS and ATKIS

vector data into a geoinformation system, whereby a railway symbol was used and the other lines were only differentiated by color. A road or a path is modeled and represented by the respective centre lines (road and path axes). The multi-lane motorway in the south is additionally broken down into several lane axes (cf. enlarged section).

A major objective in the development of the AAA model was the harmonisation of the object type catalogs of ALKIS and ATKIS. The semantic harmonisation was aimed at matching term and content, in particular identical object definitions, identical attribute types and values. Incorrect data were also corrected in the process.

However, geometric harmonisation is conceptually and scale-related not possible (cf. Fig. 5.23). ATKIS gives a rougher representation of the actual use, e.g. of an entire building block as an area of mixed use, whereas ALKIS shows individual parcels, e.g. as residential land, area of mixed use, industrial and commercial land and agriculture (garden land). ALKIS is parcel-specific, whereas ATKIS combines several parcels and goes beyond the parcel boundary to the road center line. ALKIS identifies the actual use of a parcel as a road or path, while ATKIS stores the road axis or the path axis.

In addition to the basic DLM, which makes up the old ATKIS with its updates, further, more coarsely structured digital landscape models are planned:

- Digital Basic Landscape Model*
- Digital Landscape Model 1:50000 (DLM50)*
- Digital Landscape Model 1:250000 (DLM250)**
- Digital Landscape Model 1:1000000 (DLM1000)**

  *realisation by the Land Surveying Authorities

  **realisation by the BKG (Federal Agency for Cartography and Geodesy)

The basic DLM and the DLM50 (except for Bavaria) are completely available for all German federal states. The DLM250 and DLM1000 are also completely available for Germany.

Since the 1990s a discussion has started about a modern and sustainable map design for the German topographic map series. In 1995, the AdV adopted new design principles for

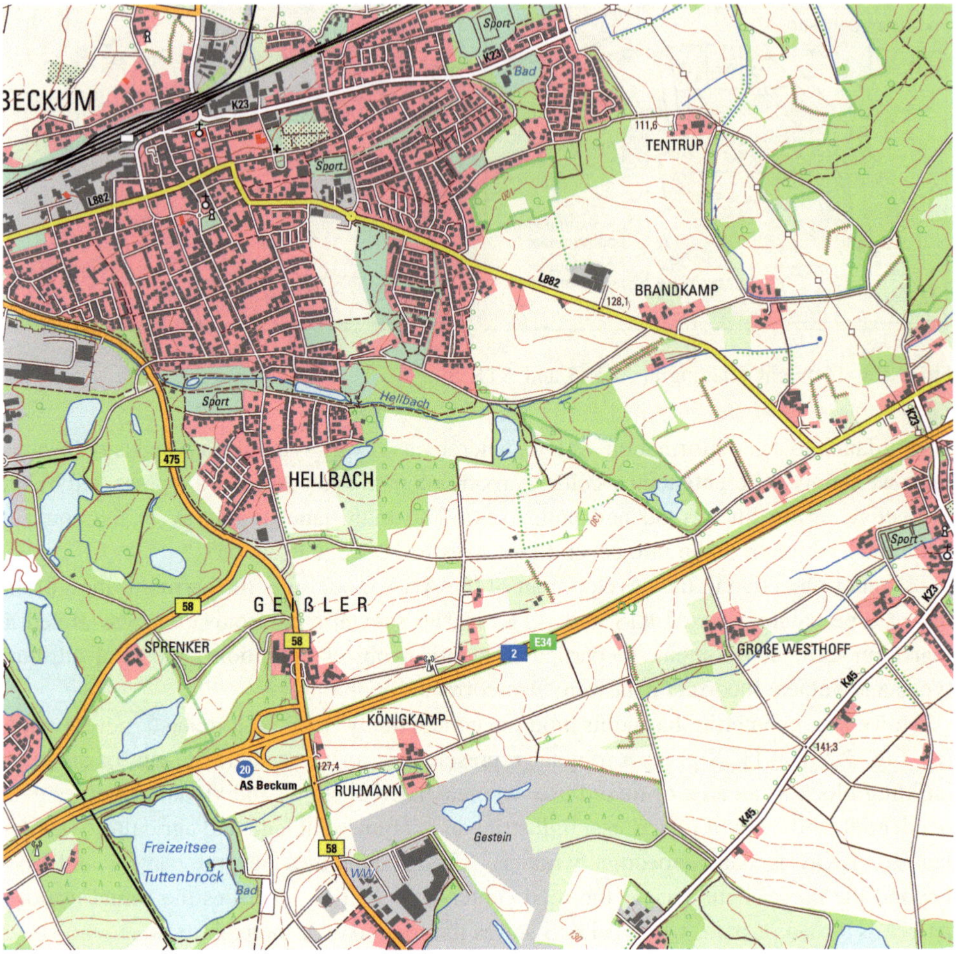

Fig. 5.24 DTK 25, Neubeckum/Kreis Warendorf. (Data source: TIM-online, Land NRW (2019) Daten Lizenz Deutschland – Version 2.0 https://www.govdata.de/dl-de/by-2-0, cf. Fig. 5.22)

the German topographic map. The new map design are characterised, among other things, by expanding the coloring (more surface colors and colored symbols), increasing the minimum cartographic dimensions (broadened line widths, increased line spacing), expansion of the symbol stock and selection of a modern map font. This new TK25 with a new color scheme and symbols is derived from digital databases of the ATKIS-DLM and from ALKIS (building information).

Figure 5.24 shows a section of a DTK25, it is identical to Fig. 5.22, so that the design principle, the modeling and the symbols become clear. The map is based on the European Terrestrial Reference System 1989 (ETRS89, cf. Sect. 4.4.3).

The German national survey provides several topographic map series of different scales for the uniform topographic description of the territory of the Federal Republic of Germany,

- Digital Topographic Map 1:10000 DTK10 or TK10*
- Digital Topographic Map 1:25000 DTK25 or TK25
- Digital Topographic Map 1:50000 DTK50 or TK50
- Digital Topographic Map 1:100000 DTK100 or TK100
- Digital Topographic Map 1:250000 DTK250 or TK250
- Digital Topographic Map 1:1000000 DTK1000 or TK1000

  *The DTK10 is available in the eastern federal states as well as in North Rhine-Westfalia and Baden-Württemberg and as a digital local map in Bavaria.

The surveying authorities of the Bundesländer are responsible for DTK10, DTK25, DTK50 and DTK100, while the BKG is responsible for the DTK250 and DTK1000. The AdV compiles an annually updated overview of the range of digital landscape and terrain models and digital topographic maps (cf. AdV-Online 2023).

The ATKIS product family also includes digital orthophotos (DOP), which are based on aerial photographs that have been georeferenced to ETRS89/UTM, projected onto the DGM5 and processed according to a uniform national standard. Thus, rectified and true-to-scale, digital vertical images are available with a ground resolution of 20 cm × 20 cm (DOP20) or 40 cm × 40 cm (DOP40) per pixel. In some Bundesländer, DOPs with a ground resolution of 10 cm × 10 cm also exist. The aerial photographs cover the entire area and are usually renewed in a 3-year cycle.

5.5.4.6 Obtaining Official Basic Geodata in Germany

In Germany the official basic geodata are made available under various licences by the respective surveying authorities or by the BKG (Federal Agency for Cartography and Geodesy). There are different forms of delivery and delivery conditions. For most of the Bundesländer, the fee is (still) subject to an official cost regulation, although individual geodata are also provided free of charge. Only the open data states of Berlin, Hamburg, North Rhine-Westfalia and Thuringia and the BKG make all their geodata available for download and online use free of charge. Within the scope of the open data offer of the BKG, maps and data that are produced and maintained by the BKG itself at a scale of 1: 250.000 and smaller are made available free of charge in accordance with the German Federal Digital Geodata Access Act (cf. Sect. 6.7.3, BKG 2023a, 2023b).

Official basic geodata can be accessed digitally within the framework of the licence condition via the geoportals of the surveying authorities of the individual federal states (cf. Sects. 6.7.3 and 6.7.5). A Web Map Service makes it easy to integrate a digital topographic map into a geoinformation system (cf. Fig. 6.4). The importance of paper maps is therefore declining sharply.

The product TopPlusOpen (TPO, cf. BKG 2023c) deserves special attention. The BKG provides a freely usable worldwide web map based on free and official data sources. TPO is available as a download of sections or tiles from the web map and presentation graphics and, above all, as a Web Map Service.

5.6 Volunteered Geographic Information (VGI)

5.6.1 Data in the GeoWeb 2.0

The term Volunteered Geographic Information (VGI) was created by Goodchild (2007), who sees it as a special form of a much more general trend on the World Wide Web: the creation of "user generated content", for which the creation, collection and distribution of geographic information is characteristic, which are provided voluntarily on the World Wide Web by mostly private individuals. This is related to the recent change in provider and user behaviour towards Web 2.0 (cf. Sect. 2.8.4), which has also changed the management and provision of geodata.

The counterpart to *Web 2.0* is often referred to as *GeoWeb 2.0,* which encompasses new forms of publishing and using geodata (cf. Kahle 2015). However, the GeoWeb should also be understood as a vision of recording spatial data in real time via sensors, integrating them with other data and using them through mobile geoinformation systems (GIS) or through GIS functions that are ubiquitously available on the World Wide Web (cf. Longley et al. 2010 pp. 275). These ultimately not yet foreseeable developments were largely initiated by offers of commercial companies such as Google, Here and Microsoft as well as formerly Yahoo and Nokia, which provide maps and user-friendly functions to present their own data on the WWW. Probably the most significant examples at present are Flickr and Google Earth. The photo portal Flickr provides access to countless pictures, which was one of the first WWW applications to allow photo to be located. In Google Earth, you can save and reload your own geodata in *KML* format (KML = *Keyhole Markup Language*), so that countless users can add a wide range of individual additional data. These are primarily georeferenced photos with brief explanations, links to further advertising sites and 3D city models.

5.6.2 Open Street Map (OSM) Project

Compared to the presentations, which are used for advertising, self-presentation and simple dissemination of information among friends, the recording and provision of geodata, such as the course of roads, by volunteers and offering them on the WWW is to be seen as a competition to official geodata. The currently most significant example is the *Open Street Map – Project* (initiated in 2004), which collects and provides geodata that can be freely used by everyone (cf. Open Street Map 2023). The Open Street Map – database has been distributed under the Open Database License (ODBL) 1.0 since 12.09.2012. They can therefore be integrated in prints, websites and applications such as navigation software by indicating the data source. To do this, the data must be recorded by volunteers themselves. As a rule, coordinates of a street or path are recorded with a GPS receiver while driving or walking, or they are also recorded by digitising freely available templates such as aerial photographs or maps. Here is a hurdle that should not be underestimated. It is not possible to simply copy maps and plans, which are often protected by copyright, as is the case with official geodata from the surveying authorities. However, many companies or organisations have now released data for import into the Open Street Map database. Microsoft has released the aerial photographs of "Bing Maps" for drawing. The so-called TIGER data set with the road data published by the Federal Statistical Office of the USA has been imported. In Germany individual municipalities have made their road data available.

The recorded, raw geodata are uploaded via a web portal and edited in a second step. The OpenStreetMap project provides various options for this. Basic editing steps are possible with the very simple program iD. The program Potlatch 2 represents the more professional version. Alternatively, the recorded data can also be adjusted using the offline editor JOSM (JavaOpenStreetMap-Editor) as a desktop application as well as using the apps Vespucci or Go Map!! for the smartphone operating systems Android or iOS. In this way the geometries are provided with attributes and, for example, a polyline is marked as a farm road. Above all, points of interest, which include sights, public facilities or bus stops, for example, are added and edited. Editing can also be done by people other than the actual data creators, so that people without GPS devices can also participate, but who have local knowledge and therefore help to correct, expand and update the database.

The invaluable advantage of OSM data is that any registered user can freely obtain geodata and use it in their own applications without restrictive licensing requirements or fees. Furthermore, various companies offer processed data. While the processing may be associated with costs, the use is free. The data has sufficient quality for many applications. Almost countless uses exist, due to the licensing model of the OpenStreetMap project. There are many examples in the World Wide Web, especially for route planners and for displaying access options and for tourism maps from travel companies (cf. Stengel and Pomplun 2011 p. 116, cf. Fig. 5.25).

5.6.3 OSM Data Quality

An argument sometimes put forward against the use of OSM data is the inconsistent and ultimately indeterminate data quality. However, it is also applies here that the quality of geodata must be assessed with regard to use (so called fitness for use, cf. Sect. 6.6.1). The official basic geodata are characterised by a precise, comprehensive and value-neutral recording within a clearly defined area of responsibility. The data have a uniform recording quality based on official cadastral surveying. Naturally, these criteria do not apply to voluntary data recording by data collectors with very different levels of training and motivation. In particular, the quality of Open Street Map data or Volunteered Geographic Information in general is not uniform. Thus, in terms of completeness, there are major differences between large cities and areas that are well developed for tourism on the one hand and rural areas on the other. Unfortunately, the positioning accuracy is not the same everywhere. Thus, in cadastral issues, when a high level of reliability is important, such as in the route planning of energy suppliers, one cannot do without official basic geodata.

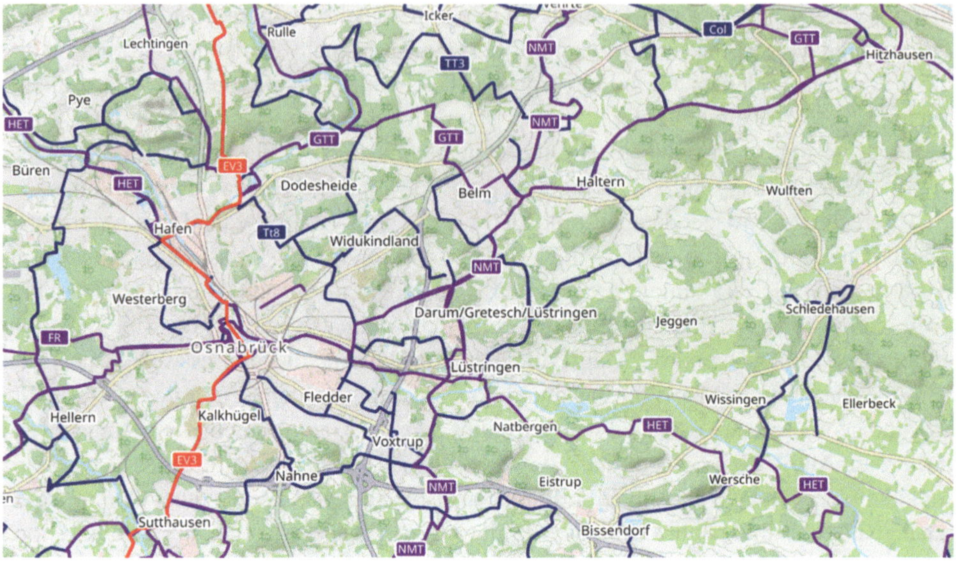

Fig. 5.25 Osnabrück cycling map based on OSM data, web mapping application in a browser (source: Open Street Map 2023)

However, the data quality of OSM data is completely sufficient for many web applications, for city and tourism portals and above all as a basis for route maps or routing systems. The Open Street Map data are a very serious competitor to the official geodata for such purposes. The official basic geodata and web services of the state surveys are mostly not used due to the restrictive, fee-based licensing policy.

References

AdV-Online (2023): AdV-Produkte. https://www.adv-online.de/AdV-Produkte/ (14.04.2023)

Bauer, M. (2018): Vermessung und Ortung mit Satelliten. Globale Navigationssysteme (GNSS) und andere satellitengestützte Navigationssysteme. Berlin: Wichmann 7. Ed.

Bayerische Vermessungsverwaltung (2023): Geländemodell. https://www.ldbv.bayern.de/produkte/3dprodukte/gelaende.html (14.04.2023)

BeiDou Navigation Satellite System (2020): Completion and Commissioning of the BeiDou Navigation Satellite System (BDS-3). http://en.beidou.gov.cn/WHATSNEWS/202008/t20200803_21013.html (14.04.2023)

BeiDou Navigation Satellite System (2023): System. http://en.beidou.gov.cn/SYSTEMS/System/ (14.04.2023)

BKG, Bundesamt für Kartographie und Geodäsie (2023a):Digitale Geodaten. https://gdz.bkg.bund.de/index.php/default/digitale-geodaten.html (14.04.2023)

BKG, Bundesamt für Kartographie und Geodäsie (2023b): Open Data. https://gdz.bkg.bund.de/index.php/default/open-data.html (14.04.2023)

BKG, Bundesamt für Kartographie und Geodäsie (2023c): TopPlus-Produkte. https://gdz.bkg.bund.de/index.php/default/digitale-geodaten/topplus-produkte.html (14.04.2023)
China Satellite Navigation Office (2018): Development of the BeiDou Navigation Satellite System Version 3.0). http://en.beidou.gov.cn/SYSTEMS/Officialdocument/201812/P020190523251292110537.pdf (14.04.2023)
China Satellite Navigation Office (2019): BeiDou Navigation Satellite System Signal In Space Interface Control Document Open Service Signal IB1I (Version3.0). http://en.beidou.gov.cn/SYSTEMS/Officialdocument/201902/P020190227601370045731.pdf (14.04.2023)
Dana, P.H. (2023): GPS. https://foote.geography.uconn.edu/gcraft/notes/gps/gps.html (14.04.2023)
de Lange, N. u. K. Wessels (2000): Aufbau eines Grünflächeninformationssystems für die Stadt Osnabrück. In: N. de Lange (Hrsg.): Geoinformationssysteme in der Stadt-und Umweltplanung. Fallbeispiele aus Osnabrück. S. 49-64. Osnabrück: Rasch. = Osnabr. Stud. z. Geographie 19.
Dodel, H. u. D. Häupler (2009): Satellitennavigation. Berlin: Springer. 2. Ed.
EGNOS (2023a): About EGNOS. https://egnos-user-support.essp-sas.eu/new_egnos_ops/egnos-system/about-egnos (14.04.2023)
EGNOS (2023b): EGNOS about OS. https://egnos-user-support.essp-sas.eu/new_egnos_ops/services/about-os (14.04.2023)
EGNOS (2023c): EGNOS about SoL. https://egnos-user-support.essp-sas.eu/new_egnos_ops/services/about-sol (14.04.2023)
EGNOS (2023d): EGNOS about EDAS https://egnos-user-support.essp-sas.eu/new_egnos_ops/services/about-edas (14.04.2023)
EUSPA, European GNSS Service Centre (2023a): Programme. https://www.gsc-europa.eu/galileo/programme (14.04.2023)
EUSPA, European GNSS Service Centre (2023b): Constellation information. https://www.gsc-europa.eu/system-service-status/constellation-information (14.04.2023)
EUSPA, European GNSS Service Centre (2023c): System. https://www.gsc-europa.eu/galileo/system (14.04.2023)
EUSPA, European GNSS Service Centre (2023d): Services. https://www.gsc-europa.eu/galileo/services (14.04.2023)
FAA, Federal Aviation Administration (2023a): Satellite Navigation – WAAS – How it works. https://www.faa.gov/about/office_org/headquarters_offices/ato/service_units/techops/navservices/gnss/waas/howitworks (14.04.2023)
FAA, Federal Aviation Administration (2023b): Satellite Navigation—GPS/WAAS Approaches. https://www.faa.gov/about/office_org/headquarters_offices/ato/service_units/techops/navservices/gnss/approaches (14.04.2023)
Flühr, H. (2010): Avionik und Flugsicherungstechnik. Einführung in Kommunikationstechnik, Navigation, Surveillance. Heidelberg: Springer.
Foley, J.D. u.a. (1996): Computer graphics: principles and practices. Reading: Addison-Wesley.
GeobasisNRW (2023): Höhenmodelle. https://www.bezreg-koeln.nrw.de/brk_internet/geobasis/hoehenmodelle/index.html (14.04.2023)
GeoInfoDok (2008a): Dokumentation zur Modellierung der Geoinformationen des amtlichen Vermessungswesens (GeoInfoDok), ALKIS Objektartenkatalog. https://www.adv-online.de/GeoInfoDok/GeoInfoDok-6.0/Dokumente/binarywriterservlet?imgUid=c9e63fd2-1153-911a-3b21-718a438ad1b2&uBasVariant=11111111-1111-1111-1111-111111111111&isDownload=true (14.04.2023)
GeoInfoDok (2008b): Dokumentation zur Modellierung der Geoinformationen des amtlichen Vermessungswesens (GeoInfoDok), Erläuterungen zu AFIS. https://www.adv-online.de/GeoInfoDok/GeoInfoDok-6.0/Dokumente/binarywriterservlet?imgUid=74b43fd2-1153-911a-3b21-718a438ad1b2&uBasVariant=11111111-1111-1111-1111-111111111111 (14.04.2023)

GeoInfoDok (2009): Dokumentation zur Modellierung der Geoinformationen des amtlichen Vermessungswesens (GeoInfoDok), Hauptdokument. https://www.adv-online.de/icc/extdeu/binarywriterservlet?imgUid=8f830072-8de8-9221-d5ad-8f138a438ad1&uBasVariant=11111111-1111-1111-1111-111111111111 14.04.2023)

GeoInfoDok (2019): Dokumentation zur Modellierung der Geoinformationen des amtlichen Vermessungswesens (GeoInfoDok), Erläuterungen zum ATKIS Basis-DLM V. 7.1. https://www.adv-online.de/icc/extdeu/binarywriterservlet?imgUid=73850d50-1d5d-0b61-7905-ba3403b36c4c&uBasVariant=11111111-1111-1111-1111-111111111111 (14.04.2023)

GeoInfoDok (2023a): GeoInfoDok 6.0, Dokumente. https://www.adv-online.de/GeoInfoDok/GeoInfoDok-6.0/Dokumente/ (14.04.2023)

GeoInfoDok (2023b): GeoInfoDok 6.0, Referenzversion bis 31.12.2023. https://www.adv-online.de/GeoInfoDok/GeoInfoDok-6.0/ (14.04.2023)

GeoInfoDok (2023c): GeoInfoDok 7.1, Referenzversion ab 31.12.2023. https://www.adv-online.de/GeoInfoDok/GeoInfoDok-NEU-Referenz-7.1/ (14.04.2023)

Goodchild, M.F. (2007): Citizens as sensors: the world of volunteered geography. GeoJournal 69 (4). S. 211–221.

GPS.gov (2023a): GPS Space Segment. https://www.gps.gov/systems/gps/space/ (14.04.2023)

GPS.gov (2023b): GPS Control Segment. https://www.gps.gov/systems/gps/control/ (14.04.2023)

Hake, G. u.a. (2002): Kartographie. Visualisierung raum-zeitlicher Informationen. Berlin: de Gruyter, 8. Ed.

Harbeck, R. (1995): Überblick über Konzeption, Aufbau und Datenangebot des Geoinformationssystems ATKIS. In: Kophstahl, E. u. H. Sellge (1995): Das Geoinformationssystem ATKIS und seine Nutzung in Wirtschaft und Verwaltung. Vorträge 2. AdV Symp. ATKIS, S. 19 – 37.

Harding, D. (2008): Pulsed Laser Altimeter Ranging Techniques and Implications for Terrain Mapping. In Shan, J. u. C. K. Toth (2008, Hrsg.): Topographic Laser Ranging and Scanning. Principles and Processing. Boca Raton Fl. S. 173-194

Heritage, G.L. u. A.R.G. Large (2019, Hrsg.): Laser Scanning for the environmental sciences. Chichester: Wiley-Blackwell

Hessische Verwaltung für Bodenmanagement und Geoinformation (2023): Airborne Laserscanning. https://hvbg.hessen.de/landesvermessung/geotopographie/3d-daten/airborne-laserscanning (14.04.203)

Hilling, F. u. N. de Lange (2010a): Vollautomatisierte Ableitung eines digitalen Solardachkatasters aus Airborne Laserscannerdaten. In: Bill, R., Flach, G., Klammer, U. u. C. Niemeyer (Hrsg.): Geoforum MV 2010 – Vernetzte Geodaten: vom Sensor zum Web. Berlin: GITO, S. 17-20.

Hilling, F. u. N. de Lange (2010): Webgestützte interaktive Solardachkataster. Ein Instrument zur Darstellung der Nutzungseignung von Dächern für Photovoltaikanlagen am Beispiel der Stadt Lage. In: Standort, Zeitschrift für Angewandte Geographie 34-1, S. 104-109.

Höfle, B. u. M. Rutzinger (2011): Topographic airborne LiDAR in geomorphology: A technological perspective. In: Zeitschrift für Geomorphologie 55, 2, S. 1–29.

Hofmann-Wellenhof, B. u.a. (2008): GNSS Global Navigation Satellite Systems. GPS, GLONASS, Galileo and more. Springer: Wien.

IAC Information-Analysis Center for Positioning, Navigation and Timing (2023): GLONASS. https://www.glonass-iac.ru/en/about_glonass/ (14.04.2023)

Jäger, E. (1995): Kartographische Präsentation aus ATKIS. In: Kophstahl, E. u. H. Sellge (1995): Das GIS ATKIS und seine Nutzung in Wirtschaft und Verwaltung. Vorträge anl. des 2. AdV Symposiums ATKIS. S. 231–241. Hannover: Nieders. Landesverwaltungsamt – Landesverm.

Kahle, C. (2015): Geoweb 2.0 – nutzergenerierte Geoinformationen. Bibliotheksdienst, Band 49, Heft 7, Seiten 762–766. https://www.degruyter.com/view/j/bd.2015.49.issue-7/bd-2015-0086/bd-2015-0086.xml (14.04.2023)

Kahmen, H. (2006): Angewandte Geodäsie: Vermessungskunde. Berlin: de Gruyter. 20. Ed.

Kohlstock, P. (2011): Topographie: Methoden u. Modelle der Landesaufnahme. Berlin: de Gruyter.

Longley, P.A., Goodchild, M.F., Maguire, D.J. u. D.W. Rhind (2010): Geographic Information Systems and Science. New York: John Wiley 3. Ed.

Lower Saxony Surveying and Cadastral Administration (2010): Basiswissen ALKIS / ETRS89. https://www.lgln.niedersachsen.de/download/126790/Basiswissen_ALKIS_ETRS89_Schulungsmaterial_Stand_12.04.2010.pdf (14.04.2023)

Mansfeld, W. (2010): Satellitenortung und Navigation: Grundlagen, Wirkungsweise und Anwendung globaler Satellitennavigationssysteme. Wiesbaden: Vieweg + Teubner. 3. Ed.

Matikainen, L. u.a. (2017). Object-based analysis of multispectral airborne laser scanner data for land cover classification and map updating. ISPRS Journal of Photogrammetry and Remote Sensing, Volume 128: S. 298 – 313

NASA (2012): Global Positioning System History. https://www.nasa.gov/directorates/heo/scan/communications/policy/GPS_History.html (14.04.2023)

NASA (2014): Global Positioning System. https://www.nasa.gov/directorates/heo/scan/communications/policy/GPS.html (14.04.2023)

Nicolaus, S. (2000): Erfassung städtischer Grünflächen auf Grundlage von Luftbildern. - In: N. de Lange (Hrsg.): Geoinformationssysteme in der Stadt- und Umweltplanung. Fallbeispiele aus Osnabrück. S. 37–48. Osnabrück: Rasch. = Osnabrücker Studien zur Geographie 19.

NMEA (2023): NMEA 0183. http://www.nmea.de/nmea0183datensaetze.html (14.04.2023)

Open Street Map (2023): About. https://www.openstreetmap.org/about (23.08.2022)

Osnabrück cycling map based on OSM data. https://www.openstreetmap.org/#map=12/52.2801/8.0890&layers=C (14.04.2023)

Pfeifer, N. u.a. (2017): Laserscanning. In: Heipke, C. (2017): Photogrammetrie und Fernerkundung, S. 431–482. Handbuch der Geodäsie. (Hrsg. Freeden, W. u. R. Rummel). Berlin: Springer.

Resnik, B. u. R. Bill (2018): Vermessungskunde für den Planungs-, Bau- und Umweltbereich. Berlin: Wichmann. 4. Ed.

SAPOS (2023a): Satellitenpositionierungsdienst der deutschen Landesvermessung. https://sapos.de (14.04.2023)

SAPOS (2023b): SAPOS. Precise Positioning in Location and Height. https://www.lgln.niedersachsen.de/download/121560/SAPOS-Broschuere_english_.pdf (14.04.2023)

SAPOS (2023c): Satellitenpositionierungsdienst SAPOS. https://www.bezreg-koeln.nrw.de/brk_internet/geobasis/raumbezug/sapos/index.html (14.04.2023)

Solarrechner Thüringen (2023): https://solarrechner-thueringen.de/#s=startscreen (14.04.2023)

Sellge, H. (1998): Zukunft der Geobasisdaten in Niedersachsen. In: Creuzer, P. (Hrsg.): Geographische Informationssysteme. Einsatz und Nutzung in Kommunen und Landesverwaltung. S. 91–96. Hannover: Landesvermessung + Geobasisinformation.

Shan, J. u. C. K. Toth (2008, Hrsg.): Topographic Laser Ranging and Scanning. Principles and Processing. Boca Raton Fl.: CRC Press

Stengel, S u. S. Pomplun (2011): Die freie Weltkarte Open Street Map – Potenziale und Risiken. In: Kartographische Nachrichten, Jg. 61, H. 3, S. 115-120.

Teledyne Optech (2023): Optech Titan Multispectral Lidar System. https://geo3d.hr/sites/default/files/2018-06/Titan-Specsheet-150515-WEB.pdf (14.04.2023)

Vosselman, G. u. H.-G. Maas (2010, Hrsg.): Airborne and Terrestrial Laser Scanning. Boca Raton Fl.: CRC Press

Wichmann, V. u.a. (2015). Evaluating the potential of multispectral airborne lidar for topographic-mapping and landcover classification. ISPRS Annals of the Photogrammetry, Remote Sensing and Spatial Information Sciences, Volume II-3/W5, S. 113–119.

Worboys, M. F. u. M. Duckham (2004): GIS: A computing perspective. Boca Raton: CRC Press.

6 Standards and Interoperability of Geodata

6.1 Standardisation and Interoperability

6.1.1 Multiple Use Through Standardisation

The simple basic idea is to use geodata that have already been collected and are therefore available for new, further questions and then not to collect them again. The focus is on increasing efficiency, since multiple use is almost always more economical. Regardless of any fees for multiple use, new acquisition is usually associated with higher costs. Multiple use avoids data redundancy, which also reduces the risk of data inconsistency.

In the meantime, large digital geodata stocks are available, so that the likelihood of reuse has increased and *multiple use* is obvious, assuming that data providers and data sources on the one hand and data demanders on the other hand are different. The desire or the need to use (geo)data multiple times leads to two differently complex questions:

- Which data are available in which quality for a specific question of a user (anywhere)?
- How is a data exchange to be carried out (technically)?

The first question aims at the (pure) knowledge of suitable data. It is often unknown that data are already available for a certain question, knowledge about the properties of existing data is often insufficient. The data itself is located in several places on different data carriers. The data are mostly based on different purposes, recording methods, data accuracies and data formats. Descriptive information about the data is therefore an indispensable prerequisite for multiple use. In particular, a mandatory minimum level of information about the data, i.e. *metadata*, must be provided. Standardised metadata catalogs, i.e. metadata on the actual subject that must be compulsorily recorded and

N. de Lange, *Geoinformatics in Theory and Practice*, Springer Textbooks in Earth Sciences, Geography and Environment,
https://doi.org/10.1007/978-3-662-65758-4_6

maintained, can be used to find suitable data and to assess the possibilities and limits of their use (including availability, provision costs and rights of use).

The second question includes sub-questions on forms of *data transfer*, *data interfaces* and *data formats*. In particular, the data must be exchanged completely and without loss. Thus, a complete exchange of geodata not only includes geometry data, but also topology data as well as attribute data and metadata. In addition information on graphical features (e.g. line width or color) can also be exchanged. An essential prerequisite for the exchange of data and the multiple use of (geo)data is the development and implementation of *standards*, which are adopted by both data providers and data consumers. However, an unhindered and effortless data exchange is not only a question of a standardised interface. Furthermore, the modeling of geoobjects in the information systems of providers and consumers must be known and standardised with regard to the greatest possible interoperability (for the definition of geoobjects cf. Sect. 4.1). In addition to the more technical aspects, it is also relevant that the meaning of a geoobject is the same for providers as well as for consumers. Thus, it is not irrelevant that both sides agree on which features distinguish a stream from a river, for example.

6.1.2 Syntactic and Semantic Interoperability

Interoperability describes the ability to access distributed data resources and to use the data, which are generated in different software systems and which are available in different, i.e. primarily proprietary data formats and may be modeled differently. Interoperability is data exchange and data use across system boundaries. In addition to the purely technical interoperability, which, among other things, refers to data transfer, e.g. via a fiber optic network, a distinction is made between syntactic and semantic interoperability.

Syntactic interoperability characterises the structure of the interfaces or data formats between the participating systems. *Semantic interoperability* in geoinformatics means that the meaning (semantics) of the terms, the data schema and the modeling of the geoobjects are known and usable in the systems involved.

Below two data sets are listed with the same content, but with different syntax. In each case, the same properties of parcels are specified within a commercial land register of a municipality. Only the second example is directly understandable due to an HTML-like syntax:

```
12-3773; osnabrueck; 2012-05-24; 1100;g;yes
<parcel>
        <number_ parcel> 12-3733</number_ parcel>
<municipality>osnabrueck</municipality>.
        < features datum="2012-05-24">
                <size>1100</size>
                <fnp>g</fnp>
                < suspected_contaminated_site >yes</
suspected_contaminated_site >.
        </features>
</parcel>
```

Syntactic interoperability cannot prevent one information system from interpreting the data of another information system differently. Such *semantic problems* occur regularly when integrating geodata from different sources. It should be emphasised that the OGC Web Services (cf. Sect. 6.4.2), which have just revolutionised the data exchange across system boundaries, offer syntactic but not semantic interoperability.

The following is an example of critical semantic interoperability for a dataset that has an attribute for a parcel that indicates the presence of a contamination ("suspected_contaminated_site" with the options "yes" or "no"). The meaning of the attribute value "no" is clear in the original context in which the data were collected. This designation should have resulted from an assessment by an engineering firm and relates to the dumping of construction waste.

```
12-3773; osnabrueck; 2012-05-24; 1100;g;no
```

In another information system, however, the local attribute with the same name ("suspected_contaminated_site") may have a completely different meaning and refer to the result of chemical soil analyses. The simple data transfer of the parcel with the attribute value "no" for "suspected_contaminated_site" can then have considerable consequences in the new context, if soils are contaminated with PAHs (polycyclic aromatic hydrocarbons), which was not checked in the original context. Nevertheless, the parcel is now listed as free of pollutants. There is no semantic interoperability for "suspected_contaminated_site" in this example. This even affects two aspects: ambiguity of the term "contamination" and multiple forms of verification.

To solve these problems, unique terminologies and metadata can be used and *ontologies* can be set up to represent knowledge. In computer science, an ontology is a formal, machine-understandable model of knowledge that systematises terms or information about the concepts for a topic or field of knowledge and semantically links the interrelated concepts. Above all, the relationships that exist between the concepts are formalised. Only when factors such as relief, runoff, precipitation or evaporation are systematised or operationalised and the relationships are shown, a possible flood risk can be automatically derived from the variable water level.

Ontologies are used, among other things, in automated knowledge processing (e.g. in expert systems) and generally in artificial intelligence application systems. The World Wide Web Consortium (W3C) has specified the Web Ontology Language (OWL) to describe and disseminate ontologies using a formal description language. Ontologies such as OWL are components for the construction of the *Semantic Web*, which is an extension of the World Wide Web in order to offer web content in a machine-understandable form and to exchange and use it more easily or automatically between different computers (for an introduction cf. Hitzler et al. 2008).

6.2 Standardisation

6.2.1 Standard and Norm

Standards in geoinformatics or with regard to geodata refer to established data formats of geoinformation systems such as the drawing exchange format (DXF) of Autodesk or the shapefile data format of the software company ESRI (cf. Sects. 9.1.5 and 9.3.3). These industry standards or de facto standards result from the importance of the software of major companies similar to the data formats of Microsoft Office or the PDF format. *Standards* denote widely accepted and used formats and rules that have gained acceptance on the market. Industry standards or proprietary data formats are always associated with manufacturer-specific properties.

Standards can be developed by standardisation institutes into a recognised, non-manufacturer-specific standard. A *norm* identifies a set of technical rules, i.e. a de jure standard. Standardisation by a manufacturer-independent authority is intended to ensure that the content of the standards is generally recognised. The *German Institute for Standardisation (Deutsches Institut für Normung e. V.)* defines in DIN 820-1 of 1994: "Standardisation is the planned activity carried out jointly by interested parties for the standardisation of tangible and intangible objects for the benefit of the general public." The standardisation process itself is also regulated in detail in DIN 820, insofar as it takes place within the German Institute for Standardisation.

6.2.2 Standardisation Institutions

Several international and national organisations are involved in the standardisation work (cf. selection in Table 6.1). The Internet addresses provide a wealth of references and standardisation criteria. However, it is often necessary to obtain authorization from the institutions for a fee in order to view the standards. This must be seen as counterproductive to the objective of achieving the most rapid possible dissemination of standards and norms. In spring 2019, a search using the keyword "geoinformation" yielded more than 630 hits on the site of the German Institute for Standardisation. In the International Organisation for Standardisation (ISO), the Technical Committee 211 (Geographic Information/Geomatics) is of central importance.

Table 6.1 Internationally important standardisation organisations

Name	Entry URL
AFNOR Association Française de Normalisation	www.afnor.org
ANSI American National Standards Institute	www.ansi.org
BSI British Standards Institution	www.bsigroup.com
CEN Comité Européen de Normalisation	www.cenorm.be
CENELEC Europ. Committee for Electrotechnical Standard.	www.cenelec.eu
DIN German Institute for Standardisation	www.din.de
ETSI European Telecommunications Standards Institute	www.etsi.org
FGDC Federal Geographic Data Committee	www.fgdc.gov
IEC International Engineering Consortium	www.iec.org
International Electrotechnical Commission	www.iec.ch
ISO International Standard Organisation	www.iso.org
OASIS Org. for the Advancement of Struct. Inform. Standards	www.oasis-open.org
OMG Object Management Group	www.omg.org/
Open Geospatial Consortium (OGC)	www.opengeospatial.org
World Wide Web Consortium (W3C)	www.w3.org

6.2.3 International Organisation for Standardisation (ISO)

The *International Organisation for Standardisation* (*ISO*), headquartered in Geneva, is the international association of standardisation organisations. The acronym is not an abbreviation of the organisation, but is derived from the Greek word "isos" (meaning equal). ISO develops international standards in all areas outside of electrics, electronics and telecommunications. The organisation is divided into technical committees, subcommittees and working groups. ISO has developed over 22,000 international standards, and more than 1000 new standards are published each year. ISO standards are developed through a precisely defined work process that includes several clearly defined steps and phases in order to achieve the broadest possible consensus among manufacturers, providers, consumer groups, government agencies, and research institutions (cf. ISO 2023a). In general, a shift from national to international standards can be seen.

The *Technical Committee TC 211 Geographic Information/Geomatics* in the ISO deals with the development of standards for geoinformatics (standards family 191xx,) in several working and special groups. Many standards have reached international standards, most of them are already in a new phase of revision. The catalogue of standards covers many aspects of geoinformatics, such as "Geography Markup Language", "Web Feature Service", "data quality", "addressing" or "calibration and validation of remote sensing imagery sensors and data" (cf. ISO 2023b).

In addition to the general objective of sustainability, TC 211 pursues further guidelines, of which the following are most prominent: independence of software and hardware, development of (interface and data) models on the conceptual level instead of data formats, communication services (e.g. data transfer) based on standardised interfaces.

6.2.4 Open Geospatial Consortium

The *Open Geospatial Consortium* (*OGC*, before 2004 Open GIS Consortium Inc.) is a non-profit standardisation organisation founded in 1994, in which more than 520 companies, authorities and research institutions voluntarily cooperate in order to develop generally available interface standards in a consensual process. The goal of the OGC can be paraphrased by: "geo-enable the Web by geospatial standards":

> The OGC Vision
>
> Building the future of location with community and technology for the good of society.
>
> The OGC Mission
>
> Make location information Findable, Accessible, Interoperable, and Reusable (FAIR).
>
> The OGC Approach
>
> A proven collaborative and agile process combining consensus-based open standards, innovation projects, and partnership building. (OGC 2023a).

The OGC produces several types of documents. The most important are the *OGC Implementation Standards* and the *OGC Abstract Specifications*, in addition to OGC reference model (ORM) best practices documents, engineering reports, discussion papers and white papers. The OGC Abstract Specifications provide the conceptual basis for most OGC developments and provide a reference model for the development of implementation specifications (cf. OGC 2023b).

Implementation standards are aimed at technical target groups and describe in detail the interface structures between software components (cf. OGC 2023c). For example, the *feature geometry model* represents an abstract specification of the OGC that forms an implementation-independent framework to which implementation specifications of software manufactures should be oriented (cf. Sect. 6.3.1). The OGC specifications (and now also ISO standards) are based on the feature geometry model or subsets thereof:

- the simple feature-geometry-object-model (for the description of two-dimensional vector geometries in two ISO standards simple feature access ISO 191251 u. 19125-2),
- the *Geography Markup Language* (*GML*, for the *XML*-based representation of geodata, ISO 19136).

Among the OGC implementation standards, the Web services (more concretely or more extensively named: the OGC-compliant spatial data services, cf. Sect. 6.4) define interoperability on the web, i.e. primarily the *Web Map Service* (*WMS*, tiled as *Web Map Tile Service WMTS*), the *Web Feature Service* (*WFS)* and the *Web Coverage Service* (*WCS*).

The important *Catalogue Service for the Web* (*CSW*, also *Web Catalogue Service*) is defined within the Catalogue Service Implementation Specifications.

OGC and ISO/TC211 have been working closely together since 1997 (cf. ISOTC211 2023). For example, the OGC's implementation specifications are submitted to ISO as a standard proposal.

Compliance with OGC standards has become the decisive criterion for the *interoperability* of geodata and software systems (cf. Sect. 6.4.2). OGC conformity is not only a general quality feature, but also a purchasing criterion for software and the basis for spatial data infrastructures. If the OGC standards are implemented in software or online services of two independently operating, i.e. also competing, software manufacturers, interoperability between the resulting components is given via this standardisation.

6.3 Standards for Modeling Geodata

6.3.1 Feature Geometry Model of the OGC

In addition to standards for data exchange, an important prerequisite for interoperability between geoinformation systems relates in particular to the modeling and storage of geodata. With the abstract specification of the feature geometry model, which now corresponds to the ISO standard 19107 (2003) "Geographic Information Spatial Schema", the OGC has developed a conceptual data model that describes spatial properties of maximally two-dimensional geoobjects (vector geometry and topology). In this way, standard spatial operations are defined for accessing, querying, managing and exchanging geoobjects (cf. Brinkhoff 2013 pp. 67).

The *feature geometry model* consists of the two main packages "geometry" and "topology", which themselves consist of five or three subpackages (cf. Fig. 6.1). Thus, the geometry package includes the "geometry root" subpackage with the geometry superclass "GM_Object", the subpackage "geometry primitive" with the definition of geometric primitives (points, curves, surfaces, solids) and the subpackage "coordinate geometry" with classes for describing geometries using coordinates, the subpackage "geometry aggregate" for aggregating multiple geoobjects into loose collections of geometries, and the

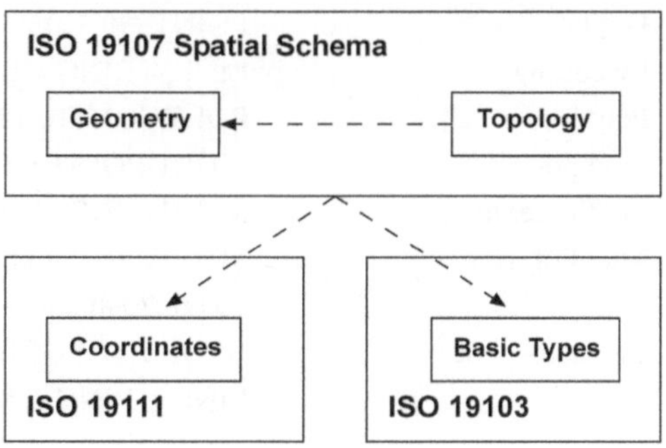

Fig. 6.1 The main packages for the feature geometry model (cf. Brinkhoff 2013 p. 67)

subpackage “geometry complex” for aggregating multiple geometric primitives describing a complex geoobject (cf. Brinkhoff 2013 pp. 68 in more detail). Furthermore, the package “coordinates” from the ISO standard 19111 “Spatial Referencing by Coordinates” and the package “basic types” from the ISO standard 19103 “Conceptional Schema Language” are used.

6.3.2 Simple Feature Geometry Object Model of the OGC

Within the OGC, the *feature geometry object model* is of central importance, which describes the spatial properties of geoobjects as an abstract, implementation-independent, conceptual data model. Version 1.2.1 from 2011 is the current version, which is somewhat extended compared to ISO standard 19125 (cf. OGC 2023d). ISO standard 19125-1 “Simple Feature Access – Common Architecture” specifies the technology-independent properties of the data model. ISO Standard 19125-2 (also from 2004) describes the implementation in an SQL database schema. The remaining two specifications for simple features developed by the OGC, i.e. for the Common Object Request Broker Architecture (CORBA) and for the Object Linking and Embedding/Component Object Model (OLE/COM) for data exchange in a Windows network, have lost technological significance.

The *simple feature geometry object model* is intended to be used here as an example to illustrate modeling principles and to introduce the procedure for modeling simple geometry structures and object formation for vector data (cf. Sect. 9.3.2). Simple features are defined as two-dimensional geoobjects, whose points are connected in a straight line. Thus, curves and circles are not simple features, but only points, lines, linear routes and polygons. Table 6.2 illustrates the implementation of the simple feature geometry object model in the PostgreSQL/Post-GIS spatial database (cf. Sect. 8.7.3). With “point”, “linestring” and “polygon” three basic geometry types are available, from which four further geometry types are derived.

Table 6.2 Geometry types of the OGC simple feature geometry object model (according to ISO 191251 Simple Feature Access Table 2 “Example Well-known Text Representation of Geometry”)

Type	Example
Point	POINT (10 12)
LineString	LINESTRING (10 12, 20 20, 30 40)
Polygon	POLYGON (10 12, 20 20, 30 40, 10 12)
Multipoint	MULTIPOINT (10 10, 20 20)
MultiLineString	MULTILINESTRING ((10 10, 20 20), (15 15, 30 25))
MultiPolygon	MULTIPOLYGON (((10 10, 10 20, 20 20, 20 15, 10 10)), ((40 40, 60 60, 70 40, 40 40)))
GeometryCollection	GEOMETRYCOLLECTION (POINT (100100), POINT (200200), LINESTRING (150 150, 200 200, 300 300))

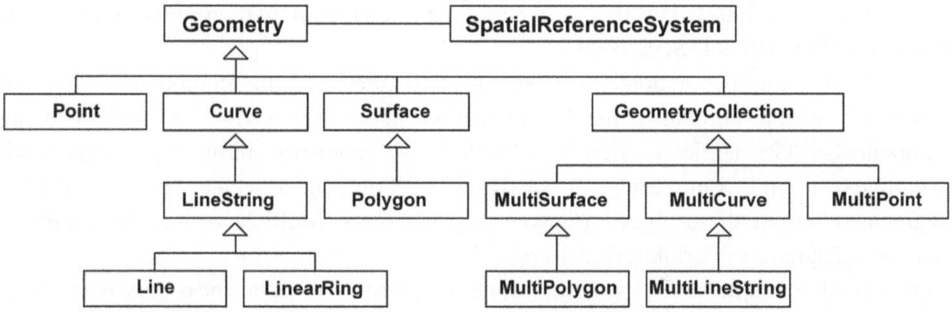

Fig. 6.2 Class hierarchy of the Simple Feature Geometry Object Model (2D)

From the superclass "Geometry", which refers to a spatial reference system (class "SpatialReferenceSystem"), more specific classes for geometric base objects (so-called geometric primitives) and for geometry collections are derived (cf. Fig. 6.2). With regard to an example (cf. listing of GML-code in Sect. 6.3.3), only lines will be discussed in more detail here. Lines are modeled in the simple feature geometry object model by the abstract class "Curve", which has only a single subclass "LineString" for storing line segments from an (ordered) sequence of (line) points. "LineString" has the subclass "Line" (line segments with exactly two line points) and the subclass "LinearRing" for closed simple line segments (rings).

The simple feature geometry object model defines methods to be provided by the classes. Examples of methods that must be provided by all geometry classes are "IsEmpty(..)" (check if the geometry is empty) or "Point.X(..)" (returns the x-coordinate of a point). Furthermore, the simple feature geometry object model offers a number of methods that express spatial relationships between geometric objects and enable spatial analysis (distance, buffer, convex hull, intersection operations such as intersection or union, cf. Sect. 9.4.4). The simple feature geometry object model is used, for example, in spatial database systems such as Oracle Spatial or PostGIS based on PostgreSQL or in program libraries such as the JTS Topology Suite, which serves, for example, as the basis for the GeoTools program library.

6.3.3 Geography Markup Language

The *Geography Markup Language* (*GML*), which is based on the feature geometry model, represents an XML-based description of geodata (cf. OGC 2023e). GML version 3.3 has been available since 2012 and complements the previous version 3.2.1. GML describes the geometric, topological and temporal features of up to three-dimensional geoobjects.

Since 2005, GML profiles have been introduced that define subsets of GML for specific application areas (e.g. GML Simple Features Profile). GML is ISO-compliant with version

3.2.1 (ISO 19136:2007). This means in particular that GML now represents an implementation of ISO 19107 (cf. Sect. 6.3.1).

The OGC describes geoobjects as so-called features with the components "Element Property", which contain general information about the geoobject, and with the components "Geometric Property", modeled by geometric basic types (geometric primitives) "Point", "LineString", "LinearRing" or "Polygon" as well as by more complex, aggregated sets of these objects (Geometry Collections, MultiCurve and MultiSurface) (on class diagrams cf. Brinkhoff 2013 pp. 339).

In a small example, a section of an information system for a business park is modeled, which consists of individual commercial parcels with attributes. As a basic geometric type, only the simple "line string" is to be presented, which is a sequence of points connected by straight line segments (cf. other types of line strings such as those with intersections or closed line strings). A geographic feature is modeled as a list of names that has geographic and non-geographic properties (cf. "featureMember"). The geographic properties are encoded by the already defined geometry types. The GML-code for modeling of objects (here: parcels) in a spatial information system is:

```
<?xml version="1.0" encoding="utf-8" ?>
<ogr:FeatureCollection
  xmlns:xsi="http://www.w3.org/2001/XMLSchema-instance"
  xsi:schemaLocation=""
  xmlns:ogr="http://ogr.maptools.org/"
  xmlns:gml="http://www.opengis.net/gml">
    <gml:boundedBy>
      <gml:Box>
        <gml:coord>
          <gml:X>434046.472</gml:X><gml:Y>5733655.540</gml:Y>
        </gml:coord>
        <gml:coord>
          <gml:X>434234.540</gml:X><gml:Y>5733770.769</gml:Y>
        </gml:coord>
      </gml:Box>
    </gml:boundedBy>

  <gml:featureMember>
  <ogr:GML3_UTM_2019 fid="GML3_UTM_2019.0">
    <ogr:geometryProperty>
      <gml:LineString srsName="EPSG:25832">
        <gml:coordinates>
          434230.018,5733742.126 434207.881,5733746.490
          434184.157,5733750.299 434193.679,5733770.769
          434234.540,5733762.597 434232.716,5733757.519
          434230.018,5733742.126
        </gml:coordinates>
```

```
      </gml:LineString>
    </ogr:geometryProperty>
    <ogr:Id>0</ogr:Id>
    <ogr:F_ID>5926918</ogr:F_ID>
    <ogr:Name>Gisbert Block</ogr:Name>
  </ogr:GML3_UTM_2019>
  </gml:featureMember>

  <gml:featureMember>
    <ogr:GML3_UTM_2019 fid="GML3_UTM_2019.1">
      <ogr:geometryProperty>
        <gml:LineString srsName="EPSG:25832">
          <gml:coordinates>
            434046.472,5733660.831 434059.802,5733739.539
            434105.715,5733751.599 434099.156,5733703.993
            434115.236,5733700.607 434108.465,5733655.540
            434046.472,5733660.831
          </gml:coordinates>
        </gml:LineString>
      </ogr:geometryProperty>
      <ogr:Id>1</ogr:Id>
      <ogr:F_ID>5926977</ogr:F_ID>
      <ogr:Name>Gisela Polygon</ogr:Name>
    </ogr:GML3_UTM_2019>
  </gml:featureMember>
</ogr:FeatureCollection>
```

The XML document is easy to understand even without broad explanations and deep knowledge of GML. Each parcel has a unique number, the affiliation to a management unit, an owner and a boundary. The geometric properties of the feature are defined by "geometryProperty". A collection of feature elements represents a "FeatureCollection". The spatial reference system must be specified for the geometry data (srsName, srs abbreviated to spatial reference system). So-called EPSG codes are used (cf. Sect. 4.5.5).

6.3.4 GeoPackage

In 2014, the OGC developed an open, non-proprietary and platform-independent standard for storing and transmitting geodata. This is intended in particular to achieve independence from industry standards (cf. the proprietary shapefile and geodatabase data formats from the software company ESRI, cf. Sect. 9.3.3). The OGC GeoPackage standard describes conventions for storing geodata in a SQLite database: vector data, image and raster maps, attributes (non-spatial data) and extensions. A GeoPackage is the SQLite container, while the GeoPackage Encoding Standard defines the rules and requirements for how content is stored in a GeoPackage container (cf. OGC 2023f).

The entire database, which may include several different themes and feature classes, is in a single, very compact, i.e. space-saving file. The data in a GeoPackage can be retrieved and updated in a "native" storage format without transformation into intermediate formats. This makes the format particularly suitable for mobile applications on smartphones or tablet computers (use of geoinformation systems on mobile devices).

6.4 Geodata Services

6.4.1 Interoperability Through Standardised Geodata Services

Data exchange is basically nothing new. Until the late 1990s, it was mostly characterised by cumbersome and time-consuming tasks. A customer, such as an energy supplying company, sends its request, e.g. for current parcel geometries, to a cadastral authority using an order form (or by e-mail). The requested data are converted from the authority's database or from the format of the information system used there into a standardised data format (e.g. EDBS or NAS in Germany, cf. in Sect. 5.5.2 or 5.5.4.1), returned on a DVD or emailed (along with a cost notification). The energy supplying company converts the data and inserts them into its own database, i.e. into the data format of the company's internal information system (cf. Fig. 6.3). The data exchange was characterised by conversion between proprietary data formats of different software manufacturers. This can lead to data loss. The process was somewhat simplified, if the data were exchanged using a uniform data format or if the information system at the requesting company had corresponding data import functions.

In the meantime, data exchange is largely automated via the Internet using standardised data (exchange) interfaces and on the basis of OGC-compliant geodata services.

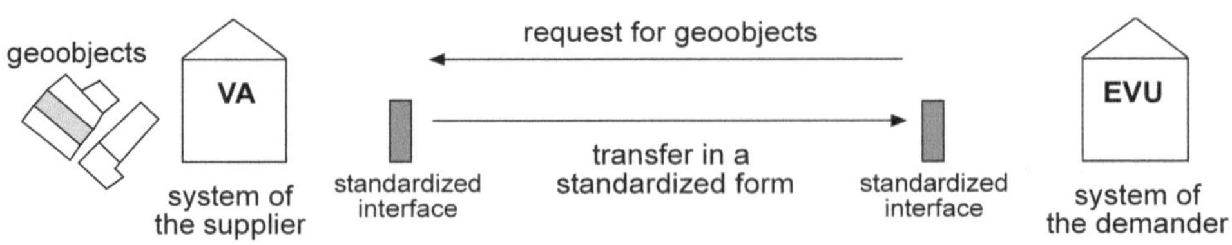

Fig. 6.3 Data exchange using standardised interfaces

6.4.2 OGC-Compliant Geodata Services

In their general form, *geodata services* are *web services* that make distributed geodata accessible via the WWW. A web service is a software application that was developed to support interoperability between (different) computers on the World Wide Web. Standardised geodata services are of great importance, when setting up a *spatial data infrastructure* (cf. Sect. 6.7), since they ultimately enable *interoperability* between distributed computers on the WWW and their databases, which are available in different data models and data formats.

Figure 2.14 can be used to illustrate the basic access and transfer of geodata on the Internet using geodata services (cf. Sect. 2.8.3), for which the term web services is commonly used in simplified terms. A webbrowser starts a request to a *webserver*, which forwards it to a so-called *mapserver,* which accesses the geodata. The webserver takes over the pure communication in the WWW. The logic layer of the mapserver is now characterised by special functionalities. The mapserver runs the web services, prepares the geodata and sends them back to the calling client via the webserver. Here, webserver and mapserver refer to software systems that usually run on the same physical server (hardware). The terms webserver and mapserver are used here to distinguish between the two systems with different functionality.

The OGC has developed several standards for geodata services for interoperability. The Web Map Service (WMS) is currently the most important and widespread. It returns the geodata in the form of a (single) raster map to the calling client (or the Web Map Tile Service (WMTS), which returns several georeferenced map tiles). The Web Catalogue Service (CSW) and the Web Feature Service (WFS), which returns descriptive metadata and vector data as a GML file follow in terms of importance (cf. Sect. 6.4.6).

Basically, calling an OGC Web Service on the mapserver is a two-step process. First, the capability is determined with the instruction "GetCapabilities". This is followed by service-specific requests (e.g. "GetMap"). In the standard case, the web services access a database directly. However, multi-level variants are also possible. For example, a WMS can call a WFS or WCS or both services in parallel and return the vector data, which is then processed by the WMS as a raster map and sent back to the calling client (cascading service).

6.4.3 Functionality of an OGC-Compliant WMS Using the Example of the UMN-MapServer

The *MapServer* is (by now) a software project of the Open Source Geospatial Foundation (OSGeo), which goes back to a development of the University of Minnesota (UMN) and is therefore often still called *UMN-Mapserver* or *UMN-MapServer*. The free software provides web services according to the specifications of the Open Geospatial Consortium (WMS, WFS, WCS). An excerpt from a mapfile of a UNM MapServer application is:

```
MAP
  NAME      OS_open_spaces
  IMAGETYPE  PNG
  EXTENT        3427000 5787000 3445000 5805000
  SIZE        800 800
  SHAPEPATH     'C:/osdaten/fnutzung/'
  IMAGECOLOR     255 255 255 # background color white
WEB
  METADATA
    'WMS_TITLE' 'sport facilities in Osnabrück WMS'
    'WMS_ASTRACT' 'Example WMS'
    'WMS_ONLINERESOURCE' 'http://localhost/cgi-bin/mapserv.exe?map=
      C:/ms4w/Apache/htdocs/os_fnutzung/freiflaechen_wms.map'
    'WMS_CRS' 'EPSG:31467'
    'WMS_ENABLE_REQUEST' '*'
  END
END
  PROJECTION
    "init=epsg:31467"
  END
# start LAYER DEFINITION ----------------------
  Layer # description Layer ONE
    NAME  'city boundary'
    DATA  'grenzen.shp'
    STATUS  ON
    TYPE  LINE
      METADATA
        'WMS_Title'   'city boundary'
        'WMS_SRS'   'EPSG:31467'
      END
    CLASS
      STYLE
        COLOR 0 0 0
      END
    END #class
  END # description Layer ONE
  Layer # description Layer TWO
    NAME  'open spaces'
    DATA  'freiflaechen.shp'
    STATUS  DEFAULT
    TYPE  POLYGON
    METADATA
      'WMS_Title'   'open spaces'
      'WMS_SRS'   'EPSG:31467'
    END
```

```
    CLASSITEM 'nutzcode'
      CLASS
        EXPRESSION '10'
          STYLE
            COLOR  0 200 0
          END
      END
      CLASS
        EXPRESSION '20'
          STYLE
            COLOR  255 0 0
          END
      END
  END # description Layer TWO
END # end mapfile
```

A UMN MapServer application has a so-called *mapfile* in addition to the data, which serves as the central layout and configuration file (text file with the extension .map). This mapfile is in a directory on a server (here in the folder C:/ms4w/Apache/htdocs/os_fnutzung) and contains parameters for controlling the WMS (i.e. for the display, information on the scale, to the legend or to the map projection). In a very common default configuration this file is accessed using the Apache HTTP server as the webserver. The WMS request "GetMap" delivers a raster graphic to the browser (cf. Sect. 6.4.4). This can only be the first step in the construction of a so-called Web-GIS, since a browser initially lacks such simple tools as e.g. for moving the map (integration of web mapping libraries, cf. Sect. 7.2.3).

A mapfile is organised into several sections. The layers specify geodata, which in this example are located on a server in the folder C:/osdaten/fnutzung (for the layer principle cf. Sect. 4.1.4). This defines the city limits and areas for sports facilities and campsites in the city of Osnabrück. The geodata are available in the proprietary shapefile data format from the software company ESRI (cf. Sect. 9.3.3, extension shp).

6.4.4 Access to Geodata Via a Web Map Service

An OGC-compliant Web Map Service can prepare raster data (i.e. graphics) and vector data, i.e. mainly vector data in proprietary data formats such as those of the software companies Oracle, AutoDesk and ESRI, which are largely industry standards. A WMS enables simple, standardised access to (almost any) geodata on the WWW via the Hypertext Transfer Protocol (HTTP). A georeferenced image is usually returned in a simple raster format (e.g. in JPEG or PNG formats). Thus, a user can call a Web Map Service via a browser, in which the requested map is then displayed after successful execution.

A WMS prepares vector data and raster images in a standardised form primarily for pure display purposes. In addition, this web service can provide information about metadata of the geodata and allow general queries of the underlying thematic data.

An OGC-compliant WMS has three functions, which should be illustrated using the example of the MapServer and the listed mapfile:

```
GetCapabilities
http://localhost/cgi-bin/mapserv.exe?map=C:/ms4w/Apache/htdocs/
os_fnutzung/freiflaechen_wms.
map&service=wms&version=1.3&request=GetCapabilities
```

This requests the capabilities of the WMS. An XML file is sent back with metadata, including details of the data provider, supported output formats (e.g. PNG) and, above all, the data layers that can be queried.

```
GetMap
http://localhost/cgi-bin/mapserv.exe?map=C:/ms4w/Apache/htdocs/
os_fnutzung/freiflaechen_wms.map&service=wms&version=1.1&request=
GetMap&layers=Siedlungsfreiflaechen&BBOX=3427066,5787000,3445000,
5805000&SRS=epsg:31467&Format=image/png&width=600&height=600
```

As a result, a georeferenced raster graphic (map) is returned by the WMS. The request includes, among other things, the desired data layer (here: Siedlungsfreiflaechen referenced to the layer freiflaechen.shp in the mapfile), the size of the map output (here 600 × 600 pixels), the output format (here PNG), the underlying coordinate system (here: EPSG code 31467, i.e. third Gauss-Krüger stripe) and the size of the data or map section. The returned PNG file can be displayed in a browser. The areas appear in green for code 10 (cf. RGB code 0 200 0 for the display specifications in the mapfile) and in red for code 20. The code designates an attribute in the original data, i.e. in the shape "freiflaechen.shp".

```
GetFeatureInfo
http://localhost/cgi-bin/mapserv.exe?map=C:/ms4w/Apache/htdocs/
os_fnutzung/freiflaechen_wms.map&service=wms&version=1.3&request=
GetFeatureInfo&layers=Siedlungsfreilaechen&CRS=
EPSG:31467&BBOX=3427000,5787000,3444200,
5801000&width=800&height=800&INFO_FORMAT=text/
plain&QUERY_LAYERS=Siedlungsfreiflaechen
```

This optional function allows a Web Map Service to return information about a mouse-clicked position on the map. The underlying thematic information is returned in XML format. It can be processed in the browser using JavaScript, for example.

It should be clearly pointed out that in this example geodata are accessed in a proprietary vector data format (shapefile data format from the software company ESRI, cf. Sect. 9.3.3) and that this data can be queried and displayed from anywhere using any browser. Furthermore, it should be emphasised that only a graphic, i.e. a rigid map, is returned. The user and demander can hardly use this information. For the efficient visualisation of the returned graphics, special software components are still required for the browser (cf. Sect. 7.2.3).

6.4.5 Access to Geodata Via a Web Feature Service

An OGC-compliant *Web Feature Service* is used for the standardised transfer of georeferenced vector data on the WWW, which is returned as an XML file in GML format. Thus, in contrast to a WMS, no raster file is returned. In the requesting browser, this XML file can initially only be viewed as a text file. To visualize these vector data, e.g. as lines, special software components such as so-called geoviewers are still required on the client. A WFS can also be integrated into a geoinformation system, which then takes over the graphical representation of the vector data. For example, the free geoinformation system QGIS can directly process vector data in GML format.

An OGC-compliant WFS (so-called basic WFS) has three mandatory functions that can be sent to the WFS as HTTP requests. The functions that evaluate the above listed mapfile should be illustrated using the mapserver as an example:

```
GetCapabilities
http://localhost/cgi-bin/mapserv.exe?map=C:/ms4w/Apache/htdocs/
os_fnutzung/frei-flaechen_wfs.
map&service=wfs&version=1.1&request=GetCapabilities
```

This requests the capabilities of the WFS. An XML file with metadata is returned: including information on the data provider, the feature types to be queried and on possible operations.

```
DescribeFeatureType
http://localhost/cgi-bin/mapserv.exe?map=C:/ms4w/Apache/htdocs/
os_fnutzung/freiflaechen_wfs.map&service=wfs&version=
1.1.0&request=DescribeFeatureType
```

This requests information on the structure of the individual feature types of the geoobjects, which can be transferred via the WFS.

```
GetFeature
http://localhost/cgi-bin/mapserv.exe?map=C:/ms4w/Apache/htdocs/
os_fnutzung/freiflaechen_wfs.map&service=wfs&version=
1.1.0&request=GetFeature&typename=Siedlungsfreiflaechen
```

This request returns vector data in GML format. Further queries can be used to create, change or delete geoobjects, to lock geoobjects for editing or, for example, to query individual elements from the GML file. The basic WFS offers read-only access. The Transaction WFS allows write access to the data.

6.4.6 Access to Geodata Via Other Web Services

In addition to WMS and WFS, the Open Geospatial Consortium is developing further standards for exchanging geodata on the Web:

The *Web Map Tile Service* (*WMTS*), which has conceptual similarities to the Web Map Service (WMS), is used to offer and retrieve digital map tiles in a standardised way (cf. OGC 2023g). These map tiles can be displayed in a browser, where many image files requested individually over the web are seamlessly connected (cf. e.g. loading OSM tiles in a web mapping application, cf. Fig. 5.25). In contrast, a WMS usually displays a single large image.

The *Web Coverage Service* (*WCS*) defines a standardised interface and operations for interoperable access to so-called coverages. The term "coverage" or "grid coverage" refers to raster data such as digital aerial photos, satellite images or other digital raster data, in which each raster cell does not have greyscale values like a photo, but data values (e.g. elevation data in a so-called elevation grid). A WCS returns the requested raster data with the associated detailed descriptions. Like a WFS, a WCS enables further processing of the data on the client side.

The *Web Catalogue Service* (*CSW*, Catalogue Service for the Web) enables standardised searches for geodata and geodata services based on metadata.

6.4.7 Processing Geodata by Standardised Web Processing Services

A *Web Processing Service* (*WPS*) provides a standard interface that simplifies the task of making computational services accessible via web services. These services include functions that are standard in GIS software such as *buffering* or *intersection* as well as specialised processes for spatial-temporal modeling and simulation (cf. OGC 2023h). Rules are specified for data input and output for functions such as buffering or intersections. The standard also specifies how a client can request the execution of a (geoprocessing) process and how the results of the process are processed. It defines an interface that enables the public provision of spatial analysis processes and supports the finding and connection of clients to such processes. In this context, a WPS can process both vector and raster data. A WPS offers three different functions that can be requested by a client:

- "GetCapabilities" provides the client with metadata that describes the capabilities of the available service (including metadata about the provider, a list of available operations or the processes offered).
- "DescribeProcess" provides a detailed description of the process, which also contains a list of the required input parameters and the expected data output.
- "Execute" triggers a process of the WPS, which must be specified in more detail by corresponding call parameters. For example, the required input data, if specified by the process, must be specified via the "DataInputs" parameter. The output can be specified via "ResponseDocument".

A typical call has the form:

```
http://path?request=Execute&service=WPS&version=1.0.0&language=de-
DE&Identifier=Buffer&DataInputs=Object=freiflaechen.shp;
BufferDistance=10&ResponseDocument=BufferedPolygon
```

The wildcard "path" identifies the path for calling the service on a server. The desired process (here: creating a buffer) is specifies by "Identifier". "DataInputs" specifies the required input data (here the shapefile to be buffered and the width of the buffer). The return of the results of this operation depends on the specified return form. In the simplest case of "RawDataOutput", the results are returned directly to the client (e.g. as a GML file). However, the GML file can also be stored temporarily on the server first. Then only the path to this file is returned. The client, such as the geoinformation system QGIS, can then retrieve the file at the appropriate time.

Web Processing Services offer great potential for the processing of geodata. Since only a browser or a small Java program is required on the client side, this new standard can be used to analyse geodata across systems, i.e. on different operating systems and hardware platforms (e.g. desktop, tablet PC or smartphone). The functions are available everywhere via the web and mostly free of charge. Until now, the analysis functions that characterise a geoinformation system and differentiate it from other software systems have been reserved for geoinformation systems on desktops (GIS as a comprehensive software system, cf. Sect. 9.3.4). With a WPS, a user can have ubiquitously access and only access the functions, which he actually needs. This approach is fundamentally more user-friendly than cost-intensive, often overloaded and confusing desktop geoinformation systems. However, the implementation of web processing services is still in its infancy. For example, for each concrete analysis the call has to be specified in a very cumbersome way, so that the handling itself is not yet user-friendly. However, some geoinformation systems support WPS.

6.4.8 Processing Geodata by a Web Map Service in a Geoinformation System

The possibility of accessing geodata with the help of a WMS has a high practical relevance in a geoinformation system. The task often arises of displaying own geodata such as biotope cadastres or representations of leisure facilities with geodata such as maps from public administration. The background information can be official geodata such as the official digital topographic map 1:25000 or official digital maps of the properties, but also user-specific maps such as maps with cycle paths that are only available locally.

A lot of geodata from the surveying offices in particular are directly available via a WMS and do not have to be procured and installed in a laborious manner. The data on the client are always as up-to-date as possible. In addition, they are seamless between adjacent map sheets. Authentication can be used to ensure that only a special group of users can access the geodata (e.g. due to data protection requirements, e.g. when displaying sensitive data such as owner information). Of course, license-based and paid access to geodata can also be realised in this way. The surveying offices of the "open-data-states" in Germany Berlin, Hamburg, North Rhine-Westfalia and Thuringia as well as the Bundesamt für Kartographie und Geodäsie (BKG, Federal Agency for Cartography and Geodesy) make all geodata freely available as WMS, among other things. This form of offer and access is exemplary. This should be the standard practice worldwide. However, it should be noted that the QGIS geoinformation system, for example, offers access to some free geodata (cf. Google Maps or Google Earth API).

Figure 6.4 illustrates how the digital topographic map 1:25000 can be integrated into the QGIS geoinformation system. The digital geodata are available as part of "Open Data -Digitale Geobasisdaten NRW" (cf. GeoBasis.NRW 2023a, cf. Sect. 6.7.5). With the help of suitable software, it is also possible to set up an own WMS for own issues in a company or an authority (see implementation with the Geoserver in Sect. 7.2.2). Further examples of official geodata can be found on the web, although some authorities find it difficult to offer their data freely as WMS for licensing reasons. The geodata supply in Great Britain should serve as an example. Parts of the digital geological map of Great Britain are made available for viewing as a WMS service, namely the 1:50000 scale geological maps for England, Wales and Scotland (cf. British Geological Survey 2023). Usually an application programming interface key (API key) is required, whereby an API key is a unique identifier used to authenticate a user. After registering and paying a fee, the user will receive an access key (cf. British Ordnance Survey 2023a). Often there is free access to a subset of the data just like to OS OpenData British Ordnance Survey (cf. British Ordnance Survey 2023b).

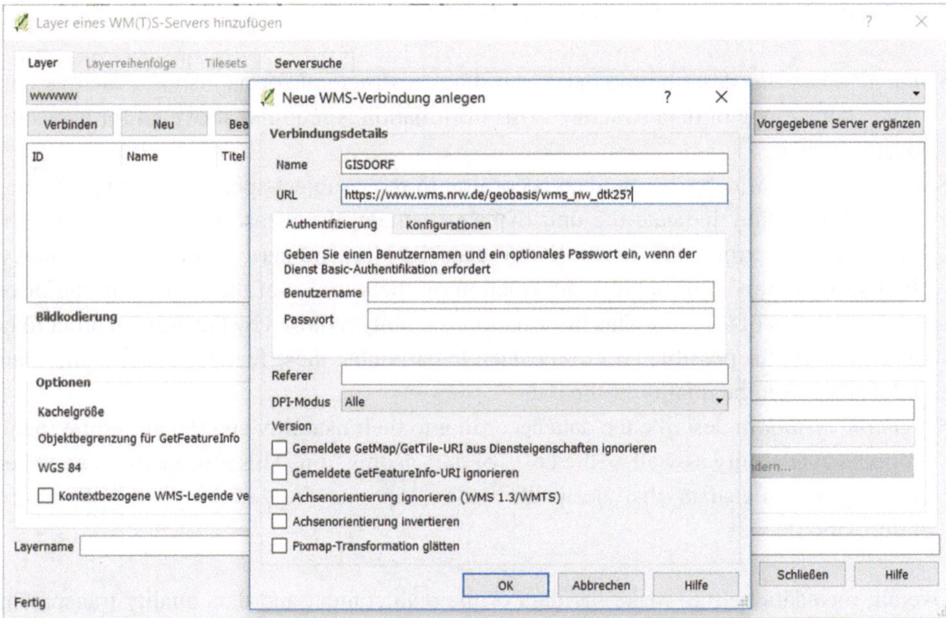

Fig. 6.4 Calling a WMS and integrating the DTK25 in QGIS

6.5 Metadata

6.5.1 From Data to Information Through Metadata

Geodata serve to solve spatial problems. These data can not only carry the most diverse technical meanings and be available for different spatial sections, but can also be stored in a variety of formats and have very different qualities. The extent and heterogeneity of the data stored quickly reach an order of magnitude that makes effective data handling impossible without additional information. Only additional data-describing information, data about data or so-called *metadata*, can then open up the information content of geodata. The availability of data alone is completely insufficient if detailed descriptions are not available, according to which procedures or accuracy specifications, for what reason, at what time and by which processor they were created. The term *metadata* refers to such information that is required for the verification and access to databases or allows the description of complex information in a formalised form. Descriptive metadata should

- inform about data and data sources,
- show data alternatives,
- identify or avoid data redundancies,
- identify or avoid data gaps,
- enable or facilitate the exchange of data.

Several conceptual levels of metadata can be distinguished:

Syntactic metadata describe the data according to structural-formal aspects such as data type, value range or data structure. This information is required above all for a specific data exchange.

Semantic metadata describe the data according to their subject-specific meaning. Among other things, they indicate the unit of measurement, the measuring devices used, the measurement errors, the location reference and the numerical recording accuracy. Furthermore, this also includes information on the purpose of the data recording or on the originators of the data. This information is mainly required by the users in order to be able to assess the possible uses of the data. In particular, these features contain important information on the quality of the data.

Pragmatic metadata describe the data according to their usability and (legal) requirements for data availability as well as the costs of data acquisition. This also includes so-called navigational metadata that designate access options (e.g. access paths, hypertext information).

Overall, metadata help to make the data or the data content and data quality transparent. Such metadata are also necessary when setting up complex environmental information systems, whereby often less new data are recorded than data collections from previous years are processed and transferred into a common structure. Several typical organisational features are present here:

- The recording of the original data was carried out by different agencies, i.e. by other departments in a municipality or by private institutions.
- The data recording was carried out independently of each other with different objectives and varying depth of information.
- The data were recorded without intending to later integrate the data into an environmental information system.

Multiple use and further processing of these data can be quite critical if knowledge of the data collection methodology is insufficient, the reasons for the collection are unknown or the data are incomplete. Without metadata, such specialised data can quickly become worthless. It is only through metadata that geodata ultimately become *geoinformation*. All in all, the simultaneous management of metadata together with the (geo)data is undisputed, when building databases. But it is not clearly mandatory, which and how much metadata should be recorded. On the one hand, hundreds of features are possible to describe data or data resources. On the other hand, in practice, the requirement to collect a lot of metadata prevents precisely this collection. Common arguments are, for example, that the recording takes too much time, a benefit that is not immediately recognizable, or a mismatch disproportion in the extent of metadata and the actual data. In addition, there are diverging ideas as to which metadata are sufficient for the description. Especially against the background of the interoperability of (geo)data, the definition of (international) standards makes sense:

- estimation of the possible uses of (geo)data by different users and for different tasks
- detecting and providing (geo)data by different users

Multiple use and exchange of data are only possible in practice, if the data producers adhere to a relatively small but binding standard to specify the data through metadata.

6.5.2 Standards for Spatial Metadata

6.5.2.1 Dublin Core Metadata Initiative

The metadata set of the *Dublin Core Metadata Initiative* (*DCMI*) forms a general standard for metadata. It includes a vocabulary of 15 properties that can be used to describe general resources. The name "Dublin" dates back to a workshop in 1995 in Dublin, Ohio. The terms "core", "core elements", or "core fields" make it clear that the elements are to be understood broadly, comprehensively, and generically in order to be able to describe a wide range of resources. This general standard is not limited to geodata. However, spatial and temporal features can be specified by the DCMI element "coverage". ISO 15386:2017 contains the 15 core metadata elements in Table 6.3 (cf. ISO 2023c).

Table 6.3 Dublin Core Metadata Initiative metadata set (according to Dublin Core Metadata Initiative 2012)

TermName	Definition
Contributor	Entity responsible for making contributions to the resource
Coverage	Spatial or temporal topic of the resource, the spatial applicability of the resource, or the jurisdiction under which the resource is relevant
Creator	Entity primarily responsible for making the resource
Date	Point or period of time associated with an event in the lifecycle of the resource
Description	Account of the resource
Format	File format, physical medium, or dimensions of the resource
Identifier	Unambiguous reference to the resource within a given context
Language	Language of the resource
Publisher	Entity responsible for making the resource available
Relation	Related resource
Rights	Information about rights held in and over the resource
Source	Related resource from which the described resource is derived
Subject	Topic of the resource
Title	Name given to the resource
Type	Nature or genre of the resource

Table 6.4 FGDC's 1998 Content Standard for Digital Geospatial Metadata (cf. Federal Geographic Data Committee 2023b)

Groups	Selected attributes
Identification information	E.g. citation, description, time period of content, status, access constraints, keywords
Data quality information	E.g. attribute accuracy, positional accuracy, logical consistency report, completeness report
Spatial data organisation information	E.g. point and vector object information (e.g. types and numbers of vector objects), raster object information (e.g. row count)
Spatial reference information	Horizontal and_vertical_coordinate_system definition (e.g. map projection)
Entity and attribute information	Detailed description among others entity type, attribute definition, attribute domain values
Distribution information	e.g. distributor, digital transfer option, available time period
Metadata reference information	Metadata date, metadata use constraints, metadata security information
Citation information time period information	Publication date, title, series information, publications information single date/time, range of dates/times
Contact information	Contact person primary, contact position, contact electronic mail-address

6.5.2.2 Content Standard for Digital Geospatial Metadata

The *Content Standard for Digital Geospatial Metadata* (CSDGM), which was developed by the US Federal Geographic Data Committee (FGDC), served as a model for the structure of metadata (cf. Table 6.4). On the basis of ten categories and a total of more than 200 attributes, the content and context of origin of spatial data are described and rules for documentation are established. Since 1995, every U.S. federal agency has had to document all new spatial data that it directly or indirectly collects or produces according to this standard (cf. Federal Geographic Data Committee 1998). U.S. federal agencies are now being encouraged to use international standards. The FGDC advocates the use of ISO metadata standards. US federal authorities and actors involved in setting up a National Spatial Data Infrastructure (NSDI) are encouraged to switch to ISO metadata (cf. Federal Geographic Data Committee 2023a).

6.5.2.3 ISO 19115

The standard of the International Organisation for Standardisation (ISO) *ISO 19115* "*Geographic Information – Metadata*"is the most important international metadata standard. It defines a comprehensive set of over 400 metadata elements. The core data set of ISO 19115:2003, which is required to identify a dataset, consisted of only 22 elements, which are not all specified as mandatory. The revised ISO 19115-1: 2014 standard defines the schema required to describe geographic information and services using metadata (cf. ISO 2023d). It provides information about the identification, scope, quality, spatial and temporal aspects, content, spatial reference, representation, distribution and other features of digital geodata and services. The standard now takes into account the increasing use of the web for metadata management (cataloguing all types of resources). Core metadata are no longer defined.

ISO 19115-1: 2014 defines:

- metadata sections, entities and elements
- a minimum set of metadata required for most metadata applications
- optional metadata elements to allow for a more comprehensive standard description of the resources, if required
- a method to extend metadata for special needs.

The older compilation of the core metadata fields of ISO 19115-2003 forms the basis for the compilation of metadata, when setting up the German spatial data infrastructures (cf. Table 6.5).

Table 6.5 ISO 19115:2003 Core Metadata for Geographic Datasets of ISO 19115 (GDI-DE Koordinierungsstelle 2023)

Dataset title (m)
Dataset reference date (m)
Abstract describing the dataset (m)
Dataset topic category (m)
Dataset language (m)
Metadata point of contact (m)
Metadata date stamp (m)
Geographic location of the dataset (by four coordinates or by geographic identifier) (c)
Dataset character set (c)
Metadata language (c)
Metadata character set (c)
Dataset responsible party (o)
Spatial resolution of the dataset (o)
Distribution Format (o)
Spatial representation type (o)
Reference system (o)
Lineage statement (o)
On-line resource (o)
Metadata file identifier (o)
Metadata standard name (o)
Metadata standard version (o)
Additional extent information for the dataset (vertical and temporal) (o)

m mandatory, *o* optional, *c* mandatory under certain conditions

6.6 Data and Geodata Quality

6.6.1 Quality Features

Metadata are primarily used to describe the content of data. In this way, quality features are also recorded and described, but less so the quality of data. This can only be assessed on the basis of these quality features with regard to a specific question or suitability for a clearly defined purpose. Although the geometric resolution of raster data is (far) lower than that of vector data, raster data are not therefore "worse" than vector data. Although raster data are not suitable for building real estate or pipeline cadastres, for example, they are the common data model for emission cadastres.

In general, quality can be defined as the sum of all characteristic properties of a product. *Data quality* can be described as the set of data features that enable the data to be used for a specific task ("quality = fitness for use"). Metadata provide evaluation features and therefore ultimately enable the multiple use of data (cf. data quality information as part of ISO 19115-1:2014).

Quality specifications can be given quantitatively using suitable measures or parameters such as standard deviation, RMS error (cf. Sect. 4.2.5.5), confidence intervals or probabilities, or they can be named using qualitative, purely textual descriptions. Table 6.6 lists the quality parameters according to the ISO standard ISO 19113 adopted in 2002. They can be differentiated according to quantitative and non-quantitative features. In the meantime, this old standard has been replaced by the new standard ISO 19157:2013, which specifies the principles for describing the quality of geographic data. The standard

Table 6.6 Quality features according to ISO 19113:2002 (meanwhile revised)

Non-quantitative quality features	
Purpose	Reasons for collecting the data, including information on the intended use of the data
Usage	Use of the data
Lineage	History of the data set, life cycle of the data
Quantitative quality features	
Completeness	Data scrap (too much data), data failure (missing data)
Logical consistency	Consistency in the conceptual, logical, and
	And physical data structure, adherence to value ranges, topological consistency
Positional accuracy	Absolute (or external), relative (or internal) accuracy, accuracy of raster data
Temporal accuracy	Accuracy of a time measurement, validity of temporal indications of dates and temporal relations, consistency of chronology
Thematic accuracy	Correct classification of geoobjects, correct recording of non-quantitative and accuracy of quantitative attributes

- defines components to describe the data quality;
- specifies components and content structure of a registry for data quality measurements;
- describes general procedures for assessing the quality of geographical data;
- specifies principles for reporting on data quality.

In particular, data quality indicators are defined for use in evaluating and reporting data quality (cf. ISO 2023e).

6.6.2 Spatial Resolution, Generalisation and Positional Accuracy

Of particular importance for the quality assessment of geodata are *spatial resolution* and *positional accuracy*. Spatial resolution can include the differentiation of vector and raster data and, in the case of the latter, above all the resolution of the raster image in pixels, i.e. the size of a pixel in a real size specification such as 30 × 30 m. It can also name the extent of a geometric data generalisation, with the transitions to positional accuracy being fluid. The digital representation of a transport network by manual recording of road centre lines on the basis of map at a scale of 1:5000 inevitably leads to a different resolution than the corresponding data recording on the basis of an analog map at a scale of 1:50000. The reason for this is the increasing geometric as well as thematic generalisation of the recording basis as the scale decreases.

The recording and presentation of geoobjects as well as related questions of generalisation are highly relevant in practice if, as is usually the case, several layers of different scales or generalisations are superimposed in a geoinformation system. In municipal planning, for example, the positional accuracy of geoobjects with regard to a regional plan (in Lower Saxony, the so called RROP on a scale of 1:50000) and with regard to a real estate map (scale 1:1000) cannot be compared with one another. The RROP identifies, among other things, areas in which the mining of raw materials such as clay is permitted. In an approval process, it cannot be decided unequivocally whether a plot or part of a brickworks is still in the area designated for mining. A technical overlay of the digital property map and the RROP is possible without any problems in a geoinformation system. The data layers can easily be brought to a uniform scale. However, the different degrees of generalisation in the original scales make it difficult or even impossible to make precise statements about the extent of the designated mining area and therefore about the admissibility of the project. The exact course of a boundary may be simplified by omitting individual intermediate points of the spatially finer map template. Furthermore, simple enlargement or reduction effects must be taken into account. The boundary, represented in the RROP by a line width of 0.1 mm, fills a width of 5 m in reality or 5 mm at the scale of 1:1000.

The *positional accuracy* quantifies the accuracy of coordinates of the position of geoobjects, which essentially depends on the recording method. For example, the engineering surveys of parcels using a theodolite or high-precision DGPS (cf. Sect. 5.3.5) will inevitably produce more accurate results than recording the position of geoobjects using the GPS sensor of a smartphone or from a georeferenced paper map. Generalisation and geometric resolution of a map as well as distortions due to aging of a map can lead to positional deviations.

The use of the data determines the size up to which a position error can be tolerated. While, for example, a positional error of 50 cm is acceptable when recording biotopes when setting up a biotope cadastre, a real estate cadastre or a cadastre of gas and electricity pipelines require higher accuracies. Last but not least, the costs of data acquisition, which generally increase with higher accuracy requirements, play a significant role in economic considerations. The question often arises as to whether the additional benefit justifies the additional expense. In addition to the pure acquisition costs, costs must also be taken into account that arise due to more extensive and therefore more time-consuming data post-processing, data storage and data continuation. Additional costs due to increasing storage requirements are negligible compared to personnel costs.

Overall, no general information can be given about the positional accuracy. Metadata can be used to decide on possible uses. However, a careful handling of positional accuracies is advisable. Thus, errors due to inaccurate point determinations are propagated in both the distance calculation and the area calculation.

6.7 Spatial Data Infrastructures

6.7.1 Spatial Data Infrastructures: Concept and Objectives

A *spatial data infrastructure* (*SDI*) is characterised by a (data) technical and an organisational perspective. On the one hand, a spatial data infrastructure includes spatial databases that are usually linked via the Internet and contain geodata and geospecific data as well as functionalities for exchanging these data. On the other hand, a SDI includes legal, organisational and technical regulations that promote and secure the development and expansion as well as maintenance. The German Federal Digital Geodata Access Act defines: "Spatial data infrastructure is an infrastructure consisting of spatial data, metadata and spatial data services, network services and technologies, agreements on sharing, on access and use as well as coordination and monitoring mechanisms, processes and procedures with the objective of making spatial data of different origins available in an interoperable way." (§ 3(5) Federal Digital Geodata Access Act 2009, cf. Sect. 6.7.3). In general, the optimisation of access and exchange of geodata is the core objective for the development and operating a spatial data infrastructure. This is intended to achieve better, i.e. more reliable, faster, barrier-free information retrieval.

The development of spatial data infrastructures is now a concern in many countries, such as the European Community, the USA (National Spatial Data Infrastructure NSDI) or Canada (Canadian Geospatial Data Infrastructure CGDI). In Germany, as in the other EU countries, the development of a national spatial data infrastructure is closely linked to

European initiatives (cf. Sect. 6.7.2). Similarly, a fundamental prerequisite for the functioning of spatial data infrastructures is compliance with international norms and standards. Only then is it possible for individual, independent and thoroughly heterogeneous components to work together smoothly (cf. above all distributed data offerings in proprietary formats).

6.7.2 INSPIRE

The first activities to set up a European spatial data infrastructure date back to 2001, when the first INSPIRE or E-ESDI expert group (E-ESDI for European Environmental Spatial Data Infrastructure) met and a first action plan was published (cf. INSPIRE 2012). A key objective was to provide the EU's environmental policy with the necessary information from the member states (for an introduction to the policy background of INSPIRE cf. INSPIRE 2023a).

Directive 2007/2/EC of the European Parliament and of the Council of 14 March 2007 for the creation of a spatial data infrastructure in the European Community (*INSPIRE Infrastructure for Spatial Information in the European Community*), which came into force on 15 May 2007, was a landmark in the development of a European spatial data base and had a far-reaching impact on the importance of geoinformation in the EU Member States. The aim of the Directive 2007/2/EC is to facilitate the cross-border use of data in Europe, based on the national spatial data infrastructures, and therefore to establish a European SDI. In particular, the INSPIRE directive obliges all member states of the European Union to provide standardised services for searching, visualising and retrieving data (cf. INSPIRE 2007a, for an introduction to the legal basis of INSPIRE, cf. INSPIRE 2023b).

The creation of a European Spatial Data Infrastructure, which is based on the Spatial Data Infrastructures (SDI) to be set up nationally, should enable the use of interoperable spatial data and spatial services across the various administrative levels. So-called implementing rules are intended to ensure that the spatial data infrastructures of the member states are compatible with one another and can be used throughout the Community and across borders. In this way, INSPIRE supports decision-making with regard to political concepts and activities that may have a direct or indirect impact on the environment (cf. entry into topics on INSPIRE cf. INSPIRE 2023c).

The INSPIRE Directive is addressed to public authorities, i.e. more precisely to the geodata-keeping bodies that perform public tasks (cf. Art. 3(9)), and refers to geodata that are available in electronic form and that relate to one or more of the topics listed in the annexes to the directive (cf. Table 6.7).

Table 6.7 INSPIRE thematic fields in Annexes I to III of the Directive (cf. INSPIRE 2007b)

Themes Appendix I	**Themes Appendix III**
Addresses	Agricultural and aquaculture facilities
Administrative units	Area management/restriction/regulation zone & reporting units
Cadastral parcels	Atmospheric conditions
Coordinate reference systems	Bio-geographical regions
	Buildings
Geographical grid systems	Energy resources
	Environmental monitoring facilities
Geographical names	Habitats and biotopes
Hydrography	Human health and safety
Protected sites	Land use
Transport networks	Meteorological geographical features
Themes Appendix II	Mineral resources
Elevation	Natural risk zones
Geology	Oceanographic geographical features
Land cover	Population distribution and demography
Orthoimagery	Production and industrial facilities
	Sea regions
	Soil
	Special distribution
	Statistical units
	Utility and governmental services

The core requirements of the initiative relate to:

- the generation and regular updating of metainformation on geodata and geoservices
- the provision of geoservices, which are used, among other things, for searching with the help of metadata (discovery services), for displaying geodata (presentation services), for downloading geodata (download services) and transformation services for converting geodata in order to achieve interoperability.

Search services and presentation services are to be made available to the public free of charge. In this context of INSPIRE the recording and collection of new geodata is not explicitly prescribed.

The directive had to be transposed into national law by the member states within two years (i.e. by 15 May 2009). Furthermore, a schedule is laid down which provides for a gradual development of the European spatial data infrastructure (cf. INSPIRE 2023d).

In Germany the coordination office SDI-DE has published a recommendation for the verification to be carried out for the uniform identification of INSPIRE-relevant geodata, with which each geodata-holding agency can verify its own data step by step and report them for INSPIRE.

In addition to the INSPIRE directive, which provides the framework, the 37 Implementing Rules (IR) form the basis for the development of an EU-wide spatial data infrastructure. They are intended to ensure that the individual national SDIs of the EU are compatible with one another and can be used together. A total of five areas have been defined for INSPIRE (cf. INSPIRE 2023e):

- metadata (describing the data and services)
- data specification (defining the data specifications of the topics listed in the annex necessary for interoperability)
- network services (specifications on the performance of the services)
- data and service sharing (development of licenses or copyrights, among other things)
- monitoring and reporting (definition of indicators for quality management of geodata and geodata services and definition of reporting obligations).

Much geodata are now available via the INSPIRE geoportal. As one example from Portugal shows, the current administrative borders of northern Portugal can be downloaded as GML data (cf. INSPIRE 2023f). This situation is similar to other European countries.

6.7.3 Spatial Data Infrastructure in Germany: A Trend Setting Example

Due to the federal structure, the legal implementation of the INSPIRE directive in Germany meant a codification both at the federal level and within the federal states (just like Bavaria or North Rhine Westfalia). At the federal level, the Digital Geodata Access Act 2009 was enacted, which creates the legal framework for the development of a national spatial data infrastructure. In addition to definitions, e.g. of geodata services or geoportals, and the definition of responsibilities and topics (according to the annexes of the INSPIRE directive), it regulates the provision of geodata, geodata services and networks and, above all, the development of a geoportal at federal level. According to § 6 of the Geodata Access Act, data relating to topics in Annexes I to III of the INSPIRE directive must be provided as services (including search services, presentation services, download services and transformation services). In this context, the authorities that provide geodata are obliged to comply with the specifications for all INSPIRE representation services (for more information, cf. GDI-DE 2023a).

The mission of the *SDI-DE* aims at an efficient and innovative provision of public geodata within the framework of a web-based, networked and standards-based spatial data infrastructure (cf. GDI-DE 2023b). In order to achieve this goal, the SDI-DE steering committee and its coordination office were set up at the Bundesamt für Kartographie und Geodäsie (BKG, Federal Agency for Cartography and Geodesy).

In the meantime, the current missions of the SDI-DE are derived from the decisions of the steering committee on the basis of the National Geoinformation Strategy (NGIS). Furthermore, the SDI-DE acts as the national contact point for the European Commission for the implementing of directive 2007/2/EC (INSPIRE). In this important and legally binding function, it formulates the mandatory task for all geodata-keeping agencies: The INSPIRE-relevant geodata and services, after they have been described with metadata and made accessible via web services, must be made available in an interoperable manner in a final step. In doing so, they must meet the requirements of the regulation for implementing the INSPIRE directive with regard to the interoperability of spatial data sets and services. Among other things, the geodata must be transformed into the format specified by the regulation for the 34 INSPIRE themes. By the end of 2020, all spatial data must be made available in an interoperable manner. Different deadlines apply depending on which INSPIRE theme the geodata set is to be assigned to and whether it is an existing or newly recorded geodata data set (cf. GDI-DE 2023c).

In order to identify INPIRE-relevant geodata, the SDI-DE coordination office developed recommendations for geodata-holding agencies in 2018 (cf. GDI-DE Koordinierungsstelle 2020) and developed a so-called SDI test suite for quality assurance for geodata and geodata services (cf. GDI-DE 2023c). In addition, the surveying authorities of the Bundesländer (in Germany federal states just like Bavaria or North Rhine Westfalia) provide extensive documents and support (for Lower Saxony cf. GDI-NI 2023a and 2023b).

The core of the SDI-DE is *Geoportal.de*, which opens up a view of the contents of the SDI-DE and provides central access to the data and services of the SDI-DE (cf. Geoportal. de 2023a). Geoportal.de is linked with the Geodatenkatalog.de, which is a powerful metadata information system for geodata and geoservices (cf. Geoportal.de 2023b).

The federal government implements the basic idea of INSPIRE and the creation of a (national) spatial data infrastructure in an exemplary manner. This should be considered in particular against the background that tax funds were usually used to capture the data. Therefore the data (should) be part of the generally available infrastructure.

The amendment to the Federal Digital Geodata Access Act 2012 accomplish the basis for providing geodata and geodata services, including associated metadata, available free of monetary benefits for commercial and non-commercial use. However, this amendment only affects federal geodata and geodata services. It serves to reduce bureaucracy by regulating the terms of use in a uniform and binding manner. This is intended to activate the potential for added value in the geodata of the federal government. However, this release only concerns databases that are the responsibility of the federal government. As part of the open data offer of the Bundesamt für Kartographie und Geodäsie (BKG, Federal Agency for Cartography and Geodesy), maps and data produced and maintained by the BKG itself from a scale of 1:250000 and smaller are made available free of charge in accordance with the Federal Digital Geodata Access Act (cf. BKG 2023).

The basic idea of INSPIRE has also been implemented in other countries. In the Netherlands, cadastral data are freely available. In Austria, a free high-resolution orthophoto WMS service is available for non-commercial use. In Great Britain, the British Geological Service with its OpenGeoscience project and the Ordnance Survey with its OS Open-Data project offer free data as well as services and interfaces for free integration into their own applications. But in most cases, as with the British Ordnance Survey, you have to register and to purchase an access key, which is subject to a fee, exactly for special or high-resolution data (cf. Sect. 6.4.8).

6.7.4 National Geoinformation Strategy (NGIS) in Germany

The National Geoinformation Strategy was adopted in 2015 by the SDI-DE steering committee. Politicians have recognised that a broad strategic approach is necessary in order to make German geoinformation policy sustainable and forward-looking and to take into account the entire complexity of the topic: Geoinformation as an essential raw material of a digital society (cf. GDI-DE 2023e). This political declaration of intent is now clearly superimposed by the development of a European spatial data infrastructure.

6.7.5 Spatial Data Infrastructure Within the Federal States in Germany

Due to the federal structure of the Federal Republic of Germany, the INSPIRE directive was implemented in state law through 16 state laws, e.g. for North Rhine-Westphalia, Saxony or Bavaria. Despite different names, the contents of the individual laws are strongly based on the Federal Digital Geodata Access Act (cf. Sect. 6.7.3).

The central component of the SDI-NI for Lower Saxony, as well as for other federal states, is a geodata portal, whereby all 16 geoportals differ in structure and clarity, but ultimately not very much in the offers (for an overview of the geodata portals of the federal states, cf. GDI-NI 2023c). Options such as "data offer", "metadata", "GDI standards", "GDI-NI", "INSPIRE" and "general information" are available (cf. GDI-NI 2023d). Searches for geodata, geodata services and specialised information systems are possible via metadata as well as the visualisation of geodata for the public. Usually, geodata viewers are offered that enable visualisation of geodata, i.e. simplified display of maps in a webbrowser. However, the geodata viewers differ considerably in terms of layout and clarity, but above all in the depth of the data offered (cf. the BayernAtlas 2023, the map viewer in GeoPortal.rlp 2023 or the Sachsen-Anhalt-Viewer 2023).

As a further example, the components of the *SDI-NRW* should be listed, which represent the idea of INSPIRE in an exemplary manner and which can serve as an inspiring model for other countries:

- The GEOportal.NRW is the central information platform and interface between users and providers of geodata in NRW (cf. GEOportal.NRW 2023a).
- The GEOviewer as a map viewer offers extensive functions and tools for the research and visualisation of geodata (cf. GEOportal.NRW 2023b). This is the display component of GEOportal.NRW.

– The GEOkatalog.NRW together with the metadata information system of the state of North Rhine Westfalia represents the central component of the GEOportal.NRW (Geodateninfrastruktur NRW 2023).

The geodata available in the GEOviewer, which are systematised according to several subject categories, are exemplary. For example, it enables the visualisation of extensive databases up to large-scale property maps. Above all, these data can be downloaded directly for free.

Particularly noteworthy is the possibility of downloading extensive databases free of charge as part of the initiative "OpenGeodata.NRW" (cf. OpenGeodata.NRW 2023). The following are available in the category "Geobasisdaten":

– 3D building models in LoD1 and LoD2
– 3d-gm digital terrain models
– digital landscape models
– digital elevation models
– digital orthophotos
– digital topographic maps
– real estate cadastre NRW
– special topographic maps

Furthermore, an Open Data download client can be found in the GEOviewer. In addition, there are various web services that are described in the GEOkatalog.NRW and associated catalogs with metadata.

Special attention should be drawn to the extensive standardised geodata services of GeoBasis NRW, which enable the web-based use of the geodata of the NRW state survey and real estate cadastre, regardless of manufacturer and platform (WMS/WMTS, WFS and WCS, cf. GeoBasis.NRW 2023b).

Another visualisation component for NRW is TIM-online (cf. TIM-online 2023). TIM-online is an Internet application of the state of North Rhine-Westfalia to visualise the geodata of the surveying and cadastral administration of NRW (cf. Figs. 5.21 and 5.24). Depending on the zoom level, the service offers all topographic maps from the overview map 1:500000 to more detailed maps 1:10000 and 1:5000 and orthophotos. The property map is also available (cf. Fig. 5.20), including information on the size of the property.

References

BayernAtlas (2023): BayernAtlas. https://geoportal.bayern.de (14.04.2023).

Brinkhoff, T. (2013): Geodatenbanksysteme in Theorie und Praxis. Einführung in objektrelationale Geodatenbanken unter besonderer Berücksichtigung von Oracle Spatial. Heidelberg: Wichmann. 3. Ed.

British Geological Survey (2023): BGS Geology 50K WMS. https://www.bgs.ac.uk/technologies/web-map-services-wms/web-map-services-geology-50k/ (14.04.2023).

British Ordnance Survey (2023a): Welcome tot he Ordnance Survey Data Hub. https://osdatahub.os.uk/plans (14.04.2023).

British Ordnance Survey (2023b): OS Maps API: Overview. https://osdatahub.os.uk/docs/wmts/overview (14.04.2023).

Dublin Core Metadata Initiative (2012): Dublin Core Metadata Element Set. https://www.dublincore.org/specifications/dublin-core/dces/ (14.04.2023).

BKG, Bundesamt für Kartographie und Geodäsie (2023): Open Data. https://gdz.bkg.bund.de/index.php/default/open-data.html (14.04.2023).

Federal Geographic Data Committee (1998): FGDC-STD-001-1998. Content standard for digital geospatialmetadata (revised June 1998). https://www.fgdc.gov/standards/projects/metadata/base-metadata/v2_0698.pdf

Federal Geographic Data Committee, FGDC (2023a): Geospatial Metadata Standards and Guidelines. https://www.fgdc.gov/metadata/geospatial-metadata-standards (14.04.2023).

Federal Geographic Data Committee, FGDC (2023b): Content Standard for Digital Geospatial Metadata. https://www.fgdc.gov/metadata/csdgm/ (14.04.2023).

GDI-DE (2023a): Einstiegsseite zu INSPIRE. https://www.gdi-de.org/INSPIRE (14.04.2023).

GDI-DE (2023b): GDI-DE. https://www.gdi-de.org/GDI-DE (14.04.2023).

GDI-DE (2023c): Interoperabilität. https://www.gdi-de.org/INSPIRE/technische%20Umsetzung/Interoperabilit%C3%A4t (14.04.2023)

GDI-DE (2023d): GDI-DE Testsuite. https://testsuite.gdi-de.org/#/ (14.04.2023) Informationen zur Testsuite.

GDI-DE (2023e): NGIS Nationale Geoinformations-Strategie. https://www.gdi-de.org/NGIS (14.04.2023).

GDI-DE Koordinierungsstelle (2023): German translation of ISO 19115 Geographic Information Metadata. http://docplayer.org/7769550-Deutsche-uebersetzung-der-metadatenfelder-des-iso-19115-geographic-information-metadata.html (14.04.2023).

GDI-DE Koordinierungsstelle (2020): Handlungsempfehlung zur Identifizierung INSPIRE -relevanter Geodaten für geodatenhaltende Stellen. https://www.gdi-de.org/download/GDI-DE_Handlungsempfehlung_Identifizierung_relevanter_Geodaten.pdf (14.04.2023).

GDI-NI (2023a): Metadaten im Geodatenportal Niedersachsen. https://www.geodaten.niedersachsen.de/startseite/metadaten/metadaten-integraler-bestandteil-einer-geodateninfrastruktur-25492.html (14.04.2023).

GDI-NI (2023b): Schritt für Schritt zu perfekten Metadaten. https://www.geodaten.niedersachsen.de/download/26323/Schritt_fuer_Schritt_zu_perfekten_Metadaten.pdf (14.04.2023).

GDI-NI (2023c): Die Geodatenportale der Länder. https://www.geodaten.niedersachsen.de/gdini/geodatenportale_laender/die-geodatenportale-der-laender-25533.html (14.04.2023).

GDI-NI (2023d): Startseite Geodatenportal Niedersachsen. https://www.geodaten.niedersachsen.de/startseite/ (14.04.2023).

GeoBasis.NRW (2023a): Datenangebote. https://www.bezreg-koeln.nrw.de/brk_internet/geobasis/ (14.04.2023).

GeoBasis.NRW (2023b): Geodatendienste NRW. https://www.bezreg-koeln.nrw.de/brk_internet/geobasis/webdienste/geodatendienste/ (14.04.2023).

Geoportal.de (2023a): Geoportal.de, suchen finden, verbinden. https://www.geoportal.de/ (14.04.2023).

Geoportal.de (2023b): Suche nach Metadaten. https://geoportal.de/search.html?q=Metadaten (14.04.2023).

Geodateninfrastruktur NRW (2023): GEOkatalog.NRW - eine Komponente der GDI-NRW. https://www.gdi.nrw/komponenten/geokatalognrw-eine-komponente-der-gdi-nrw (14.04.2023).
GEOportal.NRW (2023a): GEOportal Nordrhein-Westfalen. https://www.geoportal.nrw (14.04.2023)
GEOportal.NRW (2023b): geoviewer. https://www.geoportal.nrw/?activetab=map (14.04.2023).
GEOportal.NRW (2023c): GEOkatalog.NRW. https://www.geoportal.nrw/metadaten (14.04.2023).
GeoPortal.rlp (2023): Kartenviewer im GeoPortal Rheinland-Pfalz. https://www.geoportal.rlp.de/map?LAYER[visible]=1&LAYER[querylayer]=1 (14.04.2023).
Hitzler, P., Krötzsch, M. Rudolph, S. u. Y. Sure (2008): Semantic Web – Grundlagen. Berlin/Heidelberg: Springer.
INSPIRE (2012): INSPIRE Expert Group and Working Groups. http://inspire.jrc.ec.europa.eu/index.cfm/pageid/4 (18.11.2012, no longer available).
INSPIRE (2007a): Directive 2007/2/EC of the European Parliament and of the Council of 14 March 2007 establishing an Infrastructure for Spatial Information in the European Community (INSPIRE). https://inspire.ec.europa.eu/documents/directive-20072ec-european-parliament-and-council-14-march-2007-establishing (14.04.2023).
INSPIRE (2007b): INSPIRE knowledge base. https://inspire.ec.europa.eu/Themes/Data-Specifications/2892 (14.04.2023).
INSPIRE (2023a): INSPIRE Policy Background. https://inspire.ec.europa.eu/inspire-policy-background/27902 (14.04.2023).
INSPIRE (2023b): INSPIRE Legislation. https://inspire.ec.europa.eu/inspire-legislation/26 (14.04.2023).
INSPIRE (2023c): About INSPIRE. https://inspire.ec.europa.eu/about-inspire/563 (14.04.2023).
INSPIRE (2023d): INSPIRE Roadmap. https://inspire.ec.europa.eu/inspire-roadmap/61 (14.04.2023).
INSPIRE (2023e): INSPIRE Implementing Rules. https://inspire.ec.europa.eu/inspire-implementing-rules/51763 (14.04.2023).
INSPIRE (2023f): INSPIRE Geoportal. https://inspire-geoportal.ec.europa.eu/results.html?country=pt&view=details&legislation=all%20 (14.04.2023).
ISO (2023a): Developing Standards. https://www.iso.org/developing-standards.html (14.04.2023).
ISO (2023b): Standards by ISO/TC 211. https://www.iso.org/committee/54904/x/catalogue/p/1/u/0/w/0/d/0 (14.04.2023).
ISO (2023c): ISO 15836–1:2017. Information and documentation – The Dublin Core metadata element set – Part 1: Core elements. https://www.iso.org/standard/71339.html (14.04.2023).
ISO (2023d): ISO 19115:2014 Geographic information – Metadata – Part 1: Fundamentals. https://www.iso.org/standard/53798.html (14.04.2023).
ISO (2023e): ISO 19157:2013 Geographic Information – Data quality. https://www.iso.org/standard/32575.html (14.04.2023).
ISOTC211 (2023): ISO/TC 211 and OGC collaborate on geoprocessing standards. https://www.isotc211.org/press/1997/05/14/tc211-ogc-collaborate-on-standards.html (14.04.2023).
OGC (2023a): About OGC. https://www.ogc.org/about-ogc/ (14.04.2023).
OGC (2023b): OGC Abstract specification. https://www.ogc.org/standards/as/ (14.04.2023).
OGC (2023c): OGC Standards. https://www.ogc.org/standards/ (14.04.2023).
OGC (2023d): OGC Simple Feature Access – Part 1: Common Architecture. https://www.ogc.org/standards/sfa (14.04.2023).
OGC (2023e): OGC Geography Markup Language (GML): https://www.ogc.org/standards/gml (14.04.2023).
OGC (2023f): OGC Geopackage. https://www.geopackage.org/ (14.04.2023).

OGC (2023g): OpenGIS Web Map Tile Service Implementation Standard. https://www.ogc.org/standards/wmts (14.04.2023).

OGC (2023h): OGC WPS 2.0.2 Interface Standard Corrigendum 2. http://docs.opengeospatial.org/is/14-065/14-065.html (14.04.2023).

OpenGeodataNRW (2023): Datenangebot OpenGeodata.NRW. https://www.opengeodata.nrw.de/produkte (14.04.2023).

Sachsen-Anhalt-Viewer (2023): Sachsen-Anhalt-Viewer. https://www.lvermgeo.sachsen-anhalt.de/de/startseite_viewer.html (14.04.2023).

TIM-online (2023): TIM-online 2.0. https://www.tim-online.nrw.de/tim-online2/ (14.04.2023).

Spatial Information: Visualisation

7

7.1 Interdisciplinary View of Cartography

7.1.1 Digital Graphic Presentation of Information

Graphic information processing refers to all graphic, i.e. non-alphanumeric, representations of information using a computer and special graphics-capable input and output devices. The term *graphic presentation of information* characterises very comprehensively both the acquisition and the presentation. In particular, it includes digital, graphics-oriented information technologies and forms of communication such as *2D* and *3D computer graphics*, *computer animation*, *multimedia techniques* and *augmented reality* as well as *virtual reality.*

In graphic information processing, information without spatial reference is often also graphically presented, such as pure numerical data that is to be presented in the form of bar charts, or e.g. photos or images that are to be prepared for advertising purposes. Such forms of graphic presentation, which include so-called business graphics, are not characteristic of geoinformatics. Instead, *graphic forms of presentation of geoobjects* are of interest here (for the definition of geoobjects cf. Sect. 4.1).

7.1.2 Computer-Assisted Scientific Visualisation

Computer-assisted scientific visualisation is a relatively new field of research that has existed since the late 1980s, involving computer science, cognitive science, psychology, and communication research, and which can be referred to as *Visualisation in Scientific Computing* (ViSC). The beginnings are often traced back to McCormick et al. (1987) (for a

N. de Lange, *Geoinformatics in Theory and Practice*, Springer Textbooks in Earth Sciences, Geography and Environment,
https://doi.org/10.1007/978-3-662-65758-4_7

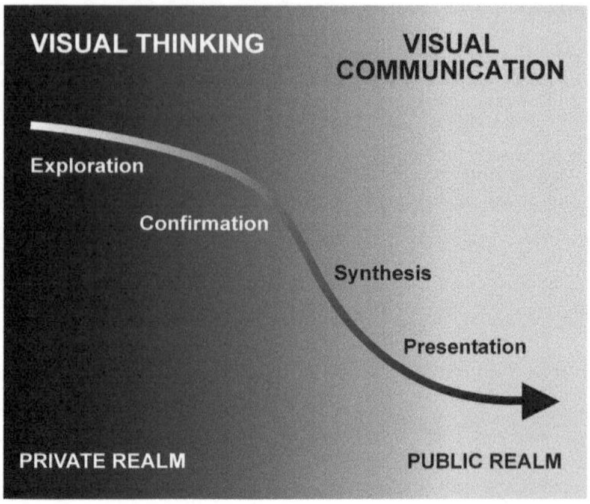

Fig. 7.1 Visualisation as a tool of scientific research according to DiBiase 1990. (Quoted from MacEachren 1994 p. 3)

brief overview cf. Wood and Brodlie 1994). This refers to research methods in computer science that aim at the interactive visual exploration of large scientific data sets in order to stimulate the visual thinking of specialists. The data sets are presented graphically to improve their readability or to reveal structures or regularities. In this context, computer-assisted scientific visualisation does not (primarily) aim at methods of how existing specialist knowledge can be "descriptively" implemented using graphical methods and representations (explanative communication). Rather, the focus is on the actual cognitive process (explorative communication).

Fundamental to this understanding of visualisation as a tool of scientific research is DiBiase's model, which clarifies the various functions of (computer-assisted) visualisation in the research process, whereby a fundamental distinction is made between two areas. In the internal domain, the scientist only communicates with himself. Visualisation here means "visual thinking", which serves to explore and structure data as well as to find and verify hypotheses (based on data analysis). In the external domain, the scientist communicates with others. Visualisation here means "visual communication". Figure 7.1 summarises the aspects in a synthesis.

The so-called Anscombe quartet is often used to prove that only a graphic presentation reveals structures in data sets. A still very small and clear data set of four pairs of characteristics for 11 units is to be evaluated (cf. Table 7.1). In the original form, the data are abstract and of academic interest. However, one can definitely create a reference to reality in order to counter the arguments that these number series are only imaginary and cannot occur at all in this form. The data could represent the altitude (in 100 m) above sea level (features x_1, x_2, x_3), the monthly mean temperatures in April, May and October (features y_1, y_2, y_3) as well as the number of frost days in January (x_4) and the snow depth in mm (feature y_4) for 11 climate stations.

An evaluation using methods of descriptive statistics provides an image of equality or homogeneity of the distributions of x_i and y_i, respectively, as well as of the correlations

Table 7.1 The Anscombe Quartet (source: Anscombe 1973)

I		II		III		IV	
X_1	Y_1	X_2	Y_2	X_3	Y_3	X_4	Y_4
10.0	8.04	10.0	9.14	10.0	7.46	8.0	6.58
8.0	6.95	8.0	8.14	8.0	6.77	8.0	5.76
13.0	7.58	13.0	8.74	13.0	12.74	8.0	7.74
9.0	8.81	9.0	8.77	9.0	7.11	8.0	8.84
11.0	8.33	11.0	9.26	11.0	7.81	8.0	8.47
14.0	9.96	14.0	8.10	14.0	8.84	8.0	7.04
6.0	7.24	6.0	6.13	6.0	6.08	8.0	5.25
4.0	4.26	4.0	3.10	4.0	5.39	19.0	12.50
12.0	10.84	12.0	9.13	12.0	8.15	8.0	5.56
7.0	4.82	7.0	7.26	7.0	6.42	8.0	7.91
5.0	5.68	5.0	4.74	5.0	5.73	8.0	6.89

For these dates applies in each case:

Arithmetic mean values of x_i	$= 9$
Variances of the x_i	$= 11$
Arithmetic mean values of the y_i	$= 7.5$
Variances of the y_i	$= 4.12$
Correlation of x_i and y_i	$= 0.816$
Linear regression	$y_i = 3.0 + 0.5\ x_i$

between x_i and y_i. All variable x_i as well as y_i have the same arithmetic mean and variance, the correlation coefficient is relatively high. Then an evaluation with the help of parameters of statistical methodology is finished. An editor will not even think of questioning this result, but will present exactly this image of homogeneity, obtained with the help of objective statistical parameters, as the final result. Only the representation of the feature pairs in a diagram, i.e. the visualisation, breaks up this procedure. The four pairs of features show very different correlation structures, which are not resolved by the statistical parameters. This analysis of a still very manageable amount of data must plunge scientists into a deep depression. What is the situation with large data sets? Can parameter-based statistical analysis deliver "correct" results? Is there a risk that these methods, which are considered to be objective, will obscure facts?

The calculation of statistical parameters alone is often not sufficient. Therefore, scatterplots (so-called correlograms) should always be created in correlation and regression calculations (cf. de Lange u. Nipper 2018 p. 127). However, graphical forms of presentation by no means guarantee the desired gain in knowledge. In particular, differently scaled diagram axes can lead to misinterpretations. A "clever" choice of the scale of the horizontal axis for the feature x_2 would result in a linear progression of the points instead of an arc-shaped one (cf. Fig. 7.2). For this reason, in comparative representations, a normalisation or standardisation of the diagram axes is required (cf. de Lange and Nipper 2018 pp. 161).

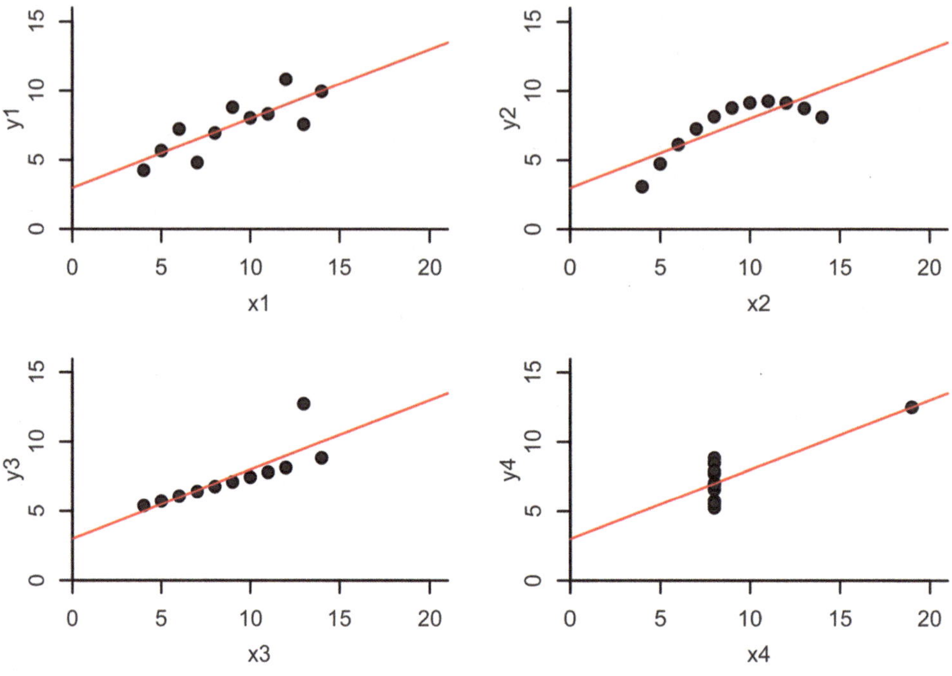

Fig. 7.2 Visualisation of the Anscombe quartet (cf. Anscombe 1973 p. 19)

7.1.3 Geovisualisation

Geovisualisation can be understood as a special form of computer-based scientific visualisation. The term "Geographic Visualisation (GVIS)" goes back to MacEachren and Ganter (1990) and was later shortened to "Geovisalisation" (cf. MacEachren et al. 1999). The 2001 definition emphasises visual exploration instead of visualisation: "Geovisualisation integrates approaches from visualisation in scientific computing (ViSC), cartography, image analysis, information visualisation, exploratory data analysis (EDA), and geographic information systems (GISystem) to provide theory, methods, and tools for visual exploration, analysis, synthesis, and presentation of geospatial data (with data having geospatial referencing)." (MacEachren and Kraak 2001 p. 3).

The conceptual development was based on the model of the map use cube according to MacEachren (1994), which highlights four different functions of map use (presentation, synthesis, analysis, exploration, cf. Fig. 7.3, for the development cf. Kraak and MacEachren 2005 and Schiewe 2013). Geovisualisation therefore pursues an integrative or interdisciplinary approach, in which cartography has an important function. Geovisualisation is continued in the more recent interdisciplinary research field geovisual analytics (geospatial visual analytics, for an introduction cf. Meng 2011 pp. 252–253, cf. G. and N. Andrienko 2005 and G. Andrienko et al. 2007 and 2010).

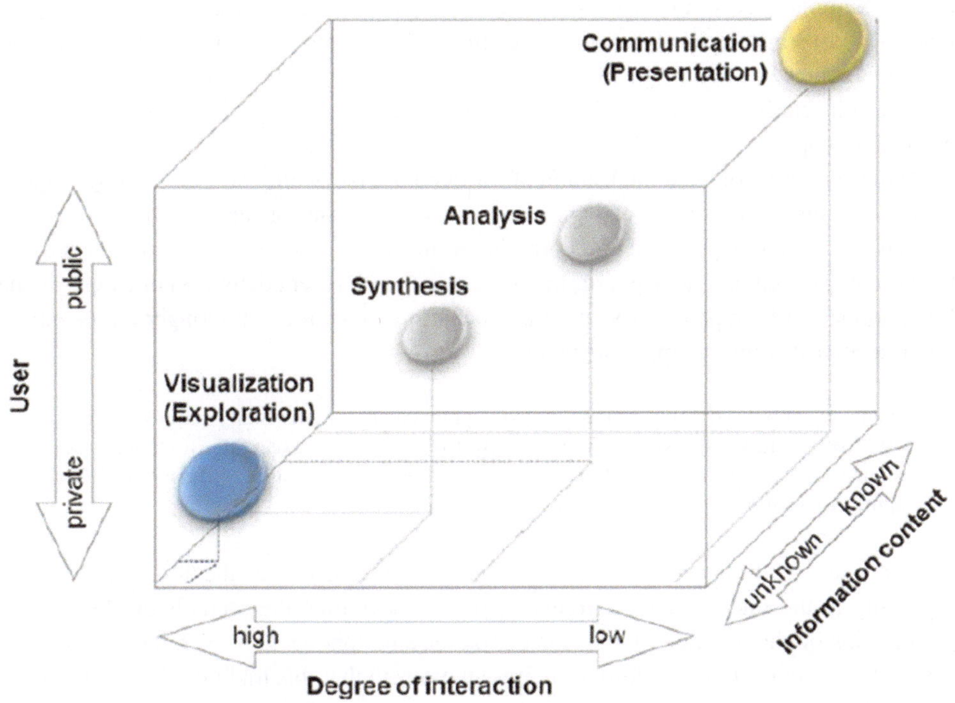

Fig. 7.3 Map use cube according to MacEachren 1994 as depicted by Schiewe 2013

The term visualisation therefore creates a new conceptualisation that is to be understood in two ways (cf. Schiewe 2013). On the one hand, the focus is on the visible product, i.e. the generation of maps to illustrate spatial data and information and make them visible (a classic concern of cartography). On the other hand, the use of map is highlighted, which leads to the generation of hypotheses and knowledge by the user and through the use of an interactive environment (e.g. through a geoinformation system). Interactive visualisation is understood as an explorative research approach to process complex and large amounts of data in the scientific knowledge process.

7.1.4 Digital Graphic Presentation of Geoobjects: Paradigm Shift Within Cartography

The new understanding of visualisation or geovisualisation is reflected in the more recent conceptual orientation of cartography and maps. The traditional view, for which static and non-interaction methods are central, is being replaced by a new view that emphasises more explorative and interactive processes.

Graphic forms of representation of geoobjects have a long tradition in *cartography*, which has developed principles of representation of spatial information that are still valid today: "Cartography is the discipline dealing with the art, science and technology of making and using maps" (International Cartographic Association 2023a). In contrast to this comprehensive definition of the International Cartographic Association, which is elegant in its clarity, older definitions were still primarily aimed at the technology and art of making maps.

At the beginning of the digitalisation of cartography, the duality between analog maps in paper form on the one hand and the model of digital data on the other hand was still emphasised, which is permanently stored in an information system, for example, and can be presented differently as required. In contrast, the most recent definition of a *map* by the International Cartographic Association is much more general and no longer refers explicitly to technical forms of implementation:

> A map is a symbolised representation of geographical reality, representing selected features or characteristics, resulting from the creative effort of its author's execution of choices, and is designed for use when spatial relationships are of primary relevance. (International Cartographic Association 2023a)

The meaning of the term "*map*" has changed significantly over the last twenty years, especially in the multimedia environment. In the meantime, the focus is on the graphic presentation on the monitor, on the tablet or on the smartphone. However, there is no other short or concise term for this form of representation, so that this traditional name has been retained. Paper maps, like digital presentations in a geoinformation system on a monitor, tablet or smartphone, have the function of an interface between the database and the user. These forms of presentation are clearly superior in conveying spatial information, due to the inherent chorographic properties of a map or two-dimensional representation to visually capture and perceive holistic and contextual spatial relationships (cf. Sect. 7.3). What is new compared to the "paper era" with unchangeable content is that the (digital) representation is also to be understood as an interactive interface that enables deeper access to various information through interaction functions and that triggers further communication processes. In particular, it is important to highlight the changes that affect the role of the actors. "In the open information society, former map users are empowered to independently generate maps from their own geospatial data." (Meng 2011 p. 250). This is precisely what characterises the real paradigm shift in cartography. Maps are no longer used passively as unmodified representations, but are "read" interactively. Moreover, the creation of (digital) representations of spatial information is no longer solely a matter of cartographic experts (for an assessment of the current situation cf. International Cartographic Association 2023a, 2023b and Virrantaus et al. 2009).

The digital presentation of geoobjects as an *interactive communication interface* can (so far) be differentiated according to three stages of development:

In the first stage, the development started primarily with the increasing spread of geoinformation systems. Thus, the presentation of geodata is a constituent feature of a geoinformation system (cf. Sect. 9.1.2). Spatial orientation in a GIS takes place via digital cartographic representations on the monitor. Visualisation is used to graphically display the

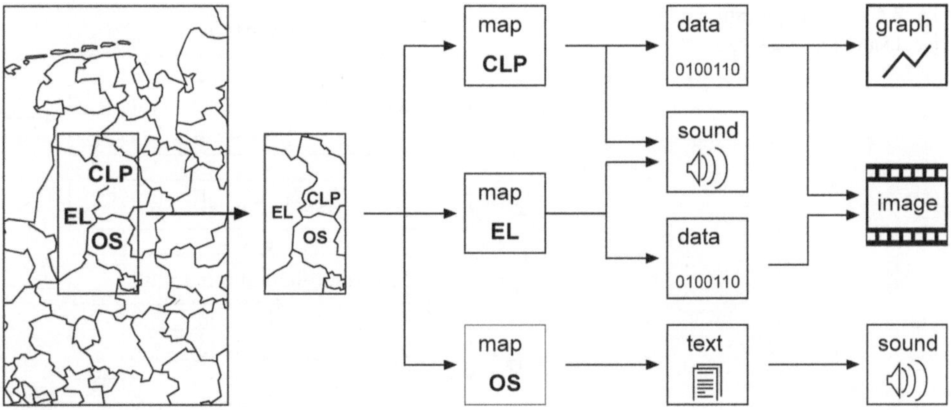

Fig. 7.4 Illustration of the hypermap concept

section of the earth's surface that is modeled in the geographic information system. The function of a digital map as an interactive communication interface between the data or the digital primary model (cf. Sect. 7.3) and the user becomes clear very early on. The GIS specialist or the GIS user, neither of whom necessarily has to have cartography knowledge, became a producer of maps at an early stage.

In the second stage, in the course of the almost exploding web applications, map applications and map services emerged on the WWW, which chronologically followed the presentation options of a desktop GIS. The term *web mapping* is used here to describe the creation and provision of maps in the World Wide Web (cf. Behncke et al. 2009, cf. Sect. 2.8.3 and for the technical realisation of an example Sect. 7.2.3). Initially, interactive map applications were offered on the WWW as ready-made products. The user was able to view the representations interactively using navigation and query functions (including zooming, but was not yet able to add its own data. Already at this stage, multimedia maps or presentations were of central importance, in which graphics, images, photos, aerial photographs, texts or sound could be called up via so-called hyperlinks. The basis for this is the so-called *hypermap concept*, which is structured similarly to the hypermedia concept (cf. Fig. 7.4).

Starting with a digital overview map on the monitor, which is called up from the WWW, a variety of information opens up one after the other, which is linked to one another according to content. The user navigates independently through the information offered, depending on his question and his prior knowledge. The context-dependent navigation through a database and the data exploration with a digital information system is characterised by a feedback process which enables the question to be made more specificly in a dialog and offers new answers that characterise the topic more precisely (cf. Fig. 7.5). This *data exploration* creates the decisive leap in quality compared to the classic use of analog maps! It should be noted that a geoinformation system also offers multimedia presentations and data exploration, but it was only the Web that provided the breakthrough

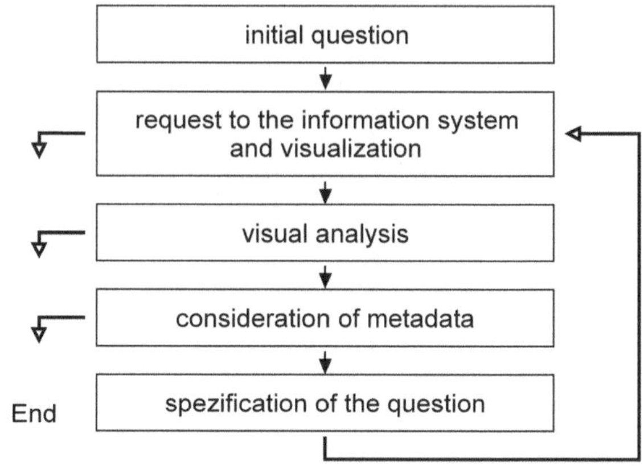

Fig. 7.5 Data exploration and visualisation as a process

for these presentation options due to the free availability of applications and data as well as easy handling.

In the third stage there are even more far-reaching options beyond the multimedia approach and data exploration, which are already "revolutionary" compared to analog maps:

In Web 2.0, users not only become consumers, but also producers of their own graphic representations.

The rapid spread of mobile devices leads to an enormous and constantly growing range of apps that use graphics or map applications as a matter of course and that are mostly connected to the web.

As the degree of interaction between the user and the digital presentation increases, applications for augmented and virtual reality open up completely new options for presentation.

7.1.5 Geoinformatics: Presentation

While geovisualisation provides the scientific framework for the presentation and exploration of geodata, far more practical or application-related tasks have to be solved in geoinformatics and in the (everyday) use of a geoinformation system. Compared to the visualisation of data without spatial reference, the presentation of geoobjects faces special challenges, since not only the attribute values have to be presented, but the geoobjects have an individual location, a neighborhood to other geoobjects, whereby a dynamic presentation should be excluded for the time being. Above all geoinformation systems allow an easy and quick presentation of geoobjects (for the definition of geoobjects cf. Sect. 4.1).

As shown in Sect. 7.1.2, the Anscombe quartet has four different sets of data pairs which have the same representation in terms of identical statistical parameters. In contrast, a single

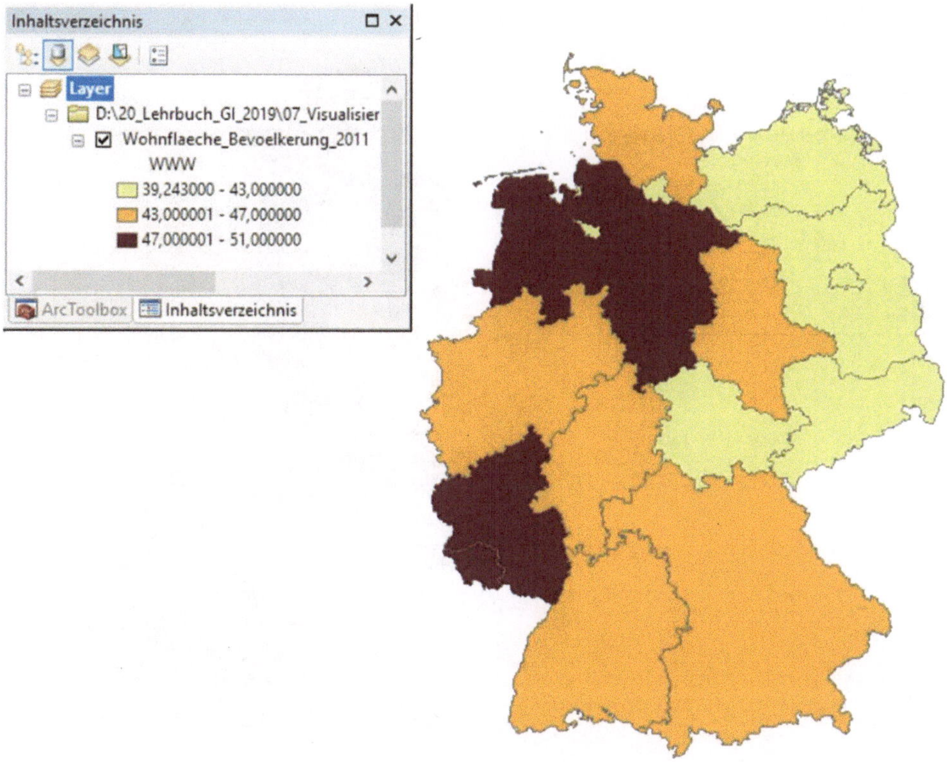

Fig. 7.6 Living space per inhabitant in m^2 for the German Federal States 2011 (simplified screenshot), draft 1

data set in cartography can be converted into different representations with differing statements (see Figs. 7.6, 7.7, 7.8 and 7.9). The question to be asked is how the topic of the geoobjects is "correctly" reproduced. Figures 7.6, 7.7, 7.8 and 7.9 show a housing indicator which is recorded in the German census and is used in official statistics in Germany. However, in international statistics housing condition are presented in form of e.g. an overcrowding rate.

The three screenshots show the indicator "living space per inhabitant in m^2", which is important for spatial planning in Germany. An identical feature is shown in each case. Figure 7.6 illustrates a stronger west-east difference (Saxony-Anhalt excepted), which could give rise to the launch of housing promotion programmes in the easterly federal states. Figure 7.7 does not show this "contrast", and finally Fig. 7.8 even shows a "balanced" picture. All three figures are based on the same data set, but the presentations are very different. Finally the question must be asked: What is "true" or "correct"?

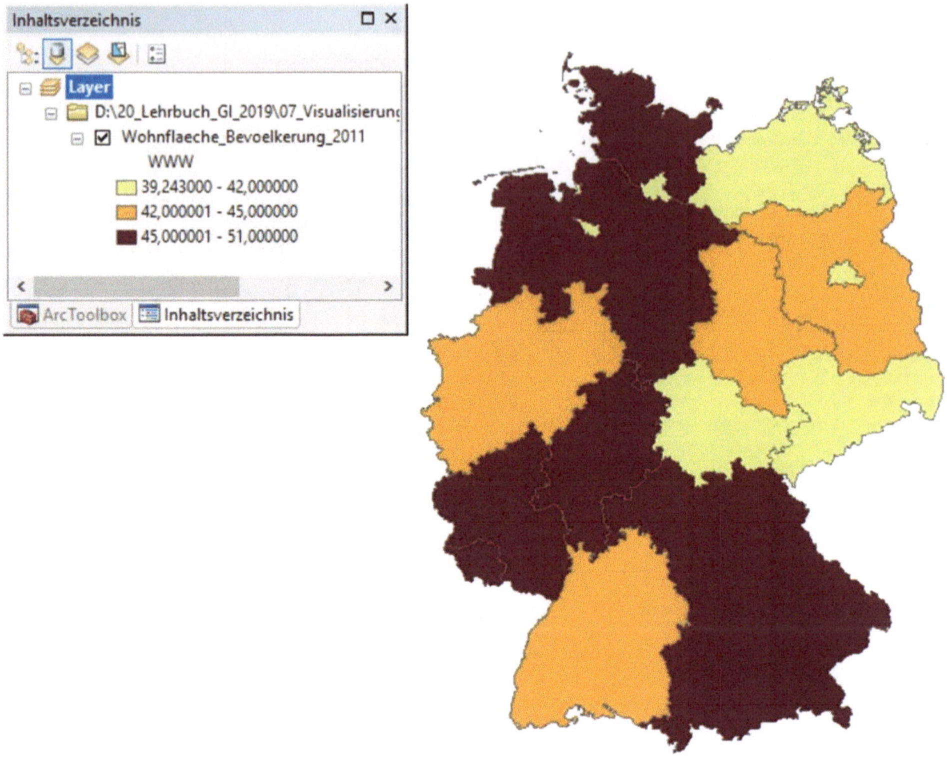

Fig. 7.7 Living space per inhabitant in m^2 for the German Federal States 2011 (simplified screenshot), draft 2

Such representations characterise almost the standard situation in geoinformatics, whereby the three illustrations even use the design tool color "correctly". Much more common are presentations such as Fig. 7.9, which contains the two cardinal errors of cartography or graphic semiology. On the one hand, the feature "living space in m^2", which consists of absolute values, is incorrectly represented by an area symbol (area filling, here the area of a Federal State with a pattern) and, on the other hand, this quantity is represented by a color. A graphically correct presentation would implement the absolute numbers using symbols of different sizes, such as circles, whose radius is proportional to the value to be presented (cf. Sect. 7.5.2). The circle symbols may also be rendered in a color graded according to lightness, but not in different colors because color cannot express values. These principles of representation belong to the standard knowledge in thematic cartography. They are based on the principles of semiology and Bertin's theory of visual variables (cf. Sect. 7.4). Unfortunately, these principles are not implemented by default in

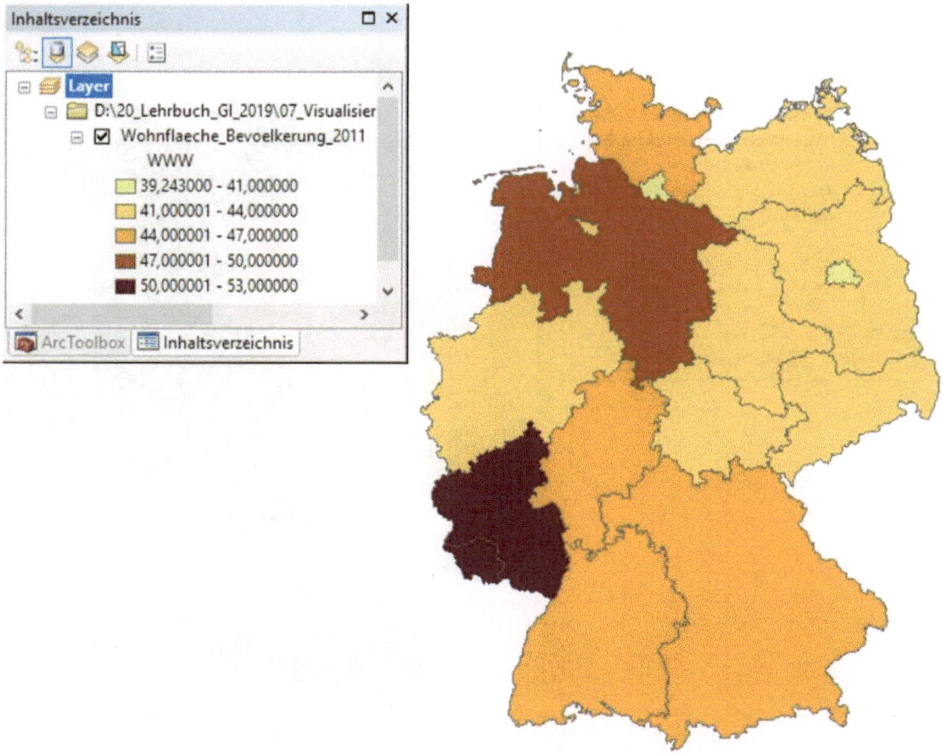

Fig. 7.8 Living space per inhabitant in m^2 for the German Federal States 2011 (simplified screenshot), draft 3

geoinformation systems. The user must intervene in order to generate graphically correct or meaningful representations.

The requirement for a correct, undistorted, non-manipulated and objective reproduction of information is fundamental. Unfortunately, this is an illusion, since a graphic representation always depends on the efforts and skills of the author to use graphic methods adequately, and always on the graphic communication process (cf. Sect. 7.3) with its many filters that prevent objective recognition. Cartographic representations are subject to the risk of manipulation or misinterpretation of the content. Especially an “inexperienced” map reader will not be aware of the limits of map interpretation that have been pointed out. The representation principles of thematic cartography or graphic semiology, which have been tried and tested for many years, also help to adequately implement graphic design in geoinformatics.

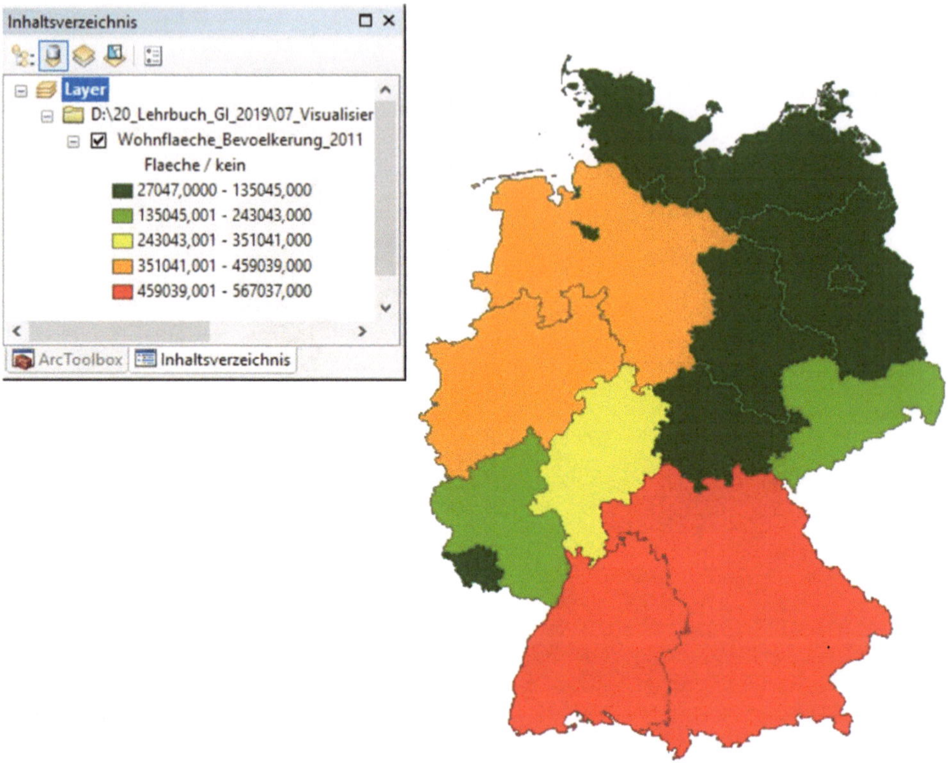

Fig. 7.9 Living space per inhabitant in m^2 for the German Federal States 2011 (simplified screenshot), draft 4

7.1.6 Augmented Reality: Virtual Reality

In geoinformatics, more and more applications of augmented reality can be found in the course of the spread of smartphones. Augmented reality (AR) or extented reality is generally understood as the computer-assisted extension of the perception of reality, which can include all human sensory perceptions. AR applications are mostly limited to visual information that supplements reality with additional computer-generated information such as text, images or videos. Almost everyday examples can be found in television broadcasts of football matches, for example, when a line or a circle is displayed to mark an offside situation or an area to be kept clear by opposing players in the case of free kicks.

More important for geoinformatics are approaches, in which additional information is superimposed on a digital image of the real environment in real time on the display of a smartphone or tablet computer. This technique has been widely used in the game Pokémon

Go for handheld mobile devices such as smartphones and tablet computers. Various sensors are required, which are now standard on modern mobile devices. The sensors for the satellite navigation signal and the gyroscope (position and rotation rate sensor) are used to position the user exactly and to display the superimposed information in the correct position in the full screen. This can be a virtual figure, as in the game, or explanatory text (for the introduction cf. Broll 2013).

An implementation of this game idea can be used to document changes of the landscape by a broad public (collection of photos through crowd sourcing). It must be ensured that the photos of an object (e.g. a section of a shore) can be taken from an almost identical recording location and from almost identical angles by different users. The user is navigated to a recording location where the camera view of the smartphone changes to an augmented reality view. The user is then instructed to align the camera to a superimposed target marker and subsequently take a photo (cf. Albers et al. 2017, cf. Kreuziger 2014).

The use of augmented reality in navigation has a promising future. In contrast to the well-known turn-by-turn navigation of common navigation systems, the route can be found using a sequence of prominent landmarks (e.g. church towers, buildings) on the way to the destination. In the smartphone camera, arrows indicate the direction to the next waypoint as well as the image (possibly with text) that reflects the next intermediate destination (cf. the prototype GuidAR in Schofeld et al. 2017). This navigation corresponds more to intuitive guidance ("from the intersection straight on to the white monument"), but it is extremely complex, as it requires the acquisition of landmarks.

Other applications can hardly be estimated: displaying virtual building within construction gaps to other buildings, superimposing data on the load-bearing capacity of floors and bridges or building information in the event of a fire, or projecting navigation instructions onto the windshield. A particularly innovative, spatial variant for navigation is the tactile orientation aid for the blind and visually impaired using vibration methods via the so-called navigation belt (cf. feelspace 2023).

Compared to applications of AR methods, implementations with virtual reality (VR) techniques can be much more (device-) complex if the user is involved in a computer-generated virtual world in real time and wants to act or move in it. While AR supports or overlays the visualisation of the real world, the possibilities of virtual reality go one step further. The real world is mapped with the user into a virtual world. This requires special output devices (VR headsets, shutter-glasses) that transmit virtual images of the environment directly to the eyes. The user no longer sees the real environment, but only images of the environment or a virtual environment. The user is embedded in the virtual world (so-called immersion) and interacts with it, e.g. via a data glove. This immersion can go so far as to see the virtual world as real. Vehicle simulators, such as those used to train pilots, are certainly the best-known area of application (for an introduction cf. Dörner et al. 2013).

7.1.7 Virtual Reality in Geoinformatics: 3D City Models

In geoinformatics, virtual reality applications with a high degree of immersion are not common. In contrast, there is an abundance of virtual representations of the environment that the user can view on the monitor. With Google Earth and Google Street View, one can virtually "walk through" many cities and regions on the monitor in a browser. Above all, virtual 3D city models are increasingly used to solve spatial tasks in urban and spatial planning (cf. the basic introduction by Coors et al. 2016, Dickmann and Dunker 2014, Edler et al. 2018a, Edler et al. 2018b).

The 3D city or building models are differentiated according to the *level of detail* (*LoD*). Level of detail 1 (LoD1) represents the buildings uniformly as simple 3D blocks with a flat roof (so-called "block model"). The next level of detail is called level of detail 2 (LoD2), which presents the buildings with standard roof shapes such as flat roof, or gable roof, as well as the outer shell with simple textures. Level of detail 3 (LoD3) characterises an architectural model that best reflects the visual impression of the original buildings (photorealistic texture). Due to the high effort, mostly only single buildings are shown in LoD3. Even more sporadically, buildings are found in level of detail 4 (LoD4), which show an interior model of the building with floors.

3D city models are available for almost every major German city. In Saxony, for example, a state-wide digital 3D city model has been set up. The buildings are available as LoD1 or LoD2 (cf. Landesvermessung Sachsen 2019). Particularly outstanding examples are available for Karlsruhe and Helsinki (cf. Karlsruhe 2023, Helsinki 2023, cf. also Deutsche Gesellschaft für Kartographie 2023). For Rhineland-Palatinate, a state-wide LoD2 model was developed and made available online (cf. Hilling and Greuler 2015). In addition to local offers from official surveying authorities, Google Earth offers a huge range of very detailed 3D maps for cities all over the world.

7.2 Graphic Presentations on the WWW

7.2.1 Web Mapping

Web mapping applications operate in the form of a client-server architecture, with the webbrowser serving as the client. The user interactively requests functions that are processed by a mapserver (cf. Fig. 7.10). The map as a result is then sent back to the client. In many cases, the function of the client are limited to visualisation and simple functions such as zooming, moving the map content or distance measurements (for the difference to Web-GIS cf. Sect. 9.2).

A distinction can be made between static and dynamic applications, even if they are not initially aligned to static and dynamic maps. In a static mapserver application, the server sends an (unchangeable) map back to the client as a raster image (static map), such as a (static) overview or route map. The user can, for example, move or zoom the map. The map

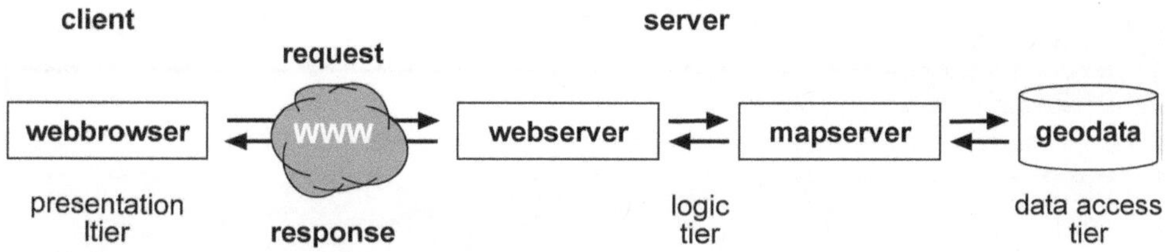

Fig. 7.10 Principle of a typical web mapping application

itself remains unchanged unless new map sections are requested from the server. In a dynamic mapserver application, a map is dynamically generated by the server based on the client's specific request and returned to the client. Common examples on the WWW are maps that change their appearance depending on the zoom level and, for example, display the symbols in a more differentiated manner as the zoom level increases (instead of one large symbol for a city, several symbols spatially resolved for individual city districts, as in Google Maps, for example). Further examples are topic-based selections, where the user can select different data levels (layers) (simple web information system).

7.2.2 Mapserver: Use Case

In practice, a frequent task is to make geodata of a company or a public authority, such as access maps, a building land cadastre or a map of land values, available to employees or a larger public. This requires the provision of a suitable server system and mapping software on the client (cf. Sect. 7.1.3 and on web mapping as a Web-GIS substitute Sect. 9.2.3). The example of the free software "GeoServer", shows how geodata are made available on the server side (cf. Fig. 7.11). The Apache Tomcat webserver is used in the standard configuration, which is included in the software download. The administrator of the GeoServer makes the geodata available by setting paths to data directories, whereby the data can be in different formats. Furthermore the layout is defined. The GeoServer provides, for example, a map service that offers a WMS layer via the WWW, which can then be presented in a browser using mapping software (cf. Sect. 7.2.3).

The GeoServer can also be used directly as a client (i.e. not in administrator mode). The (free) software can be installed as often as desired in an authority or a company, whereby the devices must have web access. Then it is possible to select different data layers, to display them in any zoom level, e.g. in Google Earth or in a client with e.g. OpenLayers. By clicking on an object, the attributes can also be queried.

Interactive webmaps and applications can be created and shared on the WWW, with both free as well as proprietary software solutions similar to GeoServer. The content created can be accessed independently of the hardware platform, i.e. desktop, in the browser, with smartphone or tablet computer. Such applications are of particular relevance for government agencies and companies that want to share the latest maps and data with different devices in different project groups. For example, the proprietary offers ArcGIS

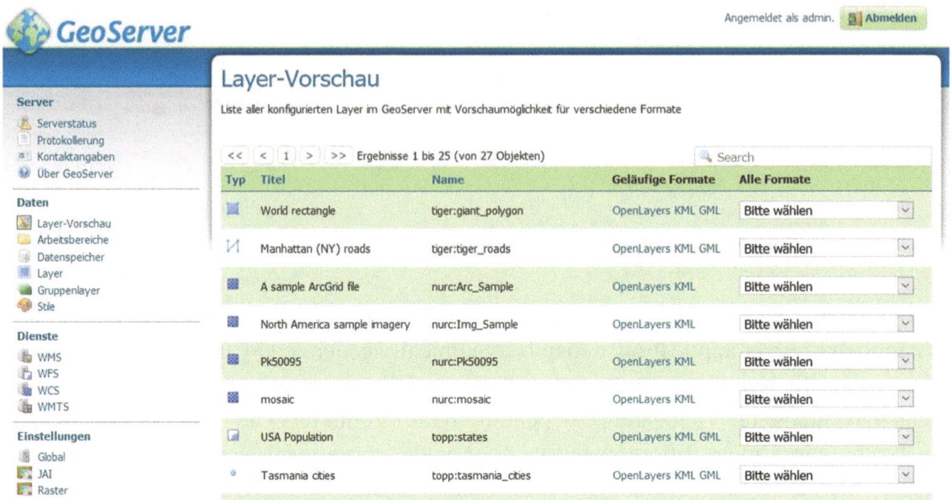

Fig. 7.11 Administrator view of the GeoServer

Online or GeoMedia WebMap are comprehensive, cloud-based GIS mapping software systems (Software as a Service, cf. Sect. 2.8.5) that connect people, locations, and data using interactive maps (cf. ESRI 2023 u. Hexagon 2023). This is particularly important for tasks that have to be done on mobile devices (e.g. mobile data recording or surveying tasks). Similarly, the free QGIS Cloud is a powerful platform for publishing maps, data and services on the WWW (cf. QGIS Cloud 2023a). With just a few instructions, one's own maps can be presented to a broad public on the WWW via qgiscloud.com. For example, a plug-in for QGIS Cloud is offered in QGIS, which simplifies putting the QGIS map online (cf. QGIS Cloud 2023b). Similar services exist, for example, from NextGIS and Gisquick (cf. NextGIS 2023 u. GISQUICK 2023).

7.2.3 Mapping Software: Example

While a mapserver provides the web map, the client must have mapping software that presents the web map. The WMS layer, offered by a free software like GeoServer or a proprietary software, is included in a HTML page with Javascript, which is retrieved from a server. Figure 7.12 shows in a very simple way the implementation with the free Javascript library Leaflet, which can be used to create web mapping applications.

Leaflet supports Web Map Service, Web Map Tile Service and the GeoJSON format for displaying geodata (cf. Sects. 6.4.1 and 6.4.4). With the JavaScript library OpenLayers, which is also widely used, geodata can be displayed in the browser independently of the server software used. Like Leaflet, OpenLayers provides typical web mapping elements, such as a scalebar for changing the displayed scale.

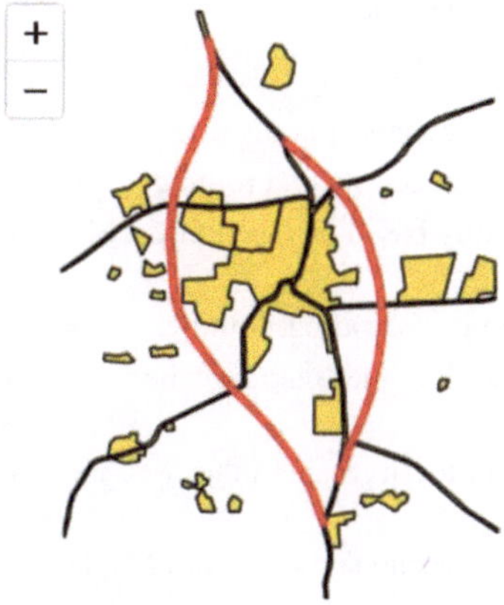


```
<html>
  <head>
    <meta charset="utf-8">
    <link rel="stylesheet" href="https://unpkg.com/leaflet@1.3.1/dist/leaflet.css"/>
    <script src="https://unpkg.com/leaflet@1.3.1/dist/leaflet.js"></script>
    <script src="jquery-3.3.1.min.js"></script>
    <title>Leaflet WMS Example</title>
    <script>
      // The loading of the pages is checked.
      $(document).ready(function(){
        var map_variable = L.map("my_map", {
          center: [52.049598, 8.032763],
          zoom: 13
        });
        // Since Leaflet cannot process the „getCapabilites" call, another way
        // via a Javascript-Object must be chosen.
        var wmsLayer = L.tileLayer.wms("http://131.173.22.70:8080/geoserver/ows?", {
          // Spezification leaflet wms options, at least property "layers"
          // Search for available layers in getCapabilites xml file
          layers : "GISdorf:siedlungUTM32,GISdorf:umgehungostUTM32,  GISdorf:
                                                        umgehungostUTM32"
        });
        map_variable.addLayer(wmsLayer);
      });
    </script>
  </head>
  <body>
     <div id="my_map" style="height:600px"></div>
  </body>
</html>
```

Fig. 7.12 Presentation of a web map with the Javascript library Leaflet

7.2.4 Graphic Presentations in Applications

Many web services provide programming interfaces (APIs, cf. Sect. 2.8.4) for programming map applications (cf. Table 7.2). Frequently, base map data such as street maps and satellite images are offered directly or can be obtained from map service providers. Various map services are available that can be used as base map data for user-specific maps (cf. Table 7.3).

This technique is used in many ways, e.g. by visualisation the location on a map on a homepage of a company and often linked to a navigation system. The web designer or computer scientist selects the base map data and supplements them with further information if required. This can be own data or references to further Internet sources. The term

Table 7.2 Selected web mapping programming interfaces

API	Features
Google Maps	Including JavaScript, static-maps, places, directions and Street-View
Bing Maps	Heatmaps, spatial math, geocoding and routing
MapQuest	Including JavaScript, partly based on Open Street Map data (e.g. directions, traffic and geocoding)
ViaMichelin	Fee-based, JavaScript (e.g. geocoding, routing, real-time traffic situation)
Map Channels Feed Maps	Easy-to-use APIs that access other map providers (esp. Google Maps, Mapbox)
Mapstraction	Free software, JavaScript for accessing other web mapping APIs like MapQuest
Mapbox	Access to map-related tools and services from Mapbox (e.g. Vector Tiles API), including geocoding, navigation, etc.
OpenLayers/Leaflet	JavaScript libraries for the integration and display of geodata and maps in a web map

Table 7.3 Selected map services

Base map data	Features
Google Maps	Different cartographic representations (street map, terrain map, aerial or satellite images, hybrid view, streetview), with own APIs (Google Maps APIs)
Bing Maps	Different cartographic representations (street map, aerial or satellite images, hybrid view, bird's eye view), with own APIs (Bing Maps SDK)
MapQuest	Various cartographic representations (map, satellite), with custom APIs (MapQuest APIs)
Open Street Map	VGI project, different predefined map styles available or own map style possible, usable e.g. with OpenLayers, MapQuest Open Maps API, Mapstraction
HERE	Emerged from Smart2Go, Map24, Navteq, Ovi Maps and Nokia Maps, online map navigation
Apple Maps geo. okapi	Map service for IOS applications freely available JavaScript library for the creation of interactive map applications for maps (BKG, Bundesamt für Kartographie und Geodäsie 2023)

mashup is used for this combination of data such as texts, images, sounds and videos from different sources on the Internet. Thus, mashup maps are created by displaying spatial data on already existing base maps (cf. Hoffmann 2011).

7.2.5 Cartography on the Web 2.0: Web Mapping 2.0

Web 2.0 is described by interactive and collaborative features (cf. Sect. 2.8.4). Accordingly, the term Web Mapping 2.0 refers to the interactive creation of maps on the Internet by the end user, who plays the important role of map designer. He no longer relies solely on

Table 7.4 Selected web mapping applications for generating maps on the web

Web mapping services	Features
Google My Maps	Drawing points of interest on Google Maps
GeoCommons	Selection of different base maps from various providers, extensive design options for displaying statistical data (for thematic maps), various data imports and exports
Scribble Maps	Selection of various base maps from different providers, diverse design options as well as data imports and exports
Mapbox	Provider of custom online maps for web presentations and applications
StepMap	Base map Open Street Map or own base maps in different styles, simple drawing functions
TargetMap	Presentation of statistical data in thematic maps

ready-made maps in a passive role, but can become an active producer himself by creating his own maps from various data sources with the help of web-based tools. He is producer and consumer at the same time, i.e. "*prosumer*" according to Toffler (1980, cf. also Hoffmann 2011). In each case, the end user essentially controls the cartographic representation process. However, often only a rudimentary cartographic expertise is implemented in the software. The web mapping software specifies the cartographic design options and often severely restricts them.

Meanwhile, the producers of their own (thematic) maps on the WWW no longer need to have programming knowledge. They do not have to program their own map application via a programming interface. Rather, relatively easy-to-use software enables them to create maps directly using a graphical user interface in a webbrowser (cf. Table 7.4). However, the creation of own thematic maps is less important than thought. There is relatively little demand from users, and correspondingly, relatively few software offerings that can be used to create your own maps. Apparently, many users would like to use Google My Maps to link their own photos and set point symbols at points of interest to them.

7.3 Visual Communication

Cartographic representations, i.e. generally graphic presentations of geodata, stand out as the best way to pass on spatial information. This principle remains even through the use of the new information and communication technologies. The main advantages are the inherent chorographic properties:

- Positional relationships are recorded directly and intuitively, which would otherwise have to be described laboriously in text form. Neighborhoods and distances can be quickly perceived visually.
- Spatial structures, i.e. the spatial distribution of data, can be recognised "at a glance". Comparisons with distribution patterns of other spatial features and processes are easier.

When emphasizing such advantages, however, it is often misunderstood that an image, a graphic, a map or a display of a geoinformation system on a computer or a mobile device is a medium for e transmitting information, through which distortions, attenuations or losses of information can also occur, as with any other information exchange. Such transfer problems are rarely considered. It is almost always assumed that the graphic presentation is "true" or that it "correctly" represents the reality depicted in the data and that the viewer can read the graphic presentation and understand it "correctly". It is assumed that the (carto)graphic communication process (cf. Fig. 7.13) runs smoothly without interference. Figures 7.6, 7.7, 7.8 and 7.9, however, already show that both basic assumptions do not have to be true. The title of Monmonier's standard work on methods of map manipulation is correspondingly provocative: "How to lie with Maps." (cf. Monmonier 1996). Both the producer and the reader of the graphic representation must have basic knowledge in order to be able to create the graphic representation correctly as well as to be able to read it and understand its contents. This general basic knowledge controls the cognitive process. Added to this is the technical knowledge of the sender as well as the receiver. Furthermore, the type and extent of the perception and storage of the information are also determined by the general context as well as the (subsequent) use of the information.

Geoinformation systems often provide graphic presentations that do not implement the traditional and proven design principles of cartography (cf. Fig. 7.9). It is not surprising that the viewer of the presentation then only perceives the information and message incompletely or distorted. The risk is that the message of the creator is transmitted in a distorted way.

Figure 7.13 schematises the *(carto)graphic communication process* that is to be transferred to geoinformatics issues:

Starting from the environment, a *primary model* of reality is created. Here, various forms of simplifications or generalisations of reality occur: filtering of information by

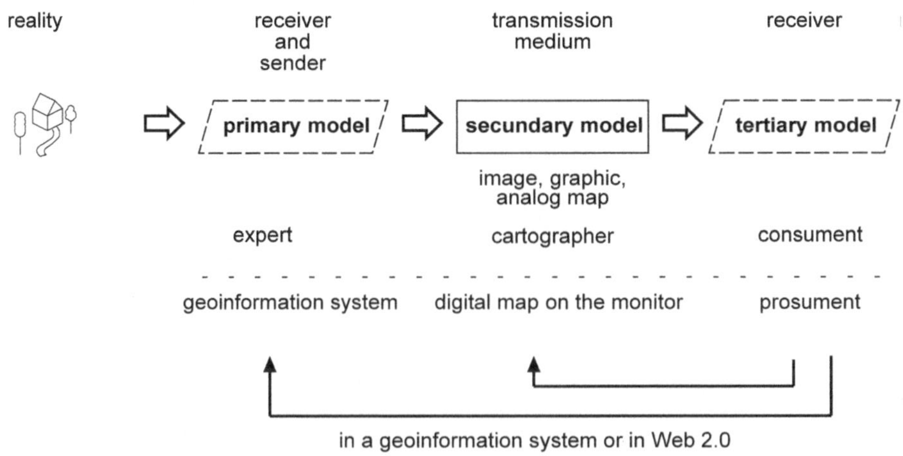

Fig. 7.13 Model of (carto)graphic communication

omitting "unimportant" information, abstraction of complex facts. For example, the road with a row of trees on both sides between two town centres is simplified to a straight connection. This creates an abstract primary model of reality, which is generally subject-related. Such a primary model exists virtually in the memory of the expert, who records and stores information. This projection of reality into the primary model is subjective, it can be incomplete, distorted or even incorrect.

In the meantime, the section of the real world is very often represented as a digital model in a geoinformation system. The pure abstraction in the form of a data model, still without graphic illustration, can then also be understood as a primary model. The mapping process is simpler here, the mapping rules are transparent, because the data acquisition is operationalised more strictly (e.g. geometries in the primary model of the geoinformation system based on official surveys).

With regard to a graphic presentation, the next step of information processing is the conversion of the primary model into a graphic description. The graphic presentation, i.e. the analog or also digital representation, is to be understood as a *secondary model* of reality. Traditionally, information processing was often carried out in such a way that the cartographer received instructions on how to graphically convert the scientist's primary model (e.g. verbal explanations or a hand-drawn sketch). Several communication problems are possible with these implementations. So it is conceivable that the cartographer may not fully understand the scientists or their concerns. The graphic design tools available in the geoinformation system (or in the API) do not allow an adequate implementation of the primary model, since, for example, the geoinformation system does not have a symbol for hardwood forest as is common in German topographic maps. Mapping errors can also occur here if, for example, colors or symbols evoke false associations (e.g. use of the color blue, which should be reserved for water areas and not commercial areas or agriculture).

In a third step of the overall communication process, the information, i.e. actually the messages encoded in the graphic, is received by the user and processed into a (new) model of the environment (*tertiary model*). Due to the user's lack of prior knowledge or inability to read and understand graphics as well as the ambiguity or lack of accuracy of the graphic presentation, reading errors may also occur here. For example, the straight line connecting two cities, represented by a symbol for a highway, should not illustrate the actual course of the road, but rather a transportation link. The exact distance is indicated by a mileage plotted on the map or by clicking on the object in a window that opens on the monitor. While these distances are usually only "read off" later, it is obvious at first glance which cities are connected by highways.

Two basic problems of graphical communication can be derived from this:

On the one hand, information that is available in the form of texts, numerical data, drawings and pictures, or even thoughts, must be transformed or translated into a graphic representation by the sender. This traditional task still applies when using the new technologies. However, now this transformation into a secondary model, in particular into a digital presentation on the monitor, is often done by a user, who usually does not have the expertise for the correct use of the (carto)graphic methods and representation tools!

On the other hand, the recipient must understand the graphic information "correctly", i.e. in the sense of the sender of the information. However, the spatial perception of a graphic, a picture or a map, the recognition of connections, the structuring of the map content or the comparison of structures create individual options of interpretation. Misinterpretations are easily possible due to the wrong use of cartographic instruments. The tertiary model then no longer corresponds to the primary model.

These problems of (carto)graphic communication – regardless of whether in analog or digital form – are given particular weight, since a graphic or pictorial presentation has a high attraction and possible sources of interference are not recognised or questioned. The frequently found, always positive sentence "A picture is worth a thousand words." is therefore clearly to be problematised or to be questioned.

In view of the changing roles of user and producer of graphic presentations in a geoinformation system or in the Web 2.0, the classic communication process, in which the user or consumer only appears at the transition from the secondary to the tertiary model, must be significantly expanded (cf. Hoffmann 2011 in more detail). The user now has considerable influence on the creation of all three models. He is particularly involved in the creation of the primary model (geodata) and the secondary model (cartographic representation). The models can now no longer be separated from one another exactly, the transitions are therefore fluid.

In *Web Mapping 2.0* (cf. Sect. 7.2.5), the base map (e.g. Open Street Map), which certainly has the function of a primary model, can also be regarded as a secondary model, since it already has a cartographic representation. Furthermore, users are able to modify the base maps with the help of suitable map services (cf. e.g. StepMap) and to add their own symbols in order to create new secondary models. Users can also actively participate in the creation of the primary model by compiling user-generated data such as GPS tracks of roads and paths, as in the Open Street Map project, or information about so-called points of interest and making them available to the basic service. This also applies in a similar form to a geoinformation system in which the geoinformatician or user supplements the data stock and creates the secondary model.

As before, viewing the graphic presentation leads to the *tertiary model* and therefore creates the user's virtual image. However, the tertiary model of the user now plays a new role, i.e. in GIS applications. Already at the beginning of the process, the user has certain ideas of his own environment, which prompt him to add or edit (basic) data (i.e. in the case of the primary model) and which influence him in the preparation of the presentation (i.e. in the case of the secondary model). The resulting tertiary model can lead to a feedback process and then also to changes in the primary and secondary models. For example, the user may update the base map data by adding data, or change the graphic presentation by selecting other symbols. This in turn can influence the tertiary model.

The user, i.e. the GIS specialist, who recently also acts as a producer of maps or graphic presentations, is often faced with a major challenge when it comes to correct graphic or cartographic presentation or communication. The user is often not even aware of this. This also shows that basic knowledge of the graphic semiology and design tools is indispensable in geoinformatics (cf. Chap. 7.4).

7.4 Graphic Semiology

7.4.1 Graphic Semiology: According to Bertin

Digital graphic presentation and cartography decisively control the emergence of the secondary and tertiary models (cf. Fig. 7.13). In this context, *symbols* are of great importance as carriers of information, i.e. in the case of a digital representation on the monitor or in a map. Thus, a map can also be understood as an abstract representation using point, line and area symbols (cf. definition according the International Cartographic Association international cartographic association in Sect. 7.1.4). In this context, a *symbol* is a graphic, image, written word, specific pattern, or sign that expresses a specific fact or topic through association, resemblance, or convention. Almost everyone knows the symbols that indicate the location of a hospital, a railway line or a lake.

Symbology is embedded in the more general framework of *semiotics*, which deals in a more epistemological way with sign processes. In semiotics, as the science and study of signs and symbols, the following dimensions can be distinguished:

The *syntactic dimension* refers to the formal creation of the symbols and to their relationships to one another. A graphic presentation is syntactically perfect, if the structure of the drawing is recognised correctly (cf. e.g. size, spacing, contrast or brilliance of the symbols). A double line, for example, is often used to represent a motorway, where the double line must be marked by a sufficient spacing.

The *semantic dimension* describes the meaning of the symbols. Thus, the meaning of the symbols at the sender (e.g. map producer) must be identical to that at the recipient (e.g. map reader). A red cross, for example, is a universally understood symbol to indicate the location of a hospital on a map.

The *pragmatic dimension* aims at the purpose and effects of the symbols. The symbols can influence behavior changes and make decisions. A double line, for example, indicating a more significant street that is believed to allow for a higher speed, decides the choice of the route.

Beyond this basic differentiation, Bertin developed the graphic semiology, which can be understood as a theory of the graphic presentation of information (cf. Bertin 1974). Bertin continued this system without making significant changes (cf. Bertin 1982). Bertin distinguishes, according to the basic geometric shapes in a geoinformation system, only

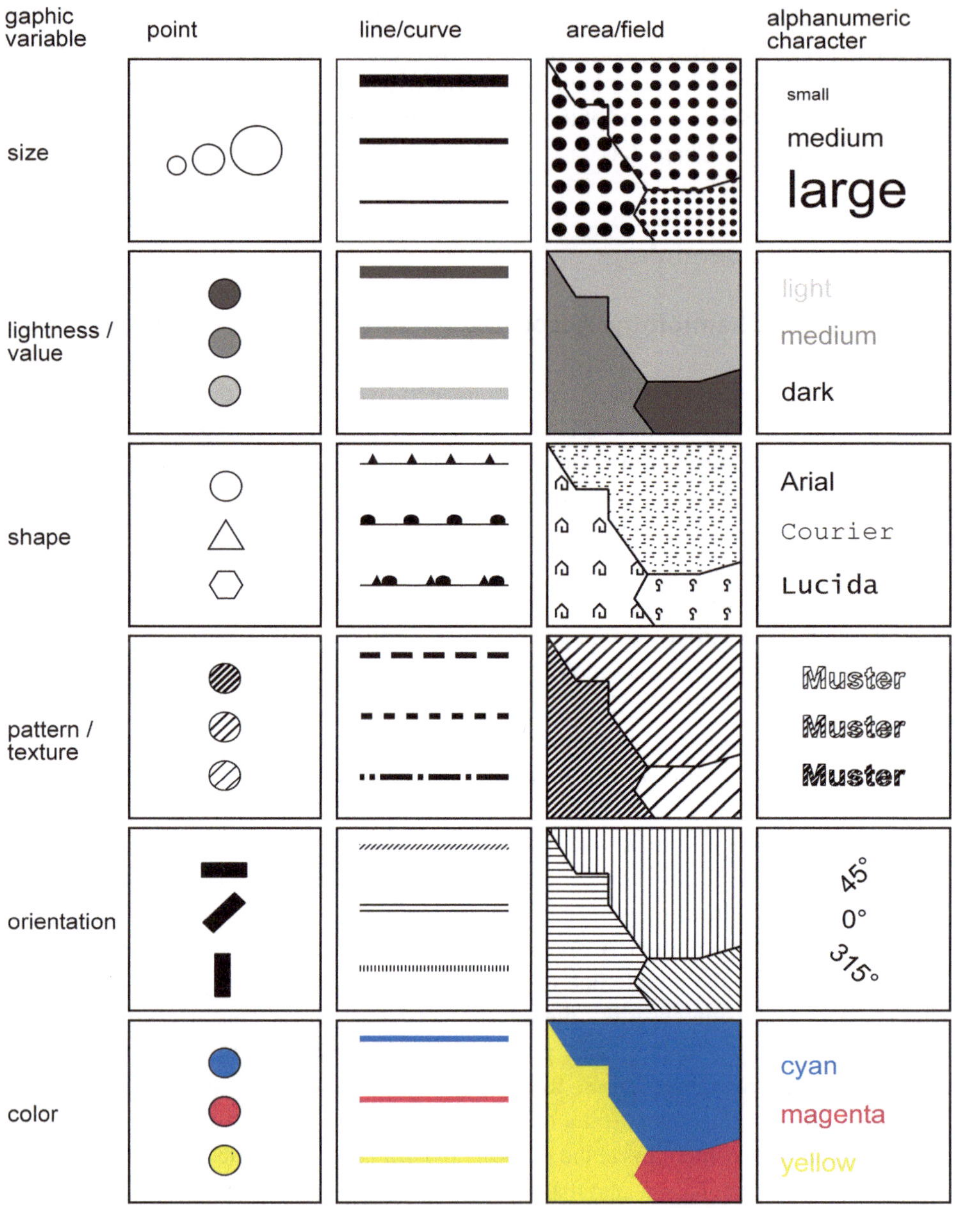

Fig. 7.14 Visual variables according to Bertin 1974 with extension of alphanumeric characters

three basic graphic elements: point, line and area. For example, in a map, cities are usually represented by point symbols, roads or rivers by line symbols, and lakes or forests by area symbols. In an extension, the alphanumeric characters, i.e. letters, numbers and special characters, should also be understood here as basic graphic elements (cf. Fig. 7.14).

Basically Bertin developed a framework of visual variables, which systemises the design tools (i.e. the symbols) for map processes and that can be used to drive the symbology and the visual appearance of the objects of a map and to differentiate them visually. A graphic representation of an object in a map always consists of at least two components. On the one hand, the position of the map objects is determined by the two dimensions of the plane (i.e. x- and y-coordinates). On the other hand, the thematic content is described by visual variable(s). Bertin identified six visual variables (besides the two dimensions of the plane): size, lightness, shape, pattern, orientation, and color.

These visual variables also allow symbols to be varied depending on the scale of the data. For example, the visual variable size with its quantitative property is well suited to display metrically scaled data. Ordinally scaled data are best represented by the visual variable lightness, because it has an ordering property.

Figure 7.14 shows ideal-typical design options. Although no generally accepted standards exist, cartography has developed conventions for the use of these basic graphic elements, which can also be transferred to graphic presentations in a geoinformation system or on smartphones. However, not every variable is equally applicable to all map objects. For example, the design of lines is very rarely varied in terms of orientation or lightness. For area symbols, the size is almost never changed in practice (in contrast to the example in Fig. 7.14).

Figure 7.14 also only shows single design variants. In a map or digital representation, however, several visual variables are usually implemented to express a single topic such as lines of different colors and width, which are intended to represent different types of road.

This systematisation makes it possible to analyse in detail the very different characteristics and means of expression of the visual variables. According to Bertin, only four special features or characteristics can be distinguished, which are not equally pronounced for all variables:

- *association* (associative = connecting, uniform visibility; dissociative = dissolving, different visibility):

  This characterises the similar perception and visibility. Thus, the ability to recognize connections and connecting structures among the objects is addressed. For example, on the one hand circles (e.g. in different sizes and color) indicate the different locations of different medical practices and on the other hand squares indicate offices of lawyers.
- *selectivity:*

  This ability expresses the property that symbols are detected more or less clearly differently. The visual variables allow a selective and separating perception of the objects. For example, the different shapes of circles and squares are clearly distinguishable as point symbols.
- *order:*

  This ability aims at the (different) performance of each visual variable in expressing a ranking between objects. For example, the different lightness (from light to dark) may indicate an increasing population density. Of course, point symbols of different sizes can express an order.

- *quantity:*

  This characterises the ability to show quantitative relationships that go beyond a pure order. For example, the size of a circle can be directly proportional to the size of the population of a city.

7.4.2 Design Rules Based on Bertin's Graphic Semiology

Figure 7.15 systematises basic rules or principles based on the stated properties of visual variables. These basic principles have lost none of their importance. They are still valid for cartographic representations in geoinformation systems and in web mapping (cf. in more detail Bertin 1974 pp. 73):

- The size of the symbols reflects a quantitative attribute (e.g. absolute number of inhabitants). The size of a symbol (e.g. size of a circle) allows an estimate of the actual numerical value. In this way, the map symbols can be compared quantitatively. Section 7.5.2 explains in more detail, how quantities can be expressed.
- Pattern and lightness have similar abilities, whereby the representation of rankings is to be emphasised by differences in lightness (ordering effect).
- Size and lightness are of different visibility. According to Bertin, they are dissociative. Several point distributions in a graphic, which only differ in size or in the lightness of the point symbols are not evenly visible. Simplified: bigger (darker) symbols are more "visible". Larger (darker) symbols are seen first and not a group of circles. Thus, the distribution pattern dissolves. In contrast, the other visual variables are evenly visible (e.g. a point distribution based solely according to the shape of the symbols). However, size and lightness also dissolve any other variable with which they are combined. For example, for point symbols that vary in size and brightness, the number of distinguishable colors decreases. Size and lightness dominate over the other variables.
- The shape (e.g. circle or square as a point symbol) only allows localisation of objects. The shape (alone) is no longer selective. This visual variable is hardly suitable for recognizing spatial relationships (cf. distributions of point symbols, which differ solely on the basis of their shape, but not, for example, according to their size or lightness).
- Orientation has little more capability than shape. It only provides selectivity on point symbols and les linear symbols.
- Using color, a high associative effect of the objects results. The recognition and structuring of the graphics are made easier.
- Colors also have good separating and selective properties.
- Strictly speaking, colors have no ordering effect, unless there is a grading, e.g. from a dark red via red and yellow to a bright yellow, in which case the lightness actually

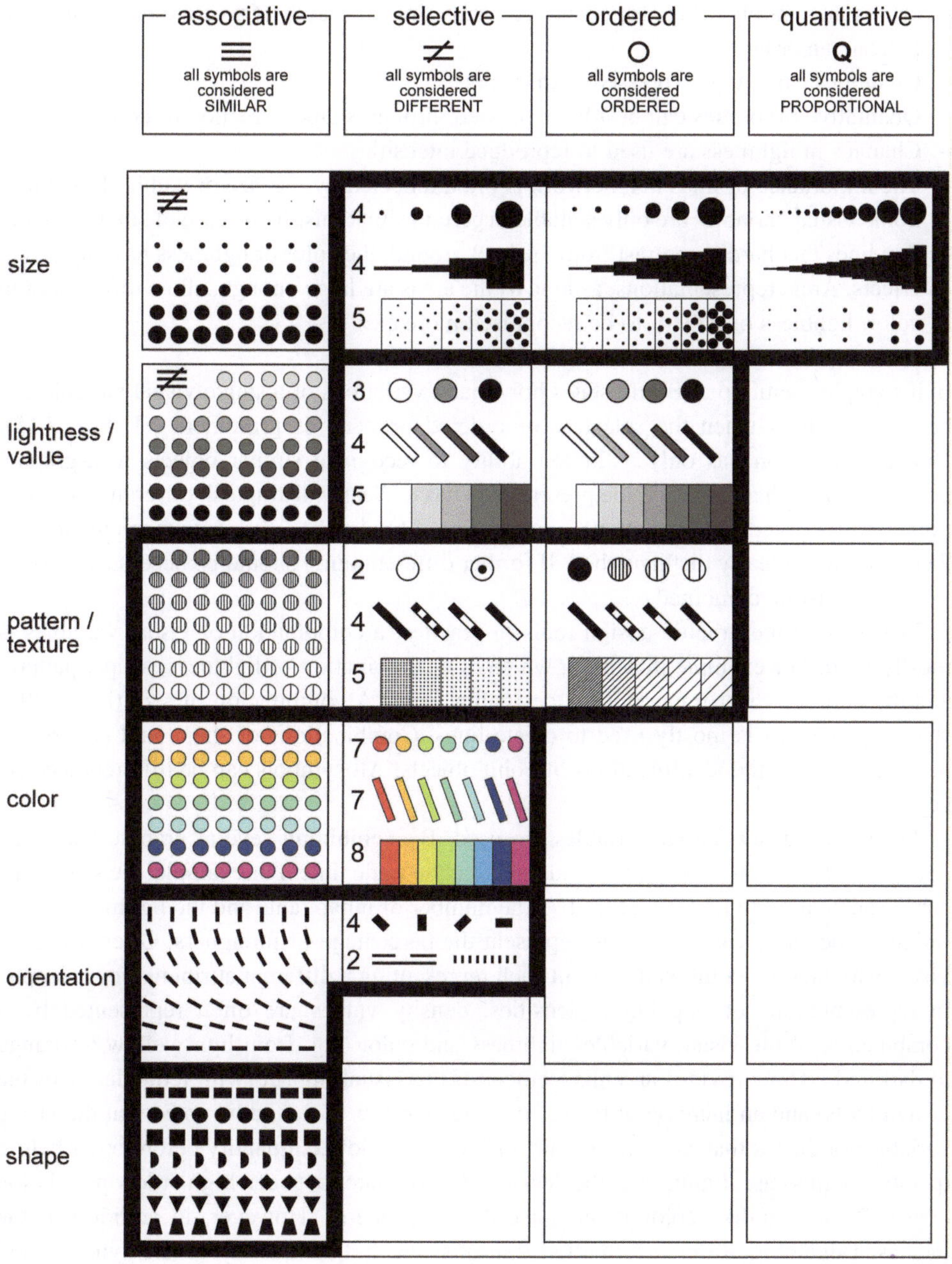

Fig. 7.15 Abilities of the visual variables according to Bertin 1974

causes the ranking. Thus, for example, evenly graded shades of grey are better suited to indicate an order.
- Quantities cannot be expressed with colors!
- Qualitative properties can best be expressed through shape, direction or color.
- Changes in lightness are used to reproduce intensities.
- The importance of lines in a network is often reflected by the size or width of the lines.
- Some visual variables are only suitable in certain situations. In the case of small dots and thin lines that hardly contrast from the background, the color or lightness have no visual effects. Areal representations, as long as the areas are large enough, allow differences in color, lightness or pattern to be recognised more easily.

In his graphic semiology Bertin states how many variations of a single visual variable can be distinguished, when the selective or ordered property is prominent (cf. Fig. 7.15). Human perception has only a limited ability to recognize similar objects in a graphic presentation, such as a map. If the presentation has to clearly distinguish different groups of point objects (selective perception), a maximum of four point sizes, or four orientations or 7 colors can be clearly distinguished. If further differentiation should be necessary, several variables must be combined.

When using the graphic design tools in practice, a combination of visual variables is mostly used. For example, hatching, which is a combination of lightness, shape, pattern, and direction, is most important for designing areas. A combination of size (i.e. width), shape and pattern is mostly used to create lines. Combinations of shape and pattern are most relevant for the identifications of point objects. All symbols can be differentiated by color.

Often only a few visual variables are used. By combining several graphic variables, several attributes can be displayed and varied at the same time. For example, the size of the circle can be used to represent the absolute number of inhabitants and the lightness for the filling of the circle can be used to represent the percentage of foreigners. Often, multiple visual variables are combined without each representing a different attribute. For example, in representations of population densities, density values are often represented by a combination of the visual variables lightness and color (e.g. from light yellow to orange to dark red). This provides an enhancement of the visual impact (with redundancy of the design tools) and an increase of the selective readability. It should be noted that the visual variable dominates that can reflect a higher level of scale (i.e. quantity before order before quality). In this case, lightness is the decisive differentiator between high or low population densities, since unlike color it has an ordering property. However, in addition to the increased unambiguity of information transmission, the complexity of the symbols and the graphic load of the presentation may increase.

Graphic design elements should always be used sparingly. This principle also applies to digital presentations, even if the design tools that are usually easily available in a geoinformation system encourage their use (cf. the many display variants of the north arrow or the use of color). A diverse mixture of pictorial and abstract symbols in bright colors, playful line symbols or textures do not lead to better readability. Rather, a clear appearance can be achieved by a cleverly combination of fewer visual variables, which can make a decisive contribution to the quick structuring of the content. For example, the shape of the point symbol should not be changed to indicate the degree of damage of a trees in a tree cadastre. The color or filling, e.g. of the circle symbol, clarifies the degree of damage (e.g. from green to red). If necessary the size of the point symbol should be varied in order to implement the size of the tree. The other visual variables such as shape, lightness are not taken into account.

In particular, the hierarchy of the objects to be presented must be transformed into an adequate *graphic hierarchy* of the symbols (e.g. preservation of size differences through correspondingly sized and graded point symbols). The group of addressees and the intended use of graphic presentations must always be taken into account, as they play a key role in deciding on the design of the graphics (including size and type of symbols) as well as the density of the content (cf. factors influencing the graphic communication).

7.4.3 Further Development and Digital Implementations

Bertin's approach has been further developed in several directions. In the 1970s, the focus was on the possible combinations of variables, in the 1980s on further differentiations and new variables, and in the 1990s on investigations within the framework of multimedia cartography (cf. Koch 2000 for a summary). Bertin's theory was two-dimensional oriented with regard to representations in (analog) maps. MacEachren was one of the first to tackle digital implementation. With regard to digital representations, i.e. visualisation on the monitor, he defined the visual variable "clarity", which is composed of the (new) variables "crispness", "resolution" and "transparency" (cf. MacEachren 1995 p. 276 and Koch 2000 p. 76).

The variables defined by Bertin exclude motion and time. *Multimedia cartography,* however, requires the addition of dynamic variables. In the early 1990s, three more fundamental variables seemed sufficient (cf. Koch 2000 p. 78): "duration" (duration of a displayed event), "rate of change" (type of change according to location and/or feature), and "order" (sequence, temporal and/or thematic). MacEachren (1995 pp. 281) proposed a six-part system of "syntactic-dynamic variables": "display date" (time of displaying the change of an object or situation), "duration", "rate of change", "order" as well as "frequency" (e.g. blinking frequency of a displayed sign) and "synchronisation" (coincidence of events). Finally, it must be pointed out that multimedia representations on the monitor include not only visual but also auditory perception.

Graphic semiology has its beginnings far before the digitalisation of cartography. In the meantime, multimedia graphic representations on a monitor are increasing or already dominating. The viewer is exposed to a different situation of perception and interpretation compared to analog maps. However, the graphic semiology founded by Bertin is fundamentally suitable to serve as a grammar of the graphic language, as a systematisation of the graphic transcription of information.

7.5 Graphic Design Elements

7.5.1 Map or Cartographic Symbols

Map or cartographic *symbols* are the most important design elements in graphics and maps – in addition to diagrams, color grading and font design. Map or cartographic symbols primarily model and create point markers, lines, areas or text. They are designed using visual variables to express information visually. For example, a street is represented on a map by a double line, i.e. by a special line symbol, which does not allow a conversion into the actual width of the street. A map symbol is understood to be an abstracted object image (cf. the sketch of an antler to mark a forest lodge) or a conventional symbol, which can be modified in many different ways. A symbol is therefore an abbreviation, ranging from abstract to pictorial, which requires less (map) space compared to a textual explanation and, especially with pictorial symbols, directly appeals to the imagination (cf. Fig. 7.16). However, symbols are not always self-explanatory, so that the use of this design tool requires a special explanation of the symbols (legend).

Figure 7.16 systematises shapes of symbols and shows examples. In the standard case, geoinformation systems only have a small selection of these display options. Symbol catalogs for the design of appealing graphics are mostly missing in the basic versions, which can often be supplemented with special symbol libraries for special applications:

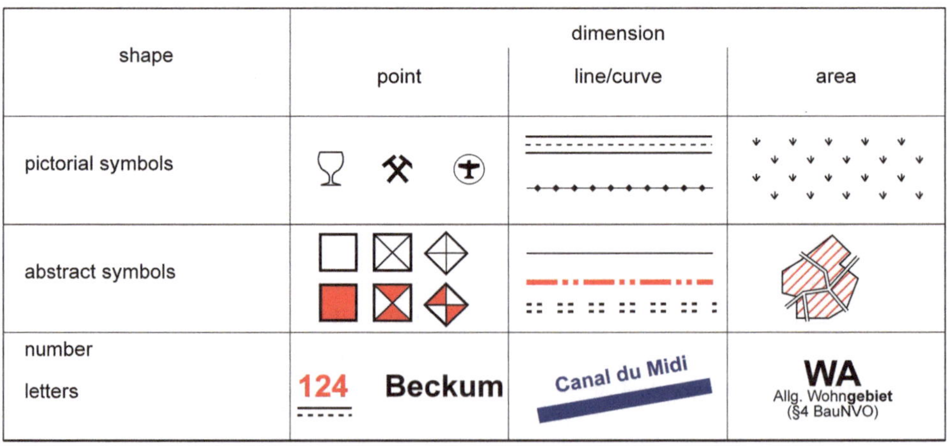

Fig. 7.16 Systematisation of symbols according to form and geometric dimension

- *Pictures* and *abstract symbols* offer individual design options. Geometric symbols offer the greatest scope for variation. In particular, individual disciplines or special applications have developed clearly defined, diverse catalogs of symbols (cf. symbols of official topographic and geological maps, the National Park standard point symbols).
- With the help of *letter or number symbols*, extensive information can be coded and visualised in a very space-saving manner (cf. e.g. nautical maps and weather maps). However, such representations may be very complex or at least not intuitively readable.

Rules for *minimum graphic sizes of symbols* are difficult to set up. Human vision has absolute limits. However, since vision can vary greatly from person to person, only general guidelines can be given. In addition, the technical options of the output devices in particular limit the smallest font size or the finest dot spacing.

With these minimum values, a graphic can just be read and evaluated by a viewer who can concentrate on the representations alone and from close up (cf. Table 7.5). These values should also be seen as reference indications. In addition to size, readability is also influenced by other factors such as lightness or contrast and shape. A squiggly font may already melt together at a reduction level, where a sans-serif, narrow font is still readable. Compared to a filigree atlas map in a scientific study, however, a graphic that is to be explained as a poster at a public hearing, for example, must be more striking.

In addition to the minimum sizes and design guidelines for paper maps, different guidelines apply to other media. Especially in the case of graphic presentations on screens, smartphones or tablet computers, the limitations resulting from the pixel matrix of the displays have to be considered. Different screen resolutions and constant technical developments make it impossible to define the minimum sizes for screen displays precisely

Table 7.5 Minimum sizes of symbols for paper maps (cf. Hake et al. 2002 p. 110)

Minimum size	Symbol for paper maps
0.05 mm	Width of a black line on a white background = maximum contrast
0.08 mm	Width of a colored line on a colored background = minimum contrast
0.15–0.25 mm	Line spacing (for thick or fine lines) at maximum contrast
0.30–0.20 mm	Line spacing (for thick or fine lines) at minimum contrast
0.3 mm	Width of an area at maximum contrast
0.4 mm	Width of an area at minimum contrast
0.15–0.20 mm	Area gap (large or small area) at maximum contrast
0.20–0.25	Area gap (large or small area) at minimum contrast
0.25 mm	Diameter of a point symbol at maximum contrast
0.45 mm	Diameter of a point symbol at minimum contrast
0.5–0.6 mm	Size of a circle/square (full or hollow) at maximum contrast
0.7–1.0 mm	Size of a circle/square (full or hollow) at minimum contrast
0.6 mm	Height of letters and digits at maximum contrast
1.0 mm	Height of letters and digits at minimum contrast

Table 7.6 Minimum sizes of map elements for screen display (cf. Brunner 2001 p. 9 and Brunner 2000 pp. 56)

Minimum size	Symbols for map elements on the monitor
3.0 mm	Point (diameter)
0.4 mm	Line width
1.5 mm	Square (side length)
10 mm^2	Area to fill with color
0.5 mm	Line spacing
10 pt. (approx. 3.6 mm)	Text

and permanently. Some basic guidelines were have been worked out by Brunner (2000 and 2001) (cf. Table 7.6).

The design tools mentioned can also be used for digital graphic representations. However, the technical features of monitors in particular (i.e. above all the number of pixels) restrict the different variants and forms of representation of symbols. The new information and communication technologies also offer further options. It is possible to reduce the amount of detail in a presentation and therefore make it easier to read and structure a figure "at first glance". After an initial overview, the user can delve deeper into the digital map or the information offered, whereby the content, the level of detail, the design and also the symbols can change depending on the scale or the so-called zoom level.

This display technique is the well-known standard principle for Google Maps as well as, for example, for the software Mapnik for the presentation of Open Street Map data. These web mapping applications generate maps from the data in a database (so-called *rendering*), whereby the design depends on the zoom level. If the user enlarges the section of the presentation, i.e. zooms more into the image, the level of detail and the design change. Then e.g. district names appear and with further enlargement street names, finally symbols e.g. for restaurants or bus stops are displayed.

7.5.2 Representation of Qualitative Features

Very often a *geoobject* has to be located on a map and at the same time a feature that characterises the object, i.e. a qualitative feature, has to be displayed. In this way, the location of a tree, a hospital or a restaurant can be identified. In this situation, a point symbol is usually used to show the position. Sometimes a small symbol or multiple symbols of different sizes are used to indicate the size. The size refers to the membership to a size category. In Germany, for example, according to the commercial code four size classes of corporations exist: micro company, small company, medium-sized company and large company. Instead of using only one symbol in various sizes for example to locate religious institutions, distinct symbols in different sizes are often used, for example for a chapel, a church or a cathedral.

The procedure for line symbols is similar. For example, the meaning of a street is indicated by different shapes (e.g. dashed line, line, double line) or additionally by colors.

Areal symbols are often used to indicate the spatial distribution of a feature on a map. Examples are forest areas or lakes or the estimated flood area in the event of a 100-year flood. Colors are mostly used for the design, but also often for a pictorial representation (e.g. symbols for coniferous forest in a topographical map).

Overall, the reproduction of a qualitative feature places special demands on the choice of symbols. The focus is on recognizing structures and legibility of the symbols. The main visual variables used are size and shape particularly for point symbols, size, shape and pattern for line symbols and pattern for area symbols, and always the color.

7.5.3 Representation of Quantitative Features

According to Bertin, only the visual variable size is suitable for representing quantities. In the further differentiation according to absolute and relative values, clear conventions for the implementation of quantitative data have emerged in cartographic practice, which should also continue to be valid for digital presentations. Quantities can be specified in several ways for point, line and area symbols (cf. Fig. 7.17).

In the case of a continuous reproduction of a quantitative feature for points and lines (e.g. population of cities or accidents per route sections), the size of the point symbol (e.g. size of a circle) or the width of a line are varied continuously (cf. Fig. 7.17). It should be noted that point symbols, such as circles or squares are perceived as two-dimensional objects. If two values are to be represented, one of which is twice as large as the other value, the symbol must be drawn twice as large. This means that the radius must not be multiplied by the factor 2, but by the factor $\sqrt{2}$ (cf. Fig. 7.17):

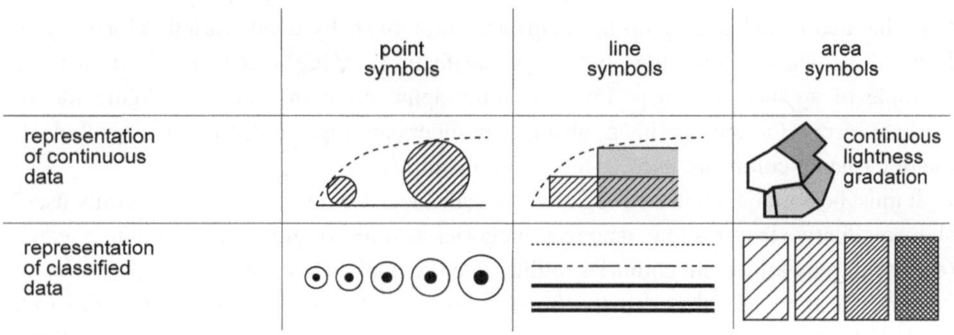

Fig. 7.17 Specifying quantities in a graphic presentation

if $F_2 = 2 \cdot F_1$ and $F_1 = \pi \cdot r_1^2$

then $F_2 = 2 \cdot F_1 = 2 \cdot \pi \cdot r_1^2 = \pi \cdot \left(\sqrt{2} \cdot r_1\right)^2$

finally $r_2 = \sqrt{2} \cdot r_1$ if $F_2 = 2 \cdot F_1$ v

If this principle is not respected, larger objects will be highlighted disproportionately. Of course, the values represented by the symbols must be explained in a legend.

For classified data for points (e.g. size classes according to the population of cities) and lines (e.g. size classes according to the number of commuters between two cities) the previous comments on the presentation of the quantitative feature apply accordingly.

A particular graphical challenge is the reproduction of a quantitative feature that is available for an area, such as for a statistical reference unit (e.g. different states or counties). There are two different approaches for an absolute feature (e.g. the number of inhabitants or the number of registered electric cars in a municipality or the number of voters for a political party) and for a relative feature (e.g. birth rate or population density).

Classic cartography offers several options for the first task to represent absolute values for (statistical) areas, the most important of which can also be implemented in a digital form:

The value or number is displayed like a name in the respective area (e.g. within the boundaries of a state or a county). However, this variant is only suitable to a limited extent for quickly capturing structures.

The numerical value is represented by a point symbol (e.g. a circle) varying in size or by a bar chart, where the size of the point symbol or the height of the column is proportional to the value.

Unfortunately, the presentation tools of a geoinformation system often automatically display absolute numbers (e.g. the number of inhabitants of a county) by filling in the area (e.g. the area of the county on the map) and, moreover, by using varied colorful colors. Figure 7.9, which shows the absolute value feature "living space in m^2", is a typical example of an incorrect map. From a cartographic point of view, this figure has two cardinal errors. On the one hand, absolute numbers are represented by area symbols. On the other hand, colors are used to represent quantities.

It must be pointed out that filling of areas with special symbols or patterns is only used to show qualitative features (e.g. land use categories as forest or grassland) or relative features (e.g. birth rate of different countries to illustrate regional difference in fertility). The spatial pattern is also perceived as density. In addition, the value is recorded spatially due to the spatial reference, an absolute value related to a small area is perceived as little, regardless of the actual value.

Classic cartography offers, in view of these facts, two options for the second task to represent relative values for (statistical) areas:

A percentage can be represented by using the corresponding degree of saturation or lightness of a color (cf. Fig. 7.17). This approach is rarely used, since the human eye can only distinguish a few lightness levels for areas (according to Bertin only 5, cf. Fig. 7.15). In addition, a saturation of for example 70% and 80% can hardly be distinguished.

In most cases, a relative feature (e.g. birth rate or population density) is classified first (cf. Sect. 7.5.4). Then the membership of a statistical unit to a certain class is represented by an area symbol. This creates a so-called *choropleth map* (cf. Figs. 7.6–7.8 using the visual variable lightness from light to dark brown). A choropleth map is probably the most common type of thematic map to spatially represent statistical data. The grading of the values, i.e. the hierarchy of the numerical values, must be translated into a visual hierarchy (cf. Fig. 7.17). The design rules of the visual variables are to be considered (cf. Sect. 7.4.2). Essentially, this boils down to using the visual variable lightness to indicate a visual hierarchy or intensity for areas. The ColorBrewer (cf. Brewer 2023) offers options to test different shading (cf. Harrower, M.A. and C.A. Brewer 2003).

7.5.4 Classification

A continuous feature usually has to be classified and therefore discretised before it can be presented graphically, particularly in a choropleth map. This characterises the standard case in a geoinformation system. Only after a classification, which implies a finite number of symbols, can an individual symbol be assigned to a class. While a qualitative feature (e.g. soil types) has clearly defined categories, several specifications have to be made for *classifying* a continuous feature: number of classes, class width and beginning of one, mostly the first class. The statistical and cartographic methodology provides first clues for this (cf. Kessler-de Vivie 1993, de Lange u. Nipper 2018 pp. 341). Three formulas for estimating the number of classes are given, whereby the rule according to Sturges has proven itself as an approximation in practice, and two variants for determining the intervals:

maximum number of classes $k = \sqrt{n}$

(n= number of objects, formula according to Witt)

maximum number of classes $k = 5 \cdot \log n$

(n= number of objects, formula according to Davis)

number of classes $k = 1 + 3,32 \cdot \log n$

(n= number of objects, formula according to Sturges)

class width $b = (x_{max} - x_{min})\ /k$

class boundaries $g_1 = x_{min}$ (first class lower limit)

$g_i = g_{i-1} + b$ für $i = 2, \ldots, k + 1$ (class upper limits)

In addition to this first variant, there is at least one other way to create equidistant intervals with a given number of classes (cf. Table 7.7):

class width $b = (x_{max} - x_{min})\ /k - 1$

class boundaries $g_1 = x_{min} - b/2$ (first class lower limit)

$g_i = g_{i-1} + b$ für $i = 2, \ldots, k + 1$ (class upper limits)

In addition to a classification into equidistant intervals, which is the most common form of classification, there are also gradings in which the class widths progressively increase. In the case of arithmetic progression, the class width increases by a constant value (cf. Table 7.7, column 3, class width increases by 7.5).

class widths $b_i = b_{i-1} + q$ $q =$ constant factor (class upper limits)

In the case of geometric progression, the quotient of two consecutive class interval boundaries is constant (cf. Table 7.7 column 4, the quotient of two consecutive class boundaries is 2.5, only if $x_{min} > 0$):

class limits $g_i = g_{i-1} \cdot q$ $q =$ constant factor (class upper limits)

Another method, also statistically oriented, uses the parameters mean and standard deviation of the data. Then each standard deviation (or part of it) becomes a class in the choropleth maps. The class boundaries may be:

Table 7.7 Different classifications in five classes

Variant 1 constant class width 20	Variant 2 constant class width 25	Arithmetic progression $q = 7.5$	Geometric progression $q = 2.5$ x
0	−12.5	0	1
20	12.5	7.5	2.5
40	37.5	22.5	6.25
60	62.5	55	15.63
80	87.5	85	39.06
100	112.5	122.5	97.66

$x_{min} = 0$, $x_{max} = 100$, with geometric progression $x_{min} > 0$

$$\overline{x} - \frac{3}{2} \cdot s,\ \overline{x} - s,\ \overline{x} - \frac{1}{2} \cdot s,\ \overline{x}, \overline{x} + \frac{1}{2} \cdot s,\ \overline{x} + s,\ \overline{x} + \frac{3}{2} \cdot s$$

In most cases, geoinformation systems also offer a classification according to so-called quantiles. The class boundaries are chosen so that each class has the same quantity. This classification principle can be useful for very scattering values, but this advantage is bought at the expense of different class widths and interpretation difficulties.

Another method tries to determine "natural" classes or "natural" class boundaries (so-called *natural breaks*). Based on the distribution of the data values, an attempt is made to identify data gaps or coherent clusters of data. The classification is done in such a way that there are few differences in each class. This method usually results in classes of different widths. Further values can quickly change the classification.

Content concerns can speak against a rigid, schematic approach. Classification can also result from logical aspects and, for example, based on threshold values. The classification is then largely independent of the distribution of the data. For example, when classifying the total fertility rate, a class break should have a value of 2.1, because this value reflects the average number of children a woman would give birth to over a lifetime and indicates broad population stability, assuming that there are no migration flows and that mortality rates remain unchanged.

Sometimes meaningful class boundaries or simple class widths will be used (e.g. in steps of 10, 100 or 1000 value units). This is particularly necessary, if two classifications are to be compared. Ideally, the class boundaries should be chosen in such a way that the essential properties of the facts to be represented and the characteristic distribution of the data are preserved.

In particular, a reasonable number of classes should not only be determined according to formal or content-related aspects, but also with regard to the maximum number of perceptible differences of a visual variable. Bertin has worked out indications for this. As shown in Fig. 7.15, a maximum of 8 colors can be distinguished. If more than 8 classes are required, several visual variables must be combined (e.g. color and pattern). Classification is to be used in particular as a method of generalizing the content. In general, the procedure to be chosen depends not least on the purpose of the presentation.

Thus, a classification requires many subjective decisions. Even with the same number of classes, many variations are possible by changing class width and a class boundary. In particular, very different statements can be made with the same data but different classifications, so that a wide field for conscious and unconscious manipulations can open up (cf. for examples de Lange and Nipper 2018 pp. 342–351). Geoinformation systems offer a wide variety of classification options and open up many opportunities for experimentation. Unfortunately, these systems lack guidance and assistance for classification.

7.5.5 Diagrams

Cartography has developed a wide range of diagram forms, some for very specific purposes: e.g. population pyramids, polar diagrams for the representation of wind directions, climate diagrams, structure triangles, concentration curves (cf. Arnberger 1997 p. 109). Such diagrams are rarely integrated in geoinformation systems by default. In most cases, only simple bar charts (histograms) or pie charts are implemented.

7.5.6 Cartograms

Numerical values of geoobjects such as population densities or unemployment rates are almost always presented in their administrative reference units as choropleth maps. A representation of population density for municipalities in North Rhine-Westphalia then also illustrates this data on a map with correct municipal boundaries. A cartogram, refers to an intentionally distorted map whose size and extent do not correspond to administrative boundaries, but rather correlate with the value of a numerical variable, while retaining neighborhoods and roughly the shape of the reference areas.

Geoinformation systems provide tools to produce cartograms. These representations have in common that they inflate areas with high values like a balloon. Figure 7.18 was

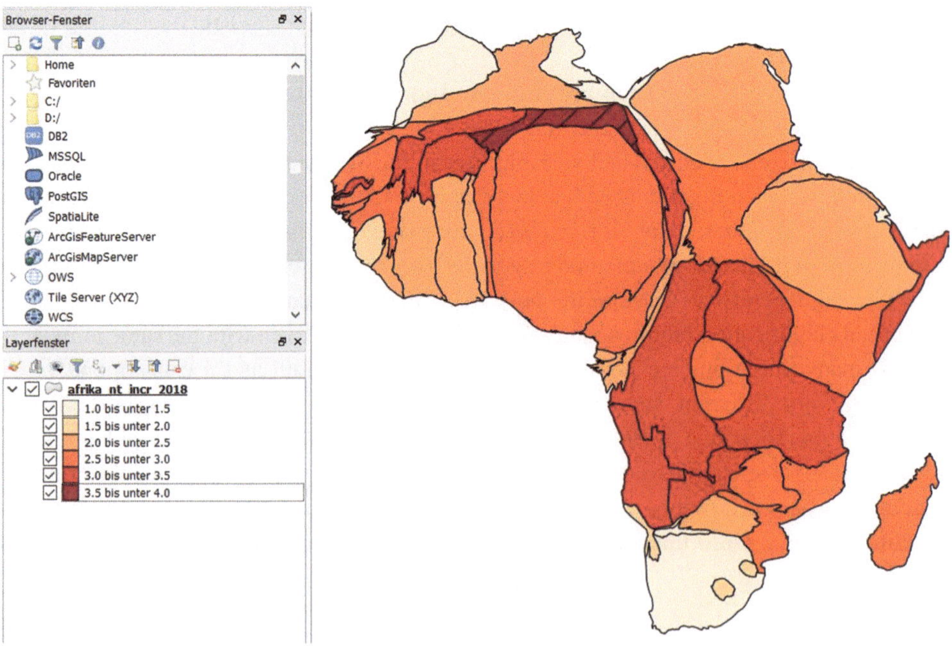

Fig. 7.18 Isodemographic map of Africa: population size and natural growth rate 2018. (Data source: World Population Datasheet 2018)

created with a QGIS plug-in (cf. Sect. 3.1.5) that implements Dougenik's algorithm (cf. Dougenik et al. 1985). The maps in the Worldmapper project with almost 600 cartograms are based on the algorithm of Gastner and Newman (2004) (cf. Worldmapper 2023, for further algorithms cf. Burgdorf 2008, cf. also Rase 2016).

Figure 7.18 shows the African countries according to their (absolute) population size. The three most populous countries in Africa, Nigeria, Ethiopia and Egypt, dominate. The color grading illustrates the annual natural population growth in percent. It is only in the combination of the distorted size of the area and the second feature, which is usually plotted into the statistical reference area, that the development dynamics of Africa's population become visible. The most populous countries no longer have the highest natural growth rates (per year) in percent, which is also due to a statistical effect with large reference sizes (increase of 1000 to the base of 10000 equals 10%, increase of 1000 to the base of 100000 equals 1%). Nevertheless, they will show largest absolute increases. For example, the highest annual natural growth rate of 3.8% for the relatively small country of Niger will not result in comparably large increases. Thus, cartograms of this kind can provide a variety of insights and added value as a supplement to traditional maps and tables.

7.6 Printed Maps and Posters: Design Features

7.6.1 Printed Maps and Posters: Formal Design

Presentations in a geoinformation system do not have to follow the traditional representation principles of cartography. In most cases, using coordinates, size information such as reduction factors are calculated automatically at each zoom level and are usually displayed at the bottom of the screen. A scale bar may be absent, a separate legend is not necessary, since necessary information results using the tools of a geoinformation system (exploration of the associated attribute tables and of menu options related to file and map properties).

If a paper printout is to be produced (e.g. a large-format poster), there are clear and compelling design instructions. It must be noted that a printout or a poster is a collection of various map elements, the actual graphic, then title, scale, map projection, legend, source, data source, author and editor. The arrangement of these elements is challenging:

- A printed map or poster that stands "alone" without any further explanatory text must have a meaningful title and a legend that explains all the symbols used, listing size relations and, if necessary, explaining the structure of diagrams. The legend must also include information on the author or on the person responsible for the map content as well as on the cartographer or creator of the graphics. In particular, references to the source of the base map and the data used should be given (especially the reference date of the data).

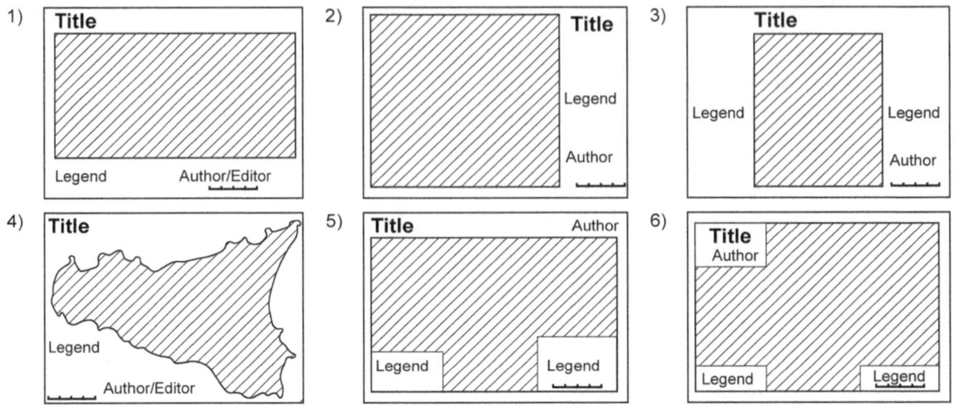

Fig. 7.19 External design of a map print

- A north arrow can be omitted if the map is oriented to the north. This rule should always be strived for, so that the range of (playful and squiggly) north arrows sometimes offered in geoinformation systems becomes almost superfluous.
- Indication of a scale is indispensable, although it is less useful to state the reduction scale in the form of a mathematical fraction (e.g. 1:12,375). In contrast, a scale bar that shows an example of a length is much more meaningful.

Figure 7.19 shows examples of the *layout of a map*. Above all the printed map or poster should appear as balanced and clearly structured as possible. The viewer's gaze should not jump back and forth between the individual explanations and the contents of the map. A single legend should structure the explanations. In view of these rules, cards 1 and especially 2 represent optimal examples of arrangements. Variant 3, i.e. the splitting of a legend, is only useful in the case of a thematic separation. The fourth example shows a recommended layout for a map, in which the margins are to be used as best as possible. The other examples represent unbalanced segmentations.

7.6.2 Printed Maps and Posters: Texts and Labels

Presentations on the WWW or in a geoinformation system usually only use a single font, which is changed according to size, color and font style (e.g. italic). In contrast, when creating printed maps or posters, there is a variety of fonts, which can be varied also by font size, font weight, font width (so-called run length) and font color, depending on the performance of the graphic system. However, there are clear recommendations. Above all, the text should appear clear, correct and concise:

Arial 12p narrow	*Arial 12p narrow italic*
Arial 12p regular	*Arial 12p regular italic*
Arial 12p bold	***Arial 12p bold italic***

Fig. 7.20 Variations of a font style

- The available fonts should be used as sparingly as possible and should be limited to what is necessary.
- The graphics should be captured visually by the graphic design options, but should be read less by labels.
- Since text in all capital letters is more difficult to read, upper and lower case should also be used.
- A maximum of two fonts are to be used in one map. Exotic fonts should be avoided. Clear, simple, sans-serif fonts such as Helvetica or similar fonts are recommended, especially with regard to a small font size (cf. Fig. 7.20). In most cases, a single font is quite sufficient, which should be further varied according to size and font weight.

Automatically placement text as well as masking text against the map background as well as of point or line symbols from each other is (still) a problem in existing geoinformation systems or graphics systems, which has not yet been solved satisfactorily. The labels usually overlap each other. Manual rework and shifts become necessary.

7.7 Use of Color

7.7.1 Color as a Simple but Also Disputable Tool for Presentations

In geoinformatics, *color* plays a very important role in the presentation in a (geo)-information system on the monitor, tablet or smartphone. The software systems for digital image processing as well as geoinformation systems require colors as an indispensable tool of expression. The advantages of color and the resulting popularity of the use of color are mainly due to two factors:

- color itself is a carrier of information.
- color simplifies and accelerates the transmission of information.

In addition, a color presentation can have a special aesthetic effect. A color presentation is generally more appealing than a black and white presentation. Another characteristic is that color is a particularly good selective variable (cf. Sect. 7.4). It has a clear stimulus effect and exerts a strong psychological attraction. Attention is aroused. Colors are very well remembered. The ability to remember is increased.

Despite many advantages of using color, it should not be overlooked that color design is not trivial and can contain many sources of error. Color design should not be done carelessly or arbitrary. With a geoinformation system, it has now become quite easy to use colors. Often randomly controlled color selection is specified by default, e.g. to indicate different lines and areas. Water surfaces then appear, for example, in green, open spaces in blue, a high population density is represented by red, a low population density by yellow (cf. as a negative example Fig. 7.6). The geoobjects are recognizable. However, the representation is only "colorful", the associative effect of colors was not used (cf. Fig. 7.15).

Last but not least it should not be forgotten that some people have anomalies in their perception of color (Daltonism). In most cases, people with color vision deficiencies look for substitute symbols or regularities in the arrangement of colors (cf. e.g. arrangement of colors at a traffic light). Graphic representations should help these people to understand the contents of the map, and therefore should not use colors exclusively, but use color in combination with textures or patterns.

7.7.2 Color Effect and Color Perception

The popularity of colors can be explained not only by their special aesthetic appearance but also by their color effect. Individual facts are inferred from the meaning of individual colors. The "correct" recognition is mostly intuitive. The *color effect* is mostly based on general *color perceptions*. However, color conventions are by no means unambiguous. For example, there are sometimes different interpretations of colors in different cultures (cf. Schoppmeyer 1993 p. 33).

With regard to the color effect, great importance is often attached to the choice of colors that are close to nature, which implements the experiences and perceptions of real objects. Recognition can simplify the reading and understanding of the map (e.g. blue for water, crimson for settlements based on red tiled roofs, yellow-green for meadows and grassland, blue-green for forests, grey for rubble). However, it is assumed that the observer has the same color perception and similar experience. Misjudgements can therefore not be ruled out.

An example of supposed color conventions and intuitive color effects is provided by elevation maps, which mostly use a dark green for lowlands and shadows of brown for low to high mountain ranges. However, this color grading is not universally valid and standardised. Elevation maps are sometimes not recognised as such, in which the elevation gradings are illustrated by a multilevel color scale ranging from rich green for lowlands, lighter shades of green for low elevations, to yellow or white for the highest elevations. This lack of standardisation makes it difficult to transfer empirical values. More problematic is that viewers often associate white with snow, yellow or brown with deserts, or green with rich vegetation. Given that deserts also extend into lowlands (usually represented by a shade of green in an elevation map) or grassland occurs in highlands, the color would incorrectly indicate land use.

The meaning of the traffic light colors red-yellow-green has been memorised: red indicates danger, yellow is associated with caution and green with safety. The distinction between warm and cold colors is common, e.g. to express warmth or coldness or to express repulsion.

7.7.3 Color Gradings

Strictly speaking, quantities and orders cannot be expressed by colors (cf. Fig. 7.15)! Yellow does not express "less" than red. It is not possible to implement a classification of population densities using colors due to the representation principles of graphic semiology (cf. Sect. 7.4.2). Instead, the lightness is usually varied (unipolar scale from light to dark) or a color transition is selected (e.g. from light green to dark blue tones). Brewer (1994) used individual color examples to develop suggestions of color gradings depending on the scale levels (qualitative, binary or sequential grading of a feature) and the number of features (cf. the ColorBrewer to try different shadings cf. Brewer 2023). However, there is no simple, memorable or easy-to-use color grading. Always the legend has to be consulted for color matching. However, few simple rules can be mentioned that must be observed when implementing quantities or orders and that also apply to presentations in geoinformation systems:

- A grading according to the spectral colors, whose lightness just does not increase sequentially, is generally not suitable.
- Good results are generally achieved by grading the lightness of a color according to the visual variable lightness.
- A lightness grading is often used in combination with a color transition (e.g. from low temperatures in blue to high temperatures in red or from light yellow to orange to dark red).

7.7.4 Color Mixing and Color Models

In the technical representation and reproduction of colors, a distinction is made between additive and subtractive color mixing. With only three primary colors each, all other colors can be represented or mixed together (cf. Fig. 7.21). Computer science uses the resulting color models.

The *additive color mixing* is based on a mixture of a red, a green and a blue light source. Additional colors result from the superimposed projecting and the resulting additive mixture of three light sources in the three (additive) *primary colors* red, green and blue, result in further colors: e.g. yellow = red + green, magenta = red + blue, blue-green

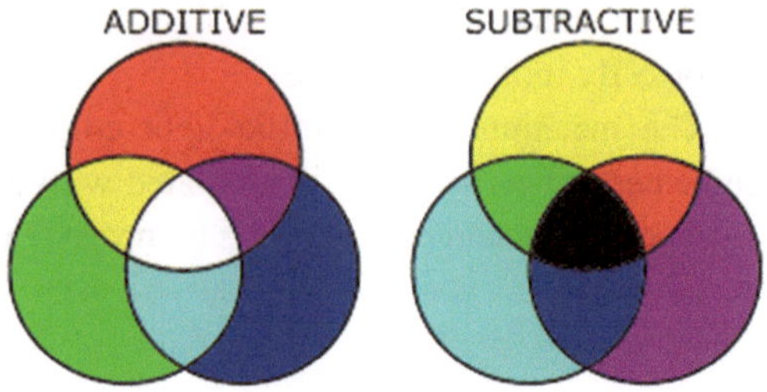

Fig. 7.21 Additive and subtractive color mixing

(i.e. cyan) = green + blue. White is created by adding the three colors together, black by missing all colors. Any bright color can be represented by varying the tonal value (lightness) of the individual light sources. Color images can therefore be produced by accurately overlaying three images from three projectors, each projecting an image in an additive primary color.

Different colored light can also be generated by placing several small light sources separately but directly next to each other. If they are no longer perceived by the eye as separate light sources, the colors mix additively (additive color mixing as a physiological process in the eye, cf. Sect. 2.5.7). This principle determines how color monitors work, in which colors are formed from red, green and blue phosphor dots.

While additive color mixing is based on the superimposition of light rays (self-illuminating) of the three primary colors red, green and blue, new color tones can also be created by mixing dyes. In this case, color impressions are created by reflecting. The basic principle is that certain color components are removed from the white light that illuminates a colored surface, i.e. "subtracted" (*subtractive color mixing*). The remaining non-absorbed color components are then reflected and perceived as color. If, for example, the blue component is absorbed by a color filter or pigments so that only the red and green components are reflected, yellow is perceived (subtractive color mixing as a physical process, additive color mixing of the reflected color components as a physiological process in the eye). The use of filters or dyes that transmit or reflect the *complementary colors* of the additive primary colors is important for printing technology. Color and complementary color additively mixed together result in white. Thus, yellow is the complement of blue, cyan of red, and magenta of green (cf. Fig. 7.21).

If an area has been filled with cyan alone, no red light is reflected from the surface, cyan "subtracts" red from the reflected light (since: white – red = (red + green + blue) – red or cyan = green + blue). Correspondingly, magenta absorbs green, yellow absorbs blue. Different grades of color are created by subtractive color mixing of the pigments. An area filled with cyan and yellow absorbs red and blue and only reflects green. An area filled with cyan and magenta absorbs red and green and is therefore blue (cf. Fig. 7.21).

The English color designations enable a clear identification CMY (for cyan, magenta and yellow) and RGB (for red, green and blue).

Any color can be created by additive or subtractive color mixing of the additive or subtractive primary colors. Thus, each color can be formally represented as a vector in a

Fig. 7.22 RGB and CMY color models or color cubes

three-dimensional color space (cf. color cube in Fig. 7.22). The *RGB color model* uses a three-dimensional Cartesian coordinate system, whose axes represent the primary colors red, green and blue. Colors are each encoded as points in the color cube. The main diagonal presents the shades of grey from black (0,0,0) to white (255,255,255). The RGB color model is based on additive color mixing. With the colors cyan, magenta and yellow, the complementary colors of red, green and blue, the *CMY color model* can also be represented as a color cube. The CMY color model is based on subtractive color mixing.

The relationships between the two color models and the representations as points or vectors in both color cubes are illustrated by two simple equations. For cyan applies, for example: $[(255,255,255) - (0,255,255)]_{\text{RGB}} = (255,0,0)_{\text{CMY}}$ or $[(255,255,255) - (255,0,0)]_{\text{CMY}} = (0,255,255)_{\text{RGB}}$:

$$\begin{pmatrix} \text{C} \\ \text{M} \\ \text{Y} \end{pmatrix} = \begin{pmatrix} 255 \\ 255 \\ 255 \end{pmatrix} - \begin{pmatrix} \text{R} \\ \text{G} \\ \text{B} \end{pmatrix} \text{ and } \begin{pmatrix} \text{R} \\ \text{G} \\ \text{B} \end{pmatrix} = \begin{pmatrix} 255 \\ 255 \\ 255 \end{pmatrix} - \begin{pmatrix} \text{C} \\ \text{M} \\ \text{Y}. \end{pmatrix}$$

With a scaling of 0 to 255 per primary color (8 bits each), a total of $256 \cdot 256 \cdot 256 = 2^{24} = 16{,}777{,}216$ combinations or colors can be encoded. Thus, a true-to-life reproduction is given, which explains the name true color. According to this grading, the representation accuracy of a color is reflected by the so-called *color depth,* which is the maximum number of colors that can be presented. The color depth is given as an exponent of the number 2. Often 8-bit color and 24-bit color have the same meaning. The 8-bit refers to the grading of each red, green, and blue primary color, while 24-bit means all three $2^{24} = 2^8 \cdot 2^8 \cdot 2^8 = 16{,}777{,}216$ color combinations

In contrast to the RGB and CMY color models, which are based on the technical options of color reproduction, so-called *HSL model* (for Hue, Saturation, Lightness) or *HSV model* (for Hue, Saturation, Value) or *IHS model* (for Intensity, Hue, Saturation) is based on color perception. Thus, not red, green or blue components are perceived, but an overall color impression. Figure 7.23 illustrates the model as a six-sided pyramid with a common color arrangement:

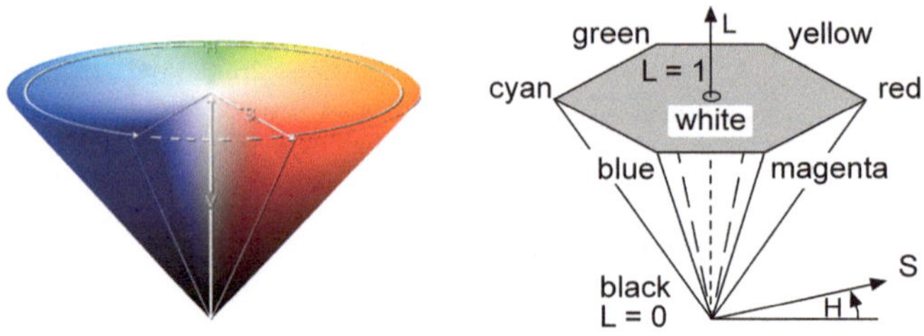

Fig. 7.23 The HSL color model

- Starting with red at 0°, the hue (H) is given as an angle.
- The ratio of the purity of a color to its maximum purity characterises the saturation S. It varies from S = 0 (at the pyramid axis) to S = 1 (maximum purity).
- The lightness of the color varies with the axis of the pyramid. It is lowest at the bottom of the pyramid (L = 0) and highest at the top (L = 1).
- The purest colors differ only in the hue angle, L = S = 1 applies to them. The color selection is based on the pure colors. Then white or black is added.

There are several approaches for the transformation of the RGB model into the HSL or HSV models (and vice versa), since HSL or HSV models are not generally defined (cf. color arrangements, representation also as a cylinder). Starting from the fact that the HSL or HSV model is based on a rotation of the RGB coordinate system and the representation with cylindrical coordinates, the following transformations result, which are of greater importance in digital image processing (cf. e.g. Abmayr 1994 pp. 177 and Fig. 10.25 in Sect. 10.6.5.3). For pure red H = 90° applies, furthermore $0 \leq S \leq \sqrt{(2/3)}$.

The following applies to the *transformation* from the *RGB color system* to the *HSL color system,* (in each case based on the unit cube):

$$\begin{pmatrix} R \\ G \\ B \end{pmatrix} \text{ to } \begin{pmatrix} H \\ S \\ L \end{pmatrix} \text{ with } \begin{matrix} H = \arctan(M_1/M_2) \\ S = \sqrt{M_1^2 + M_2^2} \\ L = L_1 \cdot \sqrt{3} \end{matrix} \quad \text{from } \begin{pmatrix} M_1 \\ M_2 \\ L_1 \end{pmatrix} = \begin{pmatrix} \frac{2}{\sqrt{6}} & \frac{-1}{\sqrt{6}} & \frac{-1}{\sqrt{6}} \\ 0 & \frac{1}{\sqrt{2}} & \frac{-1}{\sqrt{2}} \\ \frac{1}{\sqrt{3}} & \frac{1}{\sqrt{3}} & \frac{1}{\sqrt{3}} \end{pmatrix} \cdot \begin{pmatrix} R \\ G \\ B \end{pmatrix}$$

The following applies to the *transformation* from the *HSL color system* to the *RGB color system,* (in each case based on the unit cube):

$$\begin{pmatrix} H \\ S \\ L \end{pmatrix} \text{ to } \begin{pmatrix} R \\ G \\ B \end{pmatrix} \text{ with } \begin{pmatrix} R \\ G \\ B \end{pmatrix} = \begin{pmatrix} \frac{2}{\sqrt{6}} & 0 & \frac{1}{\sqrt{3}} \\ \frac{-1}{\sqrt{6}} & \frac{1}{\sqrt{2}} & \frac{1}{\sqrt{3}} \\ \frac{-1}{\sqrt{6}} & \frac{-1}{\sqrt{2}} & \frac{1}{\sqrt{3}} \end{pmatrix} \cdot \begin{pmatrix} M_1 \\ M_2 \\ L_1 \end{pmatrix} \qquad \begin{aligned} M_1 &= S \cdot \sin H \\ M_2 &= S \cdot \cos H \\ L_1 &= L \cdot \sqrt{3} \end{aligned}$$

In addition to these classic color models, several color systems are in use internationally, which are relevant for the printing industry. The most important models include the Munsell system in the USA and the DIN color system in Germany (cf. Lang 1993 pp. 741 and Pérez 1996 pp. 624). The Commission Internationale de l'Eclairage (CIE) has developed an internationally valid standard color system for evaluating and classifying colors. With this system, the color spaces realised in the various technical systems can be converted into one another (cf. Abmayr 1994 pp. 173 and Schoppmeyer 1993 pp. 32).

References

Abmayr, W. (1994): Einführung in die digitale Bildverarbeitung. Stuttgart: Teubner.

Albers, B., de Lange, N., Fuhrmann, B. u. M. Temmen (2017): The PAN project – environmental monitoring with smartphones and augmented reality. In: AGIT Journal für Angewandte Geoinformatik, 3-2017, pp. 190–199. https://gispoint.de/gisopen-paper/4164-das-pan-projekt-umweltmonitoring-mit-smartphones-und-augmented-reality.html (14.04.2023).

Andrienko, G. u. N. Andrienko (2005): Exploratory Analysis of Spatial and Temporal Data. A Systematic Approach. Berlin: Springer.

Andrienko, G., et al. (2010): Space, time and visual analytics. In: International Journal of Geographical Information Science, 24(10), pp. 1577–1600.

Andrienko, G. et al. (2007): Geovisual analytics for spatial decision support: Setting the research agenda. In: International Journal of Geographical Information Science, 21(8), pp. 839–857.

Anscombe, F.-J. (1973): Graphs in Statistical Analysis. In: The American Statistician. 27, S. 17–21.

Arnberger, E. (1997): Thematische Kartographie. Braunschweig: Westermann. 4. Ed.

Behncke, K., Hoffmann, K., de Lange, N. u. C. Plass (2009): Web-Mapping, Web-GIS und Internet-GIS – ein Ansatz zur Begriffsklärung. In: Kartogr. Nachrichten H. 6 2009, pp. 303–308.

Bertin, J. (1974): Graphische Semiologie. Diagramme, Netze, Karten. Translated from the 2nd French edition. Berlin: de Gruyter.

Bertin, J. (1982): Graphic representations and the graphic transmission of information. Berlin: de Gruyter.

Brewer, C.A. (1994): Color Use Guidelines for Mapping and Visualization. In: MacEachran, A.M. and D.R. Taylor (eds.): Visualization in modern Cartography. Modern Cartography, Vol. 2. pp. 123–147. Oxford: Pergamon.

Brewer, C.A. (2023): ColorBrewer 2.0. http://www.ColorBrewer.org (14.04.2023).

Broll, W. (2013): Augmentierte Realität. In: Dörner, R. u.a. (Eds.) Virtual and Augmented Reality. Grundlagen und Methoden von VR and AR. Wiesbaden: Springer Vieweg 2013, pp. 241–294.

Brunner, K. (2000): Neue Gestaltungs- und Modellierungsaufgaben für den Kartographen. In: Kelnhofer, F. u. M. Lechtthaler (Eds.): Interaktive Karten (Atlanten) und Multimediaapplikationen. Vienna. (= Geowissenschaftliche Mitteilungen, H. 53).

Brunner, K. (2001): Kartengestaltung für elektronische Bildanzeigen. In: Kartographische Bausteine, Vol. 19, TU Dresden..

BKG, Bundesamt für Kartographie und Geodäsie (2023): Online Karte API geo.akapi. http://sgx.geodatenzentrum.de/geo.okapi/ (14.04.2023)

Burgdorf, M.: Verzerrungen von Raum und Wirklichkeit in der Bevölkerungsgeographie. In: Kartographische Nachrichten, Heft 5 /08, Kirschbaum Verlag, Bonn 2008, pp. 234–242.

Coors, C. et al. (2016): 3D-Stadtmodelle. Konzepte und Anwendungen mit CityGML. Berlin: Wichmann.

de Lange, N. u. J. Nipper (2018): Quantitative Methoden in der Geographie. Grundriss Allgemeine Geographie. Paderborn: Schöningh.

Deutsche Gesellschaft für Kartographie (2023): Commission and Working Group 3D City Models. https://www.3d-stadtmodelle.org/index.php?do=rue (14.04.2023).

DiBiase, D. (1990): Visualization in the Earth Sciences. In: Earth and Mineral Sciences, Bulletin of the College of Earth and Mineral Sciences. Vol 59, No. 2, pp. 13–18.

Dickmann, F. u. S. Dunker (2014): Visualisierung von 3D-Gebäudemodellen – Welche Ansprüche stellt die Planung an dreidimensionale Stadtansichten? In: Cartographic News 1/2014, pp. 10–16.

Dörner, R. et al. (2013): Einleitung. In: Dörner, R. u.a. (Eds.): Virtual and Augmented Reality. Grundlagen und Methoden von VR and AR. Wiesbaden: Springer Vieweg 2013 pp 1–31.

Dougenik, J.A. et al. (1985): An Algorithm to construct continuous area cartograms. In: Professional Geographer, 37(1), pp. 75–81.

Edler, D. et al. (2018a): Virtual Reality (VR) and Open Source Software. In: Cartographic News 1/2018 pp. 5–13.

Edler, D. et al. (2018b): Potentials of spatial visualization in virtual reality (VR) for social constructivist landscape research. In: Cartographic News 5/2018 pp. 245–254.

ESRI (2023): ArcGIS Online https://www.esri.com/en-us/arcgis/products/arcgis-online/overview (14.04.2023).

Feelspace (2023): Navigürtel. Tactile Information. https://feelspace.de/ (14.04.2023).

Gastner, M.T. and M.E.J. Newman (2004): Diffusion-based method for producing density equalizing maps. In: Proceedings of the NAS, 101(20), pp. 7499–7504.

GISQUICK (2023): Let's share GIS much quicker. https://gisquick.org/ (14.04.2023).

Hake, G. et al. (2002): Kartographie. Visualisierung raum-zeitlicher Informationen. Berlin: de Gruyter, 8. Ed.

Harrower, M.A. and C.A. Brewer (2003): ColorBrewer.org: An Online Tool for Selecting Color Schemes for Maps. In: The Cartographic Journal 40(1): 27–37.

Helsinki (2023): Helsinki 3D+. https://kartta.hel.fi/3d/#/ (14.04.2023).

Hexagon (2023): GeoMedia WebMap, Geospatial Portal and Geospatial SDI. https://www.hexagongeospatial.com/products/power-portfolio/geomedia-webmap (14.04.2023).

Hilling F. u. H. Greuler (2015): 3D-Web-GIS: Online-Darstellung eines landesweiten LOD2-Modells für Rheinland-Pfalz. In: AGIT – Journal für Ang. Geoinformatics, 1-2015, p. 322–327.

Hoffmann, K. (2011): Nutzergenerierte Karten und kartographische Kommunikation im Web 2.0. In: Kartographische Nachrichten, Jg. 61, H. 2, pp. 72–78.

International Cartographic Association (2023a): Mission. https://icaci.org/mission/ (14.04.2023).

International Cartographic Association (2023b): Research Agenda. https://icaci.org/research-agenda/introduction/ (14.04.2023).

Karlsruhe (2023): Karlsruhe 3D. https://geoportal.karlsruhe.de/3d/karlsruhe_3d/ (14.04.2023)

Kessler-de Vivie, C. (1993): Ein Verfahren zur Steuerung der numerischen Klassenbildung in der thematischen Kartogra-phie. Beiträge zur kartogr. Informationsverarbeitung Vol. 6. Trier.

Koch, W.G. (2000): Kartengestaltende Variablen – Entwicklungslinien und ihre Ergänzung im multimedialen Umfeld. In: Lechthaler, M. u. G. Gartner (Eds.): Per aspera ad astra. Festschrift for Fritz Kelnhofer. Vienna. (= Geowissenschaftliche Mitteilungen, No. 52), pp. 72–82.

Kraak, M. -J., a. A. M. MacEachren (2005): Geovisualization and GIScience. Cartography and Ge-ographic Information Science 32 (2), pp. 67–68.

Kreuziger, U. (2014): Augmented Reality – Geodaten, fast zum Anfassen touch. In: Vermessung Brandenburg 1/2014, https://geobasis-bb.de/sixcms/media.php/9/vbb_114.pdf pp. 31–36. (14.04.2023).

Landesvermessung Sachsen (2019): Digitales 3D Stadtmodell. Höhen- und 3D-Stadtmodelle. Download, Dienste und Testdaten. https://www.landesvermessung.sachsen.de/download-offene-geodaten-und-testdaten-8647.html?_cp=%7B%22accordion-content-8677%22%3A%7B%223%22%3Atrue%7D%2C%22previousOpen%22%3A%7B%22group%22%3A%22accordion-content-8677%22%2C%22idx%22%3A3%7D%7D (14.04.2023).

Lang, H. (1993): Farbmetrik. In: Niedrig, H. (Ed.): Optics. Bergmann Schaefer Lehrbuch der Experimentalphysik Bd. 3. Berlin: de Gruyter. 9. Ed.

MacEachren, A.M. (1994): Visualization in modern cartography: Setting the agenda. In: MacEachren, A.M. u. D.R. Taylor (Eds.): Visualization in modern cartography. Modern Cartography, Vol. 2. p. 1–12. Oxford: Pergamon.

MacEachren, A.M. (1995): How maps work. New York: Guilford Press.

MacEachren, A.M. et al. (1999): Virtual Environments for Geographic Visualisation. Potential and Challenges. Workshop on New Paradigms in Information Visualisation and Manipulation. S. 35–40.

MacEachren, A.M. and J. H. Ganter (1990): A pattern identification approach to cartographic visualisation. In: Cartographica 27(2) pp. 64–81.

MacEachren, A.M. and Kraak, M.J. (2001): Research challenges in geovisualization. In: Cartography and geographic information science 28-1, pp. 3–12.

McCormick, B.H. et al.(1987): Visualization in Scientific Computing. New York: ACM Press.

Meng, L. (2011): Kartographie für Jedermann und Jedermann für Kartographie – Warum und Wie? In: Kartographische Nachrichten, Jg. 61, H. 5, S. 246–253.

Monmonier, M. (1996): Eins zu einer Million: die Tricks und Lügen der Kartographen. Basel: Birkhäuser.

NextGIS (2023): Web GIS made easy. https://nextgis.com/ (14.04.2023).

Pérez, J.P. (1996): Optics. Heidelberg: Spektrum Akademischer Verlag.

QGIS Cloud (2023a): QGIS Cloud Hosting. https://qgiscloud.com/en/pages/quickstart (14.04.2023).

QGIS Cloud (2023b): QGIS Documentation. https://qgiscloud.com/en/pages/quickstart (14.04.2023).

Rase, W.-D. (2016): Kartographische Oberflächen. Interpolation, Analyse, Visualisierung. February 2016. Norderstedt: BoD – Books on Demand.

Schiewe, J. (2013): Geovisualisation and Geovisual Analytics: The Interdisciplinary Perspective on Cartography. In: Cartographic News. Special Issue. S. 122–126.

Schofeld, J., F. Hillen u. N. de Lange (2017): GuidAR – Augmented Reality in der Fußgängernavigation. In: AGIT Journal for Applied Geoinformatics, 3-2017, pp. 217–222.

Schoppmeyer, J. (1993): Farbgestaltung und Farbbehandlung vor dem Hintergrund der digitalen Kartographie. Kartographische Schriften 1. pp. 32–38. Bonn: Kirschbaum.

Toffler, A. (1980): The Third Wave: The Classic Study of Tomorrow. New York: Bantam Books.

Virrantaus, K., Fairbairn, D. U. M.-J. Kraak (2009): ICA Research Agenda on Cartography and GIScience. In: The Cartographic Journal Vol. 46 No. 2 pp. 63–75 May 2009 Cartography and Geographic Information Science 36/2, pp. 209–222. https://icaci.org/files/documents/reference_docs/2009_ICA_ResearchAgenda.pdf (14.04.2023).

Wood, M. u. K. Brodlie (1994): ViSC and GIS: Some fundamental considerations. In: Hearnshaw, H. u. D. Unwin (Eds.): Visualization in Geographic Information Systems. Chichester: John Wiley and Sons.

Worldmapper (2023): Rediscover the World as you've never seen it before. https://worldmapper.org/ (14.04.2023).

Data Organisation and Database Systems

8

8.1 Data Organisation

8.1.1 Data Organisation: Basic Concepts

The storage, management and processing of extensive data are of great significance for all areas of application and therefore also for geoinformatics. The increased demand in processing data, which concern e.g. interactive access, multiple access, user-friendliness, data security or data protection, go far beyond the performance of file systems (cf. Sects. 3.2.5 and 8.1.2). These requirements have led to the development of complex *database systems*. Such systems are of great importance in geoinformatics. On the one hand, they are used to manage thematic data, i.e. the attributes of geoobjects, within geoinformation system (for the definition of geoobjects cf. Sect. 4.1). On the other hand, as so-called spatial databases, they themselves can take over many tasks of geoinformation systems. In German, these spatial databases are very aptly called "Geodatenbanken", but the English translation "geodatabase" contradicts a proprietary data structure of ArcGIS, a commercial geoinformation system, and is therefore not used here (cf. Sects. 8.7 and 9.3.3). Spatial databases can manage and process not only thematic data, but also geometric data and, in addition, provide many functions for the spatial analysis of geoobjects, such as spatial overlays. These spatial databases could therefore also be assigned to the software group of geoinformation systems to a certain extent. However, they are presented in more detail here in the context of databases, since they are based on database concepts that have been considerably extended.

In many applications, extensive amounts of data are collected which have to be optimally stored, managed and analysed. *Storage* concerns the physical form of storage on the one hand and on the other hand the (more important) logical organisation of the data.

N. de Lange, *Geoinformatics in Theory and Practice*, Springer Textbooks in Earth Sciences, Geography and Environment,
https://doi.org/10.1007/978-3-662-65758-4_8

Data management refers to the updating of datasets, which primarily involves deleting, changing and supplementing individual data or extensive data records as well as creating completely new datasets with new logical structures. The analysis of the data according to different issues is primarily aimed at sorting and searching as the main tasks. The analysis can be an individual evaluation according to a time specification, a group evaluation based on a complex query condition with preparation in table form or a statistical evaluation.

The basis of representing data in computer science is coding a number or a letter as well as multimedia information using bit sequences (cf. Sect. 2.5). At a further level of abstraction, the data type concept enables relatively elegant processing of different data such as integer or character (cf. Sect. 3.2.2). The further logical data organisation is hierarchical (cf. Fig. 8.1 and Sect. 3.2.5). Individual *items* (*data fields*), each containing attribute values, build up a *record.* Several logically related items within a record are often called a *data segment.* Similar data records that belong together based on content criteria are referred to as a file. Several files with logical dependencies or relationships between them form a file system or even a *database* (cf. Sect. 8.1.3).

The logical organisation of data is easily illustrated using tables, which together form a database, with a single table corresponding to a file, a single record to a table row, a single data field to a table column, and a single attribute value to a date in the table. The items can represent different types of data, whereby further data types occur in databases that go beyond the classical data types of computer science (cf. Sect. 3.2.2) (cf. time, currency or special field types for the integration of objects such as graphics or sounds).

Compared to technical access and storage mechanisms the logical identification of data and data records by means of (logical) keys is more important in terms of content related, conceptual side. This refers to the attribute or combination of attribute that uniquely identifies an individual data record. However, an attribute combination is only a key if all attributes are sufficient and necessary for identification, that is, if the key property is lost after one of these attributes is excluded (so-called minimum property of a key). Several keys can exist. For example, in a tree cadastre used to manage urban trees each data record is uniquely defined by the x and y coordinates (composite key) (cf. Fig. 8.1). The key which uniquely specify a record (in Fig. 8.1 the data field ID) is called the *primary key.* All fields which are suitable as key fields are called *key candidates.* A file may also contain data fields which are primary keys in other files. For example, the data field "cost center" in Fig. 8.1, which is used to charge maintenance measures such as tree pruning, can be a primary key of another file containing accounting data. Such fields are called *foreign keys.*

Own, independent numeric attributes, which represent identification numbers e.g. article or customer numbers, are usually used as key fields. Such numbers (so-called "ids") are placed in front of the attributes, which describe the object. However, the use of numeric keys is not mandatory (cf. car license plates as a combination of letters and numbers). Thus, key fields can be formed from several data fields (e.g. from name and date of birth), however, separate numeric key fields are usually clearer and enable faster data access. Sorting by a numeric field, for example, is faster than sorting by a (longer) text field, and linking two tables is also faster due to identical values in a numeric data field.

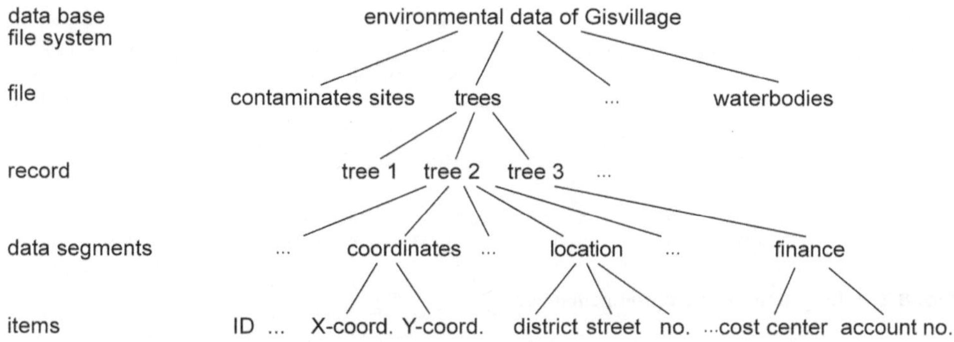

Fig. 8.1 Basic terms of data organisation

8.1.2 File Systems

File systems are the precursors of database systems. In most cases, however, the file management system of the operating system was not sufficient to evaluate the data in the files. Therefore, separate programs were developed in a high-level programming language for analysing the data, which accessed the files and processed the information. The essential characteristic of such file systems, which have to be evaluated by user programs, is the static assignment of evaluation programs to the data. Each of these user programs contains its own description of the file, which is exclusively determined by the processing of the data in the respective program. This close connection enables very individual and efficient evaluation programs. However, the programs are only suitable for this data and for exactly this use. Program maintenance in case of changed requirements is costly.

A mandatory task in an environmental agency in Germany, which can be applied to many similar problems, should make it clear that file systems are ultimately hardly suitable for managing, processing and evaluating thematically differentiated data (cf. Fig. 8.2). In a German municipality facilities for storing, filling, manufacturing and treating water-polluting substances (e.g. petrol stations or oil tank facilities) are to be managed and monitored in a register. For example, the operator is obliged to have a storage container with water-polluting substances checked by an expert at least every five years – or every two and a half years in the case of underground storage in protected areas (according to legal regulations in Germany: Wasserhaushaltsgesetz). A water agency manages the data of the individual storage tanks that are subject to inspection, the data of the test results and the relevant inspection dates. The controls of the proper determination of the test results and any necessary measures are monitored as well as missed deadlines or non-elimination of defects. To process this administrative task, a file system should be set up that shows several typical structural features (cf. Fig. 8.2):

In this example, files 1 and 2 are managed as master files that rarely need to be changed (cf. Fig. 8.3). The data of the facilities (file 1) and the operators (file 2) are kept in separate files. This already meets some of the requirements of higher database systems. For

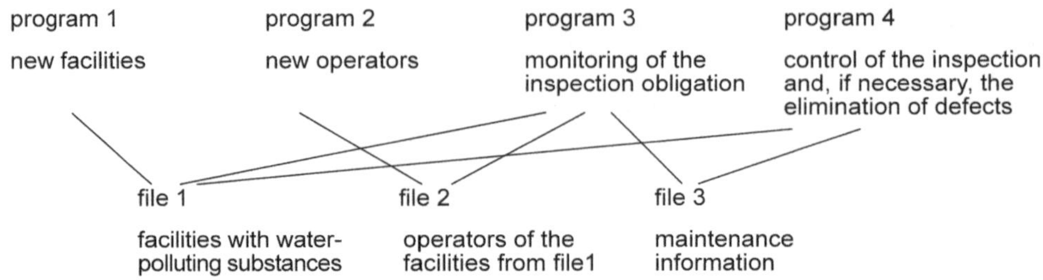

Fig. 8.2 File system in an environmental agency

file 1 facilities	file 2 operators	file 3 maintance
location (address)	name	location of the facility
operator (code-no.)	address	operator of the facility (code-no.)
type	code-no.	due date for inspection
content		defects
quantity		elimination of defects
manufacturer		
year of construction		
date of last inspection		
status		

Fig. 8.3 File structure for the example of Fig. 8.2

example, the (detailed) name and address of the operators are stored only once in file 2, making the system less redundant. If an operator is responsible for several facilities, it is easier to update the operator data. The third file contains the essential transaction data. Here, for example, information is stored, which facilities are due for inspection (e.g. after the expiry of the due examination date) and which results are available.

This file system is processed by programs of a high-level programming language. Programs 1 and 2 update the master data. Program 3 calculates from the date of the last check in file 1, when a new inspection is necessary. The data is then written to file 3, which also records the status of the inspection, among other things. Program 4 controls the status of the check and any necessary rectification of defects. For example, the operator is informed about the proper completion of the inspection or of a reminder with a new deadline. After completion of the inspection, the new inspection date is written to file 1. Extensions of this approach are conceivable: File 1 could contain a reference to an inspection file that contains the analog test report. In a later stage of development of the system, a link to the digital test report could be made. This creates a file system that has grown over time with a linked program system.

While parts of the approach make sense, the system as a whole contains weaknesses. For example, the maintenance and rectification of defects can extend over a longer period of time, during which the operator or his address has changed. In this case, the maintenance file may still refer to the old operator of a facility, while program 2 has already recorded the new operator and deleted the old operator. Such problems can be solved by clever programming or by adding more information on operators in file 1. More serious problems may arise, if the programs were created at different times and by different programmers (hopefully with sufficient documentation). When structuring the individual databases, attention was not and could not necessarily be paid to uniform formatting. Each programmer will have chosen the formatting suitable for him. A sufficient coordination could not take place.

Such a form of organisation results in several problems, which are characteristic for a file system (cf. Vossen 2008 p. 9):

- There can be a high level of redundancy between the individual files, which results from multiple storage of the same data (parallel data sets). In the example above, the location of the facilities is stored in both file 1 and file 3.
- Since multiple or multi-user access to a single file is not possible, there is a risk of inconsistency when processing the same data set. Which processing status is recorded or ultimately saved? This means that individual programs can change files without these changes being taken into account by all programs. In the above example, when a facility operator changes, program 2 modifies file 1, but the changes are not transferred to file 3. The former and not the current operator may receive a reminder. Working with duplicates often means that files, which are not always up-to-date, are evaluated. If only one file is accessed, the changes made by the first user can be overwritten by the second user, who enters the data last, when editing at the same time.
- The combination of program and file system is relatively inflexible to changing requirements and applications. This results in quite high development costs. Thus, new requirements, such as the implementation of new administrative regulations in the above example, can only be realised with quite a lot of effort. Even minor differences require the development of a new program with a new file, which cannot necessarily be derived directly from the old program.
- The low structural flexibility is also reflected in costly program maintenance. For example, when changes are made to the existing files, all relevant user programs have to be changed (cf. conversion from four-digit to five-digit postal codes, other changes of names or formats). In addition to detailed software documentation, a longer training period for the programmers is necessary.
- Access to the individual files cannot be adequately monitored. This can lead to (significant) data protection problems, especially when dealing with sensitive data. Graduated access rights can only be implemented with difficulty and at great expense.
- Data security plays a special role. At best, file systems offer the possibility of archiving the entire data at more or less regular intervals. In the event of an error, a situation can then be restored that reconstructs a previous state. Changes that have occurred since the last backup are not taken into account. Therefore, data loss cannot be ruled out.
- Ultimately, the question of enforcement and compliance with standards arises. For example, uniform data formats are essential for data exchange e.g. between different authorities and computer systems.

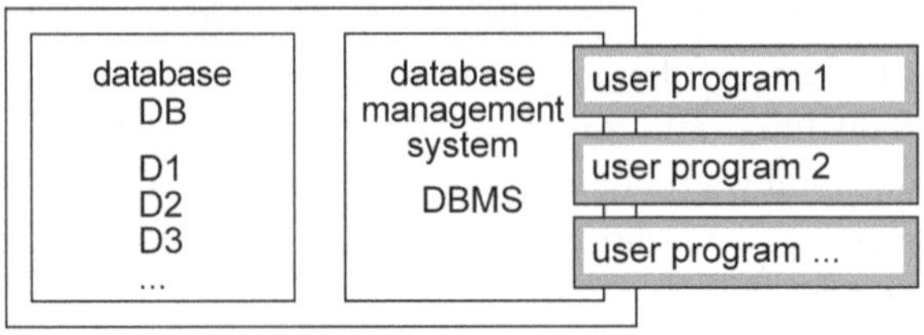

Fig. 8.4 Database and database management system

The file structure and the associated programs can certainly still be optimised. Basic problems were shown here, which almost inevitably led to the development of database systems.

8.1.3 Database Systems

A *database system* (DBS) consists of the database management system (DBMS) and (several) databases (DB) (cf. Fig. 8.4). The DBMS serves as the interface between the database and the users and ensures that data can be accessed in an efficient manner and under centralised control. Data security is therefore provided in the best possible way, offering protection against hardware and software errors and allowing individual users to have different access rights. Thus, (only) individual views of the data are released. Application programs, which are developed using tools of the database management system (for example, input editor, query macros), among others, allow efficient access to the data without knowing their actual realisation (internal data structure). However, this has the disadvantage that the export of data depends on data interfaces. Overall the database management system provides a wide range of efficient options for managing, processing and evaluating datasets.

A *database* is a structured collection of data that represents a special section of the real world in a simplified and schematised way. The data are related to one another from a logical point of view. In most cases, the database; therefore, comprises several linked files.

The specifications for the required properties of database systems result almost directly from the inadequacies of the file systems. In general, it must be possible to evaluate all data according to any characteristics or combinations of characteristics, whereby relatively simple query options with favorable evaluation times should exist. Different user rights can also be granted to individual user groups, so that individual databases do not have to be accessible to all users. The further requirements for a database system are in detail:

- *Independence of data:* In a database system, the close linkage and dependency between the data and the user programs must be resolved. This is an essential difference from a file system. Furthermore, the logical data organisation must be separated from the physical data organisation. The user only needs to know the logical data structures. The management system organises the adding, deleting, changing or searching of data

records. Finally, there must be independence from the data or information at the computer system level. Thus, the management system, together with the operating system, must control the management of the peripheral devices and the physical memories.
- *Non-redundancy data*: All information should be stored only once, if possible. This ensures optimal use of hardware resources. Above all, however, data maintenance is made easier so that the risk of data inconsistencies is reduced.
- *Optimisation of the user interface*: A powerful user interface and optimal tools should enable a simple but also comprehensive handling of the data and their analysis. Above all, this includes interactive use based on a simple user interface (e.g. menu navigation, data input via masks with checking for input errors, assistants (so-called wizards) for creating of input masks and analysis programs). Furthermore, powerful analysis tools (e.g. search and sorting procedures) are required.
- *Data integrity*: When designing the data model, inconsistencies in the data must be avoided. For example, the data stored in several tables in a relational database system must be complete, correct and therefore consistent overall (cf. Sect. 8.5). Changes to the data must be rejected in the event of integrity violations.
- *Data security*: The database management system should offer protection against hardware failures in particular. It should be possible to restore the (correct) initial state after a failure (so-called recovery functions). Furthermore, a database management system should offer protection against errors of user programs (e.g. program crash due to missing or incorrect data).
- *Data protection*: The information stored in the databases must be protected against unauthorised access. It must be possible to grant individual user groups different rights to subsets of the data. This means that, in contrast to full access with change and delete options, read-only options can also be set up that only permit querying.
- *Flexibility*: There should be the greatest possible flexibility with regard to modification and maintenance of the data and the evaluation programs. It should be possible to quickly adapt the user programs to new requirements.
- *Multiple or multi-user access*: The database management program should allow multiple, i.e. simultaneous, access to the data and the user programs (taking into account the respective access rights).
- *Good response time behavior*: The database management program should provide the data quickly, i.e. queries as well as changes to the data should be carried out quickly.
- *Compliance with standards*: The standards primarily relate to data exchange via standardised data interfaces. Furthermore, the analysis tools should correspond to a uniform standard. A standardised database language should be implemented.

Many of these requirements, and especially the technical conditions, are now met by powerful database management systems. However, not all requirements can be guaranteed equally in a specific application. In most cases, an appropriate compromise must be found. Most often, data redundancy and flexibility or efficiency are in conflict. For example, a completely normalised relational database system is non-redundant, but may be complex or confusing with regard to the linking structures (cf. Sect. 8.3.2). Therefore, in reality, only low-redundancy and not (mandatory) redundancy-free storage should to be aimed at.

8.1.4 Data Views in a Database System

The data in a database system reflect a section of the real world in a simplified or schematic way. Thus, the first question that arises is how the database can model an image of the real world and how the issues can be processed using the analysis options of the database management system. Depending on the point of view and tasks of an employee, there are three different perspectives on the data and levels of abstraction of the modeling. Thus, according to the architecture model introduced by ANSI in the 1970s, a distinction is made between three views of the dataset (so-called ANSI SPARC architecture (American National Standards Institute, Standards Planning And Requirements Committee, cf. Saake et al. 2018 pp. 47). The three levels or layers of the database design correspond to the data views (cf. Fig. 8.5):

- *external data view (level),*
- *conceptual data view (level),*
- *internal data view (level).*

This three-level architecture separates the various user's view. There are different formalisms to describe the three perspectives or levels of abstraction. While only one internal and only one conceptual schema exist for a database system, very different user views of a single dataset can exist.

Three different schema types exist that correspond to the three levels in the ANSI-SPARC architecture:

The *external level* or *external view* (*user view*) comprises all individual user views of the data. Each of this views is described by its own data structure (external data schema), which includes exactly that part of the (overall) view that a user needs or to which he should have access. The database management system provides functions for evaluating this section of the overall dataset. Users usually know neither the logical structure of the dataset nor the technical implementation of the data storage. The user view must take two perspectives into

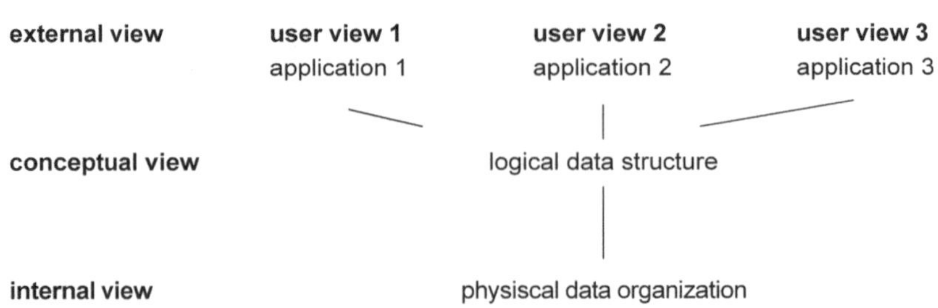

Fig. 8.5 Data views in a database system

account. On the one hand, it must be determined, which data are required in which scaling and accuracy for a specific issue. On the other hand, the user requires specific access rights and analysis options depending on the application.

The *conceptual level* or *conceptual view* develops a logical data organisation for the considered section of the real world and for the task to be solved, which is independent of hardware and software and especially of physical storage methods. The main task is to structure and organise the data set. The conceptual schema shows the logical overall view of all data in the database and their relationships to one another. It is also at this level that the algorithms or methods required to solve the problem specified by the external view must be defined.

The *internal level* or *internal view* deals, among other things, with the type and set up of the data structures (in particular their definition by data types), with special access mechanisms or with the arrangement of the data records on the data carriers (physical data organisation). This level is therefore closest to the physical memory.

8.1.5 Data Models

The conceptual level and the development of a conceptual data schema are essential for the design of a database system. In this way, depending on the issue, the required amount of information of the considered section of the real world as well as the logical data structure of the database system are described. To implement the conceptual schema, several specific data models exist, for which database systems, i.e. software solutions, are available:

The hierarchical data model (e.g. IMS/DB from IBM) and the network data model (e.g. UDS from Siemens) are also referred to as record-oriented data models. In general, they are only of historical importance and have no meaning in geoinformatics and therefore are not considered here, whereas *relational data models* are currently the most important form in commercial and also free database systems (e.g. Access from Microsoft, DB2 and Informix from IBM, Oracle Database from Oracle Corporation or Ingres, MariaDB or MySQL or SQLite).

Object-oriented or *object-relational data models* are of various importance in geoinformatics, since they can store and, in particular, process geometric as well as attribute data. With their analysis functions of geometric data, they extend classic database systems to geoinformation systems (cf. Oracle Spatial from Oracle Corporation or, as free systems, the object-relational database PostgreSQL with the extension PostGIS for the management and analysis of geodata cf. Sect. 8.7).

8.2 Database Design with ER-Modeling

8.2.1 Modeling Concepts

The development of a requirements profile for a database application is followed by the conceptual design. In this phase, the information structure is defined on a conceptual, user-oriented level. The most commonly used model at the conceptual level is the *entity-relationship model* (*ER-model*), which is an abstract model of a section of the real world. An ER model consists of (at least) three different elements (cf. Sect. 8.2.4): the so-called entities (i.e. the objects), the attributes and the so-called relationships (i.e. the relationships between the objects). As a result and especially through the graphical illustration in the form of so-called *entity-relationship-diagrams* (*ER-diagrams*) a data model is created that, before the technical implementation, first reveals all important data structures, which are often neglected by fast and rather inexperienced database programming. This supports the achievement of the stated targets for database systems.

It is often emphasised that the design step using ER modeling is independent of the database system used. The data model of the specific database management system is only taken into account in the subsequent implementation step. This means that ER modeling (also) belongs to the general methods of software engineering (cf. Sect. 3.5.2). In practice, however, ER modeling effectively determines the subsequent data model, since entity-relationship models are ideally suited for implementation in relational data models and relational database systems.

An application example from an environmental agency is used here to illustrate the terms and concepts: With the help of a database system, a well cadastre is to be modeled, which is used to monitor drinking water wells and in particular private domestic wells, and whose water samples are analysed by various laboratories (cf. Sect. 8.2.5).

8.2.2 Entity and Attribute

Entities are easily distinguishable objects of the real world (e.g. geoobjects like a specific drinking water well or a meteorological station). The individual entities that are similar, comparable or belong together (e.g. all drinking water wells in a community) are grouped together into an *entity type*, which describes a data structure, while an entity set contains the instances (the specific objects) of the given structure (for parallels to object-oriented programming cf. Sect. 3.1.4.3).

Entities have features or *attributes* (e.g. name of a business park, nitrate content of a water sample). The specific characteristics are called attribute values. The value range or *domain* includes all possible or permitted values.

The objects of the example task (cf. Fig. 8.6) can then be described using database terminology:

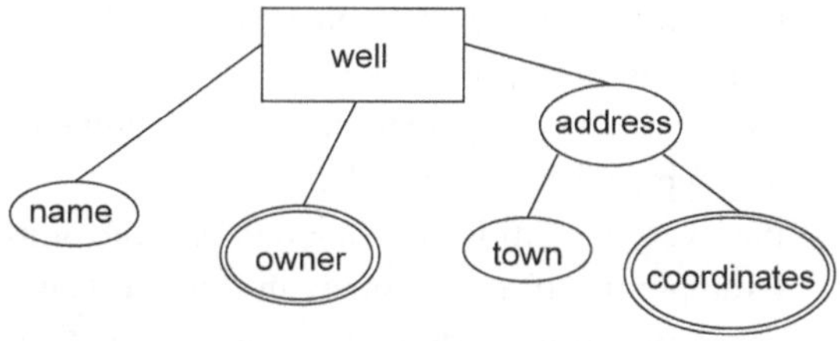

Fig. 8.6 Entity type with attributes (Chen notation)

entity-type	all drinking water wells in Gisvillage
entities	well in mountainview, well of farmer Huck Finn
attributes	name of a well, x-coordinate, nitrate content in mg/litre
domain	integers of length 5, string of length 20
attribute values	12345, waterhole 3

The name of an entity type and its attributes are time-invariant. In contrast, the content of an entity type (the entity set) and the individual attribute values are time-dependent. Attributes can be single-valued, multi-valued and composite. Instead of single-valued attributes, one also speak of atomic attributes, which have values that cannot be further decomposed. Thus, in the present example, a well can have several owners (multi-valued attribute). The address is also usually a composite attribute made up, for example, of the name of the city, the street, the house number, and the postal code. The parcel number of the building (e.g. 05529114200021) is atomic.

8.2.3 Relationship

Different entities can have relationships with each other. In the example mentioned in Sect. 8.2.1, there is a relationship between the wells and the analysis values: The wells each have special analysis values. In conceptual database design, the relationships between attributes of individual entities are less interesting than the relationships between entity types. The relationship between entity types is called a *relationship type.* According to the considerations regarding entities and their attributes, the name of a relationship and its description are time-invariant, while its specific content (e.g. "water sample from well 3 has 50 mg nitrate") is time-dependent (cf. Fig. 8.7).

A specific relation can be assigned to a relationship type. If A and B are entity types (with a and b as specific entitites) and R is a relation R(a, b), then the following applies (cf. Fig. 8.8 and application examples in Sect. 8.2.5):

- 1:1 type: one-to-one relationship
  For every a from A, there is exactly one b from B with R(a, b) (and vice versa). Example: parcels in a real estate cadastre (entity type A) and distinct related features (entity type B), each parcel has, on the one hand, information about the parcel such as

parcel name and size and, on the other hand, associated information such as property tax or rent. Both entity types (here: describing spatial characteristics and financial characteristic of a single parcel) could be combined without redundancy.

- n: 1-type: many-to-one-relationship
  For every b from B there are one or more a_i from A with $R(a_i,b)$.
  Example: n different parcels in a real estate cadastre (entity type A) and designation according to a masterplan (statutory land use plan) (entity type B), each parcel is assigned to exactly one category, e.g. housing, commercial area, green infrastructure. Several parcels belong to the category housing. This relation can be understood as forming a legend.
- n:m type: many-to-many relationship
  For every a from A there are one or more b_i from B with $R(a, b_i)$ and for every b from B there is one or more a_j from A with $R(a_j, b)$.
  Example: n different parcels in a real estate cadastre (entity type A) and m different owners (entity type B), each parcel can be owned by one or more owners, an owner can own one or more parcels.

The *complexity* or *cardinality* of the relationship indicates how many entities of the second entity set can (or may or must) be related to a specific entity of the first entity set (1:1, n:1, n: m).

Relationships can also have own attributes, which express properties of the relationship. In this example (cf. Sect. 8.2.5 and Fig. 8.10) the wells are to be maintained by laboratories, whereby the relationship "maintain" can be described in more detail, e.g. by the period of responsibility. Relationships can have more than two digits. For example, the three-digit relationship "deliver" between sampler, water samples, and laboratory would be

Fig. 8.7 Representation of a relationship in an ER-model (CHEN notation)

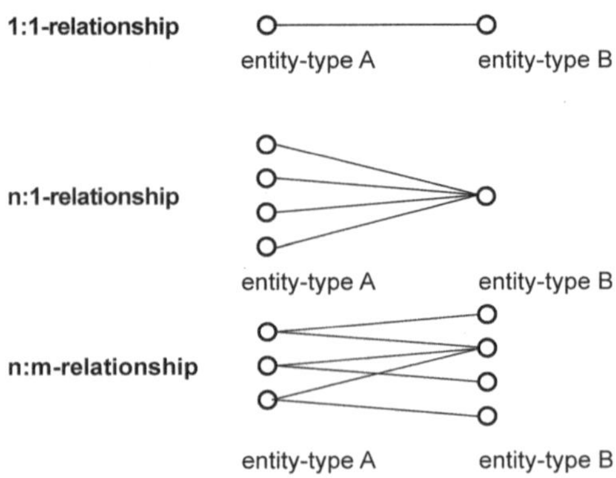

Fig. 8.8 Relationship types

conceivable. Samplers take water samples from a well and deliver them to a laboratory. The modeling takes into account that a sampler delivers several water samples (from one well) to different laboratories that have specialised in particular analytical methods. However, this more complex case is mostly excluded for reasons of simplification. Thus, often a restriction is made to the particularly important two-digit relationships, which can also be used to resolve multi-digit relationships.

8.2.4 Entity Relationship Models

Special diagram types are used for the graphical implementation of entity types and the associated relationship types. These *entity relationship models* (*ER-models*) are the graphical representation of conceptual data modeling. They describe a section of the real world. Various representations currently exist, but despite individual graphical differences, the core statements are almost identical. Figure 8.9 models or graphically illustrates the situation in which a parcel is located in exactly one town and one town has several parcels.

The so-called Chen notation (named after Peter Chen, who is known for developing the ER-model in 1976) or the modified Chen notation (extension to display attributes) is very common. In this display variant, the name of an entity type is represented as a rectangle and the attributes as circles or ellipses, which are connected with a line to exactly one entity or relationship type. Double circles or double ellipses mark multi-valued attributes, lines connect composite attributes with their components. A relationship type is represented by a rhombus, which is connected to the corresponding entity types (cf. Figs. 8.7 and 8.9). Corresponding labels explain the relationship type. Many graphical tools (i.e. programs)

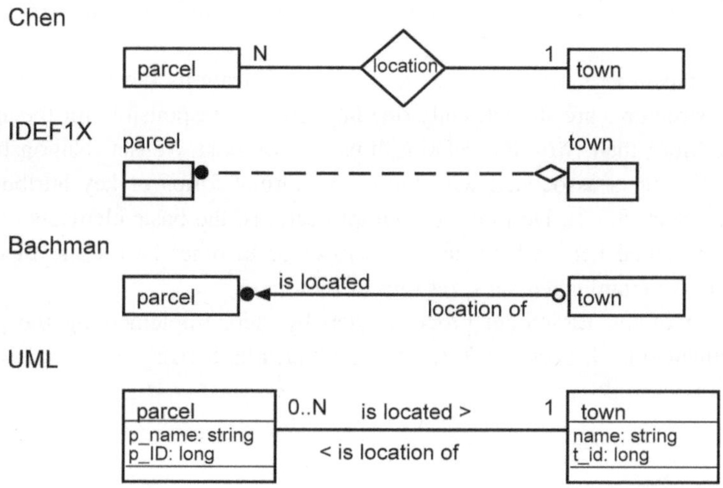

Fig. 8.9 Different methods of representing the same 1: n relationship

are available for creating ER-models, which are often integrated into programs for developing databases (cf., e.g., the free software MySQL Workbench cf. Fig. 8.11).

The IDEF1X notation characterises a long-standing standard of the US authorities. The Bachman notation goes back to a pioneer of database systems, i.e. especially of the network data model. UML (Unified Modeling Language) is now a standard for modeling software systems, i.e. for object-oriented modeling. Here, the so-called classes comprise the actual core of the modeling language. Accordingly, database models are represented by class diagrams, whereby more extensive concepts of data modeling exist (cf. e.g. specialisation). In contrast to the other forms, attributes (and methods, here called operations) are added within the class box (cf. Fig. 8.9 without methods, on UML cf. Gumm and Sommer 2013 pp. 840).

8.2.5 Conceptual Database Design: A Case Study

The conceptual database design with the help of an ER-model is to be shown on the basis of an example. In a municipality, a cadastre of the existing wells is being set up, into which new, but also existing data are to be transferred. Water samples are taken at regular intervals, but these vary in time from well to well, and are analysed in laboratories. The water sample of a single well is evaluated only by one laboratory. The individual wells are each analysed for different parameters such as nitrate or cadmium. The scope of the analysis, i.e. the number of analyses to be carried out for a water sample, and thus, the number of values to be determined for two wells can be quite different. Since the cadastral register was established, the wells may have been maintained by several laboratories one after the other. The responsibility of individual laboratories may therefore have changed. Furthermore, some laboratories have only been set up in the meantime, some have since been closed.

Although the task is very close to reality, simplifications have been made. Not all conceivable structures are shown: Only one laboratory is responsible for the analysis of a single well. More than two-digit relationships do not exist. Subset relationships are not considered. Concepts associated with the terms normal forms or key attributes are still excluded (cf. Sect. 8.3.2). Despite these simplifications, the basic elements of a database design are presented here, which can be transferred to other issues and then expanded (e.g. register of contaminated sites, register of biotopes).

The design of the ER-model proceeds step by step, implementing the principle of stepwise refinement (cf. Sect. 3.5.2, for the result cf. Fig. 8.10):

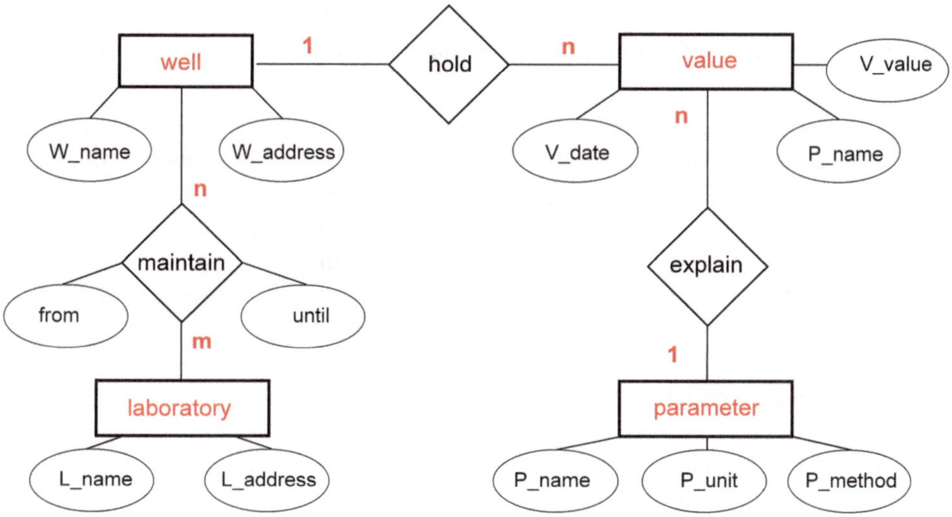

Fig. 8.10 ER-model of the well cadastre

- The first step is to only define the entity types. The model consists only of rectangles for the entity types "well", "laboratory", "value" and "parameter". At first glance, this seems somewhat incomprehensible. But: An analysis value should only have a short name of the analysed parameter. This short name of the parameter, such as "nitrate1", is explained in detail in the entity type "parameter", i.e. with specification of the measurement unit, the limiting value, further descriptions and explanation which analysis procedure is used. If a new method is used in the course of the indefinite application, the parameter "nitrate2" is introduced, for example. This avoids redundancies. Although the attributes will only be added to the model in the next step, essential conceptual considerations must already be made beforehand.
- In the second step, the attributes of the entity types are added. Now the ellipses with the attribute names are added to the model. For reasons of simplification, however, the attributes in Fig. 8.10 are only listed incompletely or grouped according to data segments. For example, the entity type "well" has the attributes W-ID (unique identifier for a specific well), the name of the well, the x- and the y-coordinate of the location, the name of the associated topographic map, the address with the postal code or also the pipe/shaft diameter of the well (cf. simplifications in Table 8.1).
- The third step introduces relationships. In the model, connecting lines are drawn between the corresponding entity types.
- In the fourth step, attributes are added to the relationships. In the present example, only the relationship between "well" and "laboratory" is to be provided with attributes, which primarily model the temporal support.
- In the fifth and last step, the cardinalities of the relations are inserted.

Table 8.1 Entity type well of the example from Sect. 8.2.5 (cf. Fig. 8.10)

W_ID	Location	W_address	W_name	L_no	L_name	L_phone
...	...	...		...		
33	10,44	Atown, X Street	springsteen	10	Aqua-pro	0123-7711
2	12,43	Atown	freshwater	28	Aquarius	0123-4567
4	14,33		blueravin	29	R-Tec	0321-8899
1314	13,35			10	Aqua-pro	0123-7711
1315	13,36			10	Aqua-pro	0123-7711
2903	11,42		foxhole	28	Aquarius	0123-4567

The entity-relationship model shows for this example:

- There is a 1: n relationship between the entity types “well” and “value”. A single well has n analysis values, but an analysis value belongs to exactly one well.
- The n:m relationship between the entity types “well” and “laboratory” defines the responsibilities, which laboratory performs the analytical procedures for which well. Here it is assumed that a laboratory is supporting several wells and that several laboratories are assigned to one well (responsibilities that change over time).
- The 1: n relationship between the entity types “parameter” and “value” explains the parameter names and therefore the analysis values. The associated limiting value can also be saved for a parameter, so that a precise evaluation of an individual analysis value can be carried out.

8.3 Relational Data Model

8.3.1 Relational Database: Structure

The relational data model, which is based on a theoretical basis by Codd (cf. Codd 1970, 1990), has become the standard of commercial database management systems since the mid-1980s. Fundamental to this model is the implementation of a relation, which here provides the only possible data structure. Formally, an n-digit relation R is a subset of the product set $A_1 \times A_2 \times \ldots \times A_n$ (here: $A_1, \ldots, A_n$ attributes).

$$R = \{(a_1, a_2, \ldots, a_n) | a_1 \in A_1, a_2 \in A_2, \ldots, a_n \in A_n\} \subseteq A_1 \times A_2 \times \ldots \times A_n$$

A tuple $(b_1, b_2, \ldots, b_n)$ of n attribute values describes exactly one entity.

Such a relation can be clearly represented by a two-dimensional table. In the relational data model, the entire database is managed by simple tables. Here, a relation or a table corresponds to an entity type, a column of a table defines an attribute, a row of a table

Table 8.2 Entity type laboratory of the example from Sect. 8.2.5 (cf. Fig. 8.10)

L_ID	L_name	L_address	L_phone
10	Aqua-pro	Atown Zlane 3	0123-7711
28	Aquarius	Atown Astreet 27	0123-4567
29	R-Tec	Btown Xroad 19	0321-8899

describes an entity and corresponds to a logical data record. An attribute or a combination of attributes is used to uniquely identify an entity, so that two identical rows can never exist. The order of the rows and columns is arbitrary.

Tables 8.1 and 8.2 explain the principles. The tables describe the location and address of the wells and the laboratories. Furthermore, Table 8.1 contains information on which of the laboratories is responsible for a well. However, it must already be pointed out that the structure of Tables 8.1 and 8.2 is not optimal (cf. non-atomic attribute “W_address” und “L_address”, repetitions of laboratory names and telephone numbers depending on the laboratory number). This is done in this context in order to show inadequacies and then their solution by introducing so-called normal forms (for the evaluation of optimality cf. Sect. 8.3.2).

The tables require *key fields* (cf. Sect. 8.1.1). Thus, the attributes must contain special features that, individually or in combination with one another, uniquely identify exactly one entity, i.e. one table row in the relational data model. *Primary keys* are the attribute “W-ID” in Table 8.1 as well as the attribute “L-ID” in Table 8.2. The attribute “L_no” is a *foreign key* in Table 8.1 (the primary key in Table 8.2). The names of associated keys do not have to be the same.

The primary keys of several tables in a relational database have another important function in addition to the unique access to an entity, i.e. to a table row. Keys are used to link different tables, i.e. ultimately to represent the relationships between the entity types of the ER-model. The links are therefore realised by software or data technology and not by specifying or managing absolute memory addresses. This results in considerable advantages. The data model is therefore invariant to changes in the system environment (i.e. change of the physical memory structure, change of the operating system). By using such links, (simultaneous) access to several relations (tables) and thus, a very efficient analysis of the data are possible.

From the user’s point of view, the relational data model is considerably simpler and more descriptive as well as more flexible than the hierarchical model or the network model. Using tables or thinking in tables is the most common way of working with data. Linking tables via key fields is almost intuitive or can be learned quickly. In particular, an ER-model can be transformed directly into the relation model (cf. Sect. 8.3.3). Thus, there is a close connection between the conceptual database design and the program-technical implementation. Furthermore, a standard database language has been established for relational database management systems (SQL = Structured Query Language, cf. Sect. 8.4.3). In particular, a large number of commercial and proprietary database systems are available for various hardware platforms, all of which can be regarded as technically well-engineered.

8.3.2 Database Normalisation

Tables 8.1 and 8.2 already illustrate relations and show the principle of how links between tables can be displayed. However, the table structure is not yet optimal. The aim should always be to avoid data redundancies. For example, multiple storage or suboptimal structuring results in high memory requirements and slower access to the data or evaluation options. In particular, problems can occur with data maintenance (multiple updates, keeping the data consistent). Since the names of the laboratories and the telephone numbers are kept in both tables, redundancies occur. The structure of Table 8.1 is particularly critical. The telephone number must be the same for the same laboratory in each table. If the telephone connection of a laboratory changes, the update would have to be carried out several times. There are considerable sources of error. The laboratory name and the telephone number are redundant in Table 8.1, which can be obtained by linking to Table 8.2. Often, however, many more features are redundantly. The concept of normal forms was developed in connection with such structuring problems, to be used as optimisation criteria for relational databases. A major concern when designing a relational database is to normalise the tables and thereby reduce redundancy.

Table 8.3 illustrates a relation which is unfortunately often occurs, but which is completely senseless from the point of view of data modeling of relational databases! It shows the analysis values for the different wells, each analysis is represented by an attribute. Here, some cells in the table remain empty, because not all analysis procedures are always performed as the respective scope of analysis is quite different for different wells. Furthermore, only a few attributes are provided for storing the analysis values. If additional parameters are to be analysed at a well, the data model must be changed. Storage space is also wasted if only a few parameters are to be measured for some wells. The

Table 8.3 Example of a poorly structured relation or table

W_ID	nitrate_ 01.07.96	nitrite_ 01.07.96	cadmium_ 01.07.96	pah_ 01.07.96	nitrate_ 03.07.97	nitrite_ 03.07.97
1	52	0.05	0.006	0.00017	54	0.05
2	50	0.051		0.00012	49	0.05
3	51		0.001	0.00009	55	
4	50				49	
5	52			0.00015		
6	49			0.00017		
7	52					

pah = polycyclical aromatic hydrocarbons

problems are multiplied, when analysis values from several analysis campaigns are to be stored. Labeling the different measurement series by dates in the attribute names is not a solution (merging of data structure and content). Above all, a table in database management systems cannot be extended to the right by any number of attributes. In addition, database queries are difficult to implement. If, for example, it is to be determined whether a parameter is recorded at a specific well, all (attribute) fields of the database must be searched. Such a table is usually not the result of conceptual data modeling with an ER-model. The concept of normal forms helps to exclude such errors.

A relation is in *first normal form* if every attribute is elementary or atomic, i.e. indecomposable. The relations or Tables 8.1 and 8.2 are not in first normal form. The attributes "location" and "W_address" and "L_address" are not atomic. Instead, there are lists of values, which consist, for example, of the components "Geogr. Longitude" and "Geogr. Latitude". It makes sense to atomize the attributes so that they can then be sorted by longitude, for example (cf. Table 8.4).

A relation is in *second normal form* if it has first normal form and, in addition, any attribute that is not itself part of the identification key is fully functionally dependent on the entire identification key. Table 8.5, which contains the analysis values of the wells with

Table 8.4 Example of a relation in first normal form

W_ID	W_name	X-coord.	Y-coord.	zip_code	town	street	no
1	rhinewine	10	44	12311	Atown	X-way	1
2	freshwater	12	43	12312	Atown	Y-lane	33
…		…	…	…	…	…	…

Table 8.5 Example of a relation in first but not in second normal form (with a composite key)

W_ID	V_date	V_value	P_name	P_method	P_unit	P_limit_value
…		…	…			
2	1.6.2000	52	nitrate	A-NO3-1	mg/l	50
2	1.6.2000	0.05	nitrite	A-NO2-1	mg/l	0.1
2	1.6.2000	0.006	cadmium	A-Cd-1	mg/l	0.005
…		…	…			
2	2.5.2001	0.00017	pah	A-PAH-1	mg/l	0.0002
2	2.5.2001	54	nitrate	A-NO3-1	mg/l	50
		…				
33	1.8.2000	35	nitrate	A-NO3-1	mg/l	50
33	1.8.2000	0.01	nitrite	A-NO2-1	mg/l	0.1
33	1.8.2000	0.00007	pah	A-PAH-1	mg/l	0.0002
…		…	…			
538	1.7.2010	49	nitrate	A-NO3-1	mg/l	50
		…				

associated explanations of the values, is in first normal form, but not in second normal form. Here is a composite identification key consisting of the attributes “W_ID”, “V_date” and “P_name”. In this relation, the attributes “P_method”, “P_unit” (unit of measurement of the measured parameter) and “P_limit value” are each non-key attributes, but depend only on the attribute “P_name”. Thus, these attributes are not fully functionally dependent on the (entire) identification key. In order to achieve the second normal form, the table must be further decomposed (cf. Tables 8.6 and 8.7).

A relation is in *third normal form* if it is in second normal form and additionally no attribute (outside the identification key) is transitively dependent on an identification key. This does not allow indirect or transferred (i.e. transitive) dependencies of an attribute on the primary key. Thus, Table 8.1 is not in third normal form. The attribute “L_name” does not belong to the identification key of the relation. The laboratory name is only dependent on the attribute “L_no”, only this attribute is dependent on the identification key (“W_ID”). Thus, there is a transitive dependency on an identification key. If several wells belong to the same laboratory, the designations are repeated. Thus, the transitive dependence indicates redundancy. In order to optimize Table 8.1 and achieve the third normal form, it must be further decomposed in addition to atomising the attributes. The solution is already provided by Table 8.2 (except for the atomisation of the address), where the attributes “L_name” and “L_phone” are removed from Table 8.1. Tables 8.6 and 8.7 normalise Table 8.5 (cf. Tables 8.8 and 8.9 in Sect. 8.3.3).

In addition to the third normal form, which is to be achieved (cf. Saake et al. 2018 pp. 179), further normal forms exist. In summary, a chain of inclusions applies, with BCNF denoting the Boyce-Codd normal form.

Table 8.6 Decomposition of Table 8.5 into “values” only (with a composite key, presence of second normal form)

W_ID	V_date	V_value	P_name
...		...	...
2	1.6.2000	52	nitrate
2	1.6.2000	0.05	nitrite
2	1.6.2000	0.006	cadmium
...		...	...
2	2.5.2001	0.00017	pah
2	2.5.2001	54	nitrate
		...	
33	1.8.2000	35	nitrate
33	1.8.2000	0.01	nitrite
33	1.8.2000	0.00007	pah
...		...	...
538	1.7.2010	49	nitrate
		...	

Table 8.7 Decomposition of Table 8.5 into "parameters" only (with a composite key, presence of second normal form)

P_name	P_unit	P_description	P_method	P_limit value	P_note
nitrate	mg/l		A-NO3-1	50	
nitrite	mg/l		A-NO2-1	0.1	
cadmium	mg/l		A-Cd-1	0.005	
chc	mg/l	chlorinated hydrocarbons		0.01	limiting value for all compounds, tetrachloromethane max. 0.003 mg/l
pah	mg/l	polycycl. aromatic hydrocarbons	A-PAH-1	0.0002	

$$5^{th}\ NF \Rightarrow 4^{th}\ NF \Rightarrow 3^{rd}\ NF \Rightarrow 2^{nd}\ NF \Rightarrow 1^{st}\ NF$$

The condition that guarantees the existence of the fifth normal form is of great practical importance (cf. Vossen 2008 p. 274): If a relation is in the third normal form and every key is simple (i.e. it consists of only one attribute), then this relation is in the fifth normal form. Thus, in practice, only simple numeric IDs are used and then only the first three normal forms are taken into account.

Overall, the normal forms represent criteria that can be used to assess a relational database system. However, the splits create many small tables, so that the data model can ultimately become quite confusing. The user should not have to look at the tables themselves at all. Access or evaluation should take place with the help of a comfortable query language, so that the type of storage should be irrelevant for the user (cf. Sect. 8.4.3).

8.3.3 Transformation of an ER-Model into a Relation Model

The ER-model of the well cadastre (cf. Fig. 8.10) is to be (partly) transformed into a relational database system. In doing so, each entity type and each relationship will be transformed into a table, since the relational model basically only has the concept of tables. While the transformation of entity types into tables has already been explained several times (cf. Sects. 8.3.1 and 8.3.2), the focus is now on the transformation of relationship types:

A 1:1 relationship is easy to implement. In the present example, the 1:1 relationship between the entity type "well" and a new entity type "well_properties" with the attributes "type of extraction point" or "thickness of filter layer" could be subsequently inserted. Each row in the table "well" (cf. Table 8.1) then corresponds exactly to one row in the new table "well_properties", the link is made via the attribute "W_ID". This could also be achieved by directly adding additional attributes to the table "well". On the conceptual side of the database design, however, it should be considered, whether time-dependent attributes should be separated from time-independent attributes such as the location definition and kept in separate tables.

In the present example, the 1: n relationship "hold" between the entity types "well" and "value" is of central importance. Each well has a different number of analysis results. An implementation, in which there is exactly one row for each well and in which the columns contain the analysis values, does not represent a solution (cf. Table 8.3). This is mainly due to the fact that it is completely unclear, how many analysis events are expected. Therefore, in the present example, the entity type "value" is the only appropriate way to convert it into a table that is unlimited "downwards" (cf. Table 8.8)! The attribute "W_ID" realises the relation "hold" between the entity types "well" and "value" (link between Table 8.1 resp. its normalised form and Table 8.8). Due to the additional introduction of the unique attribute "V_ID", Table 8.8 has its own primary key (cf. almost the same Table 8.6).

The n:1 relationship "explain" between the entity types "value" and "parameter" is used to explain the single analysis values. Thus, according to the task, several information are clearly linked to a parameter name: first, the component of the water sample that is to be analysed is named, then its measurement unit or its associated limiting value (cf. Table 8.9). Otherwise, this information would have to be listed several times, i.e. redundantly, in the "value" table. The primary key of the "parameter" table is listed as a foreign key in the table "value". Table 8.9 ("parameter") can be understood as a legend of Table 8.8 ("value"). Due to the additional introduction of the unique attribute "ID", Table 8.9 has now its own primary key (cf. almost the same Table 8.7).

The relationship "maintain" (support) between the entity types "well" and "laboratory" requires special attention, since there is a n:m relationship. A well can be supervised by

Table 8.8 Representation of a 1: n relation in third normal form for the entity type analysis values of the example from Sect. 8.2.5

V_ID	W_ID	V_date	V_value	P_name
	...		...	...
51	2	1.6.2000	52	nitrate
52	2	1.6.2000	0.05	nitrite
53	2	1.6.2000	0.006	cadmium
	...		...	...
73	2	2.5.2001	0.00017	pah
74	2	2.5.2001	54	nitrate
			...	
126	33	1.8.2000	35	nitrate
127	33	1.8.2000	0.01	nitrite
128	33	1.8.2000	0.00007	pah
	...		...	...
222	538	1.7.2010	49	nitrate
			...	

Table 8.9 Representation of a 1: n relation in third normal form for the entity type parameter of the example from Sect. 8.2.5

P_ID	P_name	P_unit	P_description	P_method	P_limit value	P_note
1	nitrate	mg/l		A-NO3-1	50	
2	nitrite	mg/l		A-NO2-1	0.1	
3	cadmium	mg/l		A-Cd-1	0.005	
4	chc	mg/l	chlorinated hydrocarbons		0.01	limiting value for all compounds, tetrachloromethane max. 0.003 mg/l
5	pah	mg/l	polycycl. aromatic hydrocarbons	A-PAH-1	0.0002	

Table 8.10 Representation of a matching table for the implementation of a n:m relationship (entity type "maintain" of the example task from Sect. 8.2.5 and Fig. 8.10)

W_ID	L_ID	from	until
1	10	01.01.96	31.12.97
2	10	01.01.96	31.12.96
2	30	01.01.97	31.12.97
3	30	01.01.96	31.12.97
4	20	01.01.96	31.12.97
5	20	01.01.96	31.12.96
1314	10	01.01.96	31.12.96
1314	28	01.01.97	31.12.97
7	30	01.01.96	31.12.97

several laboratories in succession over time. One laboratory is responsible for several wells. In the relation model, such a n:m relation is not represented by two tables, but by three tables. In addition, a so called *matching table* controls the linkage via key fields. As shown in Table 8.10, well 1314 was serviced by the lab with L_ID 10 in 1996 and by the lab with L_ID 28 in 1997. Laboratory 10 also supervised wells 1 and 2.

Overall, an ER-model can easily be converted into a relational data structure, which ultimately represents several tables linked together. There must be logical data consistency between the linked tables and in particular between the key fields that implement the link! It must be prevented, for example, that the laboratory with ID 10 is deleted in Table 8.2 as long as this laboratory is still referenced in Table 8.10. This property is referred to as *referential integrity* (cf. Sect. 8.5.2).

The present modeling (cf. Figs. 8.10 and 8.11) is already sufficient. With the database query language SQL, many issues can be analysed (cf. examples in Sect. 8.4.4). However, the modeling here is deliberately kept minimalistic in order to demonstrate the efficiency of

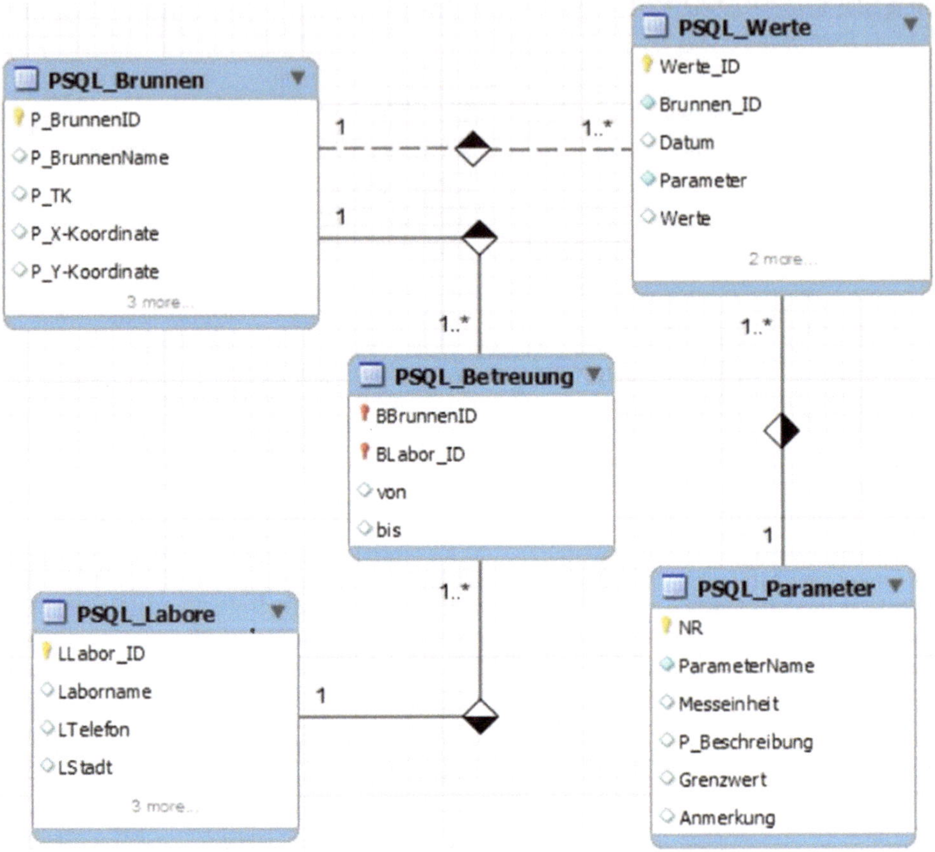

Fig. 8.11 Creating an ER-model with MySQL Workbench

the modeling and analysis. Thus, via the "detour" of the relation, which models the maintenance, the information is recorded, which laboratory has determined a specific analysis value of a well. In practice, Table 8.8, which contains the analysis values, is supplemented with an additional attribute L_ID. This is the identifier of the respective laboratory, which performed the analysis.

Some software supports this modeling, such as the MySQL Workbench tool, which is part of the free MySQL database software and which can be used to easily manage a MySQL database. After defining the table, i.e. the entities, the MySQL Workbench modeling tool expects the relationship to be specified with its cardinalities. As soon as a n:m relationship is specified, a matching table is automatically inserted immediately. In the present example, this table describes exactly the responsibilities. The relationship between wells and laboratories is also identified in more detail by attributes. Figure 8.11 shows the result of this modeling using the MySQL Workbench tool with slightly modified attribute names with regard to an application in MySQL (here still in German, the translation for the key terms is certainly helpful "Brunnen" = "well", "Werte" = "value", "Betreuung" = "maintain").

8.4 Relational Database Systems: Applications

8.4.1 Data Definition and Management Functions

In addition to the analysis functions, relational database management systems have extensive tools for defining tables or for changing a table structure or deleting tables. These data definition functions start at a higher conceptual level of working with database systems and require special access rights. For example, database management programs have control functions that can be used to define and manage user views and access rights. Standardised data interfaces allow easy data exchange. Especially in the field of commercial database software for personal computers, database management systems often offer user-friendly user environments that allow relatively easy handling. Software wizards support the creation of forms, reports or queries.

8.4.2 Data Manipulation and Data Analysis

Data manipulation primarily includes functions for inserting, modifying or deleting existing (individual) data and data records, whereby the user must be granted appropriate access rights for data manipulation. Access rights must also be granted for simple viewing or visualisation of the datasets, i.e. for read-only access. Of central importance are the functions that enable a query of individual data or data records of the entire dataset. Each analysis of the dataset of a relational database leads to a subset of the dataset, so that the result is again a relation. In extreme cases, the result is a table with exactly one column and one row. Relational database systems have three standard functions for these tasks:

- The *projection* provides only selected columns (i.e. attributes) of a table or tables, but with all rows. For example, only the attributes that are currently of interest are selected (e.g. only the laboratory name and telephone number of all laboratories).
- The *selection* provides only selected rows (i.e. entities) of a table, but with all attributes (columns). The user defines specific selection criteria (for example, all wells in a specific city). The database management system then creates a subset of the rows.
- The *join* connects the relations with each other using suitable key fields. The optimisation of a database (cf. normal forms) usually leads to several relations (tables). However, this should not affect the manageability of a relational database, whose efficiency is increased precisely by this. The tables are usually only temporarily linked to one another.

Most common is a simultaneous execution of projection and selection or a combination with selection and projection due to conditions on the attributes of several tables. For example, the search is for the addresses of all wells, whose water samples showed a nitrate content above the limiting value of 50 mg/l between March 1st, 1996 and February 28th, 1997. Queries of this complexity can be implemented relatively easily with suitable query languages (cf. examples in Sect. 8.4.3). Arithmetic operations (e.g. forming sums) or complex sorting options are also available.

8.4.3 The SQL Language Standard of a Data Manipulation Language for Relational Database Systems

With the help of a data manipulation language of a database management system, data of a database can be recorded, modified or queried in a simplified way. The *Structured Query Language* (*SQL*) represents the language standard for relational database systems. This standardised database language has been implemented in many database management systems and is therefore widely used. It was initially developed exclusively by IBM since the beginning of the 1970s. In the 1980s, a general standardisation of this language took place. The latest version SQL:2016 (ISO/IEC 9075-1:2016) is now available.

The SQL language standard consists of relatively few commands that can be entered interactively or dialog-oriented. SQL commands can also appear in an integrated form, i.e. as part of a programming language. This means that SQL queries can also be used within complex user programs. In general, the SQL standard refers to three levels:

- The *level of data description* (DDL, Data Definition Language) primarily includes language elements for defining tables, attributes (with definition of keys) or links. The associated central SQL commands are: CREATE, ALTER, DROP and JOIN.
- The *level of data manipulation* (DML, Data Manipulation Language) includes both the management and the query of data. This primarily includes language elements for inserting or deleting data as well as updating tables (i.e. deleting and modifying). The associated main SQL commands are: INSERT, DELETE, UPDATE. Database queries are also of central importance, whereby SQL has only a single, but very powerful command for formulating queries in the form of the SELECT command.
- The *level of embedding in programming language* refers to programming languages that integrate SQL commands.

All SQL commands for data manipulation have the following basic structure:

```
<Operation> indicates the type of manipulation to be performed (e.g. SELECT).
FROM        describes the affected relations.
WHERE       identifies the affected attributes of the specified relation.
```

The powerful SELECT command, which can be specified with up to six components, is important for analysing a database. The general syntax is:

```
SELECT     [ALL | DISTINCT] {columns | *}
FROM       table [alias] [table [alias]] ...
WHERE      {query | subquery}
GROUP BY   columns [HAVING {query | subquery }]
ORDER BY   columns [ASC | DESC]...;
```

Section 8.4.4 shows several examples of the SELECT command using the example of a well cadastre (cf. Figs. 8.10 or 8.11).

8.4.4 Analysis of a Database with SQL

The database built for the present example is to be analysed with regard to typical issues using the open source database PostgreSQL. Compared to the proprietary database software Access from Microsoft, PostgreSQL is of greater importance in geoinformatics, since the PostGIS extension is a powerful extension with regard to spatial data. PostgreSQL is an object-relational database management system that supports the SQL92 and SQL99 standards and impresses with extensive performance features. The October 2018 version 11 complies with at least 160 of the 179 mandatory features for SQL: 2011 (to get started cf. PostgreSQL 2023a and PostgreSQL 2023b, cf. Laube 2019).

The PostgreSQL application is based on a client-server model. A server manages the database files and regulates the links and requests from the client to the server. When installed as a package, the server and administration tools are installed at the same time. The open source software "pgAdmin" with a graphical user interface is often used for administration (cf. pgAdmin 2023, cf. Figs. 8.18 and 8.19).

The issues and the corresponding queries are based on the ER model shown in Fig. 8.11. Several tables with new unique attribute names are now used:

PSQL_well	PSQL_value	PSQL_lab	PSQL_parameter	PSQL_support
W_ID	V_ID	L_ID	P_name	S_WID
W_name	V_WID	L_name	P_unit	S_LID
W_diameter	V_date	L_phone	P_description	from
W_x_coordinate	V_parameter	L_city	P_limit value	until
W_y_coordinate	V_value	L_zipcode	P_note	
W_city		L_street		
W_zip_code		L_houseno		

The SQL statements are almost directly readable. A link is established between the tables "PSQL_well" and "PSQL_value" via the attributes "W_ID" and "V_WID" respectively, whereby this attribute represents a primary key in one table and a foreign key in the other table. This means that the condition can be evaluated in the table "PSQL_value". For the water samples selected in this way, there are no well names in the table "PSQL_value", but these are immediately apparent from the link to the table "PSQL_well". For this task there are (as usual) various implementation options. A nested query is also possible with SQL.

Search all wells in "A-town":

```
SELECT  *
FROM    public."PSQL_well"
WHERE   "W_city" = 'A-town';
```

Search all wells with a diameter greater than 1 m:

```
SELECT  *
FROM    public."PSQL_well"
WHERE   "W_diameter" > 1;
```

Search for all wells in a certain section:

```
SELECT  *
FROM    public."PSQL_well"
WHERE   (("P_x_coordinate" > 9) AND ("P_x_coordinate" <12));
```

Search all wells, whose water samples show a nitrate content >50 mg/l:

```
SELECT  "W_ID", "P_name", "V_ID", "V_value"
FROM    public."PSQL_well"
        Inner Join public. "PSQL_value" on "W_ID" = "V_WID".
WHERE   (("V_parameter" = 'nitrate') AND ("V_value" > 50));
```

Search all wells, whose water samples have a nitrate content >50 mg/l and which were sampled between 3.1.1996 and 28.2.1997:

```
SELECT  "W_ID", "P_name", "V_ID", "V_value", "V_date"
FROM    public. "PSQL_well"
        Inner Join public. "PSQL_value" on "W_ID" = "V_WID".
WHERE   ("V_date" > '3.1.1996') AND ("V_date" < '28.2.1997') AND
        ("V_parameter" = 'nitrate') AND ("V_value" > 50);
```

Search all wells, whose water samples show a nitrate content >50 mg/l and which were sampled between 3.1.1996 and 28.2.1997, with the corresponding laboratories:

```
SELECT  "W_ID", "W_name", "L_name", "P_name", "V_value", "from", "until"
FROM    (public."PSQL.well" Inner Join public. "PSQL_value" on "W_ID" = "V_WID")
        Inner Join (public. "PSQL_lab" inner Join "PSQL_support" on "L_ID" =
        "S_LID") on "W_ID"="S_WID"
WHERE   ("V_date" > '3.1.1996') AND ("V_date" < '28.2.1997') AND
        ("V_parameter" = 'nitrate') AND ("V_value" > 50);
```

The last example shows, how complex queries over several relations can be designed using the data manipulation language SQL. A query over several relations is necessary (linking with JOIN and subsequent selection). Somewhat surprising may be the result that queries e.g. between laboratories and analysis values are possible, although no direct relation was established between these entities. This shows the power of SQL. In practice, however, for reasons of simplification, the identification number of the associated laboratory is also given (cf. Sect. 8.3.3).

8.4.5 Relational Data Structures in a Geoinformation Systems

A relational database is usually integrated in geoinformation systems, in which the attribute data for the geoobjects are stored. Although the variety of functions lags behind independent database systems, there are often manufacturer-specific forms of data management and data queries. A standardised database language is usually not available. However, the general concepts of relational database systems are implemented. In particular, the thematic data of the geoobjects are stored by default in the form of tables. Before going into more detail about the specific functions of a geoinformation system (cf. Sect. 9.4), typical issues in the modeling of attribute data need to be explained.

The starting point is a management system of an industrial park, which is to be set up for the purpose of economic development. In Germany the system of the different area types is usually structured hierarchically:

- The area of the industrial park (cf. boundary in blue in Fig. 8.12 left) consists of several parcels. These parcels are also known as the smallest cadastral area units (in German: "Flurstück", a parcel within ALKIS, cf. Sect. 5.5.4.4 and Fig. 5.20, cf. the parcels separated by black lines). Tax features can be clearly assigned to each parcel and may vary from parcel to parcel. The parcels are marked in Fig. 8.12 on the left.
- These parcels are of less importance within a management systems, which is used as a marketing tool. The focus is on the commercial sites and the companies, which are located in the park. Two commercial sites are marked in Fig. 8.12 in light red and green.
- Some commercial sites do not consist of one parcel only. As shown in Fig. 8.12 the commercial site in yellow in the northwest corner of the park is subdivided into individual parcels.

- A single company may operate on different commercial sites, which do not necessarily have to be located together, but can be scattered across the park (cf. two sites marked in brown in Fig. 8.12). This means that for example production and administration sites can be physically separated from each other.
- Multiple businesses may operate on one commercial site (e.g. a forwarding company, a trading company, a storage company and an insurance office).
- The roads are marked in grey.
- In this example the master plan of the industrial park should have two sub-areas with different building regulations (not marked in Fig. 8.12). Thus, building regulations are assigned to each parcel. For example, the building height may be limited or a maximum building density must be met.

For reasons of simplification, company details such as ownership, sales and employment structure or partners for business development issues are not further differentiated. This example shows the implementation of a relational database in the context of a real task. Due to the different spatial relationships of the individual area categories, there are non-trivial requirements. The task cannot be solved by simply attaching a single attribute table to the geoobjects!

The geoinformation system usually provides two functions for modeling the relations. The "join"-operator is used to implement 1:1 and n:1 relations, while the "relate"- operator is used to implement 1: n and n:m relations.

In the associated geoinformation system each parcel is represented as a geoobject with the attributes "P_ID" and "P_size" in an attribute table, which will be referred to here as "RE_parcel_data". Furthermore, tax characteristics belong to each parcel of land, which are defined on legal basis and the municipal statutes, such as property tax or charges for street cleaning, waste water and which are stored in a separate table "RE_tax_features".

Fig. 8.12 A spatial management system of an industrial park

RE_parcel_data		
P_ID	P_size	...
...		
113	3.000	
184	4.000	
...		
221	5.000	
222	1.000	

RE_tax_features		
S_ID	P_ID	...
...		
0340	113	
...	...	
0333	184	
...	...	
03127	221	
03128	222	
...	...	

Fig. 8.13 Example of a 1:1 relationship

RE_parcel_data		
P_ID	P_size	P_reg
...		
113	3.000	I
127	5.000	I
84	4.000	I
155	2.500	M
221	5.000	M
222	1.000	S

RE_building_regulations			
B_ID	P_reg	B_type	B_building hight
...			
11	I	industrial area	15
12	R	recreation area	
20	M	mixed area	12
30	S	services	10
40	H	housing	7
50			
...	...		

Fig. 8.14 Example of a n:1 relationship

Therefore, a 1:1 relationship exists between the two tables. They can be combined into a single table, which is only temporarily available during processing, using the "join"-operator.

Then there is a n:1 relation between the geoobjects and the table of the building regulations, which explains the abbreviations (cf. Fig. 8.13).

Both tables can be combined with the "join"- operator to form a single table which is only present during processing (join via the "P_reg" attribute in both tables). The geoinformation system temporarily multiplies the entries (cf. Figs. 8.14 and 8.15).

There is a 1: n relationship between the company sites and the parcels (cf. Fig. 8.16). One company site can consist of several parcels, but each parcel belongs to exactly one site. Thus, a relation can be established with the "join"- operator.

The corresponding table of parcels is linked to the table of the company sites. This example shows, how a company site is composed of different parcels. This perspective is not the focus managing an industrial park. Rather by selecting a company site (i.e. by clicking on the area in the graphical editor of the geoinformation system, e.g. somewhere in the area marked in yellow, cf. Fig. 8.12), the user would like to have all the associated parcels displayed (and not vice versa). In this direction of the relationship, that is, between company site and parcels, there is a 1: n relationship. Both tables can no longer be merged into a single table, even if only temporarily. It is not possible to add three parcels that is

RE_parcel_data_building_regulations				
P_ID	P_size	P_reg	B_type	B_building hight
...				
113	3.000	I	industrial area	15
127	5.000	I	industrial area	15
184	4.000	I	industrial area	15
155	2.500	M	mixed area	12
221	5.000	M	mixed area t	12
222	1.000	S	services	10

Fig. 8.15 Example of an n:1 relationship resolved by the "join"- operator (cf. Fig. 8.14)

RE_company_site		
U_ID	U_size	...
...		
03122	3.000	
03412	6.000	
...		
03512	6.000	
03483	1.000	
03511	2.500	
...	...	

	RE_parcel_data		
U_ID	P_ID	P_size	...
	...		
03122	113	3.000	
03412	184	4.000	
03412	185	1.500	
03412	116	500	
03512	221	5.000	
03512	222	1.000	
03511	155	2.500	
	127	5000	

Fig. 8.16 Example of a 1: n relationship

three rows of the table "RE_parcel_data" to a single unit that is to one row of the table "RE_company_site" (e.g. parcels 184, 185 and 116 to company site 03412), and then two parcels that is two rows of the table "RE_parcel_data" respectively. In this case, the so-called "relate"- operator of the geoinformation system is used, which links both tables. When selecting a company site in the geoinformation system, i.e. by clicking on the geoobject, the associated parcels are displayed, i.e. sometimes one parcel and sometimes three parcels depending on the composition of the commercial site.

The modeling of the n:m relationship between the company sites and the companies is even more complex. Both tables cannot be directly related in the relation model, but only via a so-called matching table. The relate operator is also used here in a geoinformation system (cf. Fig. 8.17). A brief explanation shows the relation: The geoobject with "U_ID = 03112" is linked to two rows in the matching table, the geoobject "U_ID = 03512" is linked to one row in the matching table. Thus, there is a 1: n relation between these two tables. In the matching table the two rows with "U_ID = 03112" have attribute data "C_ID = 3668" and "C_ID = 3121", with "C_ID" representing a key field of table "RE_company_data". Thus, there is a m:1 relation between these two tables. The matching table also represents the case in which a company (here "C_ID = 2117" is located on several sites (here "U_ID = 03512" and "U_ID = 03555").

RE_company_site		
U_ID	U_size	…
…		
03199		
03101		
03112	3.000	
03188	4.000	
…		
03512	6.000	
03555	1.000	
…		

RE_company_data		
C_ID	C_name	…
…		
3668	PickUp transportation	
8179	TinX metal processing	
3121	PickSave insurance agency	
7112	WP wood processing	
…		
2117	Brickless building materials	
6382		
…		

1:n

matching-table	
U_ID	C_ID
…	…
03177	…
03112	3668
03112	3121
…	…
03188	7112
…	…
03512	2117
03555	2117
…	…
05331	3121

m:1

Fig. 8.17 Example of a n:m relationship

The relational data structures shown are also to be represented in geoinformation systems. At first glance, this procedure seems to be inconvient. However, the design of relational data structures serves the efficient management of the data of the geoobjects and avoids redundancies and data inconsistencies (cf. Sect. 8.5).

8.5 Data Consistency

8.5.1 Data Consistency: Concept and Meaning

In addition to storing, managing and processing data, a database management system has the important task of ensuring the consistency of the data. This very complex task includes (cf. Kemper and Eickler 2015 Chapter 5 und 9 and Saake et al. 2018 Chapter 13):

- access controls,
- maintenance of physical data integrity,
- maintenance of logical or semantic data integrity.

Access controls, which can be assigned individually for a data object or for different types of access, concern data protection against unauthorised viewing or manipulation. In contrast, integrity constraints generally denote conditions for the admissibility or correctness of databases states (so-called static integrity constraints) and of state transitions (so-called dynamic integrity constraints, e.g. when there are changes of the database). In the example of the well cadastre, there are either tube wells or shaft wells (static condition). The water samples of a well may be taken over by a new laboratory, but the well must not fall out of the supervision of the laboratories (dynamic condition when changing the relation "maintain"). Several integrity constraints are already implicit in the relationship model (static integrity):

- The definition of keys prevents two entities from having the same values in their key attributes.
- The definition of the cardinality of the relations prevents invalid relations. In the example of the well cadastre, a definition of a 1: n relationship between the relations (tables) "well" and "laboratory" would ensure that a well is not examined by two laboratories.
- By specifying a domain (a range of values) for an attribute, it is prevented that invalid attribute values are included. For example, five digits must be specified for a postal code in Germany.

The transaction concept, which is the basis for error tolerance and for parallel processing in database management systems (cf. Sect. 8.5.4), is connected with data integrity in the event of system errors and in multi-user operation.

Semantic integrity constraints can be derived from the properties of the modeled world. Such conditions are also ensured by specifying range restrictions for attributes, compliance with referential integrity (cf. Sect. 8.5.2), and by triggers (cf. Sect. 8.5.3).

8.5.2 Referential Integrity

Referential integrity ensures that data consistency exists between two linked, i.e. referenced, relations (tables) and is maintained when inserting, deleting or modification. Referential integrity characterises a property of the relationship between the primary key of a relation (table) R1 and the foreign key in another relation (table) R2. The foreign key of R2 has the same number of attributes as the primary key of the relation R1 to which the foreign key refers. The well register in Sect. 8.3.3 can serve as an example if, as suggested, another attribute "L_no" is attached that names the laboratory associated with the value. Referential integrity exists between the table "value" (now with the foreign key "L_no") and the table "laboratory" with the primary key "L_ID", if all attribute values of the foreign key of the table "value" are used as primary keys in the table "laboratories". Adhering to referential integrity ensures that when a new row is inserted or an existing row is modified in the "value" table, the foreign key must refer to an existing "L_ID" in the "laboratory" table. A primary key can only be changed, if no foreign key has referred to it. This also includes deleting a row in the table "laboratory".

8.5.3 Trigger

A *trigger* is a procedure (i.e. a program) that is defined and developed by the user and that is automatically started by the database management system, when a certain condition is met. Triggers are particularly useful when already saved data is to be changed later. Using the well cadastre as an example, it can be prevented that the determination of the pH value is (unintentionally) removed, when the scope of analysis of a well changes, if the pH determination of the water is mandatory for all wells. A trigger could also be used to ensure that entries in the "value" table are only possible with date values that are more recent than the previously stored values. This could prevent that a generally valid date like e.g. 06/25/2010 is entered, but it is (probably) wrong, because analysis values for 2012 are already stored. This problem cannot be solved by restricting the range of values, which would have to be redefined for each entry.

8.5.4 Transactions

A *transaction* is understood to be a group of several database operations, which form a unit with regard to integrity monitoring and must be executed contiguously without errors. A database system must be transferred from a consistent state into a consistent state by a transaction. Transactions consist of a sequence of elementary operations. If, in the example of the well register, the supervision by the laboratories changes, the supervision by a laboratory must not be deleted for a well (command 1) without defining a new supervision for this well (command 2). Similarly, no amount may be credited to a cost center (command 1) during a transfer without another cost center being debited by the same amount (command 2). Both commands form a transaction, which (as a unit) must fulfil four conditions (so-called *ACID principle*):

- Atomicity: A transaction is executed either completely or not at all. After a premature termination, there are no intermediate results from partially executed transactions.
- Consistency: A transaction leads from one consistent status of the database to another consistent status. The transaction is aborted and the database is left in its initial state, if integrity constraints are violated by a transaction.
- Isolation: Transactions executed in parallel are isolated from one another and do not affect one another. Each transaction has the effect that it would have caused as if it were alone in the system.
- Durability: The (new) state of the database system caused by a transaction is permanent and can only be reversed by a new transaction with the opposite effect ("chargeback").

Transaction management consists of two central components:

- *Recovery* from error situations: This so-called *recovery function* must ensure atomicity and durability. If, for example, a system crash occurs while the transaction has not yet been completed, the initial state of the transaction must be restored after the system is restarted, and effects of partial execution must be completely removed.
- Coordination of multiple simultaneous user processes or transactions: This so-called *multi-user synchronisation* must ensure the isolation of transactions running in parallel.

8.6 Extension

8.6.1 Dependent Entity Types

The simple entity-relationship model assumes that the entities are independent, are related but not dependent on each other, and are clearly identifiable in the entity set via key attributes. However, the modeling of reality and in particular many geoobjects show that dependent or weak entities can exist whose existence depends on another superior entity and which can only be clearly identified in combination with the key of the superior entity. The classic example refers to rooms in buildings. For example, the same room numbers usually exist in several buildings, and clear identification is only possible by combining the building number and the room number, such as in the University of Osnabrück, where the auditorium (E04) is identified by the ID 11/E04. If a building is demolished, the rooms will also disappear.

Dependent entities are marked in the ER-model by double bordered rectangles. A double-bordered diamond represents the relationship. The attributes are underlined with a dashed line.

The concept of dependent entity types refines and specifies the ER modeling. The implementation is done as usual by tables and by composite keys (or using new keys that represent a combination of both key fields).

8.6.2 The Is-a Relationship

With the previous constructs of the simple entity-relationship model, it is not yet possible to model a special relationship, which could be described as *specialisation* (or, in the opposite view, as *generalisation)* and which is referred to as an actual relationship or is-a-relationship (for further concepts cf. Saake et al. 2018 pp. 78). In the present example the entity type “well” is defined by several attributes:

```
well = {name, coordinates, address}
```

In addition, further information is required for individual wells. For example, municipal and private wells could be differentiated with additional attributes. Different maintenance teams are responsible for the municipal wells, for which different management keys (cost

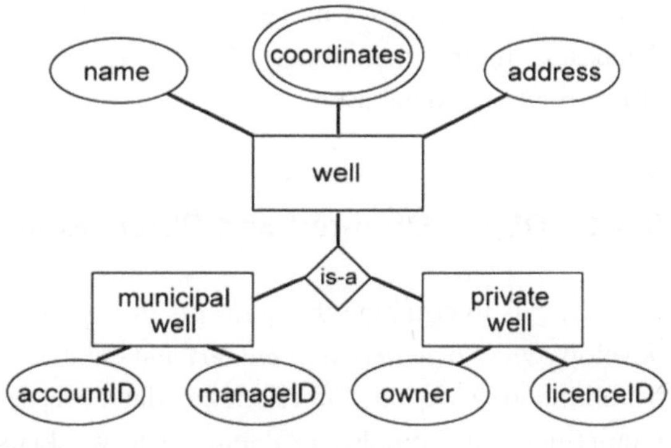

Fig. 8.18 ER-model with entities, attributes and an is-a relationship

centers) exist. The private wells are assigned to an owner (with address) and have a license number from the water company. Therefore, it makes sense to create separate entity types:

`municipal wells` = {`accountID, managementID`}
`private well` = {`owner, licenceID`}

Both entity types are specialisations of the entity type "well", the supertype is a generalisation of the subtypes. The attributes of "well" are also valid for the municipal and private wells, they are inherited by the specialisations (inherited attributes).

In the graphical representation, the is-a relationships, like the (other) relationships, are illustrated by a rhombus, in which the designation "is-a" is entered (cf. Fig. 8.18). The is-a relationship is total (opposite: partial), if there are no other entity types apart from the decomposition into the specified entity types. In the present case, the is-a relationship should be total, which means that there are no other wells owned by, for example, a utility company, which have to be modeled separately with their own attributes. Furthermore, this relationship should also be disjoint (opposite: not disjoint), since either municipal or private wells exist, and no well should to be both municipal and privately owned.

The is-a relationship is not converted into a relational database using a separate table as in the case of the relationship "maintain" in Fig. 8.10. Instead, the primary key of the more general entity type is also included in the table of the more specific entity type. In the present example, the entity type "well" has the primary key "W_ID" (cf. Table 8.1). The relation "municipal well" receives its own key and the inherited key "W_ID". The key "W_ID" in the relation "municipal well" is a foreign key with regard to the relation "well".

8.6.3 EER Model

The classic entity relationship model can be extended to the *extended ER model* (EER model). The more general concept of the so-called type constructor replaces the "is-a" relationship. Extended key concepts as well as object-valued attributes model dependent entity types. User-defined data types also extend the standard data types. Thus, these conceptual extensions can be seen as a bridge with respect to object orientation in databases. In the meantime, EER modeling has been overtaken by the concepts of object-relational databases.

8.6.4 Object-Oriented and Object-Relational Database Systems

It is obvious to combine the concepts of object orientation with database concepts and to develop *object-oriented data models* and database systems. These efforts are often justified by eliminating deficits of relational database systems and in order to better meet the requirements of complex problems. The weaknesses of relational systems are primarily (cf. with regard to geodata Sect. 8.7.1):

- limited number of available data types and limited options to model complex objects,
- cumbersome division of an object over several relations (segmentation) and merging, which requires a lot of computation time,
- artificial key attributes,
- no modeling of object- or type-specific operations (behavior) on the structures.

However, the relational model can certainly handle the tasks for which object-oriented data models offer solutions. Nor do object-oriented database systems offer only advantages. In contrast to relational database systems, object-oriented systems from different vendors have major differences. So no uniform object model or query language for object-oriented database systems has prevailed in comparison to relational database systems with SQL. In 1993, the major commercial providers of object-oriented database systems came together to form the *Object Database Management Group* (*ODMG*). After the release of version 3 of the Object Database Standard (ODMG 3.0) in 2000, the work was discontinued. The Object Management Group is currently working on a fourth-generation object database standard to incorporate recent changes in object database technology (cf. ODMG 2023).

However, standardisation efforts have not yet led to a reduction in the dominance of relational database management systems. This is due to the still rather low performance and weak distribution of mature object-oriented database management systems on the commercial market. Even more important is that a considerable effort is required for a change, which usually cannot be justified from a business point of view. Instead the classical relational model has been functionally extended to the so-called object-relational model by integrating certain concepts of object orientation into the relational model. These enhancements include multi-valued attributes, the creation of application-specific attributes, object identity (artificially generated object identifiers instead of keys created from attributes), inheritance, and class hierarchy. In particular, complex data types as well as user-defined classes can be generated. The advantages of the relational and the object-oriented data models are brought together. Central concepts such as the use of relations and the database query language SQL are retained (cf. Kemper Eickler 2015 pp. 401 and pp. 439).

The SQL standard SQL:2016 documents these developments. The standard consists of a total of nine publications and is supplemented by standardised SQL multimedia and application packages. The third part is particularly important for geoinformatics, often abbreviated to SQL-MM 3, which defines user-defined spatial data types and the associated routines (cf. ISO/IEC 9075-1:2016 and ISO/IEC 13249-3:2016).

8.7 Spatial Database

8.7.1 Managing and Processing Geodata in Relational Databases

In geoinformatics, in addition to non-geometric features that describe the topic of a geoobject, geometric data in particular must be stored, managed, analysed and presented. However, the presentation of the initially very simple sounding task of representing the borders (and areas) of the Federal Republic of Germany fails within a relational database system that requires, among other things, atomic attributes. No attribute may contain a set or list of data. The geometries of geodata, for example the borders of the federal state of Brandenburg, consist of many coordinates (attributes), whereas the borders of the federal state of Hesse have a different number of coordinates (attributes). The federal state of Brandenburg is a polygon with a "hole" for Berlin (so-called donut). The federal state of Schleswig-Holstein is a multipart polygon that describes mainland and the islands. Such multipart polygons can only be stored in a relational database system in a very cumbersome way. Thus, splitting the multipart polygons into several tables is very inefficient (for a solution approach cf. Brinkhoff 2013 p. 28).

For a long time, the standard was to store geometries and attribute data separately. This is the case with what is still the industry standard today, the proprietary *shapefile data format* from ESRI, which consists of a collection of files (cf. Sect. 9.3.3). The geometries are stored in a (manufacturer) specific format. A commercial relational database is usually used to represent the attribute data (cf. for example the dBase format in the shapefile data format, cf. Sect. 9.3.3). The coupling of both data sets is done using common keys. But using the shapefile data format has several important disadvantages: The proprietary data format can change as the manufactures makes modifications. Although there is an industry standard and the shapefile data format is widely used, the geometries cannot be used without a geographic information system or only after extensive creation of user-specific programs. Meanwhile, with the OGC GeoPackage, an open, non-proprietary, platform-independent data format exists for storing geometry data and thematic data (cf. Sect. 6.3.4).

Relational database systems now offer the option to store complex data such as raster images, audio files or videos or other arbitrary binary files as so-called *Binary Large Objects* (BLOBs). Thus, geometry data of an individual object with any structure can also be stored in binary form in the database (cf. the proprietary personal geodatabase of the software company ESRI, which stored geometry and attribute data in a Microsoft Access database file, cf. Sect. 9.3.3). Although geometry and attribute data are stored in one database (file), there are still considerable disadvantages. For example, the stored binary data cannot be interpreted and efficiently analysed by methods of the database system alone (independent of the geoinformation system) because their meaning is unknown (e.g. points or multipart polygons or audio information). Thus, no geometric functions can be triggered by the relational database management system. Elements of a query language for geometries are missing. The database management system cannot (alone) check whether consistent geometry data are available. It cannot be assessed whether relationships exist between different geodata, i.e. BLOBs. This requires functions of the geoinformation system (cf. tools for checking topologies of geometric data structures, cf. the proprietary ArcGIS of the software company ESRI in Sect. 9.3.3). Finally, such BLOBs are usually not directly usable by external users. Special programs or, in turn, geoinformation systems are required to read the data.

8.7.2 Spatial Database: Operating Principle

Spatial database systems are database systems which, in addition to the storage of non-spatial data, also enable the storage of spatial data and especially their analysis. Initially, spatial databases correspond to the general definition of database systems, which consist of a database, in which the data are stored, and a database management system, with which the data can be managed. A spatial database system must therefore meet the usual requirements of a database system, which include data integrity or data consistency and efficient query processing as well as compliance with standards. However, a spatial database also enables the storage and management of geodata, both raster and vector data, as well as their processing with regard to spatial queries and analyses.

According to Brinkhoff (2013 pp. 26), there are specific requirements for a spatial database system that:

- must offer geometric data types,
- must have methods for performing geometric functions,
- must have a suitable query language for using the methods,
- must provide suitable algorithms and data structures for the efficient execution of the methods,
- must enable the data to be used by external applications in the sense of an open geoinformation system
- must comply with common standards in order to achieve maximum interoperability.

These requirements cannot be met with classic relational database systems. Whereas object-relational database systems provide components for the management of geodata in particular. Of the commercial providers, the Oracle database should be mentioned in particular, which now includes "Spatial and Graph features" (cf. Oracle 2023, Oracle Spatial and Oracle Spatial and Graph were previously separately licensed components). IBM® Informix® is a database for integrating SQL, NoSQL, JSON, time-series and spatial data (cf. IBM 2023a). IBM DB 2 Spatial Extender is also worth mentioning (cf. IBM 2023b). The Microsoft database system SQL-Server also supports the storage and analysis of geometric data types (cf. Microsoft 2023). SAP's HANA database system also contains a spatial database engine that enables spatial data types and SQL extensions for operations with spatial data (cf. ESRI 2020). In addition, there are several open source database systems that can store and manage geoobjects, but which have quite different or limited functionalities to analyse spatial data (cf. e.g. MySQL and SpatialLite as an extension of SQLite). In contrast, there is the widely used open source database software PostgreSQL with the associated extension PostGIS for geodata (cf. PostGIS 2023a), which provide, among other things, geometric data types as well as extensive functions for the analysis and processing of geoobjects.

8.7.3 PostgreSQL/PostGIS

The PostGIS extension of the open source database system PostgreSQL provides powerful options for effectively storing, organisation and querying large amounts of geodata (cf. Refractions Research 2023). With "point", "linestring" and "polygon" three basic geometry types are available, from which four further geometry types are derived (cf. Table 6.2 and PostGIS 2023b Chapter 4). While the standard OGC data type "geometry" only supports 2D geometries, PostGIS now also offers extended formats as a superset of the OGC formats such as "3D and 4D point objects", "multicurve", "polyhedralsurface", "triangle" or "TIN". The data type "geometry" is based on specifications in a plane, where the shortest distance between two points is a straight line. Accordingly, geometric calculations (such as determining the size of an area, the length of a line, or the intersections of polylines) are performed in a Cartesian coordinate system. In addition, PostGIS provides the data type "geography", which uses a geodetic coordinate system, which supports geoobjects that are defined by geographic coordinates (longitude and latitude). The shortest connection between two points on a spherical surface is an arc of a great circle. Calculations on geographic data types (such as determining the size of an area, the length of a line, or the intersection of polylines) are performed on the spherical surface, whereby versions PostGIS 2.2 and later supporting arbitrarily defined ellipsoids (cf. PostGIS 2023b). The newest member of the PostGIS spatial type family is "raster" for storing and analysing raster data (cf. PostGIS 2023b Chapter 9).

In geoinformatics, PostGIS as an extension of PostgreSQL is of great importance and therefore widespread. The PostgreSQL database can initially only be used as a data container for storing geodata. Many open source geoinformation systems can access the database. ArcGIS, the proprietary geoinformation system of the software company ESRI, also offers a connection to PostgreSQL. This means that data can be stored in a geoinformation system independently of specifications and proprietary data formats from software manufacturers. In addition, PostgreSQL/PostGIS itself can be used to perform spatial analyses based on SQL queries. Thus, GIS operations can be carried out directly from the database (cf. Sect. 8.7.4). With the extension "pgRouting", route calculations on networks are also possible. Such functions are usually reserved for geoinformation systems.

PostgreSQL is a spatial database that can also be integrated into a geoinformation system such as in QGIS and ArcGIS, whereby this can be done quite easily in QGIS. If changes are made to the geodata in QGIS, such as adding or changing an attribute, and the changes are saved, then the data in the PostgreSQL/PostGIS database are also adopted. The mapserver can also work directly with PostGIS. Other open source geoinformation systems (e.g. GRASS, Open-JUMP, QGIS or uDIG) also usually offer convenient interfaces to integrate a PostgreSQL/PostGIS database.

In particular, greater independence from GIS software can be achieved. However, complete independence does not make sense. The geometric acquisition of geoobjects as well as the visualisation of geodata and the creation of views on the monitor or of maps via a graphic output device are easier and often more intuitive with the help of GIS software. Thus, a combination of a PostgreSQL/PostGIS database with a geoinformation system makes a lot of sense, especially for larger projects.

8.7.4 PostgreSQL/PostGIS: A Sample Application

The great performance and applicability of PostgreSQL/PostGIS in geoinformatics is outlined using a few examples. Spatial databases are no geoinformation systems since they lack (efficient) tools for recording geodata (for the EVAP model cf. Sects. 9.1.2 and 9. 1.4). The example, explained here, is based on a standard application from Chap. 9, which is presented there in connection with the tools of a geoinformation system. Here a procedure is presented that works independently of a geoinformation system using only methods of a spatial database. The command sequences refer to pgAdmin4 and postgreSQL 11.2-2, based on SQL/MM 3: 6.1.8.

As with any relational database the focus is on tables which can now contain thematic data and geometric data. The columns for thematic data are defined and created with SQL using the usual CREATE statement. However, there is only one column per table for the geometry data:

```
CREATE TABLE public.streetlamp (id_lamp integer, lname character varying
(10), geom geometry)
```

The insertion of the geometry data can be done with the following commands:

```
INSERT INTO  streetlamp
VALUES       (177,'arclamp', ST_geomFromText ('POINT(123.45 546.78)', −1))
```

The command is almost self-explanatory. Since no coordinates are specified as EPSG code, −1 is set instead.

As a rule compared to this inconvenient procedure, geometries and attribute data have been recorded using the functions of a geoinformation system, which provides suitable options for defining coordinate systems and for recording and editing geometries as well as for recording attribute data (cf. Sect. 9.4.1). Special programs (e.g. the free program shp2pgsql-gui.exe with a graphical user interface) allow easy transfer of data from the proprietary shapefile data format (cf. Sect. 9.3.3) into a PostgreSQL/PostGIS database. In this way each data layer is assigned its own table (for the layer principle cf. Sects. 4.1.1 and 9.3.6). Before saving, the so called Spatial Reference System Identifier (SRID) must be set to specify the coordinate system. The corresponding table of identifiers is created automatically, if shapefiles are imported.

The following example refers to an application that is presented in Sect. 9.4.5 in a common way with tools of a geoinformation system. Two route alternatives for a bypass are to be compared in a preliminary study, a western and an eastern bypass. It should be examined whether more areas or biotopes worth protecting are affected by the western bypass than by the eastern bypass (cf. for more details Sect. 9.4.5 and Fig. 9.18). Here, only two data layers of Sect. 9.4.5 are transferred to two PostgreSQL/PostGIS tables. The layer or table "biotopeutm32" identifies agriculturally used areas as well as forests. The layer or table "pass-westutm32" represents the planned bypass road west of a rural settlement. Both datasets are in UTM32 and fit together. Several command sequences show main analysis options:

Selection of all areas of the table "biotopeutm32" that are larger than 200,000 m^2. The result is shown in Fig. 8.19:

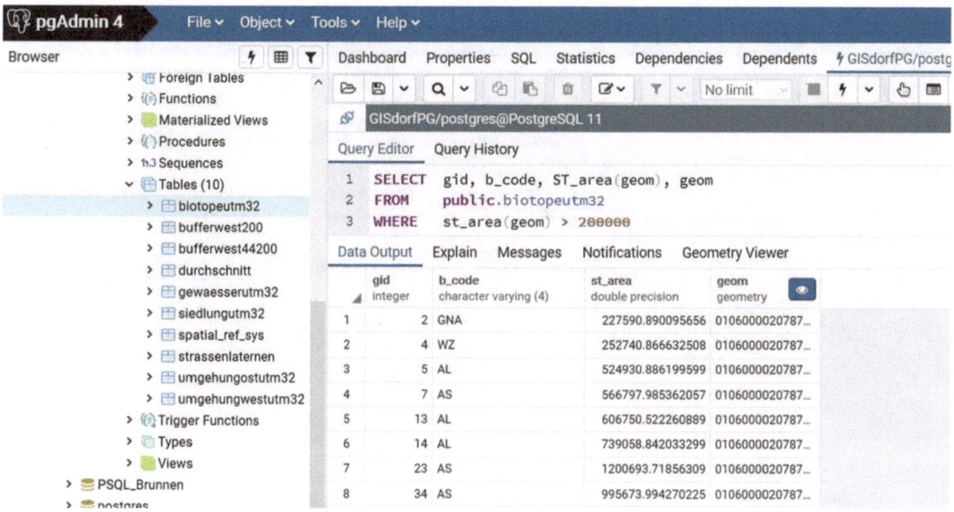

Fig. 8.19 Evaluation of a PostgreSQL database: Execution of a SELECT command on geoobjects

```
SELECT   gid, b_code, ST_area(geom), geom
FROM     public.biotopeutm32
WHERE    st_area(geom) > 200000
```

Create a buffer zone with a distance of 200 m around the planned bypass, which can be understood as an impact zone, and create a new table "bufferwest200" with the result:

```
CREATE TABLE  bufferwest200 AS SELECT ST_buffer (geom, 200)
FROM          "bypass-westutm32"
```

Create the (spatial) *intersection* of the biotopes and the just generated impact or stress zone (cf. Fig. 8.20):

```
SELECT   ST_intersection (biotopeutm32.geom, bufferwest200.st_buffer)
FROM     bufferwest200, biotopeutm32
```

As usual a database management system the query results are prepared as a table, and they are not initially visualised graphically by a map. However, with the help of the "Geometry Viewer" a very rudimentary spatial visualisation is possible (button on the far right in the line above the table in Fig. 8.19).

Finally, it should be pointed out that a PostgreSQL/PostGIS database is becoming increasingly important within geographic information systems. Instead of file-based data storage, such as with the proprietary shapefile data format, a connection to a relational

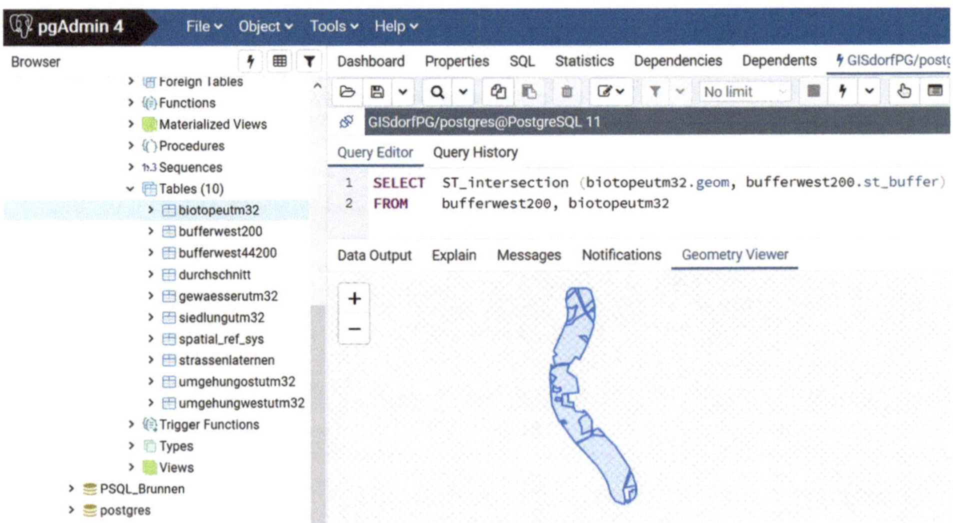

Fig. 8.20 Evaluation of a PostgreSQL/PostGIS database: Execution of a SELECT command on geoobjects (spatial average of biotope areas and impact zone)

database management system should be aimed for. The free geoinformation system QGIS, for example, offers an interface to PostgreSQL/PostGIS in order to store geometry data in addition to thematic data.

References

Brinkhoff, Th. (2013): Geodatenbanksysteme in Theorie und Praxis. Einführung in objektrelationale Geodatenbanken unter besonderer Berücksichtigung von Oracle Spatial. Heidelberg: Wichmann. 3. Ed.

Codd, E. F. (1970): A relational model for large shared data banks. Communications of the ACM, 13-6, S. 377–387.

Codd, E.F. (1990): The Relational model for database Management, Version 2. Reading, Mass: Addison Wesley.

ESRI (2020): Does the ArcGIS platform support the SAP HANA database? https://support.esri.com/en/technical-article/000012246 (14.04.2023).

Gumm, H.-P. u. M. Sommer (2013): Einführung in die Informatik. München: Oldenbourg, 10. Ed.

IBM (2023a): IBM Informix. https://www.ibm.com/products/informix (14.04.2023).

IBM (2023b): Db2 Spatial Extender. https://www.ibm.com/docs/en/db2woc?topic=data-db2-spatial-extender (23.08.2022).

ISO/IEC 9075-1:2016 (2016): Information technology -- Database languages -- SQL -- Part 1: Framework (SQL/Framework) https://www.iso.org/standard/63555.html (14.04.2023).

ISO/IEC 13249-3:2016 (2016): Information technology -- Database languages -- SQL multimedia and application packages -- Part 3: Spatial. https://www.iso.org/standard/60343.html?browse=tc (14.04.2023) (link to SQL_Spatial package).

Kemper, A. u. A. Eickler (2015): Datenbanksysteme. Eine Einführung. München: Oldenbourg Verlag, 10. Ed.

Laube, M. (2019): Einstieg in SQL: Für alle wichtigen Datenbanksysteme: MySQL, PostgreSQL, MariaDB, MS SQL. Bonn: Rheinwerk 2. Ed.

Microsoft (2023): Spatial Data Types Overview. https://docs.microsoft.com/en-us/sql/relational-databases/spatial/spatial-data-types-overview?view=sql-server-ver15 (14.04.2023).

Object Data Management Group (ODMG) (2023): Standard for Storing Objects. http://www.odbms.org/odmg-standard (14.04.2023).

Oracle (2023): Oracle Spatial and graph features in oracle Database. https://www.oracle.com/database/technologies/spatialandgraph.html (14.04.2023).

PostGIS (2023a): About PostGIS. http://www.postgis.net (14.04.2023).

PostGIS (2023b): PostGIS 2.4.10 Manual. https://postgis.net/docs/manual-2.4 (14.04.2023).

pgAdmin (2023): PostgreSQL Tools. https://www.pgadmin.org/ (14.04.2023).

PostgreSQL (2023a): PostgreSQL: The World's Most Advanced Open Source Relational Database. http://www.postgresql.org (14.04.2023).

PostgreSQL (2023b): About. What is PostgreSQL? https://www.postgresql.org/about/ (23.08.2022).

Refractions Research (2022) PostGIS. What is PostGIS? http://www.refractions.net/products/postgis/ (14.04.2023).

Saake, G., Sattler, K.-U. u. A. Heuer (2018): Datenbanken: Konzepte und Sprachen. Bonn: mitp-Verlag, 6. Ed.

Vossen, G. (2008): Datenmodelle, Datenbanksprachen und Datenbank-Management-Systeme. München: Oldenbourg, 5. Ed.

References

Geoinformation Systems 9

9.1 Concepts of Digital Information Systems and Geoinformation Systems

9.1.1 Information Systems

An information system can generally be described as a system that accesses a data pool and enables evaluation of these data in order to derive and reproduce information from them. In this first definition, the entirety of data and processing of data is already expressed, but data storage and, above all, data recording are not yet discussed or included in more detail. Thus, pure *information systems,* which only allow (possibly complex) processing of already existing data, are not counted among the information systems here, which must also permit data recording, i.e. new recording and update. Thus, an *information system* includes recording, storage, updating, processing and evaluation of information as well as its reproduction.

This very comprehensive conceptualisation also includes analog information systems. According to the type of (somehow) stored information, which then also requires special processing methods, information systems can be (alpha-)numerical, textual, visual or multimedia in nature. Various examples should be mentioned: information systems in banks (e.g. administration of the customer base and account management), in travel agencies (e.g. Instead, the focus here is on accessibility along roads and paths in a specific transport network, hotel occupancy, reservations) or in libraries (e.g. management of user data, book reservations, search options in the library stock).

Geoinformatics focuses on geoinformation systems which, in contrast to other information systems, model geoobjects of the real world and map them into a digital information system (cf. Sect. 9.3). As with all other information systems, the objects of a geoinformation system have attribute data, which describe their characteristics (theme or

N. de Lange, *Geoinformatics in Theory and Practice*, Springer Textbooks in Earth Sciences, Geography and Environment,
https://doi.org/10.1007/978-3-662-65758-4_9

topic of a geoobject). The special feature of geoinformation systems is that geoobjects also have geometry and topology as implicit components! The processing of such spatial information requires special tools or functions which are not provided by the other information systems (cf. Sects. 9.4 and 9.5).

9.1.2 Four-Component Models of an Information System

The general concept stated in Sect. 9.1.1 already contains the two fundamental perspectives according to which an information system is to be viewed structurally and functionally. According to structural aspects, the type and (physical) nature of the system and the storage media, the processing options, the information or data available and stored in any form, as well as the applications, areas of use and users must be differentiated. If these views are narrowed down to digital information systems, four structural components emerge:

hardware	computer system including processor, storage media, peripheral devices and connections
software	program systems including software tools for recording, managing, analysing and presenting of information
data	quantitative and qualitative information, which together represent a (subject-related) section of the real world
applications	applications and possible uses especially the users with their requirements and questions

According to functional aspects, four functions can be distinguished:

input	data or information recording and storage
management	data administration (i.e. management)
analysis	data evaluation and data analysis
presentation	reproduction of the information (i.e. output or presentation)

In this context, the four functional groups have different degrees of complexity and, above all, cannot be sharply separated from one another. Thus, data management can also mean, among other things, updating, when data are added or newly recorded. Sorting or selecting data records can be seen as a management function, but also as an evaluation function, which actually begins with the evaluation or analysis of the information.

The four structural or the four functional components define the HSDA or the IMAP model. In this context, it is mainly the software that determines the range of functions, i.e. all IMAP components. In this strict definition of the term geoinformation system, the analysis of data is an indispensable, constituent feature of an information system. Many data portals and web-based systems that offer data or information and that are often referred to as information systems are only data representation systems in this strict sense, since in most cases there is no collection and storage of user-specific data or analysis of the data.

9.1.3 Geoinformation System: Notion

Geoinformation systems belong to the category of spatial digital information systems. The central objects of these information systems are information about *geoobjects* (for the term geoobject cf. Sect. 4.1):

A geoinformation system is a computer-based system consisting of hardware, software, data and applications. It can be used to digitally record, store, manage, update, analyse and model spatial data and display it alphanumerically and graphically.

This definition is based on the definition established not only in German-speaking countries, which goes back to Bill and Fritsch 1991 (cf. Bill and Fritsch 1991 p. 4, and Bill 2016 p. 7). In addition, there are many different definitions, often with only minor differences (cf. e.g. Ehlers and Schiewe 2012 p. 82). In the German-speaking world, the terms *Geographic Information System*, *Geoinformation System* and *GIS* are almost always used synonymously. In English-speaking countries, the terms *Geographic* or *Geographical Information System* and *GIS* are common.

A geoinformation system is (also) defined by an HSDA or an IMAP model. As in any digital information system, the four structural components are inherently present, but here they are specific to geoinformatics applications.

The history of geoinformation systems began in the 1960s (for an overview of the historical development cf. Goodchild and Kemp 1990 Unit 23). The Canada Geographic Information System (CGIS) of the mid-1960s, developed at the Department of Forestry and Rural Development by Tomlin, the "father of GIS", was one of the first geoinformation systems. CGIS was used, among other things, to process and analyse the extensive data from the Canada Land Inventory (CLI) of rural Canada. Innovative ideas were developed in the late 1960s and 1970s at the Harvard Laboratory for Computer Graphics and Spatial Analysis. Among the pioneers are certainly the geoinformation system Arc/Info of the software company Environmental Systems Research Institute (ESRI) and the primarily raster-based GRASS GIS (Geographic Resources Analysis Support System), originally developed by the US Army Corps of Engineers (cf. Sect. 9.1.5). A milestone was the Core Curriculum Geographic Information Systems published in 1990 (cf. NCGIA 1990). Further development was favored and accelerated primarily by a variety of hardware and software improvements in the 1990s. In the meantime, geoinformation systems have become standard tools for users such as municipalities (especially cadastral and planning authorities), planning, utility, transportation and telecommunications companies. Standard textbooks are now available in several editions (cf. Longley et al. 2005 and 2015).

9.1.4 Four Component Models of a Geoinformation System

From a structural point of view, a *geoinformation system*, like any other information system, consists of four components: hardware, software, data and applications (*HSDA model*).

In terms of *hardware equipment*, there are no longer any special requirements for computer systems, which have now generally achieved a high level of performance, so that a geoinformation system can also be used on mobile devices. However, the graphical periphery is of central importance. This includes the graphic input devices such as scanners and the graphic output devices such as, above all, large-format plotters. In a longer-term and comprehensive balance that takes all components into account, the hardware is the less expensive component of an information system.

The *software* must ensure as a general task to map the *geoobjects* of the real world into a digital information system. In particular, the software must cover the four functional areas of data input, management, analysis and presentation of geoobjects. Overall, geoinformation systems are very complex software products that now cost significantly more than the hardware. However, in addition to proprietary software, there are now also powerful open source and free software available that competes with the established market leaders.

The digitally recorded and maintained *data* (geometric, topological and thematic data) make up the actually valuable component of an information system! The development of an information system leads, among other things, to systematising and completion of the previously scattered or even incomplete data and making them available to a larger number of users (for the first time). The data can survive several generations of software as well as of employees. This results in the compelling, but unfortunately often not implemented, need to clearly document the data and to describe their quality and possible uses (cf. Sects. 6.5 and 6.6). It can be stated very harshly that without a meta-information system the stored data are worthless. Particular importance is attached to the exchange of data and the multiple use of data (cf. Sect. 6.1).

The software as well as the data are only valued by the users for solving specific problems. *Applications* and *users* are inseparably linked. The users need and process the data with regard to specific applications and use the existing, derived or newly recorded information to solve their tasks. However, geoinformation systems are difficult to handle due to their complexity. Users are expected to have extensive knowledge in various areas of geoinformatics. This implies (continuous) training and further education. Above all, a successful use of new technologies requires the acceptance of the employees.

From a functional point of view, a geoinformation system, like any other information system, consists of four components: input, management, analysis and presentation (*IMAP model*). This applies to geoobjects, which are defined by spatial and attribute data. While spatial data describes the absolute and relative location of geoobjects (geometry and topology), attribute data describe the characteristics of geoobjects (topic). The *input* of real world geoobjects into a digital information system affects the geometry, topology and topic of geoobjects both on a conceptual level and on a practical input level. In this context, the problem and the application determine the type of modeling, e.g. as a vector or raster model, as a network model or as a 3D-model based on triangular meshes. A geoinformation system provides a wide range of functions for recording geodata (cf. Sects. 9.4.1 and 9.5.1).

The *management* concerns geometric and topological data on the one hand and thematic data on the other. Above all, geometrical data and subsequently topological data (depending on the data model) must be editable, i.e. it must be possible to modify and update them. This includes the conversion of geometric data into a new coordinate system (e.g. from geographic coordinates in WGS84 to UTM) or the merging of two adjacent parcels and the dissolving of a common border. Sections 9.4 and 9.5 explain essential functions. Geometric and attribute data can be stored separately in several files, in a single database and also in spatial databases (cf. Sects. 8.7 and 9.3.3). The underlying data model defines how topologies are managed.

A database management system is required to manage the attribute data and is usually integrated into the GIS software. The various functions include simple evaluation functions such as search operations in the database, reclassifications, sorting, calculation of new attributes from existing attributes or the creation of result tables and so-called frequency distributions. Thus, a geoinformation system usually offers many functions of a complete database management system for the administration of attribute data.

It should then be pointed out that an external database such as the PostgreSQL/PostGIS database or the GeoPackage, which is linked to a geoinformation system, is becoming increasingly important for data handling. Instead of file-based data storage, such as the proprietary *shapefile data format* of the software company ESRI, a connection to a relational database management system should be aimed at (cf. Sect. 8.7). Although the shapefile data format is the most widely used format in existing software packages, there are significant disadvantages: multi-file format, attribute names only up to 10 characters, only 255 attributes, limited data types, maximum 2GB file size. The free geoinformation system QGIS, for example, offers interfaces to several database systems such as PostgreSQL (cf. Sect. 8.7.3), SpatialLite, MS SQL Server or Oracle.

Spatial analyses are the central component of a geoinformation system. The most important functions include the so-called overlay functions. Sections 9.4.4 and 9.5.3 explain the central functions.

A geoinformation system also includes a variety of functions for the *presentation* of geoobjects, whereby the data is first displayed on the monitor: displaying, moving, enlarging and reducing map sections, switching on and off of different thematic layers or bringing them to the foreground, (visual) overlaying of different thematic layers, combined display of vector and raster maps and especially aerial images. In this context, the representation and generally the procedure with a GIS are map-oriented. If required, the corresponding thematic data are displayed for a geoobject. Tables or diagrams are also presented on the screen, as well as images, sound and video sequences. In addition to the presentation in the form of two-dimensional representations, a geoinformation system generally also owns functions that enable perspective, pseudo-three-dimensional views such as block images (including changes in the direction of illumination or sunlight) and allow rotations of the overall image. This presentation can also be shown on an analog two-dimensional data medium, i.e. mostly the production of a (paper) map (with an automatically generated legend and scale bar) or a poster, which may contain maps, diagrams, pictures, tables and texts. For the presentation on the screen as well as for the production of an analog map, the graphic design principles derived from traditional thematic cartography apply, as shown in Sect. 7.5.

It should be emphasised that one can only speak of a geoinformation system if all functions are available according to the IMAP model. Many software products must be judged according to this fundamental statement. Many systems that regard themselves geoinformation systems but do not, for example, record geoobjects or have no spatial analysis functions are not geoinformation systems. Even spatial databases, which have many management and analysis functions for spatial data, do not belong to the group of geoinformation systems, strictly speaking. Functions for recording geometries are only rudimentary, they do not meet the necessary requirements. Graphic presentation functions are hardly available.

9.1.5 GIS Software

Different GIS-software-systems are offered to date. On the one hand, many free and open source products are available. On the other hand, proprietary geoinformation systems are offered. These are often internationally available products. But mostly smaller companies also offer their own geoinformation systems, but then only for a smaller sales area. The extent of the latter group cannot be estimated. Frequently, special solutions are offered that are tailored to the specific needs of the users. These products often rely on open source software.

Of the geoinformation systems offered worldwide by large international software companies, dominated by US firms, the Big Six should be mentioned. The GIS products are often integrated into a larger software portfolio of the providers. The names are product names of these companies – in alphabetical order:

- As early as 1981, the software company Environmental Systems Research Institute (ESRI) internationally launched Arc/Info, a software that could run on personal computers (cf. ESRI 1995). The ArcGIS product family is ESRI's current flagship product, which is widely used worldwide (cf. ESRI 2023a). With ArcGIS Online, there is also a cloud-based offer (cf. ESRI 2023b). Important data formats have also been developed by ESRI (cf. Sect. 9.3.3).
- With the AutoCAD Map 3D toolset, which is included in the scope of services of the CAD software AutoCAD 2020, data from geoinformation systems and CAD systems (CAD, Computer Aided Design) can be merged (cf. Autodesk 2023). Autodesk also

developed the DXF format, an important data format that is still to be understood as the industry standard for the exchange of geometries.

- The Geomedia geoinformation system, originally introduced by Intergraph in the late 1990s and now offered by Hexagon Geospatial, is a powerful, flexible GIS management platform that can be used to aggregate and analyse data from a variety of sources (cf. Hexagon 2023a). GeoMedia WebMap is a server solution for web-based visualisation and analysis of geodata (cf. Hexagon 2023b).
- With MapInfo Pro, a desktop GIS and a GIS mapping software are available, which, together with other products such as MapInfo Vertical Mapper, offer a comprehensive range for the analysis of geodata (cf. MapInfoPro 2023).
- OpenCities Map and OpenCities PowerView are software products from Bentley that are engineer-oriented and enable the editing, analysis and management as well as the display and manipulation of 2D/3D geospatial data (e.g for 3D-building design, cf. Bentley 2023).
- With Smallworld, General Electric (GE Energy) distributes an internationally wide-spread geoinformation system, which is mainly used by network operators in the energy and water industries (cf. Grintec 2023).

Among the free or open source geoinformation systems are of global importance (in alphabetical order):

- GRASS GIS (Geographic Resources Analysis Support System), originally developed by the US Army Construction Engineering Research Laboratories, a branch of the US Army Corps of Engineers, is used by many US government agencies and many academic and commercial organisations worldwide. The system includes over 350 modules for analysing raster and vector data, including analysis of networks and multispectral image data (cf. GRASS GIS 2023).
- gvSIG (for Generalitat Valencia Sistema de Información Geográfica) characterises an extensive open source software family (desktop, online and mobile versions). The software is platform-independent and is compliant with current standards (cf. gvSIG 2023).
- OpenJUMP is an open-source geoinformation system written in the Java programming language that is primarily focused on vector data (cf. OpenJUMP 2023).
- QGIS (formerly Quantum GIS) is a very powerful software product for which extensive documentation, training materials and tutorials are available. QGIS is an official member of the Open Source Geospatial Foundation (OSGeo) and runs on LINUX, UNIX, Mac OSX, Windows and Android. A wide variety of vector, raster, and database formats and functions are supported (cf. QGIS 2023a). A very large range of plug-ins, which is made available by a worldwide user community, extends the basic functionalities (cf. QGIS 2023b).
- Spring GIS is an advanced geoinformation system and image processing system with an object-oriented data model that enables the integration of raster and vector data

representations in a single environment. Spring is a product of the National Institute for Space Research in Brazil (cf. Spring GIS 2023).

Although not a globally used geoinformation system, SAGA (System for Automated Geoscientific Analyses) is still worth mentioning. It is a geoinformation system originally developed at the University of Göttingen that stands out due to the implementation of spatial algorithms for hydrological modeling and analysis of terrain models as well as methods of geostatistics (including kriging) (cf. SAGA 2023).

9.1.6 Geoinformation Systems and Similar Systems

The general definition of a geoinformation system deliberately does not contain any precise statements regarding the type of (spatial) data or geodata. The further definition of the tasks and areas of application as well as the more precise determination of the data content lead to further terms:

The Fédération Internationale des Géomètres (1974) define a *land information system* (LIS): "A land information system is an instrument for decision-making in law, administration and economics as well as a tool for planning and development. It consists on the one hand of a data collection containing land-related data for a specific region, and on the other hand of procedures and methods for the systematic collection, updating, processing and implementation of these data. The basis of a LIS is a uniform, spatial reference system for the stored data, which makes it easier to link the data stored in the system with other land-related data." However, this old definition rather characterise a land information system.

Land information systems are based on a purely vector-oriented representation, which allows a high geometric accuracy of the geoobjects, so that an application in surveying and in cadastral systems is possible. In most cases, spatial analysis functions are not very pronounced. The abbreviation LIS is more frequently used for landscape information systems which manage (primarily) natural and non-administratively delimited spatial units with information mainly on natural features and which are primarily used in nature conservation and landscape planning.

An *environmental information system* can generally be understood as a special form of a geoinformation system in which environmental information is processed (cf. compilation in Fürst et al. 1996 p. 3).

Based on the terms mentioned and the associated concepts, further word constructions and combinations of terms such as *municipal* or *regional information system*, *municipal* or *regional spatial information system* or *regional environmental information system* can be found. This does not express any fundamentally new conceptual content, but merely limits the area of application and purpose.

Among the spatially oriented disciplines, a further distinction is made between spatial-oriented specialist information systems, such as e.g. contaminated sites, pipeline or traffic information systems. Here too, a geoinformation system is usually, but not necessarily, the heart of such specialist information systems.

In addition to the geographic and environmental information systems mentioned above, several types of software systems work with spatial reference units, such as database systems, systems for creating maps or cartographic visualisation and CAD systems. Such software products do not represent geoinformation systems, since they usually do not manage any topological relationships and therefore no geoobjects, and have no or only (very) limited analysis functions. However, it must be clearly emphasised that the boundaries between, for example, geoinformation systems and CAD systems, which are used for computer-aided design and interactive (technical) drawing and construction in two- and three-dimensional representation and which are used, among other things, by utility companies for planning and maintaining pipeline registers, are becoming increasingly open.

9.2 Web-GIS

9.2.1 Web-GIS: Concept and Functionality

With the growing offer of geoinformation and digital maps on the web, the terminology of geoinformation systems has also been transferred to the web (cf. Behncke et al. 2009). In addition to "Web-GIS", the terms "Online-GIS", "Internet-GIS", "Net-GIS" or "Distributed GIS" are also frequently used with often the same or similar meaning. "Google Maps has created a real hype, which (. . .) has led to the term "GIS" being misused even more often than it was before." (Rudert and Pundt 2008).

The strict definition of a geoinformation system, which can only be referred to if all four components of the IMAP model are available, must also apply to a Web-GIS or Internet-GIS:

- A *Web-GIS* refers to a geoinformation system that uses the WWW service and includes all four components of the IMAP model,
- An *Internet-GIS* represents a geoinformation system that generally uses some service of the Internet.

Similar to web mapping, a Web-GIS is technically based on a client-server architecture (cf. Sects. 2.8.3 and 7.2.1 as well as Fig. 7.10). The webbrowser serves as a client, a GIS-server operates instead of a mapserver. The user interactively calls functions that are processed on one or more servers. The result is then sent back to the client. A Web-GIS differs from web mapping applications in that more interaction options are available and the scope of specific GIS functions is greater. As a rule, access to thematic data is possible via a database, and the user can make subject-related queries. Measurement functions such as calculating the street length are supported. To justify the name "GIS", a Web-GIS must provide functions for input, management and spatial analysis and comply with the IMAP principle. At least one other function such as overlay, intersect, buffer, area selection or routing must be available.

In a typical Web-GIS, the geoprocessing is done on the server-side. In thin-client architectures, the client is only used to communicate with the server and display results. In a thick-client architecture, the client communicates with the server in the same way as in a thin-client architecture, but in addition, client-side functionalities are available, which can be realised using appropriate extensions (plug-ins and JavaScript). Geoprocessing has been extended to the client. In medium-client architectures, extensions are used both on the client-side and on the server-side.

The term Internet-GIS applies to geoinformation systems that generally use any service of the Internet and do not necessarily require the use of a webbrowser. However, the browser-independent application Google Earth, which enables data collection, processing and presentation, but has so far has only few analysis functions and therefore cannot compete with a GIS, does not constitute an Internet-GIS.

9.2.2 Web-GIS in Practice

There are only a few software offers that deserve the name Web-GIS in the strict sense. The proprietary ArcGIS Online, for example, enables paid "on demand" access to a wide range of GIS functions from a central computer via the Internet. Here, "service credits" are used as a form of payment, which must be purchased in advance (cf. ESRI 2023b and ESRI 2023c, cloud-based, software as a service, cf. Sect. 2.8.5). Such a business model is relatively new. The end products, i.e. the maps, are provided in the cloud so that many users, i.e. primarily employees of a company or an authority, can use the data at the same time. Scaling to different devices like tablet computers or desktop computers is done automatically. The software company takes care of the software (i.e. updates and maintenance) and provides the servers for software and data. This all, however, increases the dependency on the software provider. But it is not necessary to lease a comprehensive software package, which is usually too large.

More frequently, server variants are offered in which the GIS-server-software is deployed on a server at the user's premises (cf. e.g. ArcGIS Server, GeoMedia WebMap or Smallworld Geospatial Server, cf. Sect. 9.1.5) as well as many solutions from locally operating software service providers. Geodata can be made available for any internal user and optionally also externally, e.g. to citizens, via an Internet connection. On the one hand, geodata can be presented via web maps (e.g. in a company's own web portal, via browser-based web apps and native apps on mobile devices). This is by far the most frequently used variant. On the other hand, analyses can also be carried out. Thus, a Web-GIS is available that in practice is often only used for web services (i.e. presentations).

In contrast to these proprietary solutions, "Dropchop" is an approach that is still formulated as a "proof-of-concept" (cf. Dropchop 2023). Dropchop is a browser-based geoinformation system with a modular structure that is created by a group of developers spread across the web. Among other things, the powerful, free JavaScript framework Turf is used, which is a collection of small modules in JavaScript. Own geodata, which are in the "GeoJSON" data format, can be uploaded and processed on the web (e.g. execution of spatial operations such as buffer or overlay functions). This approach shows basic possibilities. There are still unanswered questions about data protection and the ongoing maintenance of the web offer. However, these are essential questions for business operations in a company. It remains to be seen whether such an approach will become established in practice.

9.2.3 Web Mapping as a Substitute for a Web-GIS?

In practice, the terms Web-GIS and web mapping become blurred. In many cases, a system with which geodata can only be visualised via the web and, if necessary, also printed out is referred to as a Web-GIS (cf. Sect. 9.2.1). In many cases these functions are completely sufficient. Often all that is needed is an easy-to-use, low-cost information system. All that is required is a web mapping solution with a mapserver that does not provide GIS functions such as spatial analysis functions. These special analysis functions are executed by only a few "power users" with a desktop GIS.

The main advantages of a client-server web mapping solution are:

- connection of a large number of users (including possible access for a broad public after granting specific access rights)
- platform independence and use from any Internet-enabled computer
- very easy and intuitive access via a webbrowser
- low client requirements and low costs.

Compared to these advantages, there are limitations, which, however, are ultimately irrelevant to the needs of the general user:

- restriction of the user only to the offered data and functions
- access speed depending on web access (availability and performance) and on number of accesses
- complex administration on the server-side.

Proprietary and free software systems are available for setting up such client-server web mapping solutions. In many cases, smaller, locally or regionally active software companies develop and maintain customer-specific client-server solutions that are tailored to the respective requirements and are based on free software. The geographical proximity and the personal or direct contact with the provider, possibly a network of similar customers (e.g. municipalities in a district) as well as manuals, training documents and support in the respective national language are the strengths of such software solutions.

9.3 Modeling Geoobjects in a Geoinformation System

9.3.1 Geoinformation System as a Model of the Real World

A geoinformation system can be seen as a model of the real world that digitally records, stores, manages, updates, analyses and models *geoobjects* and present them alphanumerically and graphically. Figure 9.1 shows the user view of a geoinformation system.

The screenshot illustrates the software that enables the presentation on the monitor and the handling of the system and that provides a variety of functions via the graphical user interface. Several data layers are visible in an extract from an environmental information system. The lakes, rivers and wetlands worthy of protection according to the old paragraphs 28a and 28b of the Lower Saxony Nature Conservation Act (in blue), the parcel boundaries of the old ALK foil 001 (in black) and the buildings of the old ALK foil 011 (in red, cf. Sect. 5.5.2) as well as the suspected contaminated sites are shown, whereby the latter layer is not activated and visualised here. Next to it, the attribute table is shown for a thematic layer. The small window shows the attribute data of a geoobject selected with the pointing device (mouse). With the help of the functions of a geoinformation system, a variety of evaluations are possible, which can be based on the attribute data in the tables or on the graphical presentation of the geoobjects.

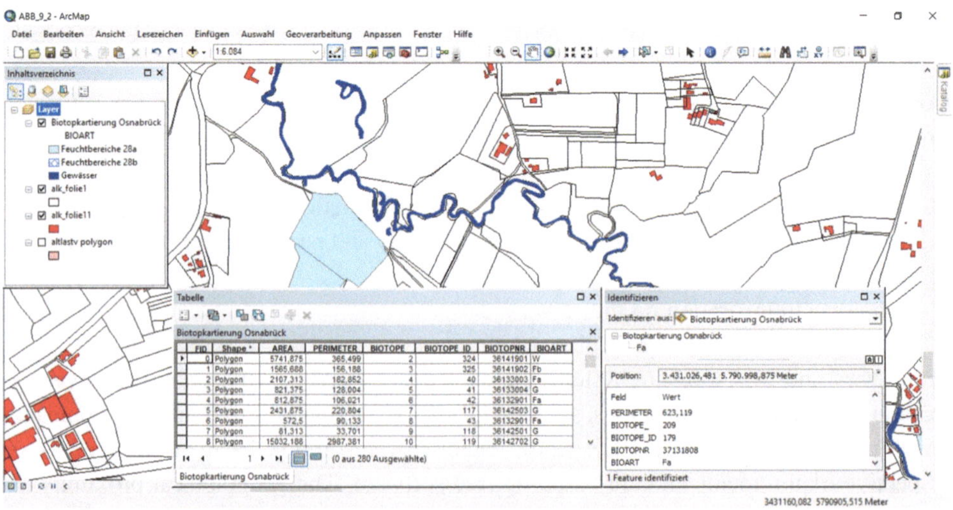

Fig. 9.1 A geoinformation system as a model of the real world

A geoinformation system enables different *technical views* of the data. For example, Fig. 9.1 could illustrate the view of an employee in an environmental agency who has to answer a request for the expansion of an industrial park. For this purpose, the current use of the existing commercial space is less of interest than the presentation of possible conflicts with areas that are still undeveloped. A business developer needs, among other things, the parceling of the commercial areas, the location of supply lines, details of the development plan and, above all, information of the current use, as well as a city map in the background for easier orientation. A complex municipal information system integrates the various data levels, such as real estate, pipelines, contaminated sites, green spaces and a tree inventory for an urban forest management. Complex issues are of interest to different users, such as:

- land registry administration, collection of property taxes, road user fees and other municipal taxes,
- management of power cables and power connections of the utility companies, maintenance of the power grid,
- management of trees worthy of protection by an environmental agency, maintenance of trees and urban forest,
- designation of land use categories for preparatory urban land use planning, site maintenance and site provision for commercial enterprises.

Often such an information system is realised by several digital cadastres in the form of individual geoinformation systems.

9.3.2 Geometric-Topological Modeling of Geoobjects in the Vector Model

In the vector model, the geometry of a geoobject is specified by coordinates on the basis of a unique spatial reference system (coordinates in a metric coordinate system, cf. Sects. 4.1.2 and 4.2). The coordinates indicate single points as well as start and end points of directed lines, i.e. of vectors. The single points are also to be understood as vectors, whose starting point is at the origin of the coordinate system (cf. Fig. 4.1). To make it very clear, when modeling geoobjects in this so-called *vector model*, ultimately only points are recorded! The entire geometric information is based on vectors or coordinates in a (Cartesian) coordinate system. Linear and planar structures must be constructed from points or vectors. As a result, all geometries are discretised: point, line and areal features.

In the vector model a polyline consists of a sequence of directed lines or segments (i.e. *vectors*). In this case, curved lines are approximated by a sequence of straight line segments (cf. Sect. 5.2.1 and Fig. 5.3 or 9.2). In the vector model, areas are modeled by polylines as *polygons* that delimit them.

Figure 9.2 shows the essential principles of geometric-topological modeling as implemented in the ArcGIS software, based on a long history dating back to ESRI's

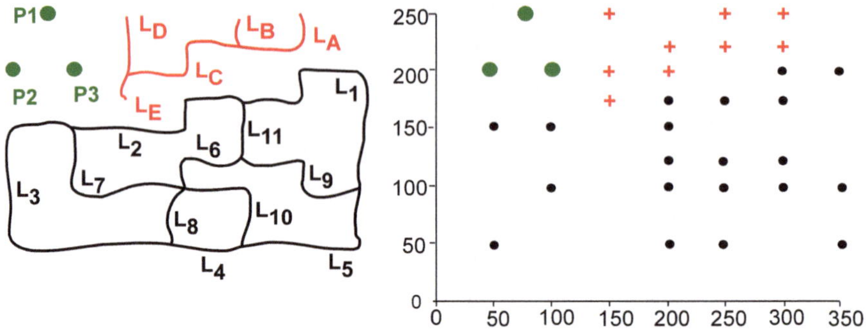

Fig. 9.2 Geoobjects of the real world: geometric modeling in the vector model

Table 9.1 Representation of the geometry of the geoobjects in Fig. 9.2 by coordinates

Point 1	(75,250)
Point 2	(50,200)
Point 3	(100,200)
Line 1	(250,175) (300,175) (300,200) (350,200) (350,100)
Line 2	(250,175) (200,175) (200,150) (100,150)
Line 3	(100,150) (50,150) (50,50) (250,50)
Line 4	(200,50) (250,50)
Line 5	(250,50) (350,50) (350,100)
Line 6	(250,125) (200,125) (200,100)
Line 7	(100,150) (100,100) (200,100)
Line 8	(250,50) (200,100)
Line 9	(350,100) (300,100) (300,125) (250,125)
Line 10	(250,50) (250,100) (200,100)
Line 11	(250,175) (250,125)
Line A	(300,250) (300,225) (250,225)
Line B	(250,250) (250,225)
Line C	(250,225) (200,225) (200,200) (150,200)
Line D	(150,250) (150,200)
Line E	(150,200) (150,175)

Arc/Info in the early 1980s. This means that the modeling of the polygons does not follow the more recent specifications of the Open Geospatial Consortium (OGC). The *simple feature geometry object model* of the OGC describes polygons using a completely closed polyline, i.e. by a sequence of segments where the starting point of the first segment and the end point of the last segment have identical coordinates, and does not store any topological information (cf. Sect. 6.3.2 and Table 6.2).

The geometry of the geoobjects shown in Fig. 9.2 is represented solely by the coordinates in Table 9.1. The consequence of this is that for the representation of line

Table 9.2 Node-edge-node modeling for the areal geoobjects in Fig. 9.3

Edge	From node	To nodes	Polygon left	Polygon right
1	1	5	–1	1
2	1	2	2	–1
3	2	3	5	–1
4	3	4	4	–1
5	4	5	3	–1
6	6	7	3	2
7	2	7	2	5
8	3	7	5	4
9	5	6	3	1
10	4	7	4	3
11	1	6	1	2

Table 9.3 Polygon modeling for the areal geoobjects in Fig. 9.3

Polygon	Edge
1	–1, 11, –9
2	2, 7, –6, –11
3	5, 9, 6, –10
4	4, 10, –8
5	3, 8,–7

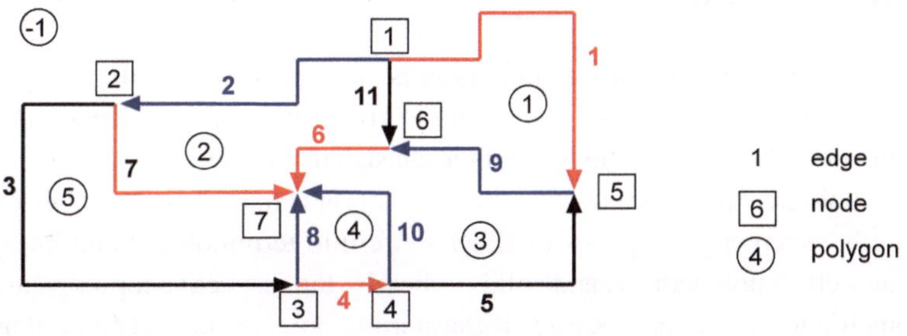

Fig. 9.3 Geometric-topological modeling of the situation in Fig. 9.2

and area geoobjects topological relationships of the coordinates must be explicitly recorded, modeled and stored in the vector model! Thus, in addition, it must be considered, which coordinates follow one another, i.e. are adjacent, and define a specific line, and which lines form a specific area, i.e. a closed polygon. Only Tables 9.2 and 9.3 represent the (topological) model of the initial situation implemented and stored in the geoinformation system (cf. Fig. 9.3). The outer area, i.e. the "infinite" area outside the contiguous subareas, is referred to here as the area with the area number "-1". The definition of the polygons is based on the mathematical positive orientation. One walks along a boundary, so that the associated polygon is on the left, and connects the polylines with the orientation corresponding to the circumnavigation.

This describes the topology of areas (polygons) that are composed of individual line segments (vectors). According to the terminology of network modeling, a polygon is defined by the sequence of its edges from node to node, so that a *node-edge-node topology* is given. This is the traditional "arc-node-topology" as ESRI's Arc/Info calls it, but neutral terminology is preferred here. It should therefore be pointed out that in the vector model the planar geoobjects are modeled as a substructure and are not directly present as in the raster model. This modeling as a connected planar graph enables a consistency check of the geometric-topological structure using network analysis methods (cf. Sect. 3.4.2).

An outer edge is defined by the fact that this edge only occurs once in the polygon edge list (cf. Table 9.3). An inner edge must appear exactly twice with a different sign according to the orientation. First of all, it must be ensured that the end node of one edge exactly matches the start node of another edge (or itself). In this example, the end node of edge 3 must be the same as the start node of edge 4 and 8. They must have exactly identical coordinates (except for a small tolerance deviation, where the GIS considers the two coordinates to be identical).

Since the following applies in Fig. 9.3 according to *Euler's theorem*, there is a connected planar graph, i.e. a correct topological structure (cf. Sect. 3.4.2)

$$k_n - k_a + p = 2 \qquad \text{here: } 7 - 11 + 6 = 2$$

k_n = number of nodes, k_a = number of edges, p = number of polygons.

The "infinite" region "-1" outside the graph is counted as a polygon.

As seen in Table 9.3 the edges 1, 2, 3, 4 and 5 (regardless of the orientation) are unique and form the outer boundary. This is another consistency check.

From a topological point of view, a strict distinction is made between nodes and points, furthermore between edges (or arcs in ESRI's Arc/Info terminology) and lines, polygons and areas as well as polyhedrons and solids, whereby the respective topological equivalent of the geometric term is named. A *node* is the starting point or the end point of an edge and then also the meeting point of several edges. An *edge* or *arc* connects two adjacent nodes, which are geometrically the start and end points of a polyline. The exact course of the polyline between the start and end nodes is irrelevant for the topology. The intermediate points that define the exact course of the polyline are called *vertices*. Topologically, the exact geometric shape of an area in the real world is abstracted (cf. Sect. 4.1.3 and Fig. 4.3).

The relational data model shown (i.e., Tables 9.1, 9.2 and 9.3) makes it possible to read graphical or topological properties of geoobjects solely through numerical evaluations. With the help of the associated database, topological questions can be answered:

- Which polygons are adjacent to polygon 3? Solution: Find the edges that define polygon 3 (solution: 5, 9, 6, −10). Find the polygons in whose formation these edges with the

inverse orientation are involved. The solution is: polygons 1, 2, 4 and −1, since edge 5 occurs only once.
- Which is the shortest path between node 2 and node 4? Solution: Determine all edge lengths using the defining coordinates and then build a weighted adjacency matrix and apply a path algorithm (cf. Sect. 3.4.2).
- Which polygon is directly adjacent to polygon 3 along edge 10? Solution: Find the polygon whose formation is involved by edge +10 or −10. Since polygon 3 is formed, among others, by edge −10, the solution is polygon 4.

The last query refers to neighborhoods. Thus, in general, two geoobjects in the vector model are adjacent, if they have at least one point in common (applies to points, polylines and polygons) or if they have at least one edge in common (applies to polygons). In Fig. 9.3, for example, polygons 2 and 3 are adjacent since they have edge 6 in common. If two polygons are adjacent only through a common point or node and through no common edge, then for a planar connected graph two edges must meet at a node, each originating from one of the two polygons (cf. node 7 in Fig. 9.3). In the node-edge table this node then also appears (at least) four times, so that these adjacent polygons can also be determined using this table (cf. Table 9.2). Polygons consisting of only one edge (i.e. islands) are excluded from these calculations.

For the definition of correct geometric structures, which reflect unique connections and relationships of geoobjects, and topologies, several rules can be defined such as (cf. the implementation in the geoinformation system ArcGIS of the software company ESRI in Fig. 9.5):

- A polyline must not have two or more identical coordinates (exception loops).
- Polygons must not overlap.
- There must be no empty spaces between adjacent polygons.
- Areas must be defined by a sequence of closed polylines, which must therefore have no gaps.
- When defining polygons from several polylines, no line segments may remain free, which do not form polygons with other polylines (so-called dangles).
- If an area has a ring-shaped structure, the inner and outer boundary lines must not cross.

If the geometries are recorded incorrectly, there are no geometrically or topologically unambiguous structures. Figure 9.4 shows three polygons of a data layer (agricultural units with grassland, arable land, corn field). The layers are intended to represent a coherent section of reality. Gaps are therefore excluded, and an area cannot be used as arable land or grassland at the same time. However, the polygons shown in Fig. 9.4, were recorded incorrectly. They partially overlap, and there is also an "empty" space between them. In this very simple example, assume a planned road that runs through the area. In a first estimate, the planner would like to know how much area of agricultural units will be lost or has to be bought from the farmer. The spatial *intersection* of these agricultural units is then formed with the polygon that represents the planned road (function "intersect", cf. Sect.

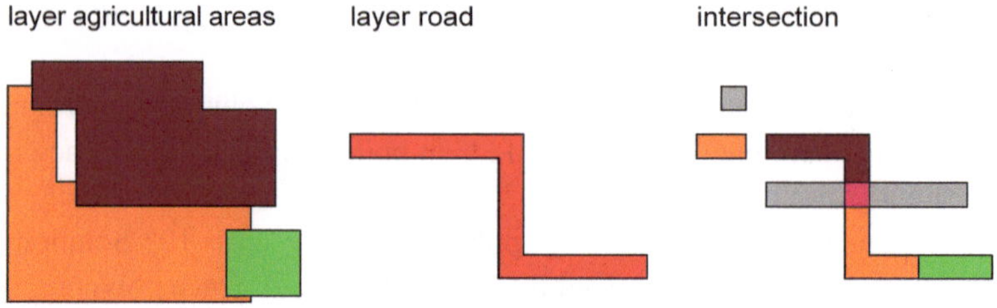

Fig. 9.4 Spatial overlays for topologically incorrectly modeled polygons

9.4.4). Artifacts have arisen due to the incorrectly modeled polygons: gaps and such areas that already exist as overlays of areas in the input layer (cf. grey areas in Fig. 9.4). The area marked in magenta was created by overlaying three areas. Such results are prevented by correct geometric-topological modeling.

The GIS software must provide tools to detect ambiguous geometrical structures and topological errors and to correct them. This requirement is independent of the underlying data model. Even if the data model itself does not store any topological information as within the simple feature geometry object model, the geometries must be clearly defined.

9.3.3 Geometric-Topological Modeling in the Vector Model: Application

Section 9.3.2 addresses geometric-topological modeling from a theoretical point of view and illustrates a certain rigor in the recording of data and particular in the creation of polygons. The example presented is based on a data model that was already implemented in the proprietary geoinformation system Arc/Info developed by the software company ESRI in the 1980s (cf. ESRI 1995). The geoinformation systems available today implement the topological requirements very differently, which primarily depends on the underlying data model.

The *simple feature geometry object model* standardised by the OGC (Open Geospatial Consortium) does not store any topological information. Polygons are modeled as closed polylines, i.e. as a closed sequence of points with identical start and end nodes (cf. Table 6.2). Common boundaries of adjacent polygons are recorded and stored twice. As a result, there is a risk of geometric-topological inconsistencies.

The proprietary *shapefile data format* of the software company ESRI must now be regarded as the industry standard that all geoinformation systems must process and that currently represents the de facto standard for the exchange of geodata. The shapefile data format is a simple, nontopological format for storing the geometric location and attribute information of geoobjects. The shapefile data format defines the geometry and attributes of geoobjects in at least three files: *.shp for storing geometrical data, *.shx for linking geometries and attribute data, *.dbf for storing attribute data in dBase format (cf. ESRI 2023d).

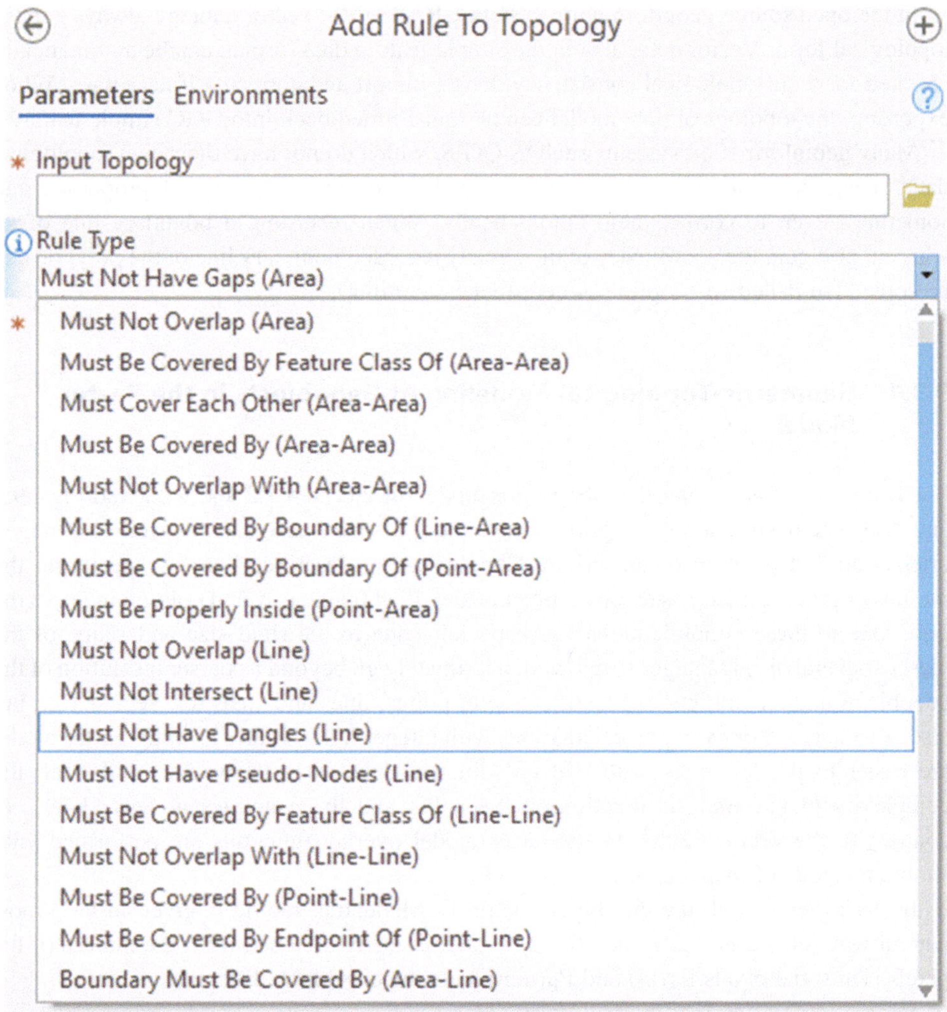

Fig. 9.5 Selected rules for defining polygons in ArcGIS Pro

A so called *geodatabase* is the native data structure of the internationally widely used geoinformation system ArcGIS of the software company ESRI. A personal geodatabase is a Microsoft Access database that can store, query, and manage both spatial and nonspatial data (phased out, support will continue until about 2025). In the long term, a multi-user relational database management system or a file system (file geodatabase) will be used in ArcGIS Pro by ESRI. Topological relationships can be defined within one or between several data layers or groups of geoobjects (cf. Fig. 9.5). With the help of suitable tools in ArcGIS Pro, these rules can be checked and errors corrected (cf. ESRI 2023e).

Figure 9.5 shows the ArcGIS Pro software wizard. It can be used to set up *topological rules* for individual types of geometry (here: definition of polygons from line segments).

In the open source geoinformation system GRASS GIS, vector data are always kept in topological form. Vector data, also in the simple feature data format, can be automatically checked for their topological consistency during import and corrected if necessary. When exporting, the topological data model can be transformed back into OGC simple features.

Many geoinformation systems such as QGIS, which do not have their own topological data model, have tools to check geometric data for their topological properties and sometimes even to correct them automatically. When recording a boundary line of an adjacent polygon, the coordinates of the already recorded boundary line of the polygon can be copied (so-called snap option of coordinate recording).

9.3.4 Geometric-Topological Modeling of Geoobjects in the Raster Model

The raster model is simpler than the vector model (cf. Sects. 4.1.2 and 5.2.2 and Figs. 9.6 and 9.7). The basis is a regular grid with a fixed grid size (mesh size) as well as arranged orientation and position of the origin. Geoobjects are described here by specifying the meshes or pixels they fill, whereby a pixel is identified by the row and column index in the grid. Due to these simplifications and especially due to the rigid size and shape of the meshes several disadvantages result, which ultimately go beyond a coarser resolution of the geoobjects and an ambiguous identification of points, lines and areas (cf. Sect. 4.1.2), but also advantages. For example, calculations with integer index values as in the raster model are easier to perform than with real coordinate values as in the vector model (cf. the complexity of the analysis functions in Sect. 9.4 and the examples in Sect. 3.4.1). In contrast to the vector model, in the raster model overlay functions are performed with simple methods of map algebra.

In the raster model, the city-block metric or Manhattan metric is given in an almost natural way (cf. Sects. 4.2.1 and 10.7.2), which is easily obtained from the indices of the pixels. Thus, the pixels $P_A(i,j)$ and $P_B(m,n)$ have the city-block distance:

$$d_{AB} = |i - m| + |j - n|,$$

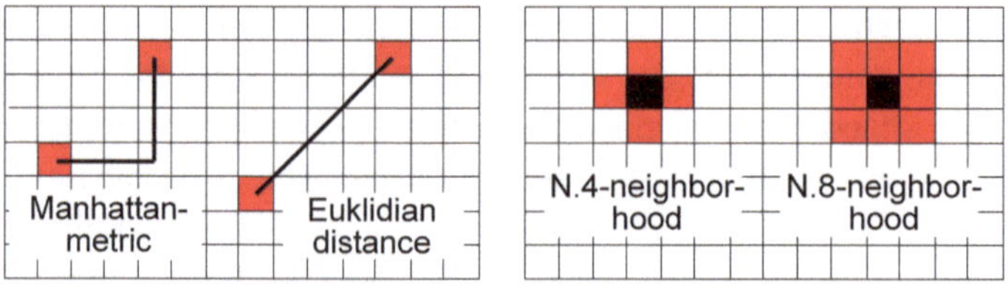

Fig. 9.6 Metrics and neighborhoods in the raster model

The distance value would have to be multiplied by the mesh size. This is almost superfluous, since then all distances have this factor. The Euclidean metric can also be used here, whereby this metric is then applied to the centers of the grid cells (cf. Fig. 9.6).

It is of great advantage that the topology does not have to be specified separately, when representing geoobjects in the raster model. The topology is already defined directly by specifying the geometry. Neighborhoods are defined relatively simply on the basis of a regular grid (cf. Fig. 9.4). Accordingly, two grid cells are adjacent if they have a common cell edge (*edge-to-edge topology*, so-called *N.4 neighbors*) or if they have at least one common cell corner (*corner-to-edge topology*, so-called *N.8 neighbors*). The neighborhood is calculated directly from the pixel coordinates. Thus, $P_A(i,j)$ and $P_B(m,n)$ are N.4-neighbors if (i = m and (j = n − 1 or j = n or j = n + 1)) or if (j = n and (i = m − 1 or i = m or i = m + 1)) holds.

9.3.5 Storing Geometries in the Raster Model

The raster model is normally based on regular square meshes (cf. Fig. 9.7). Consequently, matrices then represent the standard data model for raster data. The comparison of Figs. 9.2 and 9.7 and Table 9.1, which are basically identical but differ on the data model, shows a considerably higher memory requirement (vector model 51 coordinate pairs, raster model 384 pixels). As the resolution increases, the memory requirement increases significantly. Therefore, efficient algorithms have been developed to cope with the memory problems of raster data. They always reduce the amount of data whenever the grid matrices have larger homogeneous areas with the same attribute values.

With *run length encoding*, the matrix is scanned line by line for identical, adjacent pixels. Only the pixel value and the number of equal neighbors are saved as a pair of values (cf. Fig. 9.7, 2-0 denotes two cells with no topic, 1-L denotes one cell as a line, 12-FC denotes 12 cells

Fig. 9.7 Geoobjects in raster representation and specification of run length coding (with specification of the topic)

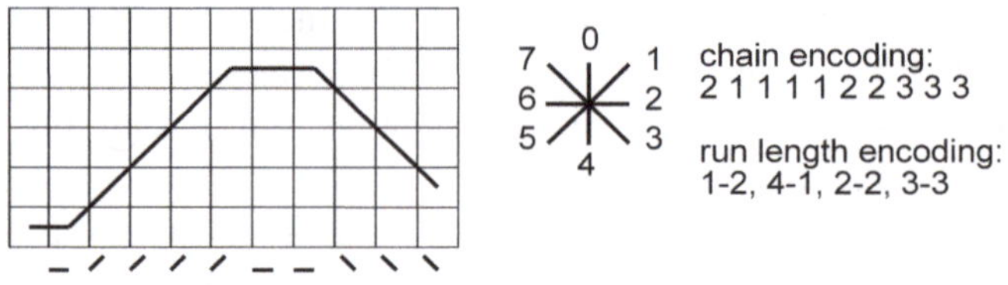

Fig. 9.8 Line in chain encoding

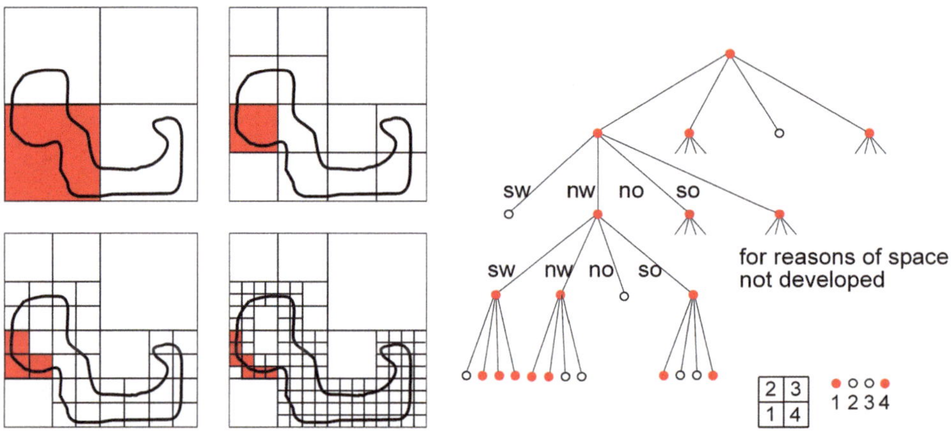

Fig. 9.9 Display of raster data as a quad-tree

as area C). The order of the rows has to be determined in advance and a one-dimensional order structure has to be defined. In Fig. 9.7, in which the origin is in the upper left corner, the procedure proceeds line-by-line with line interlace (so-called standard row order).

Chain encoding proceeds in a similar way, which is especially suitable for the storage of lines in the raster model. A line is described here by the row and column indices of the starting pixel and then by the directions R_1 to R_n to the n following pixels (cf. Fig. 9.8). A combination of both methods is possible. Efficient storage results primarily for long lines without major changes in direction.

With both methods, the original values are preserved (as with the so-called quad-tree, cf. Fig. 9.9). They are lossless in contrast to lossy data compression methods such as JPEG.

The disadvantages of the raster data model are mainly the low flexibility of the fixed mesh size, with which geoobjects can only be recorded with insufficient accuracy. Any fine graining of the grid is difficult due to the rapidly increasing memory requirements. This suggests the idea of using relatively coarse meshes for homogeneous area elements and to refine the mesh size only where the geometric data situation requires it. This idea is implemented by the quad-tree model.

The representation of raster data as a quad-tree is based on a recursive division of an in-homogeneous square into four quadrants of equal size. Thus, each quadrant has four sons. However, the quadripartition is only continued until one quadrant is homogeneous. Thus, quadrants of different sizes are often involved in the representation of an area. Figure 9.9 illustrates this successive quadripartition for the south-west quadrant. With this recursive refinement, even very small-scale structures (practically any) can be represented precisely, whereby the memory requirement is significantly lower compared to the simple raster data model. In terms of accuracy, the quad-tree model can approach the accuracy of the vector model. A quad-tree can be optimally implemented using so-called trees (cf. Sect. 3.2.4.4).

9.3.6 Topic of Geoobjects

A geoobject always has a topic or theme and carries thematic information (cf. Sect. 4.1.4). The topic is generally characterised by several attributes (features, variables) with different scale levels. The description, processing and storage of the various topics of geoobjects can be done by two basic principles: the layer concept and the object class concept (cf. Fig. 9.10).

The geoobjects with their attributes are strictly separated according to the different thematic meanings in the *layer concept* and are presented in different layers (layer principle, cf. Sect. 4.1.4). This oldest principle of presenting different topics is derived directly from the layer principle of classical cartography. Different layers with different thematic contents (e.g. layers of vegetation, soil types or rivers) are superimposed during map production. Of course, shared geometries must be identical (e.g. the boundaries of rivers and adjacent green areas on different layers). In most cases, there is not only a separation according to the different topics, but also according to the geometry (point, line, areal features). The modeling according to the layer principle has no hierarchy, all layers are formally equal.

The integration of time in a geoinformation system usually takes place according to the layer principle. Temporal information can also be included by introducing further attributes and metadata. By default, however, temporal processes are discretised by time slices that

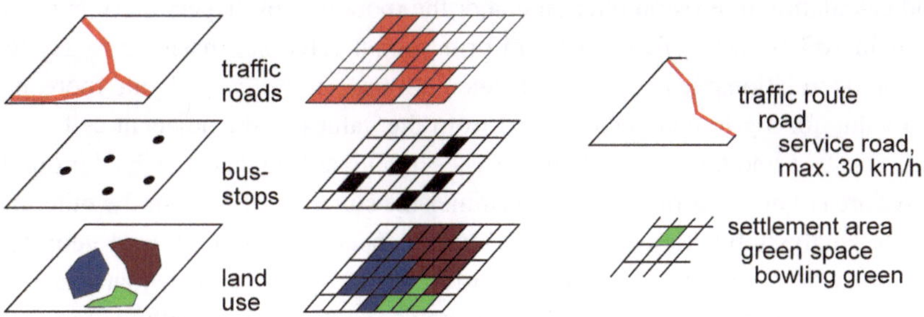

Fig. 9.10 Representation of geoobjects: vector and raster concept as well as layer- and object concept

form individual layers in a geoinformation system. Then temporal changes can be analysed using spatial overlay functions (cf. Sects. 9.4.4 and 9.5.3).

The *object class concept* assumes a hierarchical arrangement of different topics with subset relationships of the topics: e.g. hyperclass, superclass, class, subclass (cf. Fig. 9.10 and more detailed Sect. 4.1.1). In general, geoobjects with a common topic (and methods) are combined into object classes.

The different concepts of the layer and the object-oriented approach can be used both in vector-oriented as well as in raster-oriented geoinformation systems. The representation of geoobjects can easily be realised according to the layer principle. It is the standard form in a geoinformation system.

9.3.7 Vector and Raster Model: Comparison

A brief *comparison of vector and raster models* does not show any clear advantages and disadvantages (cf. Table 9.4). In general, different issues can be processed with both a vector and a raster model, although there are specific advantages and disadvantages in each case that are related to the spatial resolution of the geoobjects (coordinates versus pixel), to the memory requirement for data or to the effort required for the analysis techniques. It should be emphasised that both models do not oppose each other, but that both vector and raster models are needed.

Due to its higher accuracy and clarity, the vector model is suitable for surveying, for cadastres or in infrastructure planning as well as for large-scale studies in general. Especially in environmental planning, geoinformation systems (based on vectors) have become standard tools (e.g. biotope cadastre or register of contaminated sites).

In contrast, the raster model has become the standard for small-scale applications and for large-scale overviews as well as for applications in digital image processing, which is almost inevitable due to the database in form of raster data. Furthermore, the raster model is particularly suitable for issues concerning the modeling of spatial dispersion processes. Important areas of application are, for example, the modeling of emissions from point emitters (such as chimneys) or the modeling of water runoff (on a surface), the representation and calculation of erosion phenomena or the modeling of dispersion of environmental pollution in soil or water. Due to a uniform spatial reference basis and easy-to-handle neighborhood relationships, dispersion calculations can be carried out more easily, in which a value for a grid cell is calculated from the values of the adjacent cells.

Thus, the two model variants can only be evaluated against the background of the respective area of application, which determines the suitability of one or the other model, as well as the analysis functions available in the respective model. Hybrid geoinformation systems provide methods for data recording, data management, data analysis and visualisation for both models. In particular, geoinformatics offers methods for exchanging and transforming geodata between the two models:

Table 9.4 Comparison of vector and raster model

	Vector model	Raster model
Advantages	High geometric accuracy	Simple data structures
	Clear object description	Low effort for recording the geometry and topology
	Small data volume	
	Greater similarity between the graphic presentation and the traditional maps	Compatible with remote sensing and scanner data
		Simple overlaying of geoobjects
		Simple logical and algebraic operations
Disadvantages	More complex data structures	No shape and position accuracy of the geoobjects
	Complex recording of geometry and topology	Higher memory requirements
	Complex and computationally intensive logical and algebraic operations (including overlay and intersection)	Smaller pixel sizes for higher accuracy requirements lead to exploding amounts of data
	Parallel geometric and topological description of the geoobjects (depending on the data model)	Less satisfactory graphic presentation (depending on pixel size)
		Complex coordinate transformations

– conversion from vector to raster data (cf. Sect. 9.5.2 for spatial interpolation of thematic data at individual points to information in a raster cf. Sect. 9.7.3)
– conversion from raster to vector data (cf. Sect. 5.2.3).

Table 9.4 compares the properties of the vector and raster model. The high geometric accuracy of the vector model, based on point coordinates, requires a small data volume, but also the complex modeling of the geoobjects and, above all, the very computationally intensive execution of overlay operations. Section 3.4.1 explains the creation of an intersection of two vector layers (cf. Fig. 3.14) and outlines the associated algorithms (cf. Figs. 3.15 and 3.16) by an example.

9.4 Vector Model: Spatial Analysis of Geoobjects

9.4.1 Vector Model: Recording and Editing Geoobjects

Various technical devices are available for recording geometrical data, such as digitising tablets, which are still outdated in some places, or, more recently, smartphones, mobile geoinformation systems or GPS devices for mobile data recording. Of particular

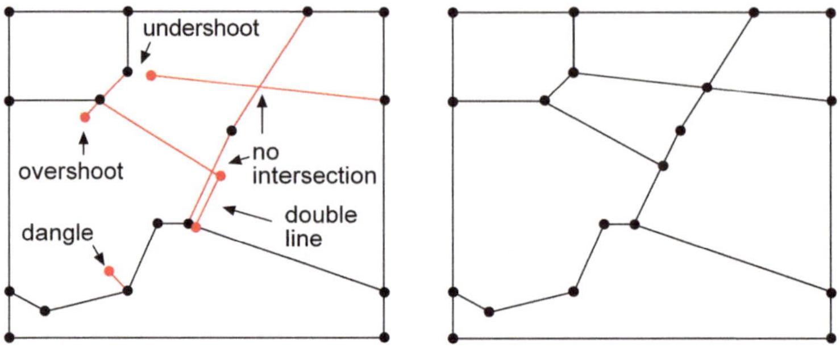

Fig. 9.11 Recording geoobjects, detecting errors and creation of clean, topologically consistent geometries

importance is the interactive data recording of coordinates with the help of on-screen data recording (cf. Fig. 5.4 and Sect. 5.2.1). The geoinformation system has functions for *georeferencing* geometries (cf. Sects. 4.2.5.1 and in particular 4.6, which shows an example of georeferencing a template). As a rule, many map projections are implemented, so that the data recording of templates with almost any projection as well as a transformation into almost any coordinate system is possible. Thus, basic knowledge of map projections and coordinate systems is essential for understanding geoinformation systems (cf. Sects. 4.2 to 4.5). Above all, very diverse functions are available for the recording of geometries: e.g. standard functions such as copying or deleting geometries as well as "snapping" coordinate values, but also special functions such as creating a parallel or perpendicular line to an existing line.

The geoinformation system provides *data interfaces*. Particular importance is attached to data exchange, i.e. the software's ability to import data in different formats or to transform and export its own data into other data formats. This also includes the transfer of data from surveying or coordinates from a GPS device.

After the data recording of the geometries, the spatial data are processed and modeled. The geoinformation system must be able to build up topological information for vector data (depending on the data model) or check the geometries for topological inconsistencies in order to correct them, define lines from the coordinate sequences and to generate polygons from polylines. Detection errors must be displayed and troubleshooting functions must be provided.

The tools for *editing geometries* mainly include:

- removing superfluous points, polylines or polygons or adding new ones (cf. Fig. 9.11)
- thinning and smoothing polylines
- splitting of polylines and polygons
- correction of geometries such as creating perpendicularity or parallelism
- dissolving a spaghetti digitisation (cf. Sect. 5.2.1)

- resolving overshoots and undershoots (cf. Fig. 9.11)
- move and copy objects
- creation of an error-free topological structure (cf. Sect. 9.3.3).

The geoobjects must not only be defined geometrically, but also through thematic information. The geoinformation system uses functions of a database management system to record the thematic information and to model the themes. In the case of relational data structures, the tables are linked to geoobjects via key attributes (cf. Sects. 8.1.1 and 8.3). The integrated database management system also allows changes of the attribute structure and the attribute values themselves: deleting and adding, copying, renaming attributes and individual attribute values, changing the type definition of attributes (e.g. reducing the number of decimal places), merging different tables via common key attributes.

9.4.2 Managing Geoobjects in the Vector Model: Data Queries

In a geoinformation system, there are many different ways to provide different views of the data and to formulate *queries.* Pure graphical illustrations and simple interactive queries by clicking on the geoobjects on the monitor are the most common forms of data queries. In addition, purely *attributive search queries* can be formed. Based on the query and search functions of the database management system, geoobjects are identified via attribute values, even from several data layers, and then marked on the monitor, e.g. by a striking color. In most cases, an intuitive query manager is offered. The display of the objects on the monitor enables an initial spatial orientation, which is often followed by a graphic-interactive query of the attribute information of individual objects (clicking on an individual object and displaying the associated attribute values).

By specifying a geometric search area *geometric search queries* can be formulated. The search area is formed, for example, by specifying a search window, a search circle or any search polygon. The search area is not calculated, but set up graphically on the monitor with the mouse (e.g. "drawing up" a window).

9.4.3 Updating Geoobjects in the Vector Model

A geoinformation system has many options for *updating*: performing various file operations (including copying and deleting), modifying data (including inserting or deleting and, above all, updating geometry and attribute values of the geoobjects) and importing or exporting data. On the one hand, these tasks only affect the functions of the integrated database management system. On the other hand, typical for a geoinformation system are the modifications that involve changes of the geometries, which according to the data model, only mean a topological review of the geometries or an update of the topology. This includes simple adding a line, which divides an old polygon, or deleting a line, which

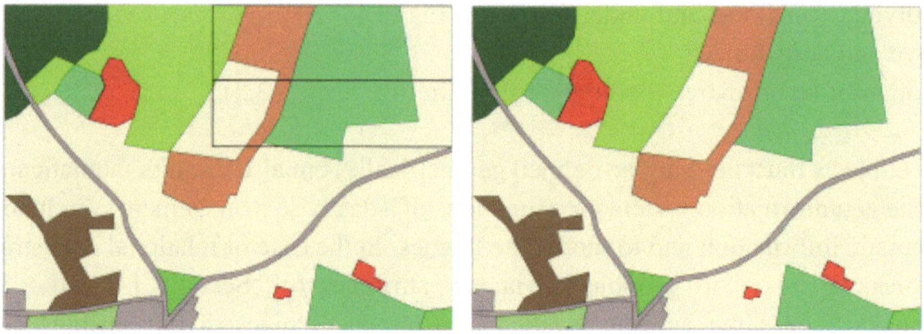

Fig. 9.12 Merging and dissolve adjacent polygons

merges two adjacent polygons. However, even simple changes of the attribute data can require changes of the geometries and then further an update of the topology. This includes, above all, the reclassification of geoobjects (e.g. merging of finely defined and then inevitably smaller subunits into coarser units).

The simple example of merging adjacent polygons illustrates very well what is special about the management of geoobjects in a geoinformation system (cf. Fig. 9.12). When merging adjacent polygons, the common polyline (the former dividing line) is omitted. A new polygon with a completely new shape (i.e. geometry) is created. In Fig. 9.12, for example, the partition of the areas in the northeast, which was installed for demonstration purposes, is removed again. In this way, the three individual green sub-areas are brought together again to form one large area (e.g. a meadow). Depending on the data model, new topological relationships of the polylines have to be established. The three data records that belong to the original polygons must be merged into one data record, since only one polygon remains. Especially this merging and updating is by no means trivial. In particular, it must be ensured in advance that there is a similarity in terms of content (here: color green, indicating a meadow), which permits the merging of adjacent polygons.

Functions for *map matching* are of particular importance. It is not uncommon for two data sets from adjacent map sheets to be merged. If the data were recorded in different coordinate systems, the coordinates must first be transformed so that both data sets are still separated in a uniform reference system, but are already "side by side". Both data sets can then be merged immediately, if exact map templates existed in each case, the data has been recorded without errors and the transformations into the common reference system have been carried out without distortion. However, if a data set was recorded on the basis of a distorted map (e.g. due to aging processes of the paper), there are considerable problems because the adjacent data sets do not fit exactly side by side. A geoinformation system usually has functions to "compress" and "tug" the geometries in a manner similar to a rubber sheet in order to achieve an adjustment.

This procedure is often described with the term "rubber sheeting". In order to carry out this function, clear connection and reference points must be available in both datasets (cf. Fig. 9.13).

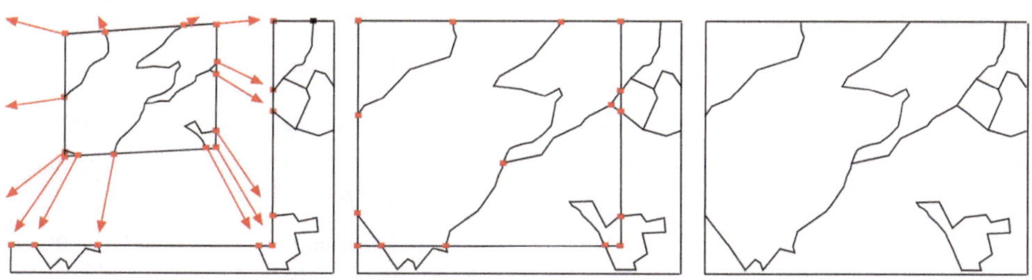

Fig. 9.13 Joining data sets from adjacent map sheets with geometrical adjustment

In the next steps, the two separate data sets will be merged into a single data set or data layer. Then the separation lines that still exist between the same characteristics will be eliminated, so that both geometrically and thematically "continuous" polygons and a dataset free of intersections are available.

9.4.4 Spatial Analyses of Vector Data

The *spatial analysis functions* can be divided into three large groups:

- generation of zones (so-called buffer functions)
- spatial selection and extraction (so-called extract functions)
- spatial overlay and intersection (so-called overlay functions).

When *generating zones,* an area is generated around the selected geoobjects (cf. Fig. 9.14). The old data layer remains unchanged, the result of *buffering* is therefore always a new polygon, which, however, has no attributes other than the size and the perimeter of the zone. Buffer zones are calculated (rigidly) according to the geometry of the geoobjects. A constant value or a numerical attribute of the selected geoobjects can be specified for the width of the buffer.

Application examples exist within the framework of urban land use planning (in North Rhine-Westphalia in Germany): maintaining a distance of 200 m between e.g. a furniture factory and a residential area or different distances form residential areas or single houses to new wind turbines. A geoinformation system is able to show distance zones. The road objects modeled as simple polylines in a geoinformation system are another application example. From this, road areas can be generated via buffer building. The width of the buffer for a line segment is determined by the associated value of the attribute "road width".

It should be pointed out that zone generation is relatively rigid and is not calculated based on modeling (such as in the calculation of flood areas of a river as a function of topography).

The attributes and attribute values are not changed when processing *extract functions.* These functions only modify the geometries of a data layer, e.g. by extracting inner parts (cf. Fig. 9.15). The associated attribute values are retained. Only the size and the perimeter of the partial areas are recalculated.

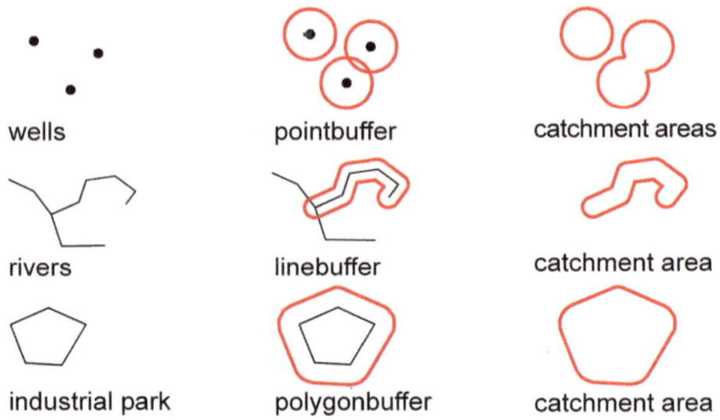

Fig. 9.14 Generating buffer zones

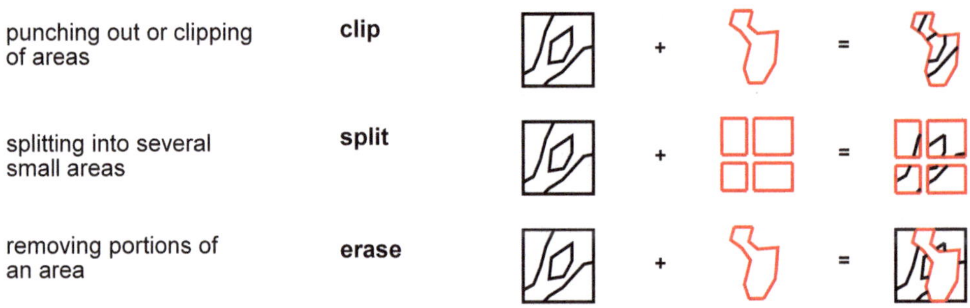

Fig. 9.15 Most important extract functions

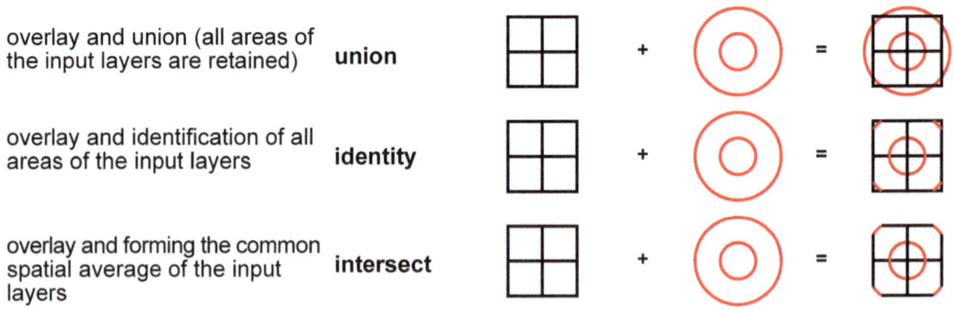

Fig. 9.16 Most important spatial overlay functions

In contrast to purely graphic overlays, *geometric-topological overlays* are of particular importance. Here, the data sets of the source data layers are linked and form a (new) data layer with geometries and attribute data (cf. Fig. 9.16). Here the functions "union" and "intersect" are of outstanding importance, which calculate the spatial union of layers or the common spatial average of layers in a result layer. The "identity" and "symmetrical difference" functions are also worth mentioning.

When intersecting areas that are not completely congruent, e.g. due to inaccurate recording of geoobjects or due to stronger generalisation of boundary lines, small residual polygons or sliver-polygons may occur. A geoinformation system should have functions to (automatically) eliminate such residuals (e.g. the function "eliminate" within ESRI's ArcGIS). For example, they can be added to the entity with which they have the largest common boundary.

In vector-based geoinformation systems, considerable effort is required to perform extract functions or to carry out the union or intersection of different layers, as the example in Sect. 3.4.1 shows (cf. Fig. 3.14). If one thematic layer represents land use types and the second layer represent parcels, a typical question is to determine the land uses on parcels of a specific owner. Then the boundaries of the land use types, which are formed by single line segments, have to be intersected with the boundaries of the parcels, which are also formed by single line segments. This job finally leads to the determination of intersection points of straight lines. So-called computational geometry has developed efficient algorithms (cf. Bentley-Ottman algorithm, cf. Sect. 3.4.1 and Fig. 3.16).

Compared to the complex processing of the geometries, the attributes and attribute values of this new data layer are only adopted from the layers involved, i.e. "inherited". Only the size and the perimeter of the new geoobjects are recalculated. Some oddities can arise here:

Figure 9.17 illustrates the type of "inheritance" implemented in many geoinformation systems. In this example, landuse information has to be intersected with ownership information (creating the common spatial average). Here, qualitative characteristics are assumed. After the intersection, the resulting entity with the identifier 4 has the properties "WL" and "X" of the two initial data layers (biotope type WL and owner X). In the attribute table of the resulting layer, the attributes of the input tables are simply "appended". For a new entity created by the intersection, the attribute values are taken from the corresponding entities of the input data layers.

Accordingly, the numerical attributes or attribute values are also "inherited". The figures are interpreted by the GIS software as qualitative data. This step easily leads to irritation: The input layer shows a total of 100 trees for parcel 20. After the intersection, the new

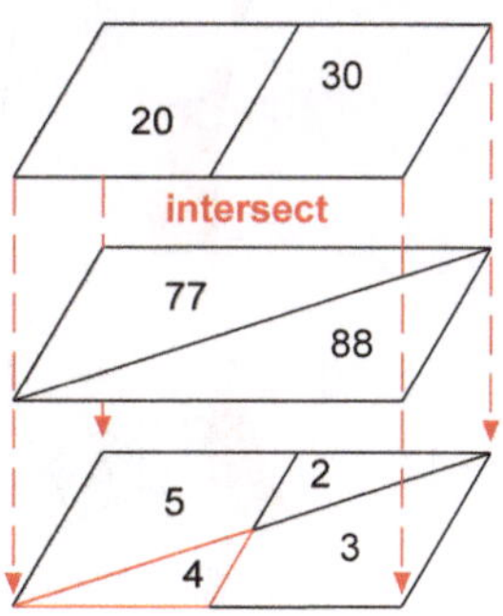

area	f-id	biotop	trees	tree density
	1			
100	20	WL	100	1
100	30	AS	10	0,1

area	f-no.	owner
	1	
100	88	X
100	77	Y

area	idd	f-id	biotop	trees	tree density	f-no.	owner
	1						
25	2	30	AS	10	0,1	77	Y
75	3	30	AS	10	0,1	88	X
25	4	20	WL	100	1	88	X
75	5	20	WL	100	1	77	Y

Fig. 9.17 Processing of the attributes of geoobjects in spatial intersections

attribute table also has the attribute "number of trees", which shows the values 100 for each of the new entities 4 and 5 in the left half. Calculated back, there would then be 200 trees in the former parcel 20, which consists of the new parcels 4 and 5. The attribute value "100" is passed on as if it were a name of the parcel. Absolute values are not converted according to their area proportions during the intersection. This is due to the fact that the conditions for such an assignment are not known. Only if it is assumed that the absolute data (e.g. the trees) are evenly distributed across the total area, an elegant solution strategy can be specified, which can be implemented automatically with GIS functions. It is not the absolute number of trees that is stored in the input data, but the "tree density". This numerical density value is correctly inherited to the corresponding entities of the resulting layer. If the trees are equally distributed across the area, this "tree density" is also present in the entities after the intersection. By multiplying this density with the respective size of the new entities, which is automatically recalculated by the software, the absolute number of trees for this entity can be estimated.

9.4.5 Spatial Analyses of Vector Data in Planning: An Example

Figure 9.18 shows a typical application of geometric-topological analysis functions. Two route alternatives for a bypass are to be compared in a preliminary study, a western and an

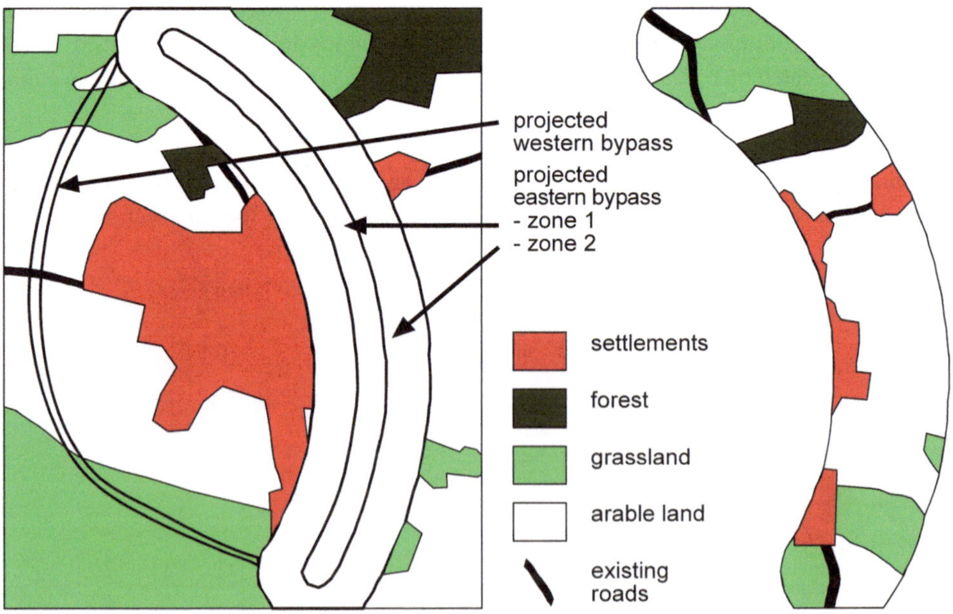

Fig. 9.18 Geometric-topological analysis in a real planning case

eastern bypass. Figure 9.18 only shows the situation in the east. On the one hand, the new road relieves the village because through traffic is routed around the village. On the other hand, valuable natural areas may be lost. Settlement areas and parts of the village population will also be affected by the traffic noise that will be caused by the bypass road. At an early planning stage the question is to be asked which of the two route alternatives is more suitable. However, it must be determined in advance and not only at the end of the analysis how the suitability is to be operationalised and, consequently, which route has the smallest impact.

In this example, the focus should first be on which areas or area categories are affected by the roads themselves and the construction project (road and hard shoulder on each side, noise barriers or cuts in the terrain, zone 1). It should be examined whether more areas in the western bypass are affected by biotopes worth protecting (i.e. the sum of areas worthy of protection). In this example only a very simple classification according to four area categories is used. In a real planning case, a biotope type survey would be carried out. In particular, the loss of natural areas and the extent of changes to existing biotopes are to be determined. One result could be that, for example, less forest areas will be lost due to the eastern bypass. Here, too, it must be defined beforehand that the loss of forest areas is an important decision-making criterion.

In this example secondly the noise propagation should be examined and analysed (zone 2). The width of this zone is determined on the basis of the traffic volume to be assumed and noise limit values that are only reached at a certain distance from the street. In Germany these values are set on a legal basis ("cf. Bundesimmissionsschutzgesetz", federal immission control law). It needs to be analysed, how many settlement areas are affected by each of the two alternatives. During more precise steps the settlement area can be differentiated according to different housing densities in order to estimate the number of residents affected.

The assessment of the environmental impact requires determining and comparing the possible changes in several impact zones along the two route alternatives. Buffer zones are generated around the routes (cf. Fig. 9.19). These buffers are intersected with the layer of the existing land uses (cf. Fig. 9.20).

It must be emphasised that with the tools of a geoinformation system not only a purely graphic overlay of the different layers or buffers is possible. A new layer is created by intersection (spatial average of the input layers), which shows the different land uses in the buffer zones (cf. Fig. 9.18). Thus, area balances can be calculated for the buffer zones on the western and eastern sides. On the basis of these area balances, it can be estimated, which of the two route alternatives is more suitable.

This quite simple procedure, which can be automated in many steps, has the advantage of providing a decision making support being relatively fast and, above all, inexpensive.

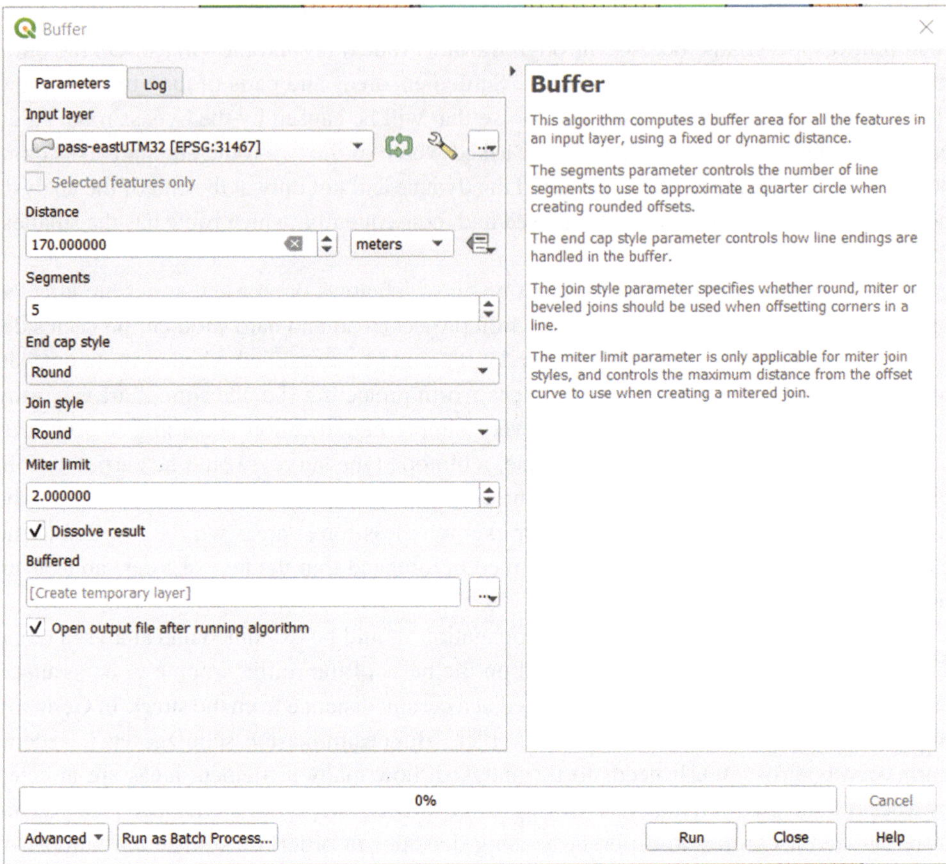

Fig. 9.19 Creating buffer in QGIS

In environmental planning this approach is applied in many ways. A common use is determining suitable areas for wind turbines. German environmental policy defines areas in which wind turbines must not be erected: e.g. legally protected biotopes or parts of open space, e.g. buffer zones to residential areas or breeding grounds of sensitive bird species. In a geoinformation system these buffer zones and all restricted areas on county basis are determined and put together (analysis function “union”). The analysis function “erase” cuts out the then newly united restricted areas from the county area. By this process the remaining area can be regarded as a potential area for wind turbines.

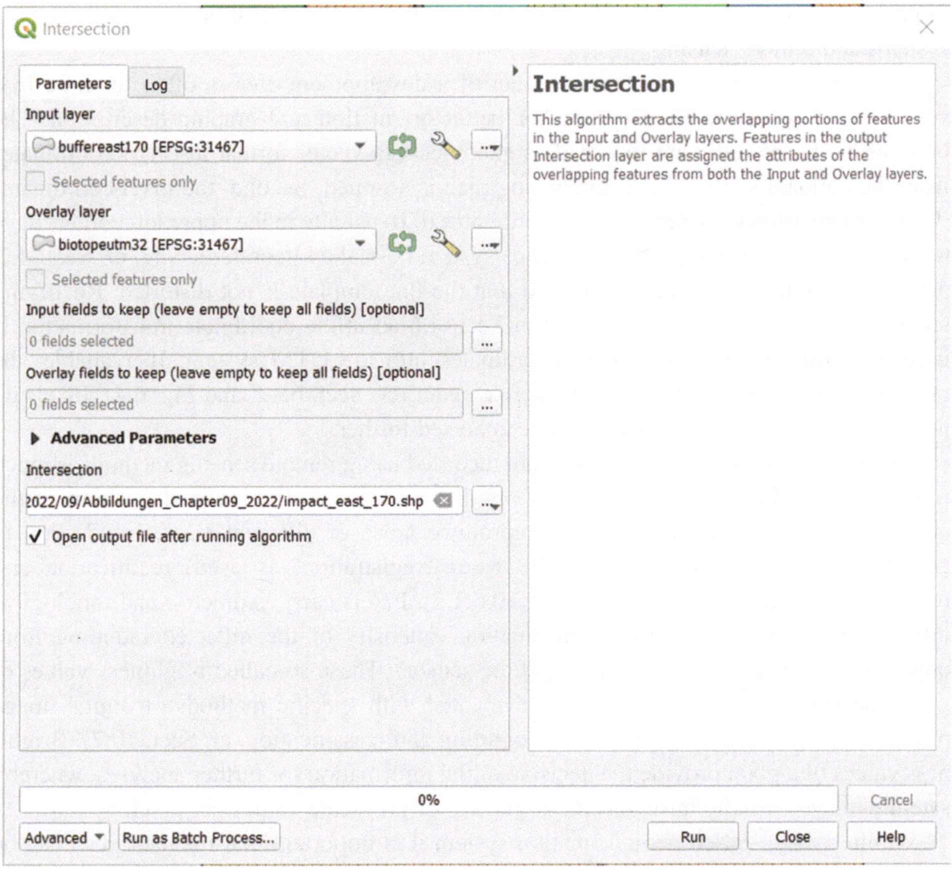

Fig. 9.20 Intersection of two input layers in QGIS

9.5 Raster Model: Spatial Analysis of Geoobjects

9.5.1 Raster Model: Recording and Editing Geoobjects

Raster data can be recorded directly in raster format via scanner systems (cf. Sect. 5.2.2). Data are imported from external systems that generate image or raster data:

- recording images with a digital camera,
- scanning analog paper templates, pictures or slides,
- recording remote sensing data with scanner systems on drones, aircrafts or satellites.

In practice there are two main groups of tasks, but related to vector based geoinformation systems and remote sensing:

A background map such as a site plan of a development area is often required in a vector-based geoinformation system for better orientation and graphic description. The building footprints only are recorded as geoobjects in vector format in UTM coordinates from the national survey. First, the analog plan is scanned. Second, the pixel coordinates (Cartesian coordinate system, origin coordinates (0,0) usually in the upper left corner) have to be transformed into a general reference system (so-called georeferencing, cf. Sect. 4.2. 5.1). In this situation it can be assumed that the flat template is not distorted. Rectifying templates are not required. As shown in Sect. 4.6 an affine coordinate transformation is used to transform the initial pixel coordinates into the UTM system. If available, the integration of a Web Map Service is more elegant (cf. Sect. 6.4.8 and Fig. 6.4). Pixels do not carry any thematic information to be analysed further.

The processing of raster data which are recorded using remote sensing methods is much more complex. For example, the image is distorted due to the inclined position of the aircraft. The pixels of the same size reproduce areas of different sizes. In addition to georeferencing (in remote sensing the word "registration" is used) rectification and resampling become necessary (cf. Sect. 10.6.1.2). Pixels carry geometric and topological information as well as thematic information (intensity of the reflected radiation from objects on the earth's surface arriving at the sensor). These so-called brightness values or greyscale values (cf. Sect. 10.5.1) are evaluated with specific methods of digital image processing (e.g. methods for pattern recognition and classification, cf. Sect. 10.7). Brightness values of a pixel provide the decisive initial information for further analyses, whereby satellite images usually have several brightness or greyscale values per pixel.

Within a raster-based geoinformation system it is important that thematic data can be assigned to a grid or calculated for a grid. Two examples illustrate typical cases of processing attribute data on a raster basis:

In an emissions inventory, the geoobjects for which emission data are available or can be calculated are usually initially available in their original spatial reference, i.e. as point or line or areal objects. Emission values can be estimated for individual groups of emitters, such as domestic heating systems, small businesses or power plants, which are then stored in a vector based geoinformation system with very different spatial references. The conversion to a uniform spatial reference system becomes necessary, if the distribution of a single air pollutant is to be represented for all groups of emitters. Aggregation can only be done on a new, "neutral" grid basis. The spread of pollutants is also modeled using a grid. The immission data are not calculated for a single point or for any area, but for a grid (cf. Berlekamp et al. 2000 as well as the particle model AUSTAL 2000 for calculating the dispersion of dust and gaseous emissions as well as odours cf. Umweltbundesamt 2023).

In order to assess the groundwater situation for planning projects and the suitability of building sites, a so-called groundwater distance map must be created, which shows the distance between the groundwater and the surface. For this purpose, the groundwater surface and the surface of the terrain are discretised by point data. The groundwater level is only available for an irregular network of individual groundwater measuring points. Elevation data can be obtained for grids of different mesh sizes from the surveying authorities. Thus, it is obvious to model both surfaces using raster data. The grid provides the common spatial reference basis. The thematic data for a grid cell are, on the one hand, the terrain heights and, on the other hand, the groundwater levels. Using the methods of map algebra (cf. Sects. 9.5.3 and 9.5.4), the difference between terrain height and groundwater level (each in m above sea level) can be calculated in pairs for two grid cells. The difference represents the distance of the groundwater from the surface of the terrain (on a grid basis). However, this procedure requires the calculation of the groundwater levels for all grid cells in the study area from the measured data at very few groundwater measuring points. For this purpose geoinformation systems offer various methods of spatial interpolation (cf. Sect. 9.7).

Both examples have in common that the initial information is available as vector data and that the problem can only be solved using raster-based methods. Due to the uniform, simple and rigid grid, preparing and editing the geometries is less relevant than the management of the thematic data on a grid basis. First of all, the thematic data available for different geoobjects (points, lines, polygons) must be assigned to individual grid cells.

9.5.2 Exchange Thematic Data Between Vector and Raster Model

When converting information of a curved line (e.g. a river) or an irregular shaped polygon (e.g. a lake) of the real world into a regular grid of a raster-based geoinformation system, the often problem arises that a grid cell contains multiple items of initial information, since the generally irregularly shaped lines or areas are difficult to approximate with a grid. These assignment problems are fundamental greater with a coarser cell structure, but a finer grid does not solve the principal problem.

Figure 9.21 illustrates a data conversion that proceeds row-by-row, where a cell retains the property of the source data layer that occupies most of the cell. For example, if a cell consists of 35% maple and 45% coniferous forest, this method assigns the cell to coniferous forest. Another method can be used to determine the property that should be assigned first or with a higher weight to a cell. If, for example, a rare plant species was found in a

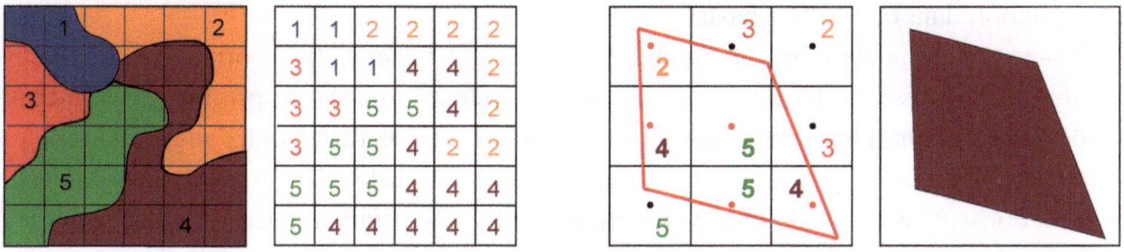

Fig. 9.21 Conversion of attribute data: vector-raster and raster-vector conversion

vegetation survey, this species can be given priority over the other plant species when converting to raster data.

A geoinformation system also provides functions for the reverse case, where thematic data are available for raster cells to which areas or polygons must be assigned in vector format. For example, a thermal image can be determined with a thermal scanner used on an aircraft that measures the radiation temperature of surfaces. From this, surface temperatures can be determined for an urban area on a grid basis. This thermal data can be assigned to building areas, whose outlines originate from a digital property map (cf. in Germany ALKIS in Sect. 5.5.4.4, cf. Fig. 10.17). To solve this conversion the so-called point method is used almost exclusively in practice, in which the geoobjects are intersected with the centers of the grid cells. If the center coordinate of a grid cell is within a geoobjects (e.g. a building), the value of this cell is used to calculate the attribute value for this geoobjects (e.g. to determine a mean surface temperature by averaging the grid-related attribute data). Figure 9.21 shows in a simplified way how the new value for the right brown area can be calculated by averaging the attribute data of the grid cells involved (mean of the values 2, 4, 5, 5 and 4).

9.5.3 Spatial Analyses of Raster Data

On the one hand there are methods for the spatial analysis of raster data that refer to the raster geometry and are comparable to the vector-based spatial analysis methods discussed in Sect. 9.3.4 (cf. Figs. 9.14, 9.15 and 9.16). On the other hand, methods are available which only refer to attribute values in matrix form and thereby abstract from the specific shape and size of a grid cell. It should be pointed out that many functions for raster data originate from image processing, which often explains the terminology (cf. filter operations, naming attribute values as brightness or greyscale values, cf. Sect. 10.6).

The focus here is not on typical image processing and image analysis functions (cf. Sect. 10.6), but rather on functions that emulate methods of spatial analysis in vector-based geoinformation systems. However, individual functions cannot be clearly restricted to one field. For example, thresholding is a method for image improvement in which brightness or greyscale values below a limit are assigned the value zero and interpreted as image errors, and another constant value is assigned above this limit. The application of such local operators is also used to classify raster data such as the spatial distribution of elevation or precipitation data on a raster basis.

Zones are generated by thickening (or thinning) raster cells. Corresponding to the operation in a vector-based geoinformation system, this is a primarily geometric-topological function in which the attribute values of the raster cells may be used for selection, but are otherwise not further considered. Likewise, spatial overlays and intersections as well as boundary functions can to be implemented by simple logical operations.

To illustrate these functions, several conventions are to be made: The neighborhood of pixels is defined by common edges (cf. Sect. 9.3.4). The input and the result data layers are denoted by InGrid(i,j) and OutGrid(i,j), respectively. The attribute values of a grid cell have the value 0 or a value ≥1, which are assigned the logical values "false" or "true", respectively. Then applies:

$$\begin{aligned} &\text{external buffer(blow)}:\\ &\text{OutGrid}(i,j) = \text{InGrid}(i,j) \quad \text{or} \quad \text{InGrid}(i+1,j) \quad \text{or} \quad \text{InGrid}(i-1,j)\\ &\qquad\qquad\qquad\qquad\qquad\quad \text{or} \quad \text{InGrid}(i,j+1) \quad \text{or} \quad \text{InGrid}(i,j-1) \end{aligned}$$

$$\begin{aligned} &\text{internal buffer(shrink)}:\\ &\text{OutGrid}(i,j) = \text{InGrid}(i,j) \quad \text{and} \quad \text{InGrid}(i+1,j) \quad \text{and} \quad \text{InGrid}(i-1,j)\\ &\qquad\qquad\qquad\qquad\qquad\quad \text{and} \quad \text{InGrid}(i,j+1) \quad \text{and} \quad \text{InGrid}(i,j-1) \end{aligned}$$

If other neighborhoods are used as a basis, further thickening and thinning operations can be defined with little additional effort.

As Fig. 9.22 shows, these functions have special applications in image processing. Thus, gaps in the original image can be closed by blowing and subsequent shrinking. Due to the reverse procedure (first shrinking, then blowing) generalisation effects of the geometry can be achieved. In this case, however, attribute values have to be recalculated by suitable functions.

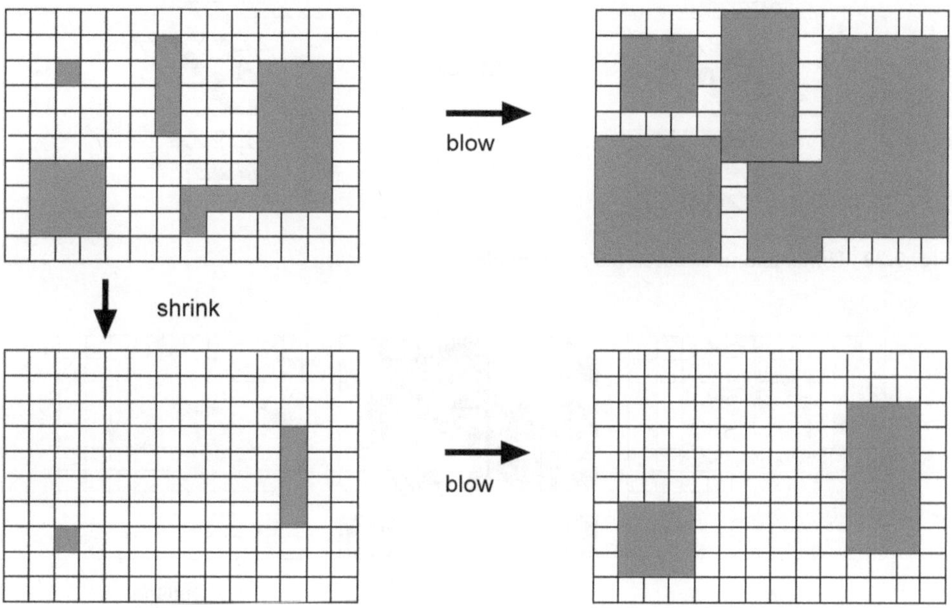

Fig. 9.22 Blow and shrink

The following applies for spatial overlays and intersections (cf. Fig. 9.23):

union of pixels of several layers (union):

$$\mathrm{OutGrid(i,j) = InGrid1\quad(i,j)\qquad or\qquad InGrid2(i,j)\qquad or\quad \ldots}$$

spatial average of pixels of several layers (intersect) (cf. Fig. 9.23):

$$\mathrm{OutGrid(i,j) = InGrid1(i,j)\qquad and\qquad InGrid2(i,j)\quad and\quad \ldots}$$

The “punching out” of areas or the “clipping” of thematic layers at the edge (for similar functions cf. Fig. 9.15) is done using local logical operators. A grid which has only the logical values “true” or “false” as attribute values serves as a mask, which covers the input grid (so-called masking, cf. Fig. 9.24):

masking:

$$\mathrm{OutGrid(i,j) = InGrid(i,j) if\ \ Mask(i,j)}$$

Fig. 9.23 Union and intersect in the raster model

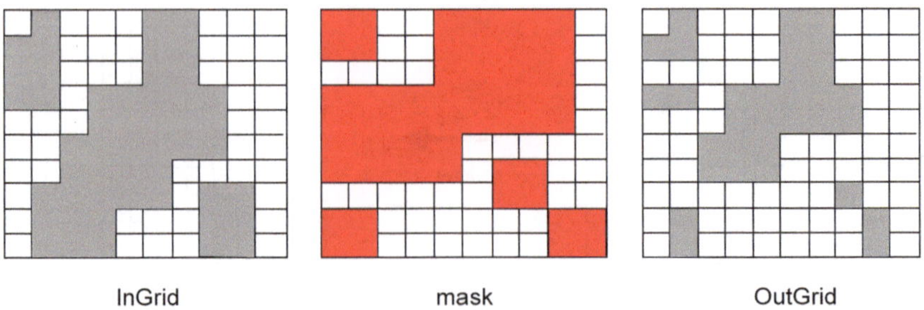

Fig. 9.24 Masking

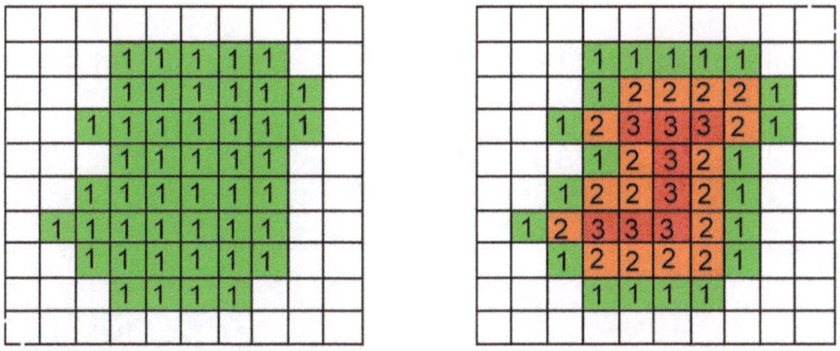

Fig. 9.25 Forming the distance transform

Another important function of processing raster geometries is the calculation of the *distance transform*. This is based on (a group of connected) raster cells that have the same value, e.g. to have the same topic. According to the underlying metric, the distance to the edge of the area is calculated for each of these grid cells. They form the distance transform, whose greyscale value are to be interpreted as distances. Therefore new values are created, which become larger and larger from the outside to the inside. The maximum values of this distance mountain are also called the *skeleton of* the raster structure.

The *distance transform* can be determined very easily by several logical operations to be carried out one after the other. For this purpose, the group of raster cells is gradually reduced by one cell width until it dissolves. The single thinnings are then added together by the logical "or" operation. Figure 9.25 shows the principle of thinning and the addition of the thinnings (cf. Zhang Suen algorithm in Sect. 3.4.3 and Fig. 3.22).

9.5.4 Map Algebra

The processing of raster data, i.e. attribute values on a raster basis, is characterised by the fact that several data layers have an identical spatial reference basis. Any spatial sections and their attribute values can now be easily linked to one another. It is irrelevant whether brightness values of an image matrix or thematically specified attribute values are available on a raster basis. Ultimately, processing this raster information boils down to processing matrices of numbers. Certain operators (links) are executed on matrices of numbers so that the result is again a matrix of numbers (cf. Fig. 9.26).

According to Tomlin (1990), the calculus is called *map algebra* in analogy to number algebra. It should be emphasised that only a limited number of operators is necessary to cover all processing options of raster data. Tomlin speaks of 59 functions like "ZonalSum", "LocalRatio" or "IncrementalDrainage" and of additional operators, e.g. to operationalize the distance or the direction of the focal functions. Figure 9.26 illustrates the underlying principle, where two data layers are linked with an arithmetic operator. Logical links can also be represented in a similar way, e.g. implementing intersections of several data layers

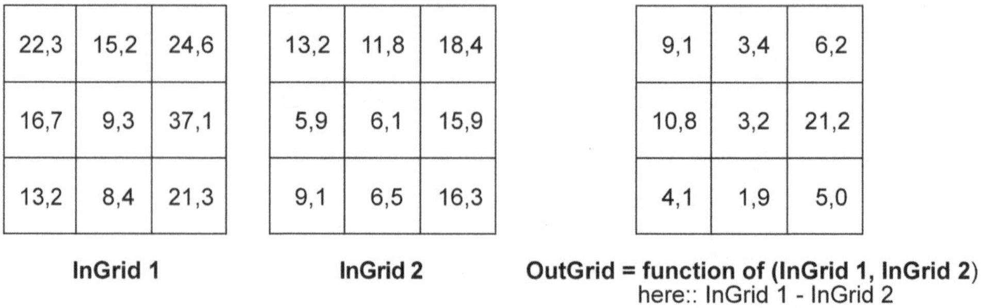

Fig. 9.26 Principle of raster data processing with Map Algebra

(cf. Fig. 9.23). The example given in Sect. 9.5.1, which explains how to create a groundwater distance map, illustrates a use case.

In the original version, Tomlin (1990) addresses four groups of functions: local, focal, zonal, and incremental functions, but no global function as is often called:

- *local operators* consider exactly one cell, possibly in several thematic layers at the same position. The values of adjacent cells have no influence. Examples are logical or algebraic operators such as comparisons or addition of attribute values and recoding or reclassification (cf. Fig. 9.26). For example, the command "newlayer = LocalRating of firstlayer with 0 for 5. . ." assigns the value 0 to all cells in the new layer that have a value greater than or equal to 5 in the source layer.
- *focal operators* each refer to a fixed environment of a cell. For example, the N.4 neighbors of a cell are included (cf. Sect. 9.3.4). Neighborhoods can have the form of a circle, rectangle, ring or wedge. For example, the command "newlayer = FocalProximity of firstlayer" assigns to all cells in the new layer the distance to the cells that have a value in the source layer, i.e. a value not equal to −0 ("nodata"). This can be used to implement the accessibility of a cell from other cells and the creation of buffer zones. Other important examples are the so-called filter functions, which play a major role in image processing (cf. Sect. 10.6.4), as well as the calculation of slope and aspect on the basis of elevation data from raster data.
- *zonal operators* calculate values from cells of a second thematic layer within a predefined area (i.e. a zone) of a first thematic layer.
- *incremental operators* proceed along given one-, two- or three-dimensional geoobjects (e.g. along a cell chain or across a terrain). The calculations of drainage directions and drainage paths based on (terrain) elevation data in raster form are important examples.

The implementation in many geoinformation systems is very different. The original commands according to Tomlin are often replaced by graphical tools or a so-called "raster calculator" (cf. Fig. 9.27).

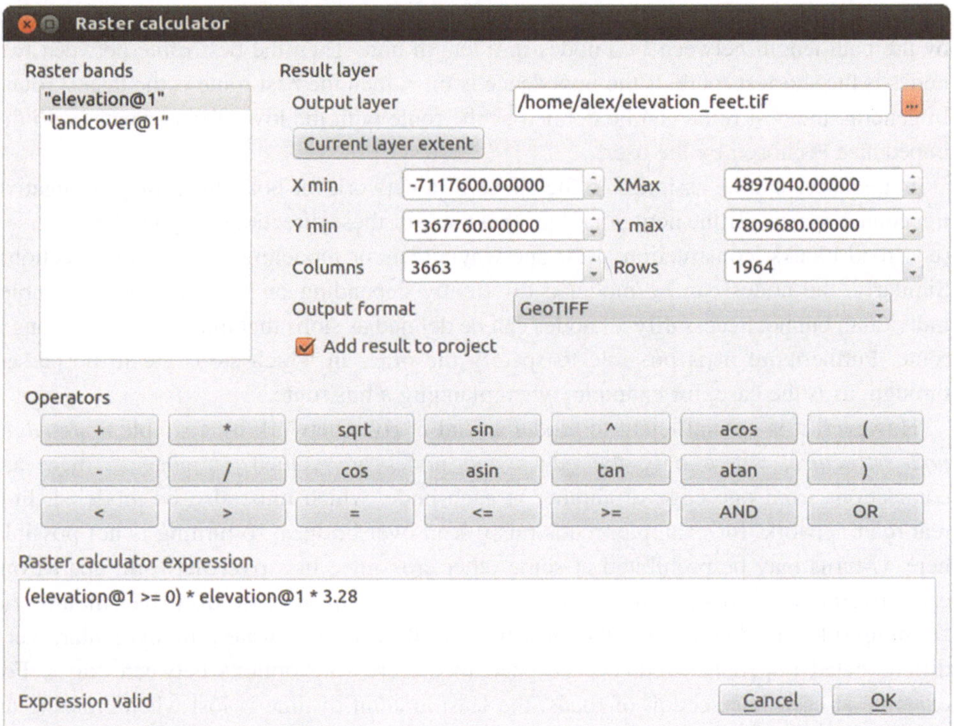

Fig. 9.27 Raster calculator in QGIS

9.6 Network Analyses

9.6.1 Network Data Model

The analysis of networks is one of the central applications of geoinformation systems. Networks play an important role in very many tasks, especially in transportation. They model traffic systems such as road or rail networks, but also line networks such as pipeline networks of supply and disposal companies or telecommunication line networks.

Formally, *networks* are defined as sets of nodes and edges. They belong to graphs, although in practice mostly only asymmetric and weighted (or valued) graphs occur (cf. Sect. 3.4.2). The modeling and analysis of networks is based on graph theory. Networks have a *node-edge-node topology* (cf. Sect. 9.3.2). They are based on the vector model.

The edges in a network represent linear entities such as roads, railways or shipping lines for a transportation network as well as traces of an electrical transmission network or the flows of a river network. The nodes of the network are, for example, bus stops or general connection points such as crossings. The edges, i.e. the segments, carry so-called cost attributes, which are to be regarded as impedances. Usually these impedances are defined by the path length between two nodes in a length unit. Then the best route between two nodes is the shortest route. If the impedance is time, then the best route is the fastest route. In general, the best route can be defined as the route with the lowest impedance, with the impedance is chosen by the user.

Impedances can be assigned to the edges of a network for both directions. A negative impedance means for the network edges that these or these directions may not be traversed (e.g. road blocks, construction sites, one-way streets or modeling of the flow direction). Similarly, the nodes can be modeled differently depending on the topic. For example, individual, but not necessarily all nodes can be defined as stops that must be visited along a route. Furthermore, it is possible to specify the order in which stops are to be passed through, as is the case, for example, when planning a bus route.

However, it is not sufficient to model a road or river network by a simple *node-edge-node structure*, where only the edges and nodes are valued. In general, there are connections from one edge to another at each node, which must also be modeled. In a real road network, for example, roads (also) lead over bridges, so turning is not possible here. U-turns may be prohibited at some other crossings. In a river network, apart from canal overpasses, crossings over bridges do not occur. In the case of a river mouth, the "turning rule" is dictated by the direction of flow of the water. In particular, very differentiated impedances can be assigned to the crossing options between edges. For example at each intersection of roads, the least amount of time is lost when crossing an intersection with a given right of way, but the most time is lost when a traffic light with a long red time delays driving on. At each intersection, n^2 possible crossing options must then be quantified, where n represents the number of edges connected at an intersection (per edge $n - 1$ turns and one u-turn). Overall, traffic networks often have very complex situations that are difficult to model: single-lane motorways, each allowing only one direction of travel, overpasses, multi-lane roads with different turning characteristics, motorway junctions, roundabouts. It has to be pointed out that the modeling of these connections is very complex, but indispensable.

9.6.2 Analysis of Optimal Paths in a Network

The standard tasks include determining the shortest paths between two nodes in a network and the solving the so-called *traveling salesman problem*, in which an optimal route is determined through several locations that leads back to the starting point. Such tasks are solved on the basis of path finding algorithms from graph theory (cf. Sect. 3.4.2).

Figure 9.28 shows the shortest route between two points in the road network of the city of Osnabrück. The freely available OSM data are the basis of the modeling (cf. Sect. 5.6.2). The evaluation was carried out according to travel time and (only) taking into account turning regulations. In addition, the network model can be further refined. For example, one-way streets or "speed 30 zones" can also be modeled, as well as the turning times. The integration of the current traffic situation is also possible.

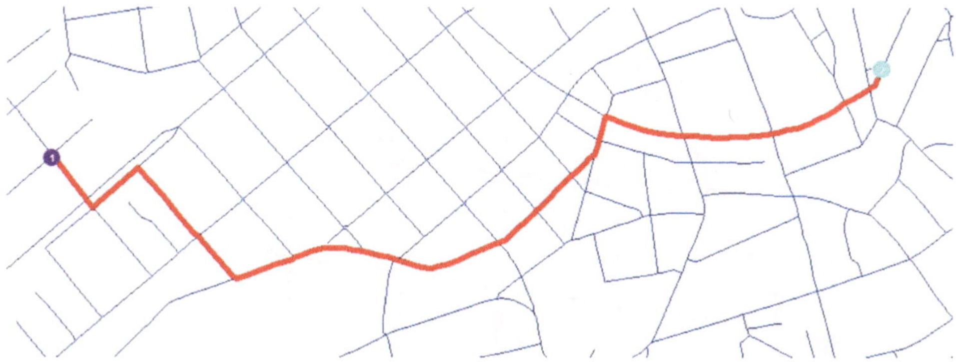

Fig. 9.28 Shortest path between two points

9.6.3 Generating Service Areas

With network analysis methods so-called *allocation problems* can be processed, which assign specific network sections to a node that can be reached from this node at a certain distance. A service area is a region that can be reached from a location within a specified travel time or distance. In a case study, the accessibility of basic food supply facilities is to be analysed.

First, roads and paths are to be determined via which a given point in the network can be reached in less than 500 m (as the maximum walking distance to be assumed for basic suppliers, cf. Fig. 9.29). A simple circle with a radius of 500 m should not be drawn around a location. Instead, the focus here is on accessibility along roads and paths in a specific transport network. For each location, the accessibility is determined on the edges of the network (so-called supply network), whereby the connected paths or edges are determined, which together do not lead further than 500 m away from a point. In the easiest case, only the footpaths are considered, so complex modeling of junctions or one-way streets is not required. Subsequently, the catchment area (service area) can be determined by a polygon that encloses the calculated access routes. To do this, all end points of the supply network can be connected by straight lines, this is the convex hull.

Once a service area is defined, the next step is to calculate the size of the population living there. To do this, the service area is intersected with the data layer that identifies residential buildings with the associated resident population. The intersection of both thematic layers is a standard job of a geoinformation system. In contrast, the allocation of population data to residential buildings is not a simple task, although in practice there are many applications. Therefore, standard solutions should be available for many planning tasks: analysis of the supply level of households in a city or their accessibility with regard to food suppliers (cf. Hackmann and de Lange 2001), locations of family physicians or the fire brigade. In a scenario, it could be analysed and quantified how the supply level would change, if an acute care hospital is closed. In order to solve such tasks, data from the

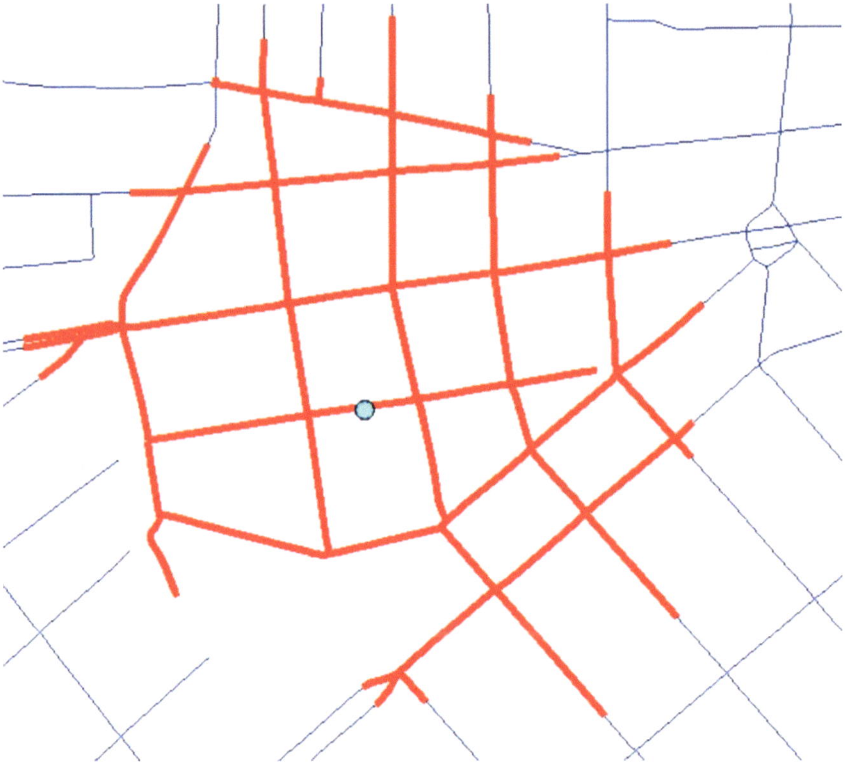

Fig. 9.29 500 m service area of the Weidencarree in Osnabrück

residents' registration office would have to be digitally accessible, ideally differentiated according to various population characteristics. This contradicts data protection, so that as a rule this sensitive data can only be accessed in emergencies (e.g. evacuation of the population in the event of a bomb find). One can try to estimate individual population data such as number or (rough) age composition based on the number of buildings and an assumed average composition of households per building. These estimates are often sufficient for marketing purposes.

9.6.4 Further Analysis Options in a Network

Geoinformation systems offer further tools for network analysis. Thus, in accordance with the traversal of a tree (cf. Sect. 3.2.4.4) all connected sequences of edges starting from a node can be determined. One application arises when, following the discharge of a pollutant, all river segments downstream in a river network are to be identified into which the pollutant could enter. Furthermore, not only the optimal route between two given nodes can be calculated. Using a similar methodology, it is also possible for a single node, e.g. an accident site, to determine the location of an emergency vehicle that can reach the accident site as quickly as possible.

Further analyses are possible, if evaluations of the nodes are taken into account. An application arises, when several locations with different demand have to be supplied and an optimal route of a transporter with limited loading capacity has to be calculated. In another use case, the value of a node can quantify an offer available there. Network analysis tools continuously allocate network edges to such a node that are connected and lead to the node. At each step, the edge that adds the smallest increase in distance from the node is added. This process continues until demand at the edges meets supply. In an application, a node may represent a primary school with an offer for a certain number of pupils. Street segments with elementary school pupils living there are assigned to this school in order to work out new school district delimitations.

9.7 Spatial Interpolation and Modeling of Surfaces

9.7.1 Initial Issues

The range of tools of a complex geoinformation system includes methods for *spatial interpolation* and for *modeling 3D-surfaces*. In principle, the associated methods aim at similar questions: Starting from a few points (x_i,y_i), which are spatially distributed with attribute values z_i (e.g. precipitation or elevation data), attribute values z_k are to be determined for further points (x_k,y_k). The unknown z_k values at these points are to be obtained from the existing z_i values. Within raster data processing, the similar task starts from a few values for individual raster cells and then determines attribute values for the remaining raster cells. Finally, an unknown continuous spatial distribution is discretised.

Interpolation procedures are often used for this purpose. As a basic assumption, it is usually implicitly assumed that those locations (or the associated values) have a stronger influence on the value sought at a new location, which are closer to it. The interpolation procedures lead to the determination of weighted mean values. The interpolation methods presented here, however, do not have any assumptions made about the underlying distribution model of the data, i.e. the z_i values. Instead of, for example, interpolating the groundwater level from a few measured data, a spatial model for simulating the groundwater level and the flow direction of the water would alternatively have to be developed. However, such procedures are very complex, so that mostly simplified interpolation procedures as presented here are used.

In another approach, it can be assumed that all z_i values are located on a surface. This procedure is obvious for elevation data, from which a *digital terrain model* (*DTM* or *DEM* for *digital elevation model*) is to be developed. This modeling of a 3D-surface can be transferred to other data with x_i,y_i coordinates with associated attribute data z_i and therefore to general value surfaces: spatial distribution of e.g. precipitation, groundwater levels or land prices. Here, a spatial process is presupposed or assumed which leads to continuous or steady changes of the values.

9.7.2 Spatial Approximation and Trend Surface Analysis

The *trend surface analysis* computes an nth-order polynomial that best approximates the z_i values of the observation points:

$$f(x, y) = a_o + a_1x + a_2y + a_3xy + a_4x^2 + a_5y^2 + a_6x^2y + a_7xy^2 + a_8x^3 + \ldots$$

Such a polynomial represents a 3D-surface, so that the approach is quite clear. An attempt is made to approximate the observed values z_i at the points (x_i,y_i) using a continuous surface. The coefficients are determined in such a way that the sum of the squares of the deviation between the observed values and the calculated values at the observation points is minimal (global fit, for the calculation cf. Sect. 4.2.5.5):

$$f(x_i, y_i) = \widehat{z}_i \cong z_i \text{ with} \sum (\widehat{z}_i - z_i)^2 \text{ minimal}$$

The simplest case of a linear trend estimation (cf. Fig. 9.30), where a plane is placed through the points, is equivalent to a linear regression with two predictor variables (here: x-, y-coordinates). Thus, a global trend can be recorded. In the case of high-order polynomials, the sum of the squared deviations becomes smaller, but these polynomials

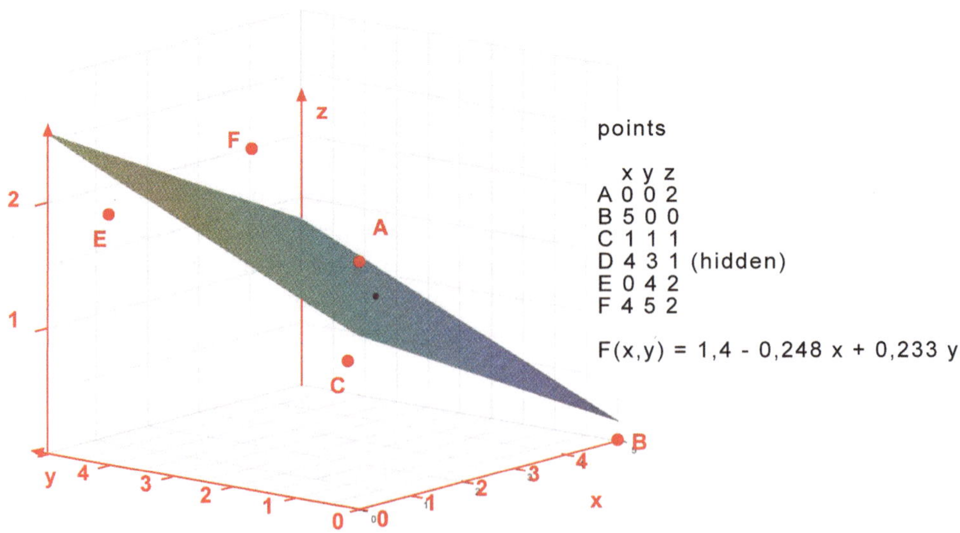

Fig. 9.30 Linear trend surface for a simple example

(or the surfaces) tend to oscillate between the observation or measurement points. Thus, mostly only trend surfaces up to the third order are used.

Trend surface analysis is based on the assumption that random fluctuations or measurement errors occur in the observed values, which can be compensated by a spatial approximation. By inserting the coordinate values of any points (x_k,y_k) into the equation, z_k-values can be calculated. However, the trend surface analysis is less used for interpolation. Rather, a general relationship between the location and the observed values can be determined. Then, similar to the analysis of residuals in the context of a regression analysis, local singularities (e.g. spatial precipitation anomalies) can be worked out.

9.7.3 Spatial Interpolation: Weighted Average

The interpolation methods, which estimate an unknown value for a point P_k by a weighted average of the measured or observed values at the nearest points P_i, are of great practical importance:

$$z_0 = \frac{\sum_i f(d_{0,i}) \cdot z_i}{\sum_k f(d_{0,k})} \quad \text{with a simplified weighting function f} : z_0 = \frac{\sum_i d_{0,i}^{-\alpha} \cdot z_i}{\sum_k d_{0,k}^{-\alpha}}$$

with

z_0 unknown but estimated value for the point $P_0(x_0,y_0)$
z_i value for the observation points $P_i(x_i,y_i)$
$d_{0,i}$ distance between $P_0(x_0,y_0)$ and $P_i(x_i,y_i)$
α weighting factor

The spatial interpolation approaches using weighted averaging differ with regard to the determination of the weights and the selection of the neighboring points. As a rule, not all initial values are included in the averaging, but only the values within a predefined radius to the point P_0. In this way, spatially distant outliers can be excluded. The weights are determined from the distances to the observation points. The most common weight functions are:

$$f(d) = d^{-\alpha} \quad \text{inverse distance}$$

$$f(d) = e^{-\alpha \cdot d \cdot d} \quad \text{Gaussian bell curve}$$

The weighting function as well as the values for the coefficient α in the exponent as well as the number of influencing values or the size of the catchment area are estimated or given due to the specific issue. It is easy to see that as α increases, the influence of more distant points decreases and that of the neighboring points increases.

In the simplest case, the inverse distances of the observation points to the point at which a value is to be estimated are used as weights (*inverse distance weighting*, *idw-interpolation*). Very common is the inverse distance with $\alpha = 2$, i.e. the inverse squared linear distance. The points that are further away are therefore given a lower weight.

The formula that appears complex can be written much more clearly for this standard case:

$$z_0 = \frac{\frac{1}{d_{0,1}^2} \cdot z_1 + \frac{1}{d_{0,2}^2} \cdot z_2 + \frac{1}{d_{0,3}^2} \cdot z_3 + \frac{1}{d_{0,4}^2} \cdot z_4 + \ldots}{\frac{1}{d_{0,1}^2} + \frac{1}{d_{0,2}^2} + \frac{1}{d_{0,3}^2} + \frac{1}{d_{0,4}^2} + \ldots}$$

It is assumed that each input point has a local influence that decreases with increasing distance. Thus, it is assumed that there is a similarity between spatially adjacent points. These methods are therefore only useful, if this basic assumption can be accepted. The calculation example for Fig. 9.31 illustrates the estimation of the daily amount of fine particles at an inaccessible location in a large city on the basis of four observation points. In order to classify the z_i values, it must be taken into account that since January 1, 2005 (Europe-wide) the value of 50 micrograms of fine particles per cubic meter of air may be exceeded on a maximum of 35 days per year.

Most likely, the weighted arithmetic mean comes closest to the unknown value, since the site is closest to the two sites P_3 and P_4 with higher fine particles values. However, the value for the arithmetic mean does not seem to be such a bad estimator, which in this case only slightly underestimates the presumably true value by averaging.

With these interpolation methods, which are not based on geostatistical models, the operationalisation of the spatial correlation or similarity is to a certain extent independent of the data, namely solely from the geometric position (distance) of the points to one another. Although the weighting function is determined on the basis of the specific issue (e.g. idw with with $\alpha = 2$), the actual spatial correlation of the z_i values is not taken into

P_i	x_i	y_i	z_i	$d(P_0, P_i)$	$d^{-1}(P_0, P_i)$	$z_i \cdot d^{-1}(P_0, P_i)$
1	1	1	34	5	0,2	6,8000
2	3	7	28	3,61	0,2770	7,7560
3	6	2	54	2,24	0,4464	24,1056
4	8	5	40	3,16	0,3165	12,6600
sum			156		1.2399	51.3216

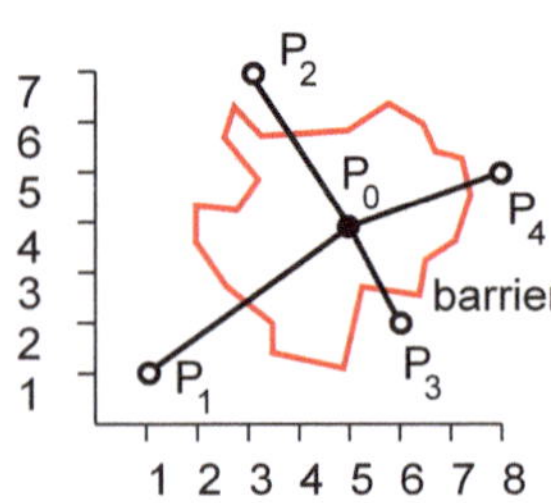

arithmetic mean for P(5,4) = 156/4 = 39.0

weighted arithmetic mean for P(5,4) = 51.3216 / 1.2399 = 41.39

Fig. 9.31 Spatial interpolation by weighting with inverse (reciprocal) distances

account, which can be different for various topics or variables and even e.g. for different precipitation events. Furthermore, in the standard case, it is assumed that the distance dependency is the same in all directions. In contrast, interpolation based on the *Kriging concept* represents a complex advancement. Here, too, weighted arithmetic mean values are calculated, but the weights are redetermined for each interpolation point using the z_i values, so that the spatial relationship of the data itself is taken into account and not just the location of their observation points. In particular, the statistical accuracy (specification of the estimation error) can be given (cf. Armstrong 1998, Wackernagel 2003 and Burrough and McDonell 1998 pp. 133–161).

9.7.4 Thiessen Polygons

The *polygon method* uses a fundamentally different approach than the other interpolation methods. Here, no new estimated values are calculated from existing values at a few observation points. Instead, if n observation points are available, the study area is divided or decomposed into n polygons, so that each point in a polygon is closest to the associated observation point. All points in this polygon receive its observation value. This results in a homogeneous distribution of values within the polygons, but there are discontinuities at the edges, which result solely from this calculation method.

For the construction, the perpendicular bisectors (cf. blue lines in Fig. 9.32) of the connections between the adjacent points are determined. Due to the laws of plane geometry, these bisectors intersect at the center of the circumcircle of the corresponding triangle. These clearly defined centers form the corners and the perpendicular bisectors the edges of so-called *Voronoi* or *Thiessen polygons* or *Dirichlet regions.*

Within a Thiessen polygon, all z_i values are the same. Differences exist (only) between different polygons. This approach therefore does not assumes any correlation between the values from different polygons. However, the discontinuities of the z_i-values at the edges cannot usually be justified in terms of content, so that the use of this technique for interpolation is limited.

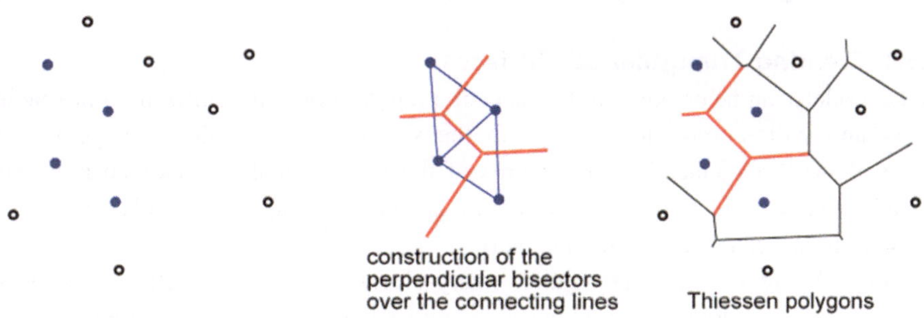

Fig. 9.32 Construction of Thiessen polygons

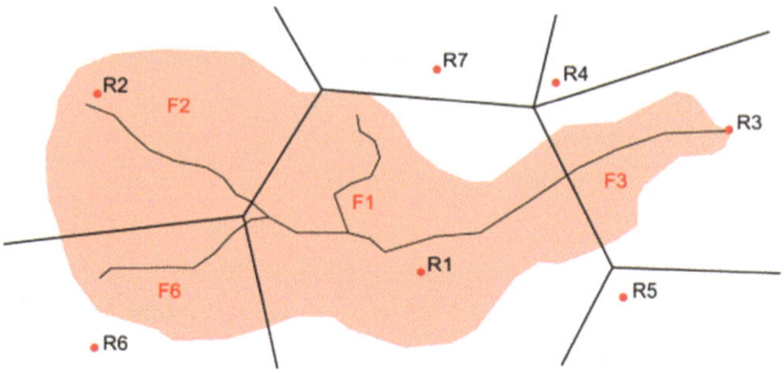

Fig. 9.33 Calculation of the area precipitation on the basis of Thiessen polygons

The widely used method for determining area precipitation (p) uses the areas of the Thiessen polygons (a_i), which belong to the stations with precipitation p_i. Precipitation is always recorded using a network of observation stations, so that ultimately only selective measurements are available at observation stations. When drawing up water balances, e.g. in a river basin, the area precipitation must be calculated from the precipitation at the stations in the catchment area. For this purpose, the weighted arithmetic mean of the station precipitation is calculated by taking the size of the subareas as weights, i.e. the size of the catchment area lying in the respective Thiessen polygon (in Fig. 9.33: $F_1 = 108\ km^2$, $F_2 = 74\ km^2$, $F_3 = 25\ km^2$, $F_6 = 36\ km^2$). The sizes of these areas are determined in an intermediate step using the intersect function of a geoinformation system (cf. Fig. 9.16). For the example shown in Fig. 9.33, an area precipitation in the river basin is then given by:

$$\frac{108 \cdot 657\text{mm} + 74 \cdot 757\text{mm} + 25 \cdot 929\text{mm} + 36 \cdot 828\text{mm}}{243} = 741\ \text{mm}$$

Only the annual precipitation at stations R_1 (657 mm), R_2 (757 mm), R_3 (929 mm) and R_6 (828 mm) are taken into account (cf. Fig. 9.33).

9.7.5 Triangulated Irregular Network

9.7.5.1 Creating Triangular 3D-Surfaces

Surface models can be created on the basis of a regular grid with uniform point spacing in both x- and y-directions and heights. In the simplest case, a block image is created (cf. Fig. 9.34). By connecting the heights of the centers of the blocks, an approximate surface is created. Due to the rigid specifications, the image appears greatly simplified and usually quite angular and coarse. The surface model is not adapted to the real elevation conditions. Where strong relief requires many elevation points, modeling uses the same number of elevation points as for planes that require few elevation points. Better results are obtained, if the surface is modeled by irregular triangles that adapt to the relief (cf. Fig. 9.35).

Fig. 9.34 Block diagram based on a regular elevation grid

Fig. 9.35 Three-dimensional network of irregular 3D triangular surfaces (terrain model)

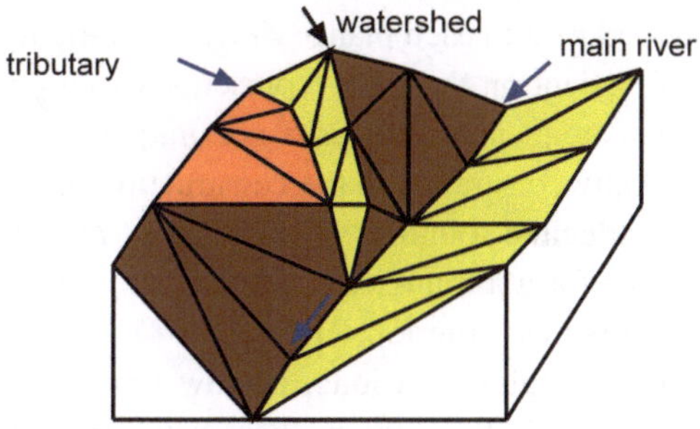

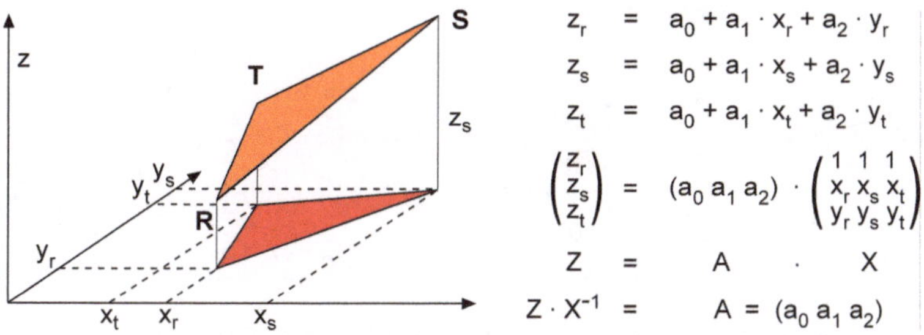

Fig. 9.36 Construction of a triangular surface in a surface model

The creation of a surface model similar to Fig. 9.35 is done in several steps. First, a triangulation of the coordinates (x_i,y_i) in the plane results in a mesh of triangles (i.e. in the xy-plane). The attribute values z_i at the locations (x_i,y_i), i.e. in the corners of the plane triangles, can be represented three dimensionally as heights. It is then obvious to place a plane through three adjacent z_i values, i.e. through the z_i heights in the corners of a triangle (cf. Fig. 9.33). This results in the three-dimensional *triangulated irregular network* (*TIN*). A three-dimensional impression of an elevation model is created with an appropriate viewer, in which the location of the viewer and the viewing angle are determined and then the hidden lines are suppressed (cf. Fig. 5.1).

Formally, the construction of a plane using three adjacent z_i values means calculating the equation $z(x,y) = a_0 + a_1 \cdot x + a_2 \cdot y$, which is to be determined exactly for each triangle. Fig. 9.36 shows the system of equations for determining the coefficients a_i and the solution.

The z_i-values for any point (x_i,y_i) within the triangle are calculated by substituting the (x_i, y_i)-coordinates into the equation. The observed values, i.e. the z_i values of the corners of the triangles, lie exactly on the surface and are not approximated as in the other methods.

9.7.5.2 Delauny Triangulation

The representation of the elevation model is primarily determined by the selection of the observation points or the points with elevation data, but also by the purely formal, technical procedure of the triangular meshing in the plane. Figure 9.37 shows the differences that alternative planar meshes of the same original data can lead to. Perspective views are placed next to each planar view. The variant on the left results in a pointed pyramid, while the variant on the right represents a valley with two slopes. Thus, the assignment of the points to planar triangles is not unique.

Only constraints in the construction of the triangles in the xy-plane lead to unique and reproducible solutions. The so-called *Delaunay triangulation* is common, in which three points form a triangle, in whose circumcircle no further observation point lies (so-called circumcircle criterion, cf. Fig. 9.38). This modeling rule leads to the formation of small wide triangles than long, narrow triangles. This makes it easier to model elevations or depressions. In addition, these small triangles are less likely to cross terrain edges. The resulting triangular mesh is unique. Irrespective of the processing sequence, the same

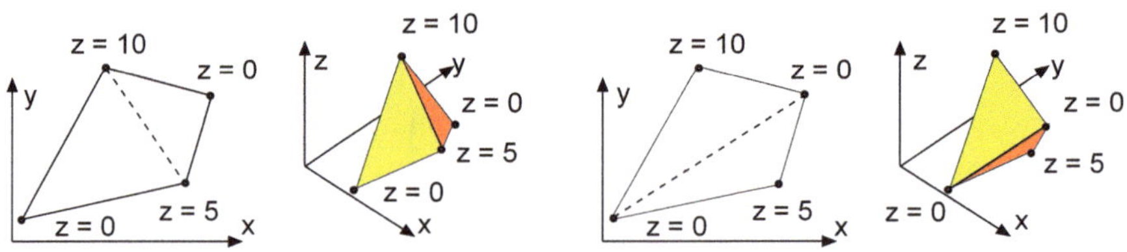

Fig. 9.37 Different surface models based on different triangular meshes in the plane

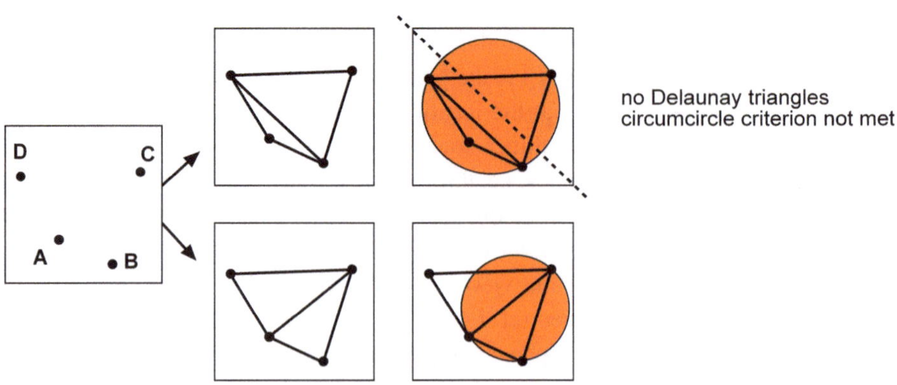

Fig. 9.38 Delauny triangulation of four points

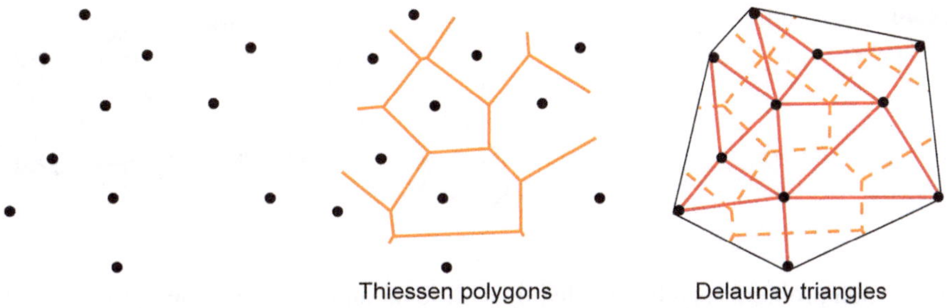

Fig. 9.39 Construction of Delaunay triangles over Thiessen polygons

triangular meshing always results (for algorithms cf. Worboys and Duckham 2004 pp. 202).

When constructing Delaunay triangles, use is made of the fact that the Delaunay triangulation represents the dual graph of Thiessen polygons. The corners of the Thiessen polygons are the centers of the circumcircles of the triangles of the Delaunay triangulation (cf. Fig. 9.39). The assignment of the points to the Delaunay triangles follows directly from this. Thus, in Fig. 9.38 the points A, B and C form a Delaunay triangle and not, for example, the points A, B and D.

The modeling of surfaces using a three-dimensional network of irregular triangular surfaces, i.e. abbreviated a TIN, is usually based on a planar triangular meshing according to Delaunay. However, as a result, a surface may not be reproduced true to the original. In reality the right variant from Fig. 9.37 may exist (representation of a valley), which just does not meet the Delaunay criterion. This situation can be checked relatively quickly with an elevation model, but more difficult in the case of a groundwater surface, for example. The 3D-data can often only be verified by plausibility considerations. In case of obvious deviations, additional points with attribute values have to be included in the meshing.

In the case of a surface model, significant points such as high and low points as well as vertices of the terrain (change from concave to convex shapes) can or must be selected, which play an important role as turning points of inclination and direction of inclination. The possibility of scattering point data according to the relief proves to be advantageous. In this way, many points can be selected in terrain with a strong relief, while flat areas are represented by fewer points. The amount of data can therefore be adapted to the relief. Above all, in terrain models, so-called fault lines and rib lines must be preserved and represented by edges (e.g. rivers, coastlines, step edges, groundwater barriers, valley floors). Such fault lines can be (pre-)determined as edges in triangulation, which are always maintained independently of the Delaunay criterion and form triangle edges.

Unwanted *plateau effects* can sometimes occur during triangulation. If mountain spurs are present, three points along a contour line may be closer to each other and then form a triangle than a third point on an adjacent contour line is included in the triangulation. This creates flat surfaces, where the terrain is uphill (cf. Fig. 9.40). Similar critical situations are peaks, depressions, hollows, saddles, ridges and valleys where triangles are equal in

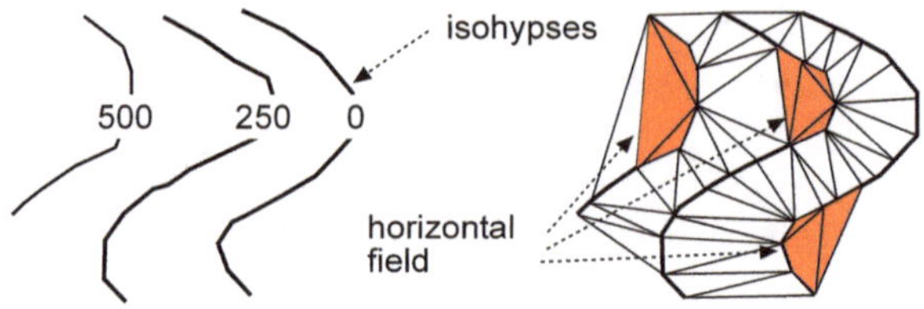

Fig. 9.40 Plateau effects during triangulation

heights. Such problems can be avoided by additional input of single elevation points at the critical terrain locations. Thus, the model must be improved step by step.

9.7.6 Parameterisation of Surfaces

Simply viewing the surface model in a 3D viewer will provide clues as to its accuracy and design flaws. Furthermore, initial statements about surface-dependent parameters such as insolation or surface runoff can also be estimated. Beyond such visualisations, a geoinformation system offers a wide range of analysis options for 3D models:

- estimation of values, which are understood as points on a (value) surface,
- generation of contour lines or isolines (e.g. groundwater levels or isohyets),
- calculate of a cross-section of the terrain,
- conducting visibility analysis,
- area and volume calculations as well as erosion calculations.

A common task in practice is the calculation of slope and exposition. Among the infinite number of changes in height that increase from one point in all directions, the value of the greatest ascent or descent is designated as the slope (cf. Fig. 9.41). The exposure then corresponds to the direction of the greatest uphill or downhill gradient, which is determined in degrees relative to the north orientation.

In the three-dimensional case or for a point (x,y,z) on a terrain surface, the slope is defined as the norm ("length") of the gradient vector, which can be calculated relatively easily, if the surface can be described by a function z = f(x,y) of the two variables x and y. Here dz/dx and dz/dy are the partial derivatives of f(x,y), if they exist:

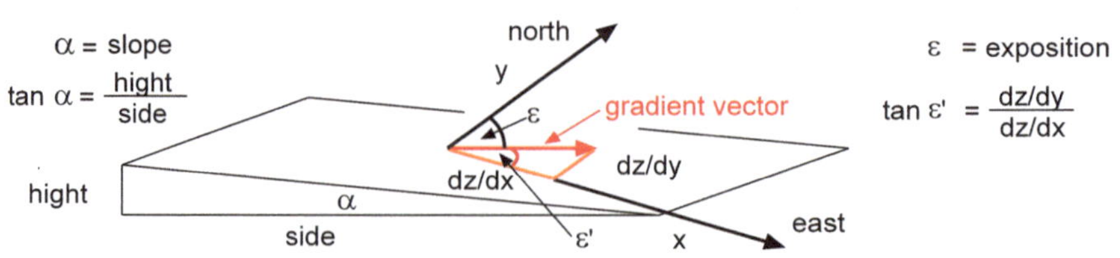

Fig. 9.41 Definition of slope and exposure

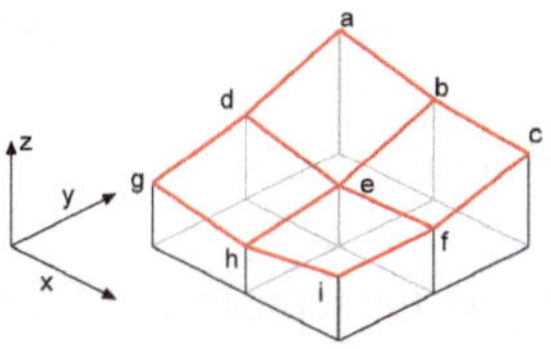

139	137	135	127	125
134	132	128	127	125
129	125	123	122	122
123	120	119	118	118
119	117	117	117	117

dz/dx = ((a + 2 • d +g) – (c + 2 • f + i)) / (8 • grid with)

dz/dy = ((a + 2 • b +c) – (g + 2 • h + i)) / (8 • grid with) and then: $\tan\alpha = \sqrt{(\frac{dz}{dx})^2 + (\frac{dz}{dy})^2}$

dz/dx = ((132 + 2 • 125 + 120) – (127 + 2 • 122 + 118)) / 8 = (502 – 489) / 8 = 13 / 8

dz/dy = ((132 + 2 • 128 + 127) – (120 + 2 • 119 + 118)) / 8 = (515 – 476) / 8 = 39 / 8

slope = tan α = 5,14 also α = 79° for grid with 1

slope = tan α = 0,10275 also α = 6° for grid with 50

Fig. 9.42 Calculation of the slope based on raster data according to Horn 1981

$$\text{slope} = \|f'(x, y)\| = \sqrt{\left(\frac{dz}{dx}\right)^2 + \left(\frac{dz}{dy}\right)^2} \quad \text{with } z = f(x, y)$$

However, a terrain surface or an elevation model can only be described by a mathematical function in the rarest of cases, so that the given formula is initially only of theoretical interest. Rather, it can be assumed that the slope (or inclination) is to be determined from a regular grid of elevation values. Here, the partial derivatives or differential quotients dz/dx (slope in x-direction) and dz/dy (slope in y-direction) are estimated using the algorithm of Horn (cf. Horn 1981, for further approaches of parameter estimation cf. Burrough and McDonell 1998 p. 190) (Fig. 9.42).

The *exposure* denotes the angle of the direction of the gradient with respect to north. It is given in positive degrees from 0° to 360°, measured clockwise from north. From the two differential quotients dz/dx (gradient in the x-direction) and dz/dy (gradient in the y-direction), ε' is first determined $\varepsilon' = \arctan[(dz/dy)/(dz/dx)]$ (cf. Fig. 9.41).

Case differentiation must be made to calculate the exposure, since the inverse function of the tangent, i.e., the arctangent, is unique only between −90° and + 90°. Figure 9.41 shows a northeast exposed slope ($\varepsilon' \neq 0$) with dz/dy > 0 and dz/dx > 0 and $\varepsilon = 90° - \varepsilon'$. If dz/dy < 0, the gradient vector would point southeast, the arctangent would be negative, and $\varepsilon > 90°$.

$$\text{Exposure} \quad \begin{cases} 90° - \arctan(dz/dy/dz/dx) & \text{if } dz/dx > 0 (\text{slope to the east}) \\ 270° - \arctan(dz/dy/dz/dx) & \text{if } dz + dx < 0 (\text{slope to the west}) \\ 0° & \text{if } dz/dx = 0,\ dz/dy > 0 \\ 180° & \text{if } dz/dx = 0,\ dz/dy < 0 \\ \text{undefined} & \text{if } dz/dx = 0,\ dz/dy = 0 \end{cases}$$

The last line illustrates a plane where the differential quotients are 0. An exposition is not clearly given and not formally defined.

Slope and exposition are used in many ways. For example, the exposure is used in issues of vegetation geography or in geomorphology and terrain analysis, e.g. for modeling potential alpine permafrost or avalanche hazards. For the construction of solar roof cadastres, which represent the suitability of roof surfaces for photovoltaics, the roof inclination and the extent of the sun (i.e. insolation) are required. The parameters are determined e.g. on the basis of LiDAR data (cf. Sect. 5.4).

References

Armstrong, M. (1998): Basic Linear Geostatistics. Berlin: Springer.

Autodesk (2023): The AutoCAD Map 3D toolset is included with AutoCAD. https://www.autodesk.com/products/autocad/included-toolsets/autocad-map-3d (14.04.2023).

Behncke, K., Hoffmann, K., de Lange, N. u. C. Plass (2009): Web-Mapping, Web-GIS und Internet-GIS – ein Ansatz zur Begriffsklärung. In: Kartogr. Nachrichten H. 6 2009, S. 303–308.

Bentley (2023): OpenCities Map. https://www.bentley.com/en/products/brands/opencities-map (14.04.2023).

Berlekamp, J., de Lange, N. u. M. Luberichs (2000): Emissions- und Immissionskataster für das Stadtgebiet Münster als Komponente eines kommunalen Umweltinformationssystems. In: Cremers, A. u. K. Greve (Hrsg.): Umweltinformatik 2000. Umweltinformation für Planung, Politik und Öffentlichkeit. 12. Intern. Symposium „Informatik für den Umweltschutz". S. 703–715. Marburg: Metropolis.

Bill, R. u. D. Fritsch (1991): Grundlagen der Geo-Informationssysteme. Bd. 1. Hardware, Software und Daten. Heidelberg: Wichmann.

Bill, R. (2016): Grundlagen der Geo-Informationssysteme. Berlin: Wichmann. 6. Ed.

Burrough, P.A. u. R.A. McDonell (1998): Principles of Geographical Information Systems. Oxford: University Press.

Dropchop (2023): Dropchop. https://dropchop.io/ (14.04.2023).

Ehlers, M. u. J. Schiewe (2012): Geoinformatik. Darmstadt: Wiss. Buchgesellschaft.

ESRI (1995): Unterstanding GIS. The Arc/Info Method. Self Study Workbook. Version 7 for UNIX and OpenVMS. New York: John Wiley.

ESRI (2023a): ArcGIS Pro. https://www.esri.com/en-us/arcgis/products/arcgis-pro/overview (14.04.2023).

ESRI (2023b): ArcGIS Online. https://www.esri.com/en-us/arcgis/products/arcgis-online/overview (14.04.2023).

ESRI (2023c): ArcGIS Online. Use the analysis tools. https://doc.arcgis.com/en/arcgis-online/analyze/use-analysis-tools.htm (14.04.2023).

ESRI (2023d): What is a shapefile? https://desktop.arcgis.com/en/arcmap/latest/manage-data/shapefiles/what-is-a-shapefile.htm (14.04.2023).

ESRI (2023e): What is a Geodatabase? https://desktop.arcgis.com/de/arcmap/10.3/manage-data/geodatabases/what-is-a-geodatabase.htm (14.04.2023).

Fürst, D., Roggendorf, W., Scholles, F. u. R. Stahl (1996): Umweltinformationssysteme – Problemlösungskapazitäten für den vorsorgenden Umweltschutz und politische Funktion. Hannover: Inst. f. Landesplanung u. Raumforschung. = Beiträge zur Räumlichen Planung 46.

Goodchild, M. F. u. K. Kemp (1990) (Hrsg.): Introduction to GIS. NCGIA Core Curriculum. Santa Barbara. Access to the digital version of the curriculum from 1990 can be found on the homepage: https://ibis.geog.ubc.ca/courses/klink/gis.notes/ncgia/toc.html (14.04.2023).

Grass GIS (2023): What is GRASS GIS? https://grass.osgeo.org/learn/overview/ (14.04.2023).

Grintec (2023): Smallworld GIS. Network information system. https://www.grintec.com/en/Smallworld (14.04.2023).

gvSIG (2023): gvSIG Association. http://www.gvsig.com/en (14.04.2023).

Hackmann, R. u. N. de Lange (2001): Anwendung GIS-gestützter Verfahren in der Stadtentwicklungsplanung – Untersuchung von Versorgungsbereichen durch Netzwerkanalyse auf der Grundlage amtlicher Geobasisdaten. In: Strobl, J. u.a. (Hrsg.): Angewandte Geographische Informationsverarbeitung XIII: Beiträge zum AGIT-Symposium 2001. S. 221–226. Heidelberg: Wichmann.

Hennermann, K. (2014): Kartographie und GIS. Eine Einführung. Darmstadt: Wiss. Buchgesellschaft. 2. Ed.

Hexagon (2023a): Leverage your geospatial data with GeoMedia for GIS and mapping. https://hexagon.com/Products/GeoMedia (14.04.2023).

Hexagon (2023b): GeoMedia WebMap, Geospatial Portal and Geospatial SDI. https://hexagon.com/Products/geomedia-webmap-geospatial-portal-geospatial-sdi (14.04.2023).

Horn, B. (1981): Hill shading and the reflectance map. In: Proceedings of the IEEE 69, S. 14–47.

Longley, P.A., Goodchild, M.F., Maguire, D.J. u. D.W. Rhind (2005, Ed.): Geographical Information Systems: Principles, Techniques, Management and Applications. West Sussex: John Wiley & Sons. 2nd Edition, Abridged.

Longley, P.A., Goodchild, M.F., Maguire, D.J. u. D.W. Rhind (2015): Geographic Information Science and Systems, 4th Edition Hoboken, NJ: Wiley

MapInfo Pro (2023): MapInfo Pro. https://www.precisely.com/product/precisely-mapinfo/mapinfo-pro (14.04.2023).

NCGIA (1990): Core Curriculum-Geographic Information Systems (1990). Access to the digital version of the curriculum from 1990 can be found on the homepage: https://escholarship.org/uc/spatial_ucsb_ncgia_cc (14.04.2023).

OpenJUMP (2023): OpenJump. http://www.openjump.org (14.04.2023).

QGIS (2023a): A Free and Open Source Geographic Information System. https://www.qgis.org/en/site/ (14.04.2023).

QGIS (2023b): QGIS Python Plugins Repository. https://plugins.qgis.org/plugins/ (14.04.2023).

Rudert, F. u. H. Pundt (2008): Standardisierte Geodienste (WMS) auf mobilen Endgeräten -ein Entwicklungsbeispiel aus dem Projekt „GeoToolsHarz". In: Strobl, J. u.a. (Hrsg.): Angewandte Geoinformatik 2008, S. 305-312. Heidelberg: Wichmann.

SAGA (2023): SAGA, System for Automated Geoscientific Analyses. https://saga-gis.sourceforge.io/en/ (14.04.2023).

Spring GIS (2023): O Sistema de Processamento de Informações Georeferenciadas - SPRING. http://www.dpi.inpe.br/spring/ (14.04.2023).

Tomlin, C. D. (1990): Geographic Information Systems and Cartography Modeling. Englewood Cliffs: Prentice Hall.

Umweltbundesamt (2023): Ausbreitungsmodelle für anlagenbezogene Immissionsprognosen. https://www.umweltbundesamt.de/themen/luft/regelungen-strategien/ausbreitungsmodelle-fuer-anlagenbezogene/uebersicht-kontakt11.07.2022textpart-1 (14.04.2023).

Wackernagel, H. (2003): Multivariate geostatistics. An introduction with applications. Berlin: Springer. 3. Ed.

Worboys, M. u. M. Duckham (2004): GIS. A computing perspective. Boca Raton: CRC Press, 2. Ed.

Remote Sensing and Digital Image Processing 10

10.1 Remote Sensing: Definition and Use

Obtaining information with the help of remote sensing systems deployed on aircraft, on satellites and increasingly also on drones (UAVs, unmanned aerial vehicles) has gained considerable importance in recent years. While initially the focus was clearly on military applications, especially espionage activities, the potential of these systems for scientific and commercial observation of processes on the earth's surface and in the atmosphere was soon recognised (on the development of remote sensing cf. Heipke 2017a pp. 8). The areas of application where remote sensing is already used as a standard procedure include above all weather observation and the broad field of environmental monitoring. Today, a worldwide distribution network ensures that data from various sensors can be obtained almost without any problems. Their evaluation is increasingly being carried out in connection with other digital geodata within hybrid geoinformation systems.

A very general definition of *remote sensing* is provided by Hildebrandt (1996 p. 1, translated): "Remote sensing in the broadest sense is the recording or measurement of objects without physical contact with them, and the analysis of data or images obtained in this way to obtain quantitative or qualitative information about their occurrence, condition, or change in condition, and possibly their natural or social relationships to one another." This broad definition also includes analog methods such as the creation of analog aerial photographs, i.e. taking photographs on film with a camera, as well as the measurement of radiant temperature by airborne measuring devices or even by hand. All methods record electromagnetic radiation such as visible light, thermal radiation and other non-visible radiation emitted or reflected by objects of investigation on earth (e.g. green areas) or in the atmosphere (e.g. clouds), whereby the objects emit or reflect in different ways depending on their type or nature (e.g. vegetation) and condition (e.g. damaged forest stock or harvested grain fields).

N. de Lange, *Geoinformatics in Theory and Practice*, Springer Textbooks in Earth Sciences, Geography and Environment,
https://doi.org/10.1007/978-3-662-65758-4_10

Table 10.1 Applications of remote sensing (here especially of satellite data)

Meteorology and climatology	Weather forecast, atmospheric and climate research
Geology	Geological mapping, prospecting, deposit development, photogeology (conclusions on rock types and tectonic structure)
Hydrology	Recording of snow cover and its melting, recording of input parameters for hydrological models (e.g. current land cover, estimation of evapotranspiration)
Forestry	Forest classification, forest damage assessment, yield estimation, forest fire assessment
Agriculture	Crop recording and crop estimation on a regional basis, precision farming, control of the allocation of EU area payments
Coastal, oceanography	Monitoring of water quality, detection of discharges, observation of coastal changes as well as (especially with radar methods) detection of waves and wind speed
Archaeology	Discovery of historical sites (esp. with radar methods)
Environmental planning	Detection of land cover, environmental monitoring, water and wetland management
Cartography	Acquisition and updating of basic cartographic information
Media	Tourism marketing, education, visually appealing background
Risk management	Volcano monitoring, flood protection

Only those remote sensing systems that provide digital image data as the result of a scanning process of the earth's surface are discussed here. These can be processed directly by digital image processing systems and easily integrated into geoinformation systems. Thus, only those techniques that use digital methods are presented. Analog photographic recording systems, analog image evaluation, methods of visual image interpretation and photogrammetric evaluation are not dealt with.

Image data from civil earth observation satellites represent a generally readily available source of information that has great potential for large-scale monitoring of processes in the geosphere and atmosphere (cf. Table 10.1). In addition to commercial missions, many remote sensing data are now available free of charge (cf. Sect. 10.5.3). These data have characteristics that no other data source can provide in this form:

- Different sensors are used to record in different wavelength ranges of the electromagnetic spectrum. Wavelength ranges which are not visible to the human eye (e.g. infrared) are also recorded.
- Remote sensing enables a synoptic recording of large areas.
- The recorded information is up-to-date (cf. weather satellites).
- Remote sensing by satellites allows a regular repetition of the survey of an area and provides comparable data in constant quality over a longer period of time.
- The integration into a geoinformation system enables a completely digital processing chain.

Remote sensing is a key technology for disaster management, resource management and environment risk research. The establishment of networked services for global environmental monitoring and the development of early-warning systems and disaster management are becoming increasingly important, especially viewing the fact of climate crisis.

The Earth Observation Center (EOC) is a group of institutes within the German Aerospace Center (DLR). It is formed by the Institute for Remote Sensing Methodology and the German Remote Sensing Data Center (DFD) and is the competence center for earth observation in Germany (for applications and projects cf. DLR 2023a).

However, with regard to the applicability of remote sensing data some limitations and problem areas must also be mentioned:

- Restrictions are linked to the weather dependency of optical systems. For example, cloud cover, haze or aerosols in the air can considerably reduce the amount of data which can be analysed.
- With many operational satellite systems, processing was only possible on scale levels of 1:25000 and smaller due to their low spatial resolution. More recent satellite systems and sensors on UAVs with a ground resolution in the meter and sub-meter range now allow more precise evaluations.
- Another problem area is the question of the reproducibility of the results, which plays an important role, for example, in the case of multiple analyses of a region at different times in the course of a monitoring process.

Even with remote sensing data and methods of digital image processing (cf. Sect. 10.2.3) it is generally not possible to obtain the "really" objective information that might be hoped for. Here, too, the researcher or analyst must intervene in a complex analysis process according to his level of knowledge and react to the respective data situation. As an example, the selection of representative training areas for a classification of land cover can be mentioned (cf. Sect. 10.7.4).

10.2 Remote Sensing and Digital Image Processing: General Approach

10.2.1 Remote Sensing: Principles

The central approach of remote sensing is based on *physical radiation processes* in the atmosphere (cf. Fig. 10.1). The starting point is a passive remote sensing system that does not emit any laser or radar beams itself:

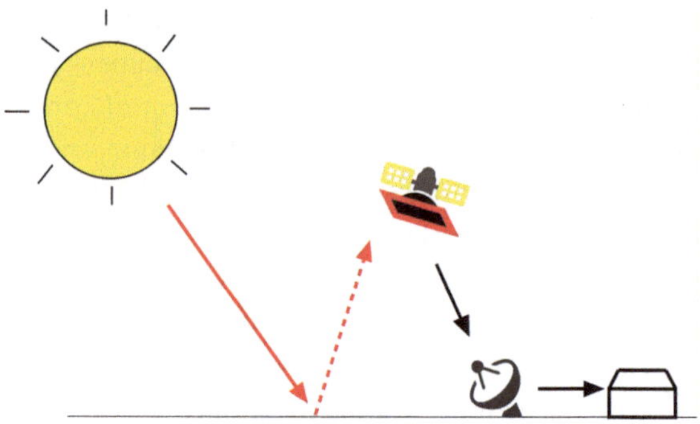

Fig. 10.1 Remote sensing approach (passive sensor systems)

- Energy radiated from the sun is predominantly in the wavelength range between 0.1 μm and 3.5 to 5 μm. This is mainly short-wavelength radiation (ultraviolet light, visible light, near infrared and middle-infrared), which is partly absorbed and partly reflected by objects on the earth's surface.
- The absorbed radiation heats up the objects on the earth's surface, so that these heated objects emit long infrared (thermal infrared) radiation back into the atmosphere (cf. Sects. 10.3.2 and 10.3.3).
- The incoming radiation is reflected and emitted back into the atmosphere differently due to the properties and conditions of the objects on the earth's surface.
- A sensor that is not in contact with the objects on earth and is mounted on a drone (UAV), an aircraft or a satellite records the reflected electromagnetic radiation.
- The intensity of the radiation arriving at the sensor is encoded and sent to a receiving and processing station, where the data are further processed.
- After processing the data, the images are interpreted visually and analysed using appropriate methods.

This basic principle is based on the fact that different objects have special spectral reflectance or emission characteristics. A water surface, for example, reflects the incoming infrared radiation in the wavelength range between 0.7 and 0.9 μm in a completely different way than a dry pasture. The specific emission or reflection of different objects ultimately enables their identification. The electromagnetic radiation detected by remote sensing allows conclusions to be drawn about the objects on earth (cf. especially Sect. 10.7). The representation here, which is still highly simplified at the beginning, does not yet take into account the influences of the atmosphere, does not yet address the differentiated reflection behavior in different wavelength ranges and still uses the word "image".

Remote sensing systems can be subdivided into a sensor segment, a ground segment and a refining segment (cf. Fig. 10.2):

- The *sensor segment* generally consists of the platform and the actual sensor. In most cases, a digital *multispectral scanner* is carried on a drone (UAV), on an aircraft or on a satellite, which detects the incoming radiation in different ranges of the electromagnetic

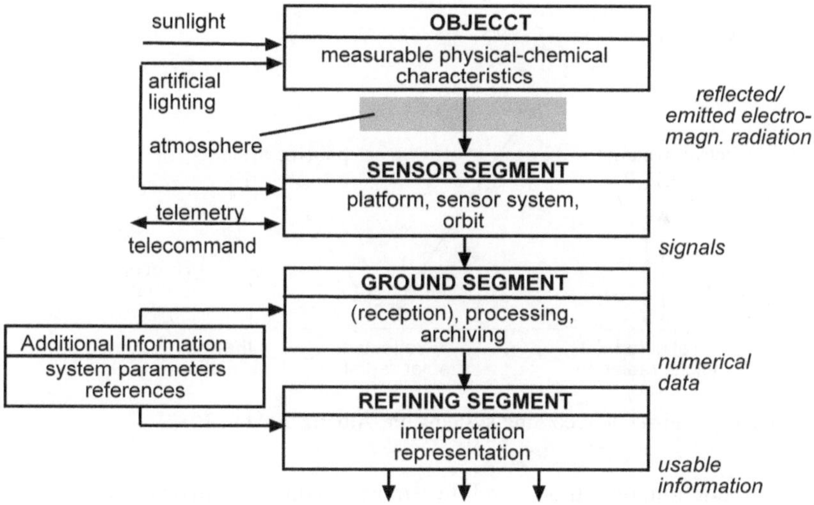

Fig. 10.2 Sensor, ground and refining segment in remote sensing (translated, cf. Markwitz 1989 p. 3)

spectrum. The tasks of a digital, satellite-based sensor segment are therefore the scanning of an area as well as, among other things, the digitisation and coding of the data and the intermediate storage with subsequent transmission to the ground station.

- The task of the *ground segment* is the reception as well as preparing and pre-processing of the data. The pre-processing includes standard data preparations (cf. Sect. 10.4.6.4), the addition of further information (e.g. recording time) as well as geometric or radiometric corrections (cf. Sect. 10.6.1).
- The actual conversion of the recorded data into interpretable output images and therefore the evaluation and use of remote sensing data takes place in the *refining segment*. For this purpose, methods of visual interpretation and digital image processing are used.

10.2.2 Sensor Systems and Platforms

With regard to data acquisition, a distinction can be made between primary passive and active as well as secondary recording systems (cf. Fig. 10.3). *Primary passive systems* can only record radiation reflected or emitted by objects (e.g. the multispectral scanners on board Landsat). In contrast, *primary active systems* emit radiation themselves and record the reflected radiation (e.g. the C-band synthetic aperture radar on board Sentinel-1, cf. Sect. 10.4.9). *Secondary recording systems* are systems with which data that are available in analog form (e.g. analog aerial photographs) are converted into digital raster form. These include, above all, special scanners that are used to digitise aerial photographs for further computer-assisted processing.

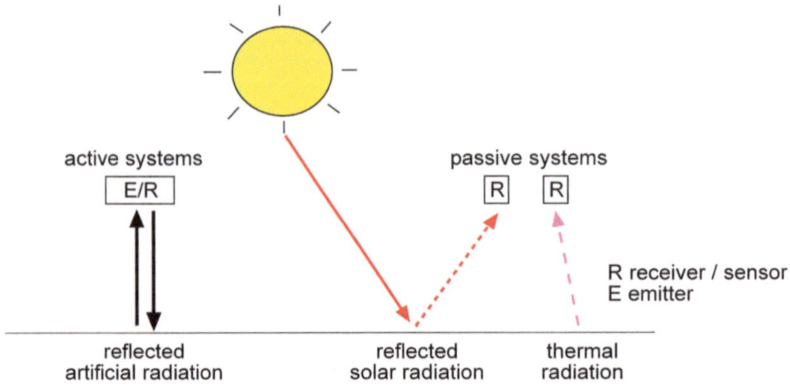

Fig. 10.3 Passive and active recording systems (cf. Albertz 2009 p. 10)

In this introduction, only those satellite remote sensing systems will be discussed which are of great importance and that can be used to illustrate basic principles. Since the 1970s, there have initially only been national or international operators of satellite systems, i.e. with their own satellites and associated sensors, for meteorological, resource-oriented and cartographic applications. For many parts of the earth, recordings are available for longer time series, so that in some cases an excellent data material for environmental monitoring exists. Recording systems such as Landsat and SPOT still belong to the standards of remote sensing and are of inestimable value due to their data continuity, although the principle of their recording was developed decades ago. Since the end of the 1990s, several purely commercially oriented satellite operators have achieved great importance. Due to the high spatial resolution of these new systems, the area coverage of their recordings is not very high. However, the strengths of these systems lie in the order-related, targeted recording (cf. Sect. 10.4.10).

Table 10.2 systematizes different platforms. Depending on the task and application for which remote sensing data are required, this provides orientation for choosing a suitable platform. For example, a satellite-based digital sensor that records visible light and near-infrared data will be used to classify large-scale land cover. The Earth Observation Portal of ESA (European Space Agency) provides a very good compilation of almost all satellite missions with extensive background information on many sensors (cf. EoPortal 2023).

10.2.3 Digital Image Processing

In general, it can be defined that *digital image processing* serves to extract information from digital image data. This includes, among other things, the recognition of structures and content in the digital images using suitable classification methods. A further distinction can be made between methods of image pre-processing (including radiometric and geometric enhancement), contrast enhancement, image enhancement (including image

Table 10.2 Typical platforms and sensors for remote sensing and their geopositioning without terrestrial systems (cf. Toth and Jutzi 2017 Table 2.1)

	Satellite	Airplane	UAV
Maneuverability	None/limited	Medium	High
Spatial extent	Worldwide	Regional	Local
Sensor diversity	MS/HS/SAR	MS/HS/LiDAR/SAR	MS (LiDAR/HS)
Surroundings	Outdoors	Outdoors	Outdoors / indoors
Cover	>10 km	Approx. 1 km	Approx. 100 m
Possible repetition rate	Day(s)	Hours	Minutes
Spatial resolution	0.30–300 m	5–25 cm	1–5 cm
Spatial accuracy	Up to 1–3 m	5–10 cm	1–25 cm
Usability	Difficult	Complex	Easy
Observability	Vertical/inclined	Vertical/inclined	Vertical/oblique/360°
Operational risk	Medium	High	Low
Costs (platform Incl. Sensors)	€€€€€	€€€	€

MS/HS Multi–/hyperspectral, *SAR* Synthetic aperture radar, *LiDAR* Light detection and ranging

transformations and calculation of vegetation indices) and thematic information extraction (including classification techniques). The selection of the necessary processes always depends on the nature and the use of the data. Procedures of thematic information extraction analyse the thematic content of the image. They are used for current information, pattern recognition and extraction thematic information such as the recognition of forests or lakes or even rivers with organic content. However, the boundaries between these groups of methods are fluid in practice. For example, methods for forming vegetation indices can certainly serve to extract thematic information to deduce plant characteristics (cf. Sect. 10.6.3.1).

10.2.4 Photogrammetry

The term remote sensing is usually distinguished from photogrammetry, which refers to image measurement, i.e. the geometric evaluation of images with the aim of determining the position, size and geometric shape of spatial objects. This requires high precision recording systems that enable the most exact possible measurement of the recorded situation. The main application of photogrammetry is the production of topographic maps. The field of application ranges from measuring the velocity of moving objects to the automated creation of digital elevation models with modern digital photogrammetric workstations (for basics and applications cf. Kraus 2012 and Heipke 2017b). Due to new developments in sensor technology and evaluation options, photogrammetry and remote sensing are increasingly converging (cf. Lillesand et al. 2008 Section 3).

10.3 Electromagnetic Radiation Principles

10.3.1 The Electromagnetic Spectrum

Remote sensing is based on electromagnetic radiation processes:

- Electromagnetic radiation is an emission of energy radiating from bodies of matter.
- An object absorbs and/or reflects electromagnetic radiation depending on its condition (e.g. heating of a body or state of growth of a plant). Part of the absorbed radiation is emitted as thermal radiation. Generally speaking, absorption refers to the complex process by which radiant energy is absorbed (i.e. picking up electromagnetic waves or particle beams) and converted into other forms of energy.
- Electromagnetic radiation transports electrical and magnetic energy in waveform at the speed of light. An electromagnetic wave is described by the wavelength λ (in meters) and frequency υ (in hertz), which determine the physical properties of the radiation.
- Several spectral ranges of the electromagnetic spectrum in the visible light, infrared and also in the microwave range are of particular importance for remote sensing of the earth. The visible light extends in the wavelength range approximately between 0.4 μm and 0.7 μm. On the short-wave side, the ultraviolet precedes, on the longer-wave side, the infrared follows. The infrared is further subdivided into the near infrared (NIR) (approximately between 0.7 μm and 1.1 μm), the shortwave infrared (SWIR, approximately between 1.1 μm and 3 μm), the middle-infrared (approximately between 3 μm and 7 μm), and the far infrared (approximately from 7 μm), which is also called thermal infrared. The different ranges cannot be sharply separated. The sub-ranges of the infrared are sometimes defined differently by different authors. It should be emphasised that remote sensing of the earth can only use parts of these spectral ranges (cf. Fig. 10.4, on the so-called atmospheric windows cf. Sect. 10.3.2).

10.3.2 Solar Radiation and Atmospheric Influences

The sun is the energy source for *solar radiation*, the wavelength range of which is to be limited between $\lambda = 0.3$ μm and about $\lambda = 3.5$ μm from the point of view of remote sensing (wavelengths from the ultraviolet through the visible to the infrared spectral range, cf. Fig. 10.4). The spectral composition of solar radiation corresponds approximately to that of a so-called blackbody with a temperature of 5900 K. A blackbody, i.e. an idealized radiator, is a physical model to which the laws of thermodynamics can be applied and spectral irradiances can be theoretically derived (on the physical principles cf. Hildebrandt 1996 pp. 14, Mather and Koch 2011 pp. 4 and especially Jensen 2015 pp. 185). On average, 35% of the incident radiation is reflected by the earth (including clouds and

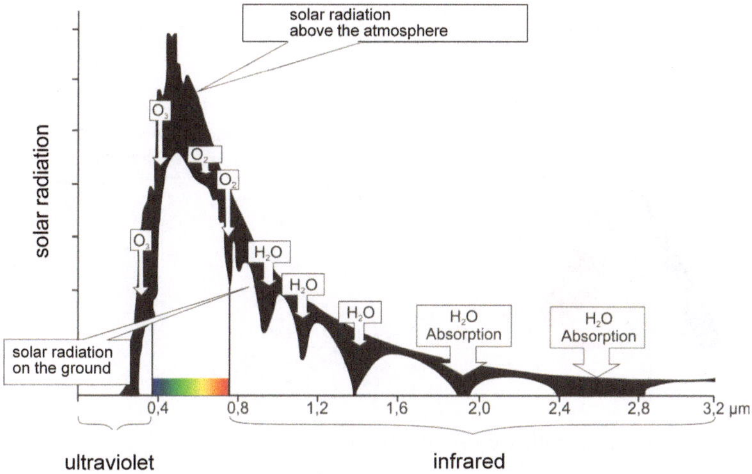

Fig. 10.4 Solar radiation and the effects of atmospheric absorption, scattering and reflection (DWD 2023)

atmosphere), 17% is absorbed by the atmosphere and 47% by materials on the earth's surface (cf. Mather and Koch 2011 p. 13–14). The sources of electromagnetic radiation to be evaluated in remote sensing are, on the one hand, the radiation reflected directly from the objects on earth and, on the other hand, the thermal radiation emitted by the objects on earth.

Direct solar radiation decreases as it passes through the atmosphere so that only part of the radiation reaches the ground. The *transmission* through the atmosphere for electromagnetic radiation depends strongly on the state of the atmosphere (aerosol content, moisture content, stratification, weather conditions), on the path the radiation has taken through the atmosphere and on the wavelength of the radiation. The different physical properties of the gases present in the atmosphere are responsible for a complex interplay of scattering and absorption. Due to oxygen or ozone, ultraviolet radiation below 0.3 μm is almost completely absorbed and to a larger extent converted into thermal energy. In the spectral range between 0.4 and 0.75 μm, only a small part of the radiation is absorbed by ozone, water vapour, aerosol and cloud particles, so that a large part of it can reach the earth's surface. In the course of evolution, the human eye has adapted to this spectral range as a special sensor for electromagnetic radiation, so that we speak of (for humans) visible light. Whereas many vertebrates can also recognize colors in the near ultraviolet below 0.4 μm. In the infrared spectral range from 0.75 to about 300 μm, radiation is strongly absorbed by water vapor, carbon dioxide, and ozone, and to a lesser extent by other trace gases (cf. Fig. 10.5).

The absorption and scattering processes are summarised by the term *extinction.* The absorption is due to the special absorption properties of gases, aerosol particles and cloud droplets, the scattering is due to interactions between wavelength and the particle sizes of

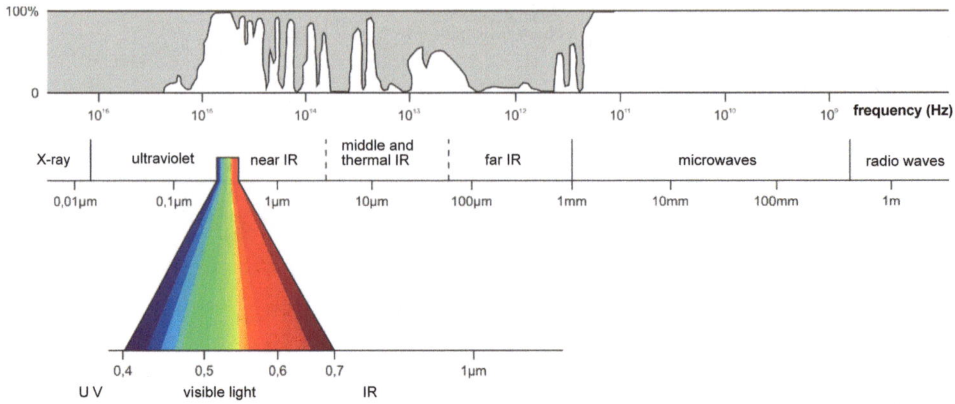

Fig. 10.5 Atmospheric transmittance (SEOS 2023)

aerosols and air molecules (cf. in further detail Mather and Koch 2011 p. 15). Thus, the *atmospheric transmittance* has a direct impact on remote sensing. The incident radiant energy can be (partially) absorbed in the atmosphere or in the terrain. The range of wavelengths (or frequencies) in the electromagnetic spectrum, within which radiant energy is absorbed, is called an *absorption band.*

The absorption bands of water vapour cause, for example, that these sections cannot be used for optical remote sensing of the earth's surface. However, there are sections in the electromagnetic spectrum, for which the atmosphere is almost transparent. These sections of the spectrum are called *atmospheric windows*. The most important of these windows are found in several sections of the electromagnetic spectrum (cf. Fig. 10.5):

In visible light (VIS)	0.4–0.7	µm
In the near infrared (NIR)	0.7–1.1	µm
In the shortwave infrared (SWIR)	1.1–1.35	µm
	1.4–1.8	µm
	2–2.5	µm
In the middle infrared (MIR)	3–4	µm
	4.5–5	µm
In the far infrared (thermal infrared, THIR)	8–9.5	µm
	10–14	µm
In the section of microwaves	>1	mm

The radiation passing through these atmospheric windows is subject to complex *scattering processes*, which in turn have wavelength-specific effects. For example, a large part of the radiation in the blue spectral range is already scattered in the atmosphere by the air molecules. This so-called *Rayleigh scattering* causes diffuse sky radiation and is reflected back to the satellite sensor. This part overlays the ground signal as unwanted "radiometric

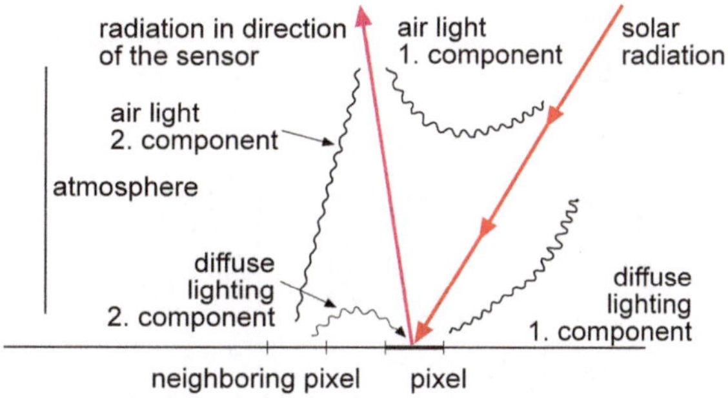

Fig. 10.6 Influences of the atmosphere (cf. Richards 2013 p. 34)

noise" and leads to a reduction in contrast. Therefore, this section is often left out of examinations or not recorded at all. For example, the ASTER sensor on NASA's Terra satellite platform does not record this wavelength range (cf. Sect. 10.4.8).

Figure 10.6 gives an impression of the complexity of the atmospheric processes that influence the signal received at the sensor. For example, both the diffuse reflection and the air light from adjacent pixels (diffuse illumination second component and air light second component, respectively) contribute to the reflection (cf. in more detail Richards 2013 pp. 33).

The transmittance of the atmosphere has a double relevance in remote sensing: on the one hand for the solar radiation on the earth's surface and on the other hand (more important) for the reception of the radiation reflected or emitted from the earth's surface by a satellite-based sensor. Therefore, the sensors of those satellites, whose main task is to record processes on the earth's surface, are adjusted to these sections or windows of the highest atmospheric transmittance.

The solar radiation is measured by the respective recording systems (sensors), whereby different sensors are used for special frequency ranges of electromagnetic radiation. These frequency ranges, i.e. sections of the electromagnetic spectrum of a certain width, are called *channels* or *bands* (as in Landsat terminology) in remote sensing. Thus, a channel or band is a discrete, clearly recorded interval of wavelengths detected by a sensor (cf. Table 10.3).

Usually, the bands are aligned with the atmospheric windows. However, various systems also carry instruments that operate within the spectral ranges of certain absorption bands and therefore make processes in the atmosphere visible and explorable. For example, the Meteosat weather satellite records the water vapour content of the atmosphere (channel 6 of the SEVIRI sensor, cf. Sect. 10.4.5).

10.3.3 Reflection of the Earth's Surface

The reflection, or the degree of reflection, as a function of the wavelength of the radiation emitted from the surface of objects on earth is crucial for the identification of these objects.

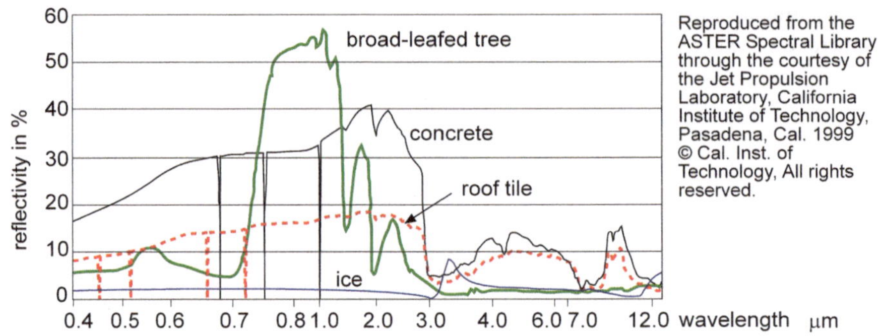

Fig. 10.7 Spectral signatures of selected surfaces

This reflectance is represented in *reflectance curves*, also known as *signature curves*, which have the meaning of "spectral fingerprints". Figure 10.7 shows selected signature curves that can be used to illustrate some differences in reflection. Further examples can be found, for example, in Mather and Koch 2011 pp. 17, Lille-sand et al. pp. 13, and especially in the spectral libraries of the United States Geological Survey (cf. USGS 2023a and USGS 2023b). The ECOSTRESS spectral library (formerly the ASTER spectral library) of the Jet Propulsion Laboratory contains, among other things, a search for spectral signatures of over 3400 surface materials (cf. NASA 2023a).

The spectral signatures of healthy green vegetation show, in addition to the chlorophyll reflection maximum in the green spectral range, a particularly noteworthy steep increase in reflection in the near infrared. This gradient, called "red edge", is of great importance when analysing data for vegetation analysis. It is used, among other things, in the development of so-called vegetation indices, which are used, for example, to identify the degree of vitality of plants (cf. Sect. 10.6.3.1, for the reflectance of plants cf. in detail Hildebrandt 1996 pp. 39).

However, there are limits to the use of signature curves. There are no generally valid signature curves for a surface type! Rather, the signature curves of objects on the earth's surface vary according to lightness or illumination, season, composition of the atmosphere, condition of the surface object (e.g. state of health, state of aggregation, moisture content) and configuration of the recording instrument. For example, the reflectance of water changes with the proportion of turbidity, among other factors. Therefore, so-called training areas are necessary in the respective investigation areas, which have a current homogeneous surface and allow a derivation or calibration of signature curves (cf. Sect. 10.7.4). Mixed forms present another problem. In particular, parts of a cultural landscape (e.g. built-up areas with roads, industrial plants, individual houses in gardens) cannot be clearly identified on the basis of a single signature curve (on the problem of mixed pixels cf. Sect. 10.7.7). The identification is primarily dependent on the geometric and spectral resolution of the recording system (cf. Sect. 10.4.1).

10.4 Major Satellite-Based Imaging Systems

10.4.1 Remote Sensing Instruments: Characteristics

With regard to the practice of image evaluation, four features are particularly important for assessing the performance of a recording system for a specific application:

- The *spatial resolution*: This characterises geometric properties of the recording system. It indicates the size of a pixel in meters, i.e. the side length of the area of the ground element that is recorded by a system at a certain flight altitude. This value is determined by the distance from the target and the aperture angle of the sensor system. This viewing angle is called *Instantaneous Field of View* (IFoV) mostly measured in milliradian (mrad, with $360° = 2\pi$ rad).
- The *spectral resolution*: The sensors used in remote sensing are designed to record different wavelengths of the electromagnetic spectrum. Many of the recording sensors used are multispectral, they record the radiation reflected from the earth's surface or atmosphere in several spectral ranges, i.e. in so-called channels or bands (cf. Sect. 10.3.2). The spectral resolution is determined by the number of channels. The location of these spectral bands in the electromagnetic spectrum and their width directly affect the discriminability of different surface types in the satellite image. Panchromatic sensors are to be distinguished from multispectral ones, whereby panchromatic denotes the broadband spectral sensitivity of a sensor. A panchromatic sensor is sensitive over the entire range of the human eye. A hyperspectral sensor system records images from a large number of wavelengths. Such systems can have 20 to 250 different bands.
- The *radiometric resolution*: Detecting objects also depends on the capability of a sensor to record the received radiation as differentiated as possible. The radiometric resolution is determined by the number of intensity levels available for the reproduction of this back radiation per spectral band (cf. Sect. 10.5.2). A common form is the reproduction in 256 levels per channel, which requires 8 bits for storage, so that the radiometric resolution (sensitivity) is specified as 8 bits. However, this is a simplified representation. For the description of the radiometric resolution one would also have to include the signal-to-noise ratio, for example (cf. Hildebrandt 1996 p. 429).
- *Temporal resolution*: The time interval within which an area can be repeatedly recorded by a particular satellite sensor determines its temporal resolution (also called repetition rate). This shortens the wider the ground track is and the further an area is towards the poles (overlapping of adjacent ground tracks in the case of polar satellite orbits).

10.4.2 Orbit Parameters of Remote Sensing Satellites

The most common orbit configurations used in remote sensing are the sun-synchronous orbit and the geostationary orbit. For a *sun-synchronous orbit,* the orbit is chosen so that the satellite always crosses a given location at the same local time. In this way, images from different years can be better compared (similar lighting conditions). Therefore, to indicate a sun-synchronous orbit, the time of crossing the equator is always given (for Landsat 8 10 a.m. ± 15 minutes). The satellite is in a near-polar orbit that is directed against the earth's rotation, and is tilted by a few degrees against the earth's axis. This angle is (usually) measured counterclockwise between the ascending direction of movement of the satellite (so-called ascending node) and the equatorial plane and is called the inclination angle (cf. Fig. 10.8, shown there as the corresponding angle for descending motion). Typical values are between 96° and 102°. Due to the rotation of the earth under its orbit, the satellite scans a new ground track at each orbit until it records the ground track again after a certain time (repetition rate). Such satellites are located at altitudes between 450 and 1.100 km and complete an orbit relatively quickly in just over 90 minutes (i.e. 99 minutes for Landsat 8).

Due to the Earth's rotation, the globe moves under the satellite orbit, so that the ground tracks of two successive orbits are offset against each other (cf. Fig. 10.9). For Landsat 8, on day M the (adjacent) orbit N + 1 runs 2.100 km west of orbit N. On day M + 1, orbit N is 120 km west of orbit N of the previous day. This results in an overlap of the images taken on both days, which is larger at higher latitudes. This sequence repeats itself every 16 days.

Geostationary satellites, or more precisely *geosynchronous satellites,* are always located at a specific position above the earth. The satellite orbit is selected in such a way that the satellite movement is synchronous with the earth's rotation, i.e. the satellite moves

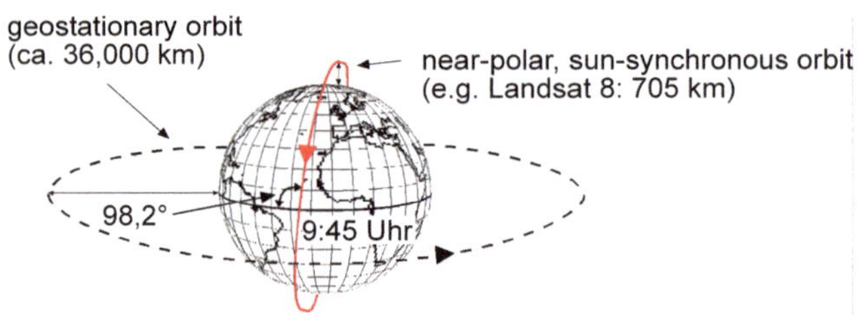

Fig. 10.8 Near-polar, sun-synchronous orbit and geostationary orbit

Fig. 10.9 Landsat orbit parameters (modified, cf. Drury 1990 p. 48)

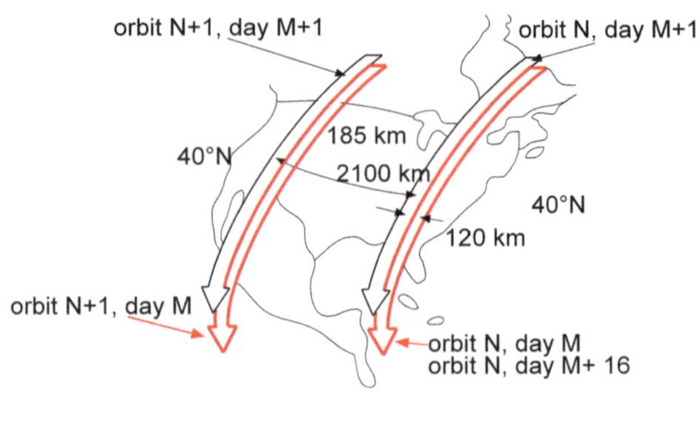

at the exact speed of the earth's rotation. This requires orbital altitudes of around 36.000 km. Such satellites are able to record the entire hemisphere of the earth at short time intervals. The best known examples are weather satellites (cf. Sect. 10.4.5).

10.4.3 Recording Principles of Scanners on Satellite Systems

The Multispectral Scanner (MSS) and Thematic Mapper (TM) as well as the more recent Enhanced Thematic Mapper Plus (ETM+) are *opto-mechanical scanners* in which one line is scanned by a mirror that oscillates back and forth (cf. Fig. 10.10). In the case of the MSS, one mirror movement recorded six lines in four spectral channels, requiring a total of 24 detectors. With the TM, 16 detectors were used to record data in each channel except the thermal channel (4 detectors). Since 6300 measurements are taken with each mirror rotation and 185 km are scanned on the ground, this results in an extent of 30 m per pixel across the flight direction.

This type of scanner of the TM is also called "whisk broom scanner" (sweeping by swinging a broom back and forth) due to the similarity of the recording principle with a broom. In contrast, there is the more recent construction principle of a "push broom scanner" (sweeping by pushing a broom).

With the older operating principle of opto-mechanical scanners, recording errors can occur due to the larger mechanical components that are generally more susceptible to failure (cf. Landsat 7). In the case of a rotating or oscillating mirror, a so-called *panorama distortion* occurs during the recording (cf. Fig. 10.10). With low-flying, airborne scanners, clearly visible distortions occur in the peripheral areas, since the distance to the recording system is considerably greater there than vertically below the recording center, i.e. in the *nadir*. In this case, the angle between the perpendicular starting from the projection center and the acquisition axis is called nadir distance or nadir deviation. The extent of the distortion depends on the aperture angle of the recording sensor, which is between 3 and

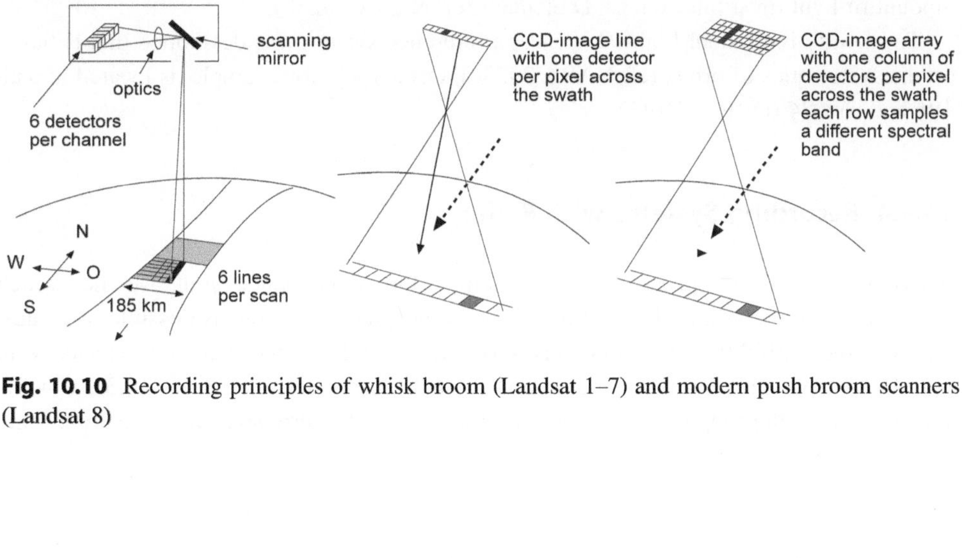

Fig. 10.10 Recording principles of whisk broom (Landsat 1–7) and modern push broom scanners (Landsat 8)

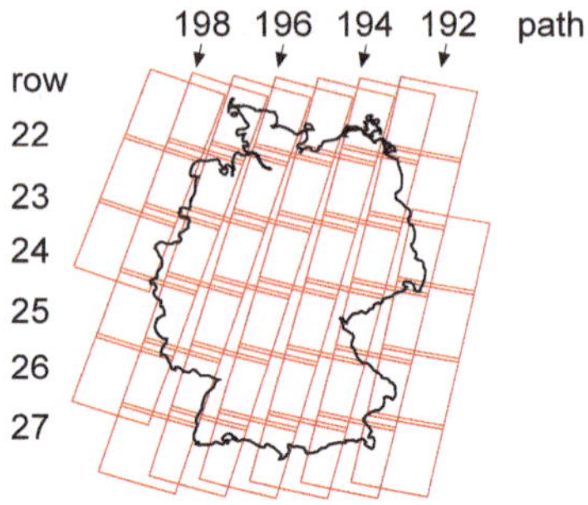

Fig. 10.11 Section of the World Reference System (WRS) from Landsat (Germany)

10 times larger for aircraft scanners than for satellites (cf. Hildebrandt 1996 p. 424). However, these distortions are not generally low in satellite systems, but only in those with a small aperture angle (such as Landsat). For example, the AVHRR sensor has a relatively large aperture angle and therefore a relatively large panorama distortion. The resolutions of this sensor are 1,1 km in the nadir and 2,4 km in the direction of flight at the edge and 6–9 km across the direction of flight (cf. Lillesand et al. 2008 p. 463). This is related to the purpose of the system to cover the largest possible area (2.400 km^2) during a swath.

In contrast, the more modern opto-electronic scanners are more powerful. In a push-broom scanner, the reflected radiation hits several photocells (CCD elements) arranged next to each other in a row, which convert the intensity of the signals into voltage values (cf. Fig. 10.10). An analog-to-digital converter then converts these values into bit sequences. All image elements of a line are recorded simultaneously. Such sensors therefore do not generate any panorama distortion. In particular, no mechanical movements are necessary. In contrast to the opto-mechanical scanning principle, which operates sequentially, the simultaneous recording of entire lines allows the received light energy to act on a CCD element for a longer period of time.

By installing many smaller CCD elements, a higher geometric resolution becomes possible. However, the CCD elements cannot be reduced in size at will, since a certain amount of light must fall on a CCD element (cf. NASA 2023b).

To identify individual Landsat scenes, a reference system was developed that is based on the systematics of orbits (cf. NASA 2023c). Osnabrück, for example, is located in path 196 and row 24 (cf. Fig. 10.11).

10.4.4 Recording Systems with Radar

A clear disadvantage of passive optical systems is their dependence on the weather, which only allows the recording of useful data if the atmosphere is as free as possible of clouds, haze or other turbidities. The relatively short wavelengths of the radiation detected with optical systems prevent the penetration of such turbidities (cf. Sect. 10.3.2), while wavelengths in the range of 1 mm to 1 m, the so-called microwaves or radio waves, are

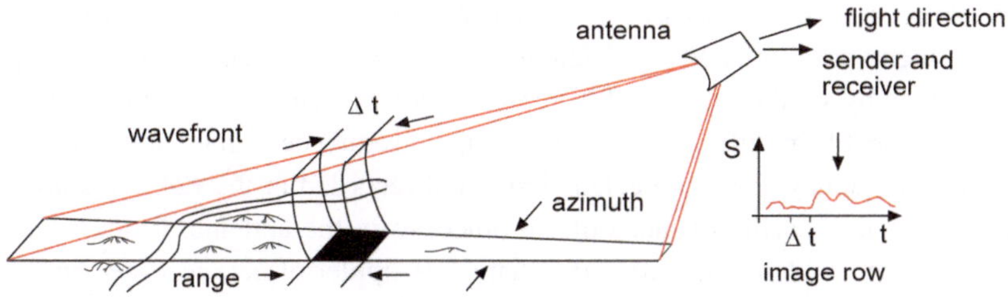

Fig. 10.12 Recording principle of radar systems (cf. Albertz 2009 p. 57)

transmitted. This is used in *radar remote sensing* (for introduction cf. Mather and Koch 2011 p. 58). Radar is an acronym for radio detection and ranging and was originally developed to detect objects using radio waves and to determine their range or position. Since objects emit radio waves naturally only to a small extent, active systems are usually used in remote sensing, which themselves emit radiation pulses and can receive the reflected radar echo again. Such a system is therefore not dependent on the natural radiation of the observation object and is therefore also capable of recording at night.

A radar system transmits short, high-energy microwave pulses to the side of the flight direction at a specific angle (angle of incidence) to the ground. The reflected signals are then recorded by the sensor on the satellite or aircraft. This approach is also based on the basic principle of remote sensing. Thus, various features of the surface such as relief or topography, microclimate, moisture, soil (including surface roughness and texture) or vegetation cover determine the reflection in the microwave range.

The radiated electromagnetic energy can vary in wavelength and polarisation. Polarisation refers to the ability to filter radar beams, so that they only send out in certain directions perpendicular to the wave motion. From the sensor's perspective, the wavelength and polarisation used as well as the angle of incidence determine the appearance of the reflected signals and therefore the characteristics of surface objects in the radar image (cf. Fig. 10.12). Furthermore, the spatial resolution of a radar system in flight direction (azimuth) is determined by the overall length of the antenna used. The temporal duration of the energy pulse influences the resolution perpendicular to the flight direction (range). Thus, with increasing antenna size (i.e. aperture), the width of the radar pulse decreases, causing the radar waves to focus more strongly and therefore resulting in a better spatial resolution. In contrast, the width of the radar pulse increases as the distance between the radar and the observed object increases. However, antennas cannot become arbitrarily large, in order e.g. to compensate the flight altitude of satellite-based systems, so that conventional radar systems (real aperture radar, cf. Hildebrandt 1996 P. 576) can only be used for small flight altitudes, which do not have a very large distance between antenna and ground surface.

Synthetic aperture radar (SAR) systems use only a short antenna, but emit microwaves in a wide lobe so that terrain points are repeatedly irradiated during flight (cf. Lillesand et al. 2008 pp. 638). The further away an object is, the more often it is detected. The long antenna, which is actually necessary for the use of microwave remote sensing from high altitudes, is therefore simulated by different positions of the antenna along the flight path. The system thereby synthesizes the effect of a very long antenna. By recording the reflected signals of an object from different positions, a signal sequence is generated for each object, in which the frequency of the reflected signal and recorded at the radar antenna changes systematically with varying distance of the object from the antenna. From the frequency shift between the emitted and received radiation (Doppler effect) recorded in this signal sequence, the signals can be reconstructed during evaluation using complex mathematical procedures as if they originated from a single antenna.

Effects caused by the oblique view of the radar and therefore by the recording technique determine the appearance of the earth's surface in the radar image and must be taken into account when processing the data. The electrical properties of the materials on the earth's surface (expressed by the so-called dielectric constant) have a great influence on the reflection and the penetration depth of microwaves. In particular, the wavelength determines the penetration depth of the radiation into certain materials. The principle applies here: the greater the wavelength, the greater the penetration depth. This relationship can be used to select specific frequency ranges for a specific examination. Overall, the processing of radar data, which goes beyond standard applications, is very complex and is not discussed in detail here.

During the Shuttle Radar Topography Mission (SRTM) in February 2000, in which two radar instruments were used simultaneously, almost the entire surface of the Earth was measured in three dimensions with high precision using radar interferometry. In 2014, the U.S. federal government announced that the generated topographical data at the highest-resolution would be released worldwide by the end of 2015, whereas previously it was only available at varying resolutions. The new data are published at a sampling rate of one arc second, which reflects the full resolution of the original measurements (cf. NASA 2023d, also USGS 2023l). The data are used to create a high-resolution digital terrain model of the earth's surface.

More recent examples of satellite-based SAR systems are the German satellite systems TerraSAR-X and TanDEM-X (TerraSAR-X add-on for Digital Elevation Measurement), which were launched in 2007 and 2010, respectively (cf. DLR 2023b). Terra-SAR-X is currently the most precise high-resolution commercial radar satellite in orbit, providing data services with a unique precision, quality and reliability. TanDEM-X and TerraSAR-X fly in close formation only a few hundred meters apart and form the first SAR (Synthetic Aperture Radar) interferometer of this type (parallel flight) in space (cf. Zink et al. 2017). The standard elevation model product (completed in 2016) has a spatial resolution of 12 m and 30 m with an absolute vertical accuracy better than 10 m and a relative vertical accuracy of 2 m (cf. DLR 2023c).

10.4.5 Weather Satellites

Weather satellites exist both in geosynchronous, i.e. geostationary, and in sun-synchronous versions. An international network of geostationary weather satellites enables recordings of the entire earth at almost any time of day. This offers optimal conditions for observing the weather development. The high temporal resolution has to be paid for by the high orbit with a low spatial resolution.

Typical geostationary weather satellites are the platforms of the US National Oceanic and Atmospheric Administration (NOAA), which are in an orbit of about 35.800 km above the Earth (cf. NOAA 2023a, 2023b and 2023c). Together, GOES-16 and GOES-17 watch over the Western Hemisphere from the west coast of Africa all the way to New Zealand (cf. GOES-R 2023). GOES-18 was launched in March 2022 and will replace GOES-17 (as of spring 2022, cf. GOES-T 2023).

These satellites carry several sensors, of which the Advanced Baseline Imager deserves special mention, which records radiation in 16 wavelength bands and operates in different scan modes (up to every 30 seconds for a 1000 × 1000 km area below the satellite, cf. NOAA 2023d).

EUMETSAT has also been operating several geostationary satellites for weather and climate observation for more than 35 years. The data for Europe are provided by the Meteosat satellites, which were developed by the European Space Agency (ESA or the former European Space Research Organisation ESRO, cf. EUMETSAT 2023a). First-generation Meteosat satellites include Meteosat-1 (launched in 1977, retired in 1979) to Meteosat-7 (launched in 1997, retired in 2017, cf. EUMETSAT 2023b). The second generation currently has four satellites in orbit with Meteosat-8 (launched 2002) to Meteosat-11 (launched 2015) (cf. EUMETSAT 2023c). They scan Europe, Africa and the Indian Ocean. Key improvements of Meteosat's second generation include a new main instrument (Spinning Enhanced Visible and IR Imager, SEVIRI) that observes the Earth in twelve channels, of which the High-Resolution-Visible (HRV) channel has a spatial resolution of 1 km in nadir and which scans half of the Earth. Meteosat-11 is the main operational satellite providing full earth disc imagery every 15 minutes. Meteosat-10 offers the Rapid Scanning Service, which even scans the northern third of the earth every five minutes (cf. EUMETSAT 2023d). Meteosat's contribution will continue into the 2040s with the introduction of Meteosat Third Generation (MTG, launch of MTG-I1 planned at the end of 2022, cf. EUMETSAT 2023e, for future satellites cf. EUMETSAT 2023f).

EUMETView is a visualisation service that allows users to view EUMETSAT imagery interactively via an online map viewer or use it as a Web Map Service (cf. EUMETSAT 2023g).

Weather satellites in a sun-synchronous orbit also enable the observation of regional processes such as snow and ice melt or surface temperature anomalies on the water surface due to their relatively high-resolution. Typical weather satellites in low, sun-synchronous polar orbit are the *POES* (*Polar-Orbiting Operational Environmental Satellites*) platforms of the *US National Oceanic and Atmospheric Administration* (*NOAA*). These satellites carry several sensors, of which the *Advanced Very High-resolution Radiometer* (*NOAA-AVHRR*) is of central importance for hydrological, oceanographic, and meteorological studies. As of NOAA-15, the AVHRR/3 sensor is used, which has six channels (cf. USGS 2023c).

The Joint Polar Satellite System (JPSS) is the successor to the POES program. Launched on 2017, NOAA-20 (formerly JPSS-1) began the next generation of polar-orbiting NOAA satellites (cf. JPSS 2023a). Of the five instruments on board, the VIIRS instrument (Visible Infrared Imaging Radiometer Suite), as a further development of the AVHHR, records the visible and infrared range between 0.412 and 12 μm in 22 spectral bands (max. 3000 km wide recording swath, resolution in nadir 400 m, cf. JPSS 2023b and USGS 2023o). The MetOp satellites also fly in a sun-synchronous orbit. They complement the Meteosat geostationary weather satellites and work in conjunction with the NOAA satellite system. The MetOp (Meteorological Operational satellite program) satellite program of the European Space Agency (ESA) offers weather data services for monitoring the climate and improve weather forecasting (launch of MetOp-A in 2006, MetOp-B in 2012, MetOp-C in 2018, cf. ESA 2023a).

10.4.6 Landsat

10.4.6.1 Mission

The Landsat program was initiated by the U.S. Geological Survey (USGS) and the National Aeronautics and Space Administration (NASA) to routinely collect land images from space. NASA develops remote sensing instruments and spacecraft, and launches and validates the performance of the instruments and satellites. The USGS assumes ownership and operation of the satellites in addition to managing all ground reception, data archiving, product generation, and distribution (USGS 2023d).

With the Landsat images, a unique collection of environmental observations of the Earth from space has been available since the first images were taken in 1972, which is also available free of charge (cf. Sect. 10.5.3). Thus, data are available with sensor technology that has remained constant for decades, making the data comparable and providing complete spatial coverage of the earth. This accounts for the great value of the Landsat mission, although more recent sensors have a considerably finer ground resolution of the image data, which are not available for such a long survey period and large spatial scope of the survey (for extensive information on the Landsat program cf. USGS 2023d). Table 10.3 lists areas of application for the Landsat sensor spectral bands. It should be noted that a single channel rarely leads to a substantive conclusion and that instead, as with classification methods, multiple spectral bands must always be considered.

Table 10.3 Application areas of the spectral bands of the TM and ETM+ sensors (cf. USGS 2023e)

Spectral bands	µm		Use
1	0.45–0.52	Visible blue	Bathymetric mapping, useful for distinguishing soil from vegetation and deciduous from coniferous forest
2	0.52–0.61	Visible green	Designed to measure peaks of reflected radiation in the green section of the electromagnetic spectrum (i.e. from vegetation)
3	0.63–0.69	Visible red	Designed to measure reflected radiation in a section of the electromagnetic spectrum characteristic of chlorophyll absorption, useful for distinguishing plant species
4	0.76–0.90	Reflected infrared	Useful for biomass mapping, distinguishing vegetation types, highlighting vegetation boundaries between land and water and landforms, and identifying plant vitality (the more "juvenile", the higher IR reflectance)
5	1.55–1.75	Reflected infrared	Useful for distinguishing the moisture content of soil and vegetation (indicator of water deficiency), separation of clouds and snow, geological and pedological mapping, penetration of haze
7	2.08–2.35	Reflected Infrared	Useful for geological and pedological mapping, useful for mapping hydrothermally altered rocks
6	10.4–12.5	Thermal-Infrared	Useful for thermal mapping and soil moisture estimation

10.4.6.2 Recording Systems of Landsat 1 to 7

Landsat 1 (originally Earth Resources Technology Satellite ERST) was the first satellite of the US Landsat program, launched in 1972. Its instruments were the *Return Beam Vidicon System* (RBV) and the *Multispectral Scanner* MSS (for geometric, spectral and radiometric properties of both instruments cf. Tables 10.4 and 10.5). Since the geometric and radiometric properties of the RBV system, which consisted of three video cameras, were inferior to those of the MSS, the MSS system became the primary imaging device on board Landsat. The choice of spectral ranges for the MSS had practical reasons: This was to provide image products that are similar to false-color infrared aerial images, for interpretation and analysis of which a wide range of experience already existed. Landsat 3 received an additional thermal channel with a geometric resolution of 240 m, which turned out to be faulty and was therefore hardly used.

The sensors of the next Landsat generation were adapted to meet the increasing demands and needs of the various geoscientific disciplines. The result was the *Thematic Mapper*, named for the purpose of providing thematic representations for the various geoscientific domains. The spatial, the spectral and also the radiometric resolution were strongly improved (cf. Tables 10.4 and 10.5). In contrast to the MSS system, the selection of the displayed ranges of the TM was based on in-depth studies of the reflection of different surface types such as vegetation and rock minerals. In addition to spectral bands in

Table 10.4 Overview of Landsat missions 1–7 (cf. USGS 2023e)

	Landsat 1	Landsat 2	Landsat 3	Landsat 4	Landsat 5	Landsat 7
Start	1972	1975	1978	1982	1984	1999
Decommissioning	1978	1983	1983	1993	2013	
Flight altitude (km)	907	908	915	705	705	705
Inclination	99.9°	99.2°	99.1°	98.2°	98.2°	98.2°
Circulation rate	103 min.	103 min.	103 min.	99 min.	99 min.	99 min.
Ground track width	185 km	185 km	185 km	185 km	185 km	185 km
Repetition rate	18 days	18 days	18 days	16 days	16 days	16 days
Equator crossing (descending path)	09:30	09:30	09:30	09:30	09:30	10:00
Scanner	RBV	RBV	RBV			
	MSS	MSS	MSS	MSS	MSS	
				TM	TM	ETM+

RBV Multispectral return beam Vidicon, *MSS* Multispectral scanner with thermal channel, *TM* Thematic mapper, *ETM+* Enhanced thematic mapper plus

the visible and near-infrared, those in the shortwave and thermal infrared were also implemented, but the latter at a relatively low spatial resolution (120 m). This system, along with the MSS system, was on board Landsat 4 and 5. The TM provided almost continuous images of the Earth from July 1982 to November 2011, with a 16-day repeat. After the TM was shut down, the MSS was briefly reactivated until January 2013. The launch of Landsat 6, which had an improved TM (*Enhanced Thematic Mapper*) instrument and still the MSS on board, failed in 1993.

The spectral sections covered by the RBV sensor on Landsat 1 and 2 were designated as channels 1 through 3. The channels of the MSS were then further numbered starting with 4 in the first generation of Landsat. TM channel 7 is not in the sequence because it was added last, after the other six channels were already fixed.

After the failure of Landsat 6, Landsat 7 was launched in April 1999. Its orbit data correspond to those of Landsat 4 and 5. The ETM instrument was further improved and expanded and was renamed *Enhanced Thematic Mapper Plus* (*ETM+*). Thus, a panchromatic channel with a spatial resolution of 15 m has been installed and the spatial resolution of the thermal channel has been increased to 60 m. Another important improvement concerns the radiometric calibration of the instruments (directly) on board the satellite. These measures allow a radiometric correction with an absolute accuracy of 5% within the different levels of data preparation (cf. Sect. 10.4.6.4).

On March 31, 2003 the scan line correction (SLC) instrument failed, making it impossible to compensate for the satellite's forward motion. The SLC instrument ensures that the lines taken transverse to the satellite's flight line (cf. Fig. 10.10) connect to one another and are parallel to one another. Landsat 7 data has data gaps (22% missing pixels). The United States Geological Survey provides information and tools concerning the data processing of

Table 10.5 Characteristics of the Landsat instruments

Instrument (platform)	Geometric resolution	Spectral resolution			Radiometric resolution
Return Beam Vidicon RBV	80 m	1.	0.48–0.58 µm	Green	
(Landsat 1, 2 and 3)		2.	0.58–0.68 µm	Red	
		3.	0.68–0.78 µm	NIR	
Multispectral scanner MSS	79 m	1	0. 0,5–0,6 µm	Green	6 bit
(Landsat 1 to 5)		2	0. 0,6–0,7 µm	Red	
		3	0. 0,7–0,8 µm	NIR	
		4	0. 0.8–1.1 µm	NIR	
Thematic mapper TM	30 m	1.	0.45–0.52 µm	Blue	8 bit
Landsat 4 and 5		2.	0.52–0.60 µm	Green	
		3.	0.63–0.69 µm	Red	
		4.	0.76–0.90 µm	NIR	
		5.	1.55–1.75 µm	SWIR1	
		7.	2.08–2.35 µm	SWIR2	
	120 m	6.	10.4–12.5 µm	THIR	
Enhanced thematic mapper plus	30 m	1.	0.45–0.52 µm	Blue	8 bit
(ETM+)		2.	0.52–0.60 µm	Green	
		3.	0.63–0.69 µm	Red	
		4.	0.77–0.90 µm	NIR	
		5.	1.55–1.75 µm	SWIR1	
		7.	2.08–2.35 µm	SWIR2	
	60 m	6.	10.4–12.5 µm	THIR	
	15 m		0.52–0.90 µm	PAN	

Landsat 7 (cf. USGS 2023f, in detail Mather and Koch 2011 pp. 89, on the missions as a whole cf. USGS 2023g).

10.4.6.3 Landsat 8 and 9

The Landsat Data Continuity Mission (LDCM) continues the 40-year history of Landsat missions under a new name. The LDCM is designed to ensure the continued collection and availability of Landsat-like data beyond the duration of the Landsat 5 and Landsat 7 missions. The launch of the Landsat Data Continuity Mission (i.e. in the continued count: Landsat 8) took place on February 11, 2013. Newly received data from Landsat 7 ETM+ and Landsat 8 OLI/TIRS will be made available for download within 12 hours. All scenes are processed as level 1 products (cf. Sect. 10.4.6.4). The objectives formulated in 2012 was: "The mission objectives of the LDCM are to (1) collect and archive medium resolution (30-meter spatial resolution) multispectral image data affording seasonal coverage of the global landmasses for a period of no less than 5 years; (2) ensure that LDCM data are sufficiently consistent with data from the earlier Landsat missions in terms of

Table 10.6 Application areas of the channels of the OLI and TIRS sensors of Landsat 8 (cf. USGS 2023h)

Channel	µm	Insert		
1	0.43–0.45	coastal/ aerosol	30 m	intensified observations of coastal zones
2	0.45–0.51	Blue	30 m	Like channel 1 TM/ETM+
3	0.53–0.59	Green	30 m	Like channel 2 TM/ETM+
4	0.64–0.67	Red	30 m	Like channel 3 TM/ETM+
5	0.85–0.88	Near IR	30 m	Similar to channel 4 TM/ETM+
6	1.57–1.65	SWIR1	30 m	Similar to channel 5 TM/ETM+, useful for detecting plant stress due to drought, delineating burned areas and fire-affected vegetation, and detecting fire
7	2.11–2.29	SWIR2	30 m	Like channel 6
8	0.50–0.68	Pan	15 m	Useful for sharpening multispectral images
9	1.36–1.38	Cirrus	30 m	Useful for better detection of cirrus clouds
10	10.60–11.19	TIRS1	100 m	To be used for thermal mapping
11	11.5–12.51	TIRS2	100 m	Like channel 10

acquisition geometry, calibration, coverage characteristics, spectral characteristics, output product quality, and data availability to permit studies of land-cover and land-use change over time; and (3) distribute LDCM data products to the general public on a nondiscriminatory basis at no cost to the user." (USGS 2012 P. 1).

The Landsat Data Continuity Mission carries two pushbroom scanners, the Operational Land Imager (OLI) and the Thermal Infrared Sensor (TIRS). The OLI collects data in nine shortwave spectral bands (cf. Table 10.6). Improved (historical) channels and the addition of a new first channel for the detection of coastlines and aerosols such as a new so-called Cirrus channel provide data with improved radiometric performance at 30 m ground resolution in the multispectral range or 15 m for panchromatic images. The TIRS acquires data in two long-wavelength thermal channels with a ground resolution of at least 100 m, which are resampled to a resolution of 30 m to match the data of the OLI sensor.

Landsat 9 will extend earth observation by Landsat over half a century. Landsat 9 was launched on September 2021. The instruments onboard Landsat 9 are improved replicas of the instruments currently collecting data onboard Landsat 8 (cf. USGS 2023i). Landsat 9 entered the orbit of Landsat 7. The older satellite is scheduled to be decommissioned, the nominal science mission ended in April 2022 (cf. USGS 2023f). Landsat 7 imaging resumed on May 5, 2022, at a lower orbit of 697 km. However, because the quality of the data remains nominal, and the health of the satellite remains steady, data acquisitions will continue into 2023. Landsat 9 overflies every point on earth every 16 days with an eight-day offset from Landsat 8.

10.4.6.4 Data Preparation

The standard data preparation of Landsat satellite data is done in three versions (besides the raw format): level L1GS (systematic correction), level L1GT (systematic terrain correction) and L1TP (precision and terrain correction). The data are radiometrically corrected (cf. Sect. 10.6.1.1). The processing levels differ in terms of geometric correction. The L1TP and L1GT products incorporate a digital elevation model for topographical accuracy, and L1TP uses ground control points for additional geometric accuracy (cf. USGS 2023j and USGS 2023k).

The importance of radiometric calibration of satellite data increases as data from different sensors are to be combined, such as within the Earth Observing System (EOS), which is a composite of multiple polar-orbiting, low-inclination satellite systems for long-term global observations of the land surface, biosphere, solid earth, atmosphere, and oceans, and in which time-series images are used to analyse change (cf. NASA 2023e).

10.4.7 SPOT and Pléiades

The satellite systems *SPOT* (*Système Probatoire d' Observation de la Terre*, also called Satellite Pour l'Observation de la Terre), which were developed by the French space agency CNES, among others, have used *opto-electronic scanners* (cf. Fig. 10.10) from the beginning. Due to the high ground resolution this system was superior to the older Landsat system. SPOT 4, for example, had an array of 6.000 detectors per line for panchromatic recordings, so that with a recording width of 60 km in nadir a ground resolution of 10 m resulted. The sensors from SPOT 1 to 3 had a multispectral and a panchromatic recording mode. The two HRVIR (High-resolution Visible and Infrared) instruments of SPOT 4, each of which also had a channel for the mid-infrared, could be operated completely independently of one another, so that, for example, one instrument was used to record in the nadir, while the other instrument used its side-capability to respond to current user requests or to record part of a stereo scene. In addition, the VGT vegetation sensor was used on SPOT 4 and 5 for large-scale detection of the vegetation. In the meantime the systems SPOT 1 to 4 have stopped their service. The data are still available in the SPOT archive (cf. Airbus 2023a).

The improvements for SPOT 5 mainly concerned the geometric resolution. An additional very high-resolution mode was able to achieve a geometric resolution of up to 2,5 m by means of a new computational process. The High Geometric Resolution Imaging Instrument (HRG) was available in duplicate. Furthermore, a Vegetation 2 Instrument (VGT) identical in construction to the Vegetation Instrument on SPOT 4 and the High-Resolution Stereoscopic Imaging Instrument (HRS) were on board, which was operated simultaneously with the other instruments and recorded panchromatic stereo image pairs with a geometric resolution of 10 m (for technical data of SPOT 1–5 cf. SPOT IMAGE 2023).

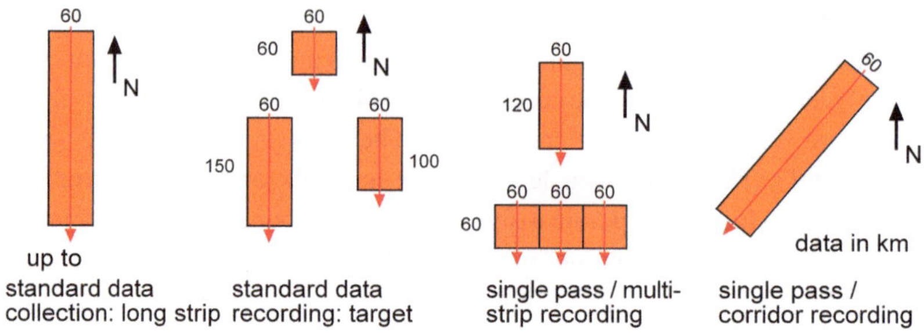

Fig. 10.13 Recording modes of SPOT 6/7 (cf. ASTRIUM 2023)

The sensor system of SPOT offers advantages due to the swivelling deflection mirror (cf. Fig. 10.13). Thus, it is possible to record image strips parallel and also sideways to the direction of flight, which also makes it possible to record stereo images. While the same area will only be recorded again in 26 days with a vertical recording direction, the same scene can be scanned much more frequently with sideways recordings in flight the direction.

SPOT 6 and SPOT 7 were launched in 2012 and 2014 respectively (for technical data cf. ASTRIUM 2023). With a large strip width, which corresponds to that of its predecessor satellites, SPOT 6 is suitable for covering large areas. The higher resolution of 1,5 m – compared to 2,5 m for SPOT 5 – enables the detection of even more precise details (using the New AstroSat Optical Modular Instrument, NAOMI). SPOT 6 is in a near-polar, sun-synchronous orbit at an altitude of 694 km with an orbital periodicity of 99 minutes. Within certain limits, the recording strips can be freely selected. Each sensor has a viewing angle that can be changed up to 30° (or extended up to 45°) in the longitudinal direction. The reacquisition capacity is one day using SPOT 6 and SPOT 7 simultaneously, and between one and three days using only one satellite (depending on the longitude of the area of interest). The sensors can be quickly oriented in all directions to cover different areas of interest in one pass (30° in 14 seconds including stabilisation time).

Pléiades-HR is a high-resolution optical imaging constellation in orbit that extends the SPOT program. Pléiades-HR 1A and Pléiades-HR 1B were launched in 2011 and 2012, respectively. While SPOT 6 and SPOT 7 scan large areas with a repetition rate of one day, Pléiades 1A and Pléiades 1B are able to record smaller sections at higher ground resolution. SPOT 6 and SPOT 7 operate in the same orbit as Pléiades 1A and Pléiades 1B, forming a 4-satellite constellation offset by 90° (cf. EOS 2023a). Pléiades provides panchromatic images with 50-centimeter resolution and four-channel multispectral images with 2-meter resolution (four multispectral channels, five near-infrared channels, one panchromatic channel, cf. EOS 2023b).

As a successor, from 2021 four Pléiades Neo satellites orbit the earth at an altitude of 620 km in a sun-synchronous orbit in order to enable twice-daily overflights anywhere on earth. The key features are: 30 cm resolution, deep blue, blue, green, red, red edge, near-infrared, panchromatic spectral bands, mono, stereo and tri-stereo acquisitions (cf. Airbus 2023b, on the overall system cf. Airbus 2023c).

10.4.8 ASTER on Terra

The earth observation satellite Terra (also known as EOS-1 or EOS-AM1) is the flagship of the *Earth Observing System* (*EOS*) launched in 1999 and is orbiting in a polar, sun-synchronous orbit. "Terra explores the connections between Earth's atmosphere, land, snow and ice, ocean, and energy balance, to understand Earth's climate and climate change and to map the impact of human activity and natural disasters on communities and ecosystems" (cf. NASA 2023f). Terra is an international mission and carries five systems:

- CERES (Clouds and the Earth's Radiant Energy System, USA), study of heat fluxes to and from the earth (2 instruments)
- MISR (Multi-angle Imaging Spectro-Radiometer, USA), study of the scattering of sunlight, clouds, aerosols and vegetation (9 cameras with different viewing directions and 4 channels each)
- MODIS (MODerate resolution Imaging Spectroradiometer, USA), study of large-scale changes in the biosphere (36 channels)
- MOPITT (Measurements of Pollution in the Troposphere, Canada), study of the concentration of methane and carbon monoxide in the troposphere (3 channels)
- ASTER (Advanced Spaceborne Thermal Emission and Reflection Radiometer, Japan), study of the earth to create detailed maps including land cover and surface temperature (3 independent sensors with 14 channels).

Channels 1 to 3 of ASTER are very similar to channels 2 to 4 of the Thematic Mapper. Among other things, the sensor supplements or replaces Landsat due to its higher resolution and ensures a certain continuity in times when Landsat instruments provide incomplete data (Landsat 7 since 2003). Channel 3 is duplicated. It scans in the nadir and backwards, so that a stereoscopic analysis and the creation of cost-effective elevation models are possible (for ASTER characteristics cf. NASA 2023g).

10.4.9 Copernicus and Sentinel

Copernicus (formerly Global Monitoring for Environment and Security-Programme, GMES) is a recent comprehensive earth observation program coordinated by the European Space Agency (ESA). The program aims to provide accurate, timely and easily accessible information to improve environmental management, understand and mitigate the effects of climate change, and ensure civil security (cf. ESA 2023b). ESA is developing a new family of satellites, Sentinel 1 to 6, specifically tailored to the operational requirements of the Copernicus program, each consists of two satellites. There are currently three complete two-satellite constellations in orbit, as well as an additional single satellite designated Sentinel-5P (cf. ESA 2023b and ESA 2023c).

Sentinel-1A (launched in 2014) and Sentinel-1 B (launched in 2016) characterise an all-weather polar orbiting radar imaging mission for land and ocean services. Sentinel-1 continues SAR earth observation (Synthetic Aperture Radar, cf. Sect. 10.4.4) in the C-band of ESA's ERS-1, ERS-2 and ENVISAT programs, as well as the Canadian RADARSAT programs. As a constellation of two satellites orbiting 180° apart, the mission images the entire Earth every six days (cf. ESA 2023d).

Sentinel-2A (launched in 2015) and Sentinel-2B (launched in 2017) is a polar-orbiting, multispectral high-resolution imaging mission for land monitoring. The Sentinel-2 mission (like Sentinel-1) is based on a constellation of two identical satellites in the same orbit, offset by 180° to ensure optimal coverage and data transmission. Together they cover the earth all five days. They have a high-resolution sensor (Multispectral imager MSI) covering 13 channels from visible blue (440 μm) to shortwave infrared (2190 μm) with a resolution of 10 m (VIS) and 20 m (SWIR). Three spectral channels are designed for atmospheric correction of cloud, water vapor and aerosol influence and provide data with 60 m ground resolution. Sentinel-2, with a ground track width of only 290 m and more frequent repetition times, continues the French SPOT and the US Landsat missions with major improvements. The mission provides basic data for agriculture (land use and cover, crop forecasts, water and fertilizer requirements), forestry (stock density, health status, forest fires), water monitoring, spatial planning and disaster management. Real-time recordings of floods, volcanic eruptions and landslides contribute to the production of up-to-date maps in the event of natural disasters (cf. ESA 2023e).

Sentinel-3A (launch 2016) and Sentinel-3B (launch 2018) is a multi-instrument mission to measure sea-surface topography, sea- and land-surface temperature, ocean color and land color with high-end accuracy and reliability. The mission is based on two identical satellites. They orbit the earth in an optimal constellation to achieve optimal global coverage and data delivery. Thus, every two days, the ocean and land color instrument will provide global coverage with a swath width of 1270 km. They complement the Sentinel-2 mission. On the one hand, the temperature, color and height of the sea surface as well as the thickness of the sea ice are recorded. On the other hand, detailed information can be provided when monitoring forest fires, for mapping land use, provide indices of vegetation state and measure the height of rivers and lakes (cf. ESA 2023f).

The Sentinel 4 and Sentinel 5 missions aim to monitor the composition of the atmosphere. The Sentinel-4 mission is to be carried out on Meteosat's third-generation geostationary satellites (including a UVN ultraviolet visible near-infrared spectrometer). The Sentinel-5 mission is to be carried out on the second-generation polar-orbiting MetOp satellite (including a UVNS ultraviolet visible near-infrared shortwave spectrometer, cf. ESA 2023g).

Sentinel-6 carries a radar altimeter to measure global sea level and will mainly be used for oceanography and climate studies. The first satellite was launched in 2020. (cf. ESA 2023h).

Sentinel-5 Precursor (also known as Sentinel-5P, the predecessor of Sentinel-5) is the first Copernicus mission dedicated to monitoring the atmospheric (launch 2018). The satellite is equipped with the Tropomi spectrometer, which can detect a wide range of trace gases such as nitrogen dioxide, ozone, formaldehyde, sulphur dioxide, methane, carbon monoxide and aerosols. Sentinel-5P was developed to reduce the data gaps between the Envisat satellite and the launch of Sentinel-5 (cf. ESA 2023i).

10.4.10 Recent Commercial High-Resolution Sensors

For a long time, satellite remote sensing has primarily characterised by the fact that first military clients and then state-funded (research) satellites were used. In contrast, several privately financed, i.e. commercially oriented missions are currently planned or active, which provide high spatial resolutions for civilian applications (for an overview of various remote sensing platforms and sensors, including aircraft and UAV platforms cf. Toth and Jutzi 2017). However, when using commercial systems, the question arises whether users also benefit adequately from this development and the new wide range of providers. So far, the space agencies, which are funded by governmental organisations, have not taken enough account of user needs. In particular, there has been too little support for data refinement. It is possible that private and commercial users want to (and now probably can) have more influence on the development of satellite systems, which has not been a priority due to the previous orientation towards research institutions. Thus, new impulses can be expected from a stronger commercial orientation.

In contrast to the "classic" satellite systems such as Landsat, the new *high-resolution satellite systems* can no longer scan the earth's surface over a wide area (here: geometrically high-resolution). The size of a scene at high-resolution creates large amounts of data that are impossible to maintain in reserve or stock. A user must order recording of an area to be covered in advance. This means that older scenes can only be accessed in rare cases, as they are often simply not available.

The older missions Early Bird 1, OrbView 2, OrbView, Ikonos, Quickbird and WorldView-4 are now terminated (for data sheets of WorldView-1 to 4 and GeoEye-1 cf. European Space Imaging 2023). The RapidEye earth observation system comprises five satellites equipped with optical cameras (launched in 2008). Within one day, the system can be pointed at any point on earth (cf. DLR 2023d).

While the Landsat or Copernicus missions, for example, are based on single, large satellites, the US-company Planet operates a large number of mostly small satellites (cf. Planet 2023a):

The PlanetScope satellite constellation (so-called flock) consists of several launches of groups of individual satellites (DOVEs). Each DOVE satellite is a small satellite in the standard CubeSat 3 U format (10 cm × 10 cm × 30 cm). Roughly 200 Doves orbit the planet every 90 minutes, providing near real-time images for time-sensitive monitoring (four bands: blue, green, red and near infrared (or five bands including an additional red-edge band) at 3.7 m ground resolution), can image the entire land surface of the earth on a daily basis (cf. ESA 2023i, ESA 2023j and Planet 2023a).

The SkySat fleet, currently consisting of 21 satellites (five channels: blue, green, red, near infrared and pan) can visit any point on Earth with a ground resolution up to 50 cm 5–7 times per day. Besides, SkySat offers a unique acquisition option to create high-resolution time-lapse and real-time videos (cf. Planet 2023b).

10.5 Digital Images

10.5.1 Recording Digital Images in Remote Sensing

The sensor systems scan the earth's surface line by line in the direction of flight, whereby each line is resolved into several raster cells. The reflected radiation as measured irradiance L at the sensor is coded using analog-digital conversion in a certain radiometric resolution (cf. Sect. 10.4.1).The sensor systems therefore provide numerical matrices (raster data) for each spectral band, the values of which represent the intensity of reflection or radiation within a specific spectral range (channel) (cf. Fig. 10.14). These values are not available in any unit of measurement and depend on the calibration of the sensor system (cf. Sect. 10.6.1.1). A higher numerical value means a higher irradiance at the recording system, i.e. higher reflection or emission at the ground. Therefore these values are called *brightness values* (*BV*) or *digital numbers* (*DN*) or *greyscale values* (cf. Sect. 10.5.2).

The possible range for the values of a number matrix depends on the recording system as well as on the issue and the topic to be represented. In remote sensing, 6- to 12-bit data types are used to represent a pixel. One channel of a scene of the Landsat 8 OLI instrument, for example, has a radiometric resolution or signal quantity (cf. Sect. 10.4.1) of 12 bits, so that $2^{12} = 4096$ values are available for storing the irradiance at the recording system for each channel and for each pixel (Landsat 5 and 7 only $2^8 = 256$ values, 1 byte).

10.5.2 Visualising Digital Images in Remote Sensing

In remote sensing, digital recording systems do not record "images" on film as in the case with analog photography with cameras. Instead, separate number matrices are recorded by the scanner systems for each recording channel (cf. Sect. 10.5.1). Images are only created, when these numerical values are reproduced by an output device. When converting the numerical values of just one channel, usually monochrome from black to white, a greyscale image is generated. With a radiometric resolution of 8 bits, black is assigned the value 0, white is assigned the value 255, and the values in between are assigned correspondingly grey levels (cf. Fig. 10.14). Therefore, the numerical values, are often referred to as *greyscale values*. However, the resulting image cannot be compared to a black and white image in photography.

Color images on a monitor or on a printer are created by additive or subtractive color mixing of three primary colors (cf. Sects. 2.5.7 and 7.7.4). Accordingly, multispectral data

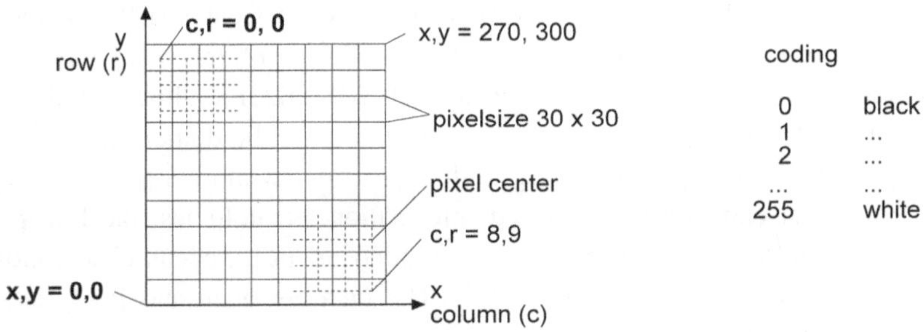

Fig. 10.14 Structure and content of an image matrix and coding with 8-bit brightness values

can be converted by assigning the recorded spectral ranges, i.e. the brightness or greyscale values of one channel, to one primary color of a monitor or a printer each, but only combinations of three channels are possible (due to the principles of additive or subtractive color mixing). With the Thematic Mapper recording system of Landsat 5, an approximate "true color" image can be generated by assigning spectral band 1 (visible blue) to the monitor color blue, band 2 (visible green) to the monitor color green, and band 3 (visible red) to the monitor color red. In addition, other channel combinations are common, so that the special characteristics of the detected objects become visible, which are reflected differently in different spectral ranges. The spectral ranges that are not visible to the human eye are therefore represented by so-called false colors. If channel 2 of the Thematic Mapper (visible green) is assigned to the monitor color blue, channel 3 (visible red) to the monitor color green and channel 4 (near infrared) to the monitor color red, the usual false-color infrared representation is created.

The color image on the monitor is therefore not to be confused with a color photograph. Thus, the term *color composite* is used here instead. It should be noted in particular that each color carries a particular piece of information. Thus, the intensity of a color represents the irradiance of a section of the electromagnetic spectrum on the sensor. The color red, for example, can then visualize the intensity of the non-visible infrared that arrives at the sensor.

10.5.3 Remote Sensing Data: Acquisition

Remote sensing data can be obtained from a variety of sources. Major suppliers have mostly been national agencies such as the National Aeronautics and Space Administration (NASA) or supranational agencies such as the European Space Agency (ESA) or major research institutions such as the German Aerospace Center (DLR). Furthermore, there are now many private missions and providers. It should be emphasised that, in addition to commercial data offers, remote sensing data are increasingly freely available. Suppliers of the probably most frequently used sensor data of the Landsat and Sentinel missions are:

The USGS Earth Resources Observation and Science Center (EROS), the primary source of Landsat satellite imagery and data products, offers vast amounts of data (cf. USGS 2023m and especially for Landsat cf. USGS 2023n). Ultimately the data were produced by a US federal agency, it is therefore in the public domain and may be used, transmitted, and reproduced without copyright restriction (with reference of course).

On the basis of earth observation and information technologies, the European earth observation program Copernicus is creating an independent European observation system with the sentinel satellites (cf. Sect. 10.4.7). The data are open and freely available to everyone (cf. Copernicus 2023).

The Earth Observation Center of the German Aerospace Center lists extensive sources on its website (cf. DLR 2023e):

- links to DFD data services (including EOWEB, the interface to the German Satellite Data Archive of DFD)
- links to data from important earth observation missions, in which DLR is involved (including TerraSAR-X, TanDEM-X, EnMap, SRTM and Envisat, the European environmental satellite)
- links to the archives of partner organisations (e.g. Astrium Geoservice, EUMETSAT)
- links to free data offered by space agencies, government authorities and universities (such as NASA, ESA (Copernicus data and products) or USGS).

The data suppliers often no longer only deliver raw data, but pre-processed data. In most cases, systematic distortions of the recordings are already corrected. In addition, the data can already be obtained at further processing stages, which include, for example, georeferencing into a desired coordinate system. In particular, there is a trend towards largely pre-processed data (so-called value added products), which is being driven above all by the new commercial systems and providers, so that the data can be used directly by the user (keyword "GIS-ready"): The data are already orthorectified with high precision on the basis of a digital elevation model to a desired map projection (cf. Sect. 10.6.1.2) and can be integrated directly as a data layer into a geoinformation system.

10.6 Digital Image Processing

10.6.1 Image Pre-Processing

The data received from the data supplier are often not yet suitable for analysis and evaluation. Then the data must be further processed and improved. For example, the data are not yet available in the desired spatial reference system or the pure brightness values first must be converted into physical units.

10.6.1.1 Radiometric Correction

The radiation received at the sensor was modified by various factors such as differences in illumination, atmospheric influences, viewing angle or characteristics of the sensor itself. The usage of the remote sensing approach determines, whether *radiometric corrections*, i.e. corrections of the received reflectance values, have to be made (on radiometric corrections cf. Chavez 1996, Hildebrandt 1996 pp. 486, Richards 2013 pp. 38, Lillesand et al. 2008 pp. 490, Mather and Koch 2011 pp. 112):

- A multi-temporal analysis of the data, such as it is necessary in the context of a regular monitoring of a region, requires minimising external influences in order to identify the temporal differences that are actually of interest.
- In the case of a mosaic formation from several image data sets, which may originate from different sensors or may have been recorded at different times, it is necessary to make the brightness values homogeneous across the different partial images, i.e. to make them comparable (so-called *histogram matching*, cf. Sect. 10.6.5.2).
- If a comparison is to be made with reflection measurements carried out on the ground or if statements about absolute reflection values are required, the brightness values must first be calibrated in absolute radiation units as measured at the sensor.
- For a conversion of the brightness values into absolute reflectance values, current calibration information for each channel of a sensor are necessary, since these values change as the progressive age of the recording instruments. This data should be taken from the file header of the image data set or from the additional information enclosed.
- Only through a sensor-specific calibration it is possible to compare recordings from different sensors.

A complete radiometric correction includes a conversion of image brightness values to the (apparent) spectral radiation L at the sensor (cf. Sect. 10.5.1), subtraction of the influence of the atmosphere, topographical normalisation and sensor calibration. The conversion of the image-related brightness values into radiation values, as measured at the sensor (at-sensor irradiance), is performed via (cf. Richards 2013 pp. 33 and 37):

$$L_{sat,\lambda} = \frac{L_{max,\lambda} - L_{min,\lambda}}{DN_{max}} \cdot DN + L_{min,\lambda}$$

with:

$L_{sat,\lambda}$	spectral radiance at the sensor ($W\ m^{-2}\ sr^{-1}$)
$L_{max,\lambda}$	maximum spectral radiance detectable by the sensor ($W\ m^{-2}\ sr^{-1}$)
$L_{min,\lambda}$	minimum spectral radiance detectable by the sensor ($W\ m^{-2}\ sr^{-1}$)
	radiance: $W\ m^{-2}\ sr^{-1}$ watts per square meter per steradian
DN	digital number (brightness value)
DN_{max}	highest possible digital count for the sensor

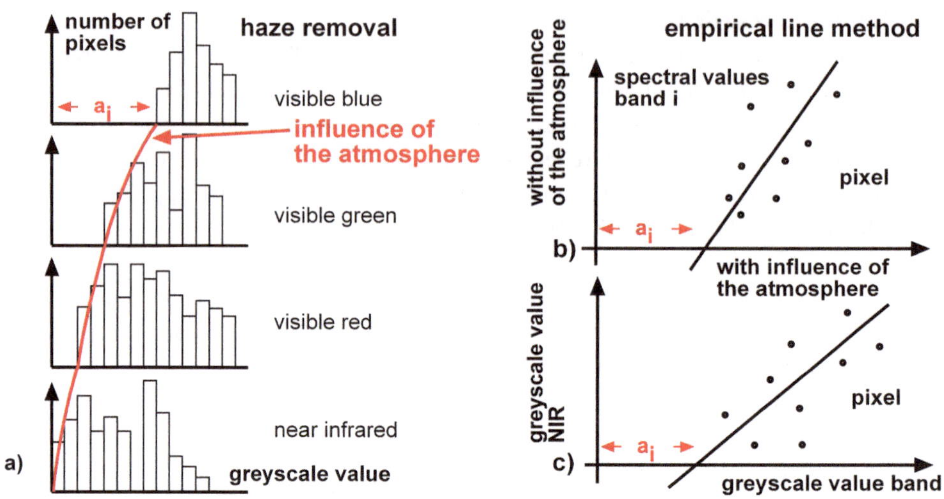

Fig. 10.15 Approximation method for the reduction of the atmospheric influence (cf. Hildebrandt 1996 p. 490 and Richards 2013 p. 45 as well as Mather and Koch 2011 p. 114)

The calibration information is subject to changes due to the age-related degradation of the instrument on the one hand, but on the other hand also as an adaptation to different reflection conditions of the recorded surface. The current parameters are available from the mission operators. However, the radiometric correction is usually already applied during data pre-processing by the data supplier. The remote sensing data are therefore (mostly) radiometrically corrected and available to the customer or user.

Taking into account the state of the atmosphere during the recording leads to the implementation of a so-called atmospheric correction. Simple approximation methods estimate correction values from the image data itself (cf. Richards 2013 pp. 44, Mather and Koch 2011 p. 114). In a very simple approach, a value a_i is subtracted from the brightness values of a scene for each channel, which is understood as haze (so-called *haze removal*). To determine these values, the histograms of the brightness values are considered for each channel. For the near infrared channel, brightness values of 0 or close to 0 already have a notable frequency, assuming clear water surfaces, i.e., free of vegetation or plant remains, or mountain or cloud shadows (i.e. reflectance close to 0). For the other channels, there may be a histogram shift by a value a_i, which marks the difference between the darkest brightness value and 0 (or the darkest brightness value of the near infrared channel) and which can be interpreted as the influence of the atmosphere (cf. Fig. 10.15a).

In another approximation method, the correction values a_i are determined by setting up regression equations with spectral values for each channel. This method (*Empirical Line Method*) requires reference objects on the ground, for which the spectral values are derived from the brightness values. The associated reflectivity is measured with a spectrometer without atmospheric influence. The distance between the zero point and the intersection of the regression line with the horizontal coordinate axis provides the correction value for the channel in question (cf. Fig. 10.15b).

Another variant of this approach sets up regression equations between the brightness values of the near-infrared channel and another channel and determines the correction quantities a_i in a similar form (cf. Fig. 10.15c).

More complex methods rely on the results of radiative transfer and aerosol models and require knowledge of various atmospheric data such as vertical stratification and visibility (for the integration of physical models cf. e.g. Richter 1996, Chavez 1996). In the case of Sentinel-2, correction information is also collected and made available so that users can perform correction calculations independently (cf. ESA 2023k and Louis et al. 2016).

Simplified, it is often assumed that the area covered by the satellite scene represents a flat surface. However, the magnitude of the signal received at a satellite sensor also depends on the relief, i.e. the illumination and the viewing angle. The correction of terrain illumination effects ultimately requires a digital terrain model, if the assumption that the surface reflects the incoming radiation equally in all directions, cannot be maintained. The so-called *cosine correction* multiplies the radiation value by the quotient of the cosine of the zenith angle of the sun (measured from the vertical) and the cosine of the radiation angle (measured from the surface normal, cf. Mather and Koch 2011 p. 117 and 123, Hildebrandt 1996 p. 496).

10.6.1.2 Rectification, Registration and Resampling

Depending on the recording system, the image data are more or less distorted. For example, airborne scanners usually provide more distorted images than satellites due to the relatively unstable flight attitude (e.g. yaw, roll and pitch, cf. Figs. 5.16 and 10.17). Therefore, a correction of the geometric distortion, i.e. a rectification of the original data, is often necessary. Since the coordinate system of the recording system can be oriented in any way due to the flight direction, referencing and adaptation of the digital original image to an analog or digital map template with a defined reference system (e.g. UTM, cf. Sect. 4.5.3) is also very important. In remote sensing it is common to speak of *registration* instead of georeferencing.

The image *rectification* is often a hybrid approach. The pixels of the input image are simultaneously rectified and assigned to a cartographic or geodetic coordinate system (registration or georeferencing). The original image has to be converted into the new coordinate system, which usually involves rotating and stretching or compressing of the

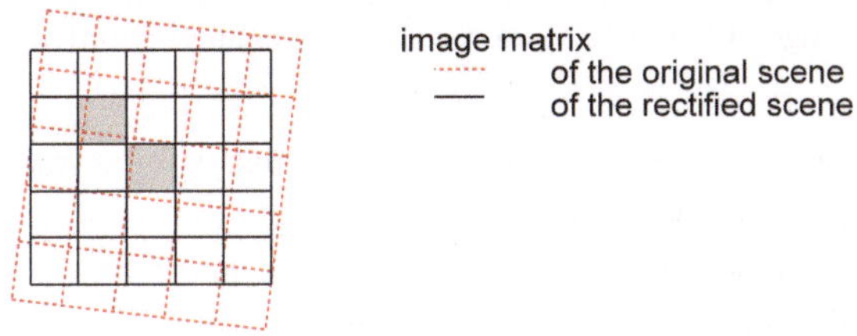

Fig. 10.16 Rectification of raster images (cf. Hildebrandt 1996 p. 480)

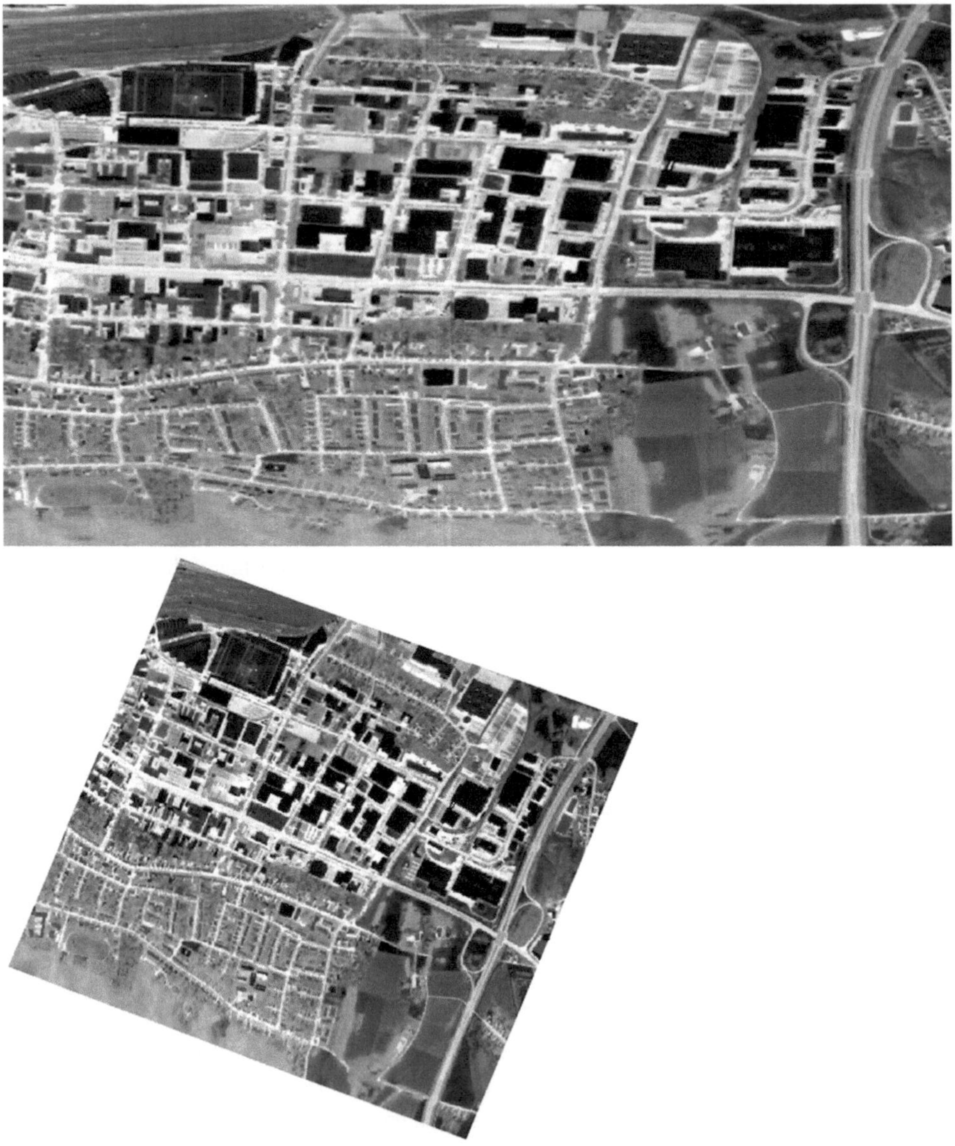

Fig. 10.17 Photographs of a commercial park in Osnabrück with an airborne opto-mechanical scanner: non-georeferenced (top) and georeferenced flight strip (bottom), thermal channel in the range of 8.5 to 12.5 μm, surface temperatures shortly after sunset (cf. Wessels 2002 p. 82)

input pixels (cf. Fig. 10.16). Thus, similar methods and procedures are used as for georeferencing (cf. Sects. 4.2.5 and 4.6 for more details on georeferencing).

In addition to the transformation into a new reference system, the brightness values of the original image must be converted into brightness values of the result image (*resampling*). Thereby several pixel values of the original image can be involved in a value for a single pixel in the rectified scene. It is required that all pixels of the result image must also receive brightness values.

There are several methods for image rectification in digital image pre-processing (cf. especially Mather and Koch 2011 pp. 94, Richards 2013 pp. 56). The central problem of polynomial rectification, which is mostly used in practice, is that the necessary transformation equations have to be set up starting from only a few control points (cf. the similar approach within georeferencing). Parametric methods, which are not discussed here, take into account the geometric properties of the sensor, which are referred to as the inner orientation in an aerial photo, and the spatial position or spatial movement of the sensor, which is referred to as the outer orientation in an aerial photo.

In *polynomial rectification*, a polynomial of the nth-order is determined in order to rectify and register the original image or to georeferenced it in one step. This requires control points in the source image with known x- and y-coordinates in the target coordinate system. In most cases, polynomials up to a maximum of third order are used. Polynomials of a higher degree hardly achieve better results in practice. This method has proven itself for the rectification of satellite data, where a uniform distortion of the overall scene can be assumed. In most cases, rectification is performed on the basis of planar control points (planar rectification).

However, ground objects at different heights or at different terrain levels show (considerable) positional shifts compared to an orthogonal image. *Orthorectification*, which requires a digital elevation model and control points with x, y and z coordinates, corrects the topographic distortions pixel by pixel (for the rather complex procedures cf. Lillesand et al. 2008 pp. 169, Schowengerdt 2006 pp. 363). In the resulting image, each pixel appears as if it had been taken at a right angle from above (orthographic projection).

There are basically two methods in digital image processing for converting the brightness values from the original scene into the (corrected) result scene (resampling) (cf. Fig. 10.18):

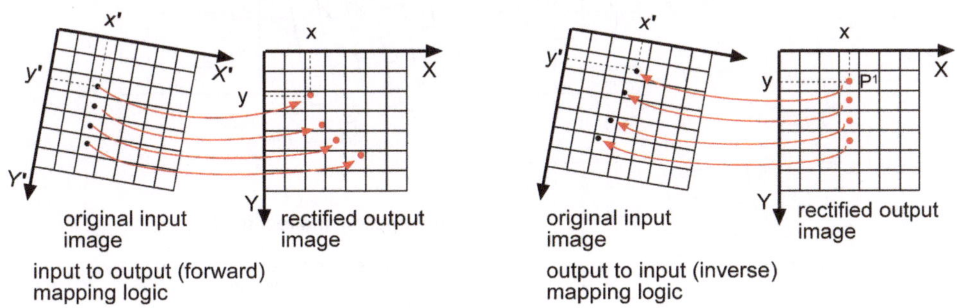

Fig. 10.18 Filling a rectified output matrix with values from an unrectified input image matrix (cf. Jensen 2015 p. 247)

- *Direct resampling* starts from a pixel in the input image for which the position in the output image is calculated. The brightness value from the input image is assigned to this pixel. In this case, all pixels of the input image are transformed. However, in the (more important) output image, individual pixels may not be assigned a brightness value, while other pixels have multiple assignments. Therefore, post-processing becomes necessary.
- *Indirect resampling* is based on the position of a pixel in the output image, for which the brightness value of the nearest pixel from the input image is determined. This means that a "re-calculation" takes place from the output image to the input image to ensure that all pixels of the output image have a brightness value. Therefore the rectified image matrix is available without further post-processing.

Resampling is usually differentiated according to three methods (cf. Fig. 10.19, for extensive example calculations cf. Jensen 2015 pp. 246):

- The *nearest neighbor method*: The pixel in the resulting image gets the brightness value of the nearest pixel in the source image. This can cause positional errors of up to half the pixel size in the resulting image. Diagonally running brightness value edges can appear stepped in the resulting image. This effect can be reduced by selecting a sufficiently small pixel size in the resulting image. The advantage of this method is that the original values of the source image are preserved. This is an important prerequisite if a classification of the pixel values is to be carried out later in order to identify objects or properties such as land cover types. Original data must be used for this.
- The *bilinear interpolation*: The required brightness value of a pixel in the resulting image is calculated as the weighted average of the brightness values of the four directly adjacent pixels of the original image. It is assumed that the four brightness values (represented as z-values in a three-dimensional coordinate system with the pixel

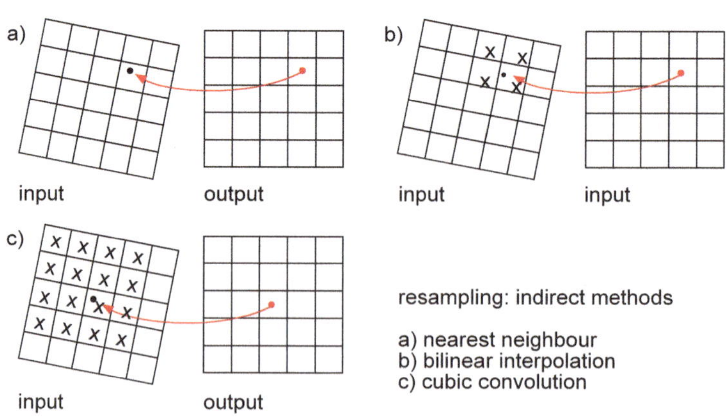

Fig. 10.19 Different forms of indirect resampling (cf. Hildebrandt 1996 p. 481)

coordinates) can be approximated by a plane on which the searched brightness value also lies (cf. Sect. 9.7.2). The resulting image then has no brightness values of the original image. Thus, the input brightness are changed, i.e. smoothed here (low-pass filtering effects, cf. Sect. 10.6.4.1). Overall, a good image quality is generated, so that this method is used for visualisation. However, the computation time is three to four times greater than of the nearest neighbor method.
- The *cubic convolution*: Corresponding to bilinear interpolation, the so-called cubic convolution calculates the desired brightness value for a pixel as the weighted average of the brightness values of the adjacent 16 pixels of the input image. It is assumed that the z-values can be approximated by a curved surface (described mathematically by third degree polynomials). The low-pass filter effect can be reduced by selecting the parameters accordingly (cf. Sect. 10.6.4.1). Here, too, the input brightness values are changed. This is the preferred method when producing satellite image maps, since it provides the best visual results. However, it also requires the most computational effort.

In practice, polynomial rectification followed by indirect resampling is of great importance. This results in a three-step workflow:

- The first step, the selection of suitable ground control points (GCP), is very time-consuming and laborious. Easily recognizable points or image elements (e.g. street crossings at small image scales) are selected as control points, which do not change over time (e.g. no shorelines). The control points should be distributed as evenly as possible over the image. The selection of suitable control points is of the utmost importance for the quality of the rectification. Control point sources can be, for example, analog maps and digital vector or raster map that have been already recorded in the desired reference system. Furthermore, image-to-image registration (co-registration) to other images is also possible, whereby the images do not necessarily have to be georeferenced to a map projection.
- In the second step, the geometric rectification equation is set up. Determining the appropriate order of the polynomial depends on the degree of distortion of the original image, the number and accuracy of the control points to be determined as well as the distribution of the control points. In practice, transformations with polynomials up to second degree have proven to be suitable for the rectification of satellite images, if the relief is not very pronounced and the aperture angle of the sensor system is not too large (cf. Hildebrandt 1996 p. 477 and Schowengerdt 2006 pp. 287 and pp. 298). Often a simpler affine transformation is taken, which follows a recommendation by Jensen (2015 p. 244): "Generally, for moderate distortions in a relatively small area of an image (e.g. quarter of a Landsat TM scene), a first-order, six-parameter, affine (linear) transformation is sufficient to rectify the imagery to a geographic frame of reference."
- The so-called *RMS error* (cf. Sect. 4.2.5.5) can be used to evaluate the quality of the rectification. A higher order polynomial usually reduces the RMS error. However, this does not automatically improve the quality of the rectification. Rather, polynomials of

higher degree usually show worse interpolation properties, since the pixels between the control points in the result image can have significant positional errors. Therefore, if possible, not only the RMS error should be considered in the rectification, as this only refers to the deviations of the control points. Systematic errors can be determined by calculating so-called residual error vectors. Residual error vectors graphically represent the different position of the measured and the calculated control point coordinates or (better) further test points. However, these new test points are not to be confused with the ground control points used for rectification.

– The third step is resampling.

10.6.2 Image Enhancement: Contrast Enhancement

With a color depth of 8 bits, a total of 256 different values for coding the radiation intensity are available per spectral band. However, this range of values is often only partially used, since the instruments have been set up in such a way that even extremely strongly or slightly reflective surfaces can still be reproduced. The remote sensing images then often appear quite low in contrast. Several methods exist for eliminating this inadequacies, which disturb a visual evaluation of the image data (for further details cf. e.g. Jensen 2015 pp. 282, Mather and Koch 2011 pp. 128 and Richards 2013 pp. 99).

10.6.2.1 Linear Contrast Stretching

In a first processing step, a histogram of the brightness values usually shows that only a few brightness levels have a significant frequency. This distribution can now be stretched over the entire range of values, resulting in contrast enhancement:

$$\text{brightness value}_{new} = \left[\frac{(\text{brightness value}_{max} - \text{brightness value}_{old})}{(\text{brightness value}_{max} - \text{brightness value}_{min})} \cdot 256 \right]$$

Linear contrast stretching reclassifies the values of the input pixels in such a way that the former minimum or maximum greyscale values are assigned the values 0 or 255 (with a color depth of 8 bits), while the brightness values in between are stretched linearly over the entire 256 scale. This reclassification is usually done with the help of a so-called look-up table, in which the new, contrast-enhanced values are assigned to the original greyscale values. In this process step, the original data remains unchanged, and the changed data are loaded from separate look-up tables for the view.

In addition to the contrast stretching between the former minimum and maximum brightness values, a purposeful stretching of the areas of particular interest can also be carried out (interactively) in order to be able to clearly differentiate and analyse them. To do this, the minimum and maximum brightness values of the surfaces of interest must first be found. Subsequently, the intermediate brightness values can be stretched over the entire

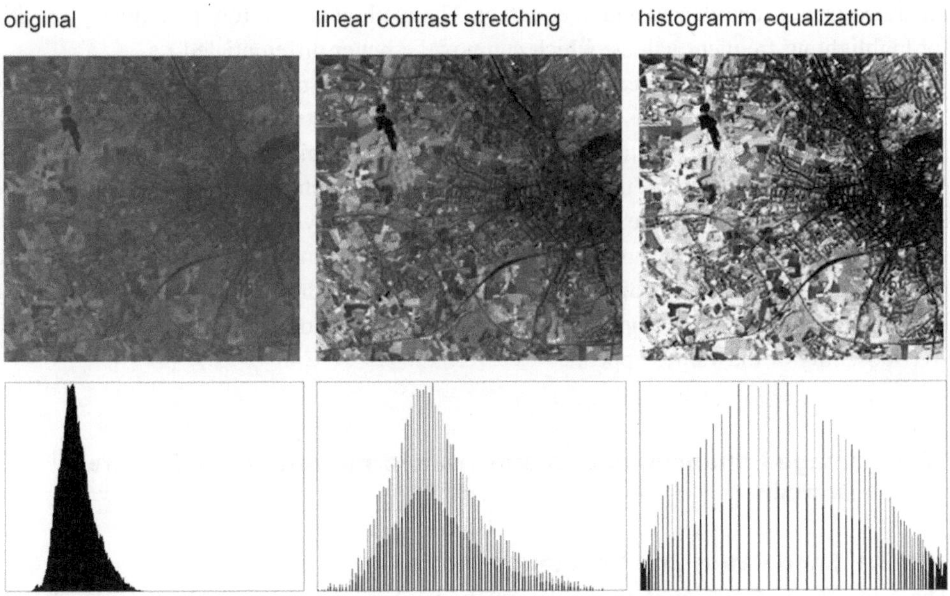

Fig. 10.20 Contrast stretching and histogram equalisation: section from a Landsat TM scene of Osnabrück, channel 4 (NIR), with associated histograms of brightness values

available value range (e.g. 256 levels), whereby the values that are unimportant for the analysis are faded out.

The linear contrast stretching method is particularly suitable for highlighting extreme values or for a more detailed examination of partial areas. Figure 10.20 shows how the contrast and therefore the recognition of objects in the image improve. Particularly low-reflective surfaces such as the water surface in the northwestern image area are assigned the lowest brightness values, while highly reflective surfaces such as meadows and pastures are assigned the highest brightness values so that they are reproduced very brightly. In between, linear scaling is used so that the relative brightness value differences are preserved. In the associated histogram, which shows the frequency distribution of the brightness values, it becomes clear that the entire dynamic range is used and at the same time the basic gradient shape of the histogram is preserved.

10.6.2.2 Histogram Equalisation

Histogram equalisation is used to stretch the brightness values depending on their frequency. Each histogram unit (mostly 0–255) should be assigned approximately the same number of brightness values in order to obtain a balanced histogram (not to be confused with histogram matching, cf. Sect. 10.6.5.2). Previously sparsely populated brightness value categories are combined, while those with a large number of brightness values are separated from each other to a greater extent. The method is therefore particularly suitable for highlighting frequent values, which can now be better differentiated.

Figure 10.20 shows the result of the histogram *equalisation* with the corresponding histogram. The effect of this method becomes clear: Between the areas which show extreme brightness values in the original (cf. flanks of the histogram), a differentiation of different brightness values is now hardly possible. This includes, for example, the sealed areas with low reflection in the east and the vegetation areas with high reflection in the west of the image. In contrast, the areas in the central area of the image can now be differentiated much better. In the corresponding histogram, it can be seen that the brightness values at the flanks are summarised or compressed, while towards the middle of the diagram a progressive loosening of the values appears.

10.6.3 Image Enhancement: Image Transformation in the Spectral Domain

10.6.3.1 Band Ratios and Vegetation Indices

The calculation of indices is one of the methods of image transformation, in which new image data can be created from a multi-band image data set at one or different recording times (mono- or multi-temporal). Special information can be emphasised more clearly through clever combination and index creation. Differential brightness value differences of the input images are amplified, which can significantly increase the interpretation options. With these operations the data values of two or more output channels are arithmetically combined pixel by pixel (for local operators of the analysis of raster data cf. Sect. 9.5.4). Finally, a new "artificial" image is created.

In practice, *vegetation indices*, which make vegetation areas more prominent, are of great importance (cf. Jensen 2015 pp. 325 and Mather and Koch 2011 pp. 152 for further details). The vegetation index VI uses the property that healthy vegetation reflects only weakly in visible red, but strongly in the infrared (cf. Fig. 10.7). Thus, vital vegetation can be distinguished relatively easily from other ground covers:

$$\mathrm{VI} = \frac{\mathrm{NIR}}{\mathrm{R}} \qquad \mathrm{NDVI} = \frac{\mathrm{NIR} - \mathrm{R}}{\mathrm{NIR} + \mathrm{R}}$$

NIR = (near) infrared channel (e.g. channel 4 for the Thematic Mapper)
R = red channel (e.g. channel 3 for the Thematic Mapper)

The more advanced *Normalised Difference Vegetation Index* (*NDVI*) is also based on the difference between the spectral signatures of unvegetated and green vegetated areas in the visible light (best in visible red) and near infrared channels. The NDVI takes values between -1 and $+1$, with positive values significantly different from 0 indicating the existence of green biomass. But the transition between animate and non-animate does not have to be marked by 0 and has to be determined anew in each case. The higher the numerical value of the NDVI, the more vital the vegetation. The NDVI is strongly correlated with the Leaf Area Index (LAI), which can be used to estimate the biomass per pixel (cf. Löffler 1994 p. 71). The area percentage of vegetation or the degree of sealing per pixel is also highly correlated with the NDVI, so that proportion values can be estimated (cf. Achen 1993 pp. 77). Cihlar et al. (1991) demonstrate a strong connection between the NDVI and evapotranspiration for areas in Canada, which, however, must not be considered independently of other growth parameters such as the availability of energy and water or soil type.

Bannari et al. (1995) and Baret (1995) point out that the standard vegetation indices, which are calculated on the basis of linear functions, should be evaluated with caution when vegetation cover is low. For example, the NDVI tends to misestimate the proportion of the surface covered by vegetation at the beginning and end of the growing season (overestimation at the beginning, underestimation at the end of the growing season, cf. Bannari and others 1995 p. 101). This led to the development of a second generation of vegetation indices, which try to take into account the interactions between electromagnetic radiation, the atmosphere, vegetation cover and soil. Based on the four bands of the Landsat MSS sensor, Kauth and Thomas (1976) developed the so-called "Tasseled Cap" transformation, resulting in four indices: Soil Brightness Index (SBI), Green Vegetation Index (GVI), Yellow Vegetation Index (YVI) and Non Search Index (NSI). A further development of these indices based on the Landsat-Thematic-Mapper bands was done by Crist and Cicone (1984). However, these indices have one major disadvantage: Since they were determined empirically, they cannot be easily transferred to other sensor systems or other regions, i.e. without new calibration.

Vegetation indices can be used in several ways. They can monitor changes in vegetation, as they provide relatively reliable information on the presence of vegetation. On a global scale, this has been done for many years using standard NDVI calculations from AVHRR data, which are used to estimate the global distribution of biomass and its development. With their help it was possible, for example, to substantiate the trend of global warming with a shift in the vegetation period in northern latitudes (cf. Shabanov et al. 2002). Vegetation indices also allow, for example, the assessment of the condition of cultivated plants (cf. e.g. Jürgens 1997 and Thenkabail et al. 1994).

10.6.3.2 Principal Components Transformation

Adjacent spectral channels are usually highly correlated with each other. The *principal component transformation* offers a way of eliminating such redundancies (for the approach of principal component analysis cf. e.g. de Lange and Nipper 2018 pp. 198, Bortz and Schuster 2010 pp. 385 and the classic textbook by Harman 1976, for the application in digital image processing cf. Mather u. Koch 2011 pp. 160, Richards 2013 pp. 163). When the number of input variables is reduced, most of the input information, expressed by the sum of the variance of the input variables, is represented by a few principal components. In this way, the information content of a six-channel Landsat TM dataset can usually be reduced to three principal components. Then, by displaying the three principal components as a color composite, much more information can be represented than would be possible with any of the possible composites of the original dataset (each based on three colors).

This method can also be used in classification approaches, where only the principal components with the highest eigenvalues are included in the classification.

10.6.4 Image Enhancement: Spatial Filtering

Spatial filter operations are used in image processing in order to emphasize or suppress certain spatial-structural features of the image (cf. Jähne 2006 pp. 111, Jensen 2015 pp. 293, Lillesand et al. 2008 pp. 509, Richards 2013 pp. 130). These filtering techniques are local operations that use the properties of the adjacent pixel. In this context, the term spatial frequency is important. This parameter refers to the variation of brightness values in a pixel neighborhood (e.g. the number of brightness value changes per unit distance for a given part of an image). Low spatial frequencies are present in the case of very few changes in brightness values (relatively homogeneous areas), while high spatial frequencies occur in the case of strong local brightness value variations over short distances as an expression of pronounced inhomogenities.

Since the spatial frequency describes differences or changes spatially, spatial methods or approaches must also be selected in order to amplify or compensate for these differences. Therefore, the adjacent pixel brightness values are considered. There are two groups of techniques: (spatial) convolution and Fourier analysis. Convolution is a neighborhood operation, where each output pixel is the weighted sum of adjacent input pixels. Convolution achieves linear filtering of an image. Basically, low-pass filters and high-pass filters can be distinguished (for edge enhancement filters such as the Sobel edge detector or the Laplacian Operator and for more advanced methods of image enhancement cf. Jensen 2015 pp. 293 and Richards 2013 pp. 133).

10.6.4.1 Low-Pass Filter

The task of the *low-pass filter* is to work out the low spatial frequencies and to suppress local extremes. This type of filter includes, for example, mean, median and modal filtering. All these image filters have in common that they result in a smoothing of the image, which may be necessary, e.g., if disturbing influences of the sensor or data transmission errors have led to local errors in the image data, or if the result of a multispectral classification has to be generalised (reduction of the so-called "salt and pepper effect", cf. Sect. 10.7.7). Low-pass filters emphasize trends in an image. The dominant image structures and typical characteristics are brought out, while high-frequency characteristics of the pixel environment (large local brightness value differences) or outliers are suppressed or smoothed.

In Fig. 10.21, a source image is transformed into an output image, where in each case the middle pixel within an environment of 3×3 pixels is replaced by a new value, which is calculated from the values of these adjacent pixels. A convolution mask (also coefficient or

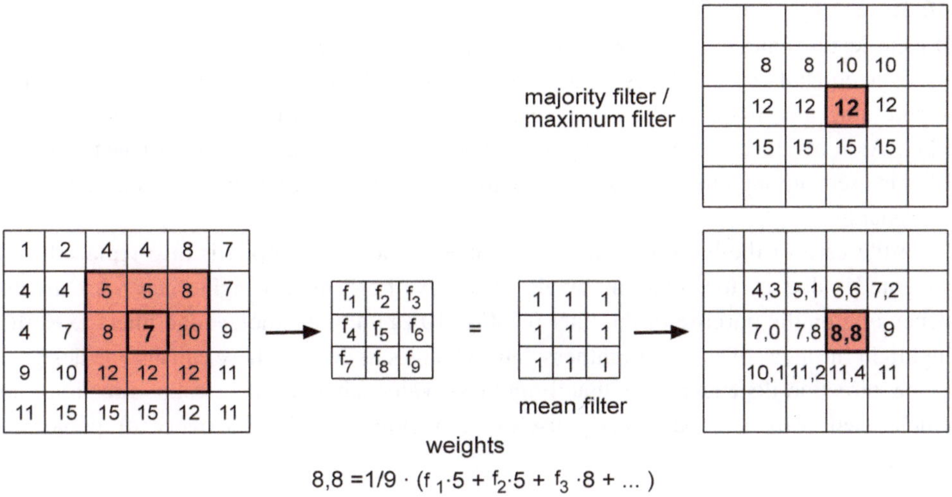

Fig. 10.21 Low-pass filter using the example of the mean filter

filter matrix) with the weights for the adjacent pixels is moved over the image (moving filter matrix). The edge pixels are not covered in the output image, since no 3 × 3 neighborhood can be created for an edge pixel in the original image. If the filter matrix is enlarged, i.e. if a larger pixel neighborhood (e.g. 5 × 5 or 7 × 7 pixels) is included in the filter operation, an even greater smoothing effect occurs. This technique also applies to qualitative data, as the example of the majority filter shows. The cells in a raster are resampled according to the majority of the qualitative feature values of the neighbor cells (corresponding to minority filter or maximum or minimum filter).

With the *mean filter*, the value of the central pixel is replaced by the (simple arithmetic) mean value of the adjacent pixel values (cd. Fig. 10.21). Due to the smoothing effect a "softer" and "blurrier" output image is obtained. The high values, i.e. high irradiance values at the recording system, are depressed (by adjacent low values). With the *median filter,* the median of the pixel values is taken instead of the arithmetic mean. With the *modal filter,* the value of the central pixel is replaced by the most frequent value of the adjacent pixel values. This filter is used, for example, to "fill in" missing pixel values (cf. the so-called *majority filter*, which uses a frequent – not necessarily the most frequent – value). It can also be used to post-process scenes that have already been classified in order to eliminate individual misclassified pixels or to adjust them to the neighborhood.

A low-pass filter can also be used for so-called contrast equalisation. For this purpose, a scene with unwanted large brightness value differences is first subjected to a very strong low-pass filtering, so that an extremely blurred image is created (cf. Fig. 10.22). Then the brightness value difference between the original and the "unsharp" mask is formed. This is amplified so that, with a suitable choice of the individual parameters, an image balanced in overall brightness with good detail reproduction can be obtained (cf. Albertz 2009 pp. 106).

10.6.4.2 High-Pass Filter

If local features and extremes are to be recognised, *high-pass filters* are used, which bring out image details by emphasising high frequencies. The high-pass filter has the effect of suppressing low frequency components. Conversely, high values can pass through the filter. As a result, contours and edges are particularly emphasised (edge detection). The filter is used, among other things, to distinguish areas with abrupt changes from those with few changes.

In the case of the high-pass filter, the value of the central pixel is also replaced by a weighted value of the adjacent pixels, whereby these filter weights can vary greatly depending on the purpose of the filter (cf. Fig. 10.22). For example, edge filters exist that highlight the areas of strong brightness value changes as edges. These can also be designed to be direction-dependent, so that brightness value changes in certain directions are emphasised, which is particularly useful in geology for the discovery of geological structures.

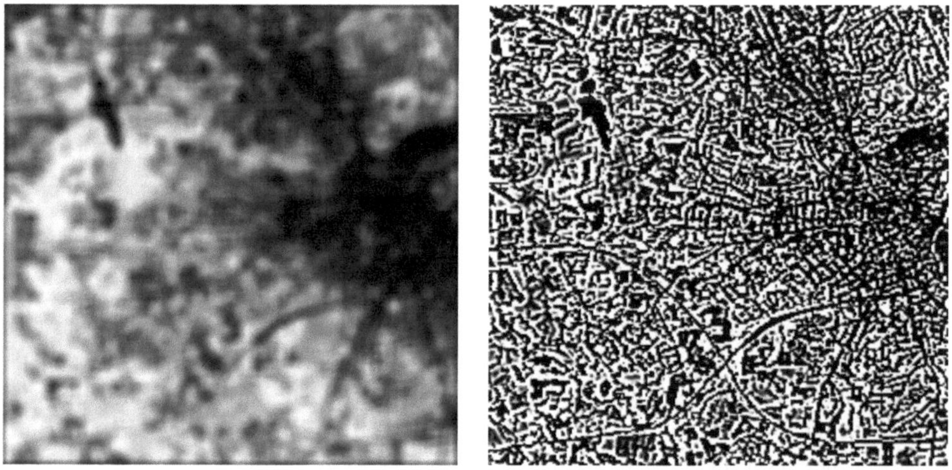

low pass filter (7x7-filter matrix)

1	1	1	1	1	1	1
1	1	1	1	1	1	1
1	1	1	1	1	1	1
1	1	1	1	1	1	1
1	1	1	1	1	1	1
1	1	1	1	1	1	1
1	1	1	1	1	1	1

high pass filter (5x5-filter matrix)

−1	−1	−1	−1	−1
−1	−1	−1	−1	−1
−1	−1	24	−1	−1
−1	−1	−1	−1	−1
−1	−1	−1	−1	−1

Fig. 10.22 Using filters

10.6.5 Merging Multiple Remote Sensing Images

10.6.5.1 Geometric Mosaicking

The task of merging several individual images into one overall image often arises in order to cover a larger examination area. There are two basic approaches for this process of *mosaicking*:

- The individual images are rectified individually and then merged into a mosaic, for which a large number of control points are required.
- The individual images are rectified together. In order to determine the transformation equations, comparatively only a few control points are required, which provide the reference to the coordinate system (e.g. UTM) and are distributed irregularly across all images. It is advantageous that only for the entire image a sufficient number of control points must be available, while there may not be a sufficient number of control points to register a single image. This approach also requires control points in the overlapping area of neighboring individual scenes, via which the scenes are merged. These control points do not require coordinates (e.g. in UTM), the control points only have to be able to be clearly identified in the scenes involved (cf. Fig. 10.23).

10.6.5.2 Radiometric Mosaicking

After geometric mosaicking, it may be necessary to adjust brightness, contrast or color differences, since they appear differently in each image depending on the conditions of the recording and the sensor used. A common procedure is the (iterative) adjustment of the brightness value histograms of the overlapping areas of adjacent individual scenes. From this, correction values can be determined so that the remaining brightness values can also be adjusted (so-called histogram matching, cf. Fig. 10.24). This results in the two images having approximately the same brightness value properties.

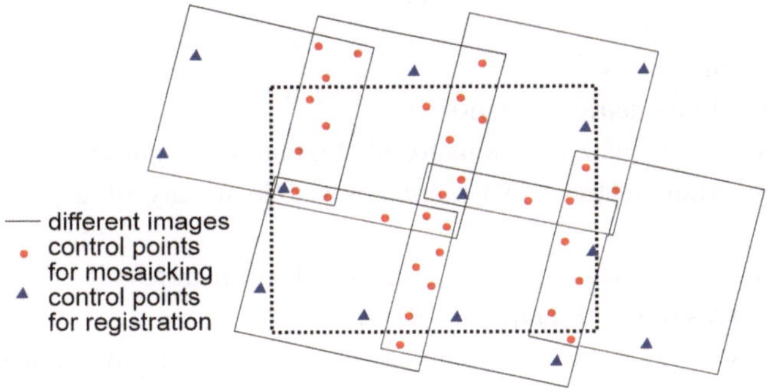

Fig. 10.23 Principle of geometric mosaicking (cf. Albertz 2009 p. 117)

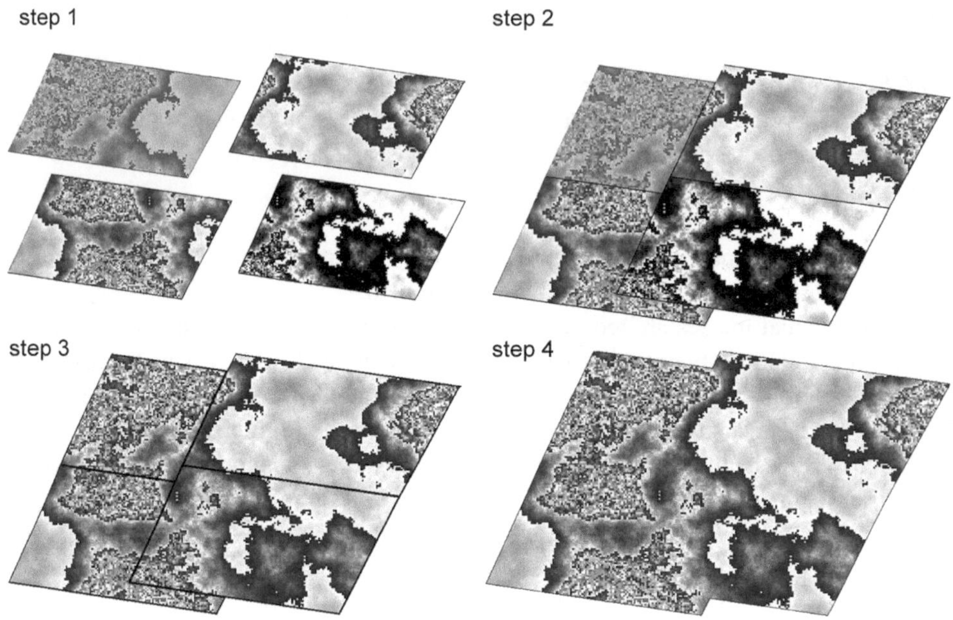

Fig. 10.24 Mosaicking and histogram matching

10.6.5.3 Image Fusion

Image fusion goes one step further than mosaicking. It aims to increase the information content by merging image data from different sources. Such an image fusion can be generated from different data sets, the most common application being the combination of multi-sensor data sets, i.e. image data from different recording instruments. Furthermore, the combination of multi-temporal data into one data set or the integration of additional data from topographic maps into an image data set for joint analysis is significant. In general, image data fusion can have different purposes (cf. Pohl and van Genderen 1998 p. 827 and for an overview Jensen 2015 pp. 168 and Mather and Koch 2011 pp. 196):

- image sharpening
- improvement of geometric correction
- creation of stereo evaluation options for photogrammetric purposes
- highlighting certain features that were not visible in any of the individual images involved
- improvement of classification, highlighting of object features
- detecting changes in multi-temporal datasets
- replacing missing information of one image by signals of another image
- replacing damaged data.

This data fusion takes place pixel by pixel. Thus, the most accurate georeferencing of the involved data sets in a single coordinate system is mandatory. However, the associated techniques cannot be used in every case. It can become problematic, for example, if the recorded spectral ranges of the involved image data differ too much from one another. Then disturbing artifacts can come into the image. In general, the methods of image fusion can be divided into two large groups (cf. Pohl and van Genderen 1999):

- color space-based techniques
- statistical and numerical techniques.

Two methods that are often used to sharpen images are to be explained here as examples. Image sharpening aims at merging the high geometric resolution of a panchromatic data set with the high spectral resolution of a multispectral data set (cf. further Pohl and van Genderen 1998 and Vrabel 2000 or on the fusion of optical data and radar data Pohl and van Genderen 1999).

The so-called *IHS* or *HSI transformation* is used for the *color space transformation*, which separates spatial (intensity) and spectral information (hue, saturation) (cf. Fig. 10.25). In a first step, the RGB representation of a color composite (e.g. from the spectral bands 3, 2 and 1 of the Landsat TM) is transformed into the IHS color space. Here, the intensity component representing the spatial properties of the multispectral data set is replaced by an image of a panchromatic sensor with a higher geometric resolution. This image is usually subjected to further image enhancement measures (e.g. contrast stretching) beforehand. After an inverse IHS transformation, the data set is again available in the RGB color space and can be viewed as a normal color composite with improved detail recognition (for conversion of the RGB representation into the IHS representation cf. Sect. 7.7.4 cf. Mather and Koch 2011 p. 172).

The group of numerical methods includes fusion using high-pass filters (cf. Sect. 10.6.4.2). Here, first the geometrically high-resolution panchromatic data set is subjected to a high-pass filtering with a small filtermask (3 × 3 matrix), which emphasizes the image details. Then, the result of this filtering is added to each individual multispectral band, which means that this method is also suitable for multispectral images with more than three

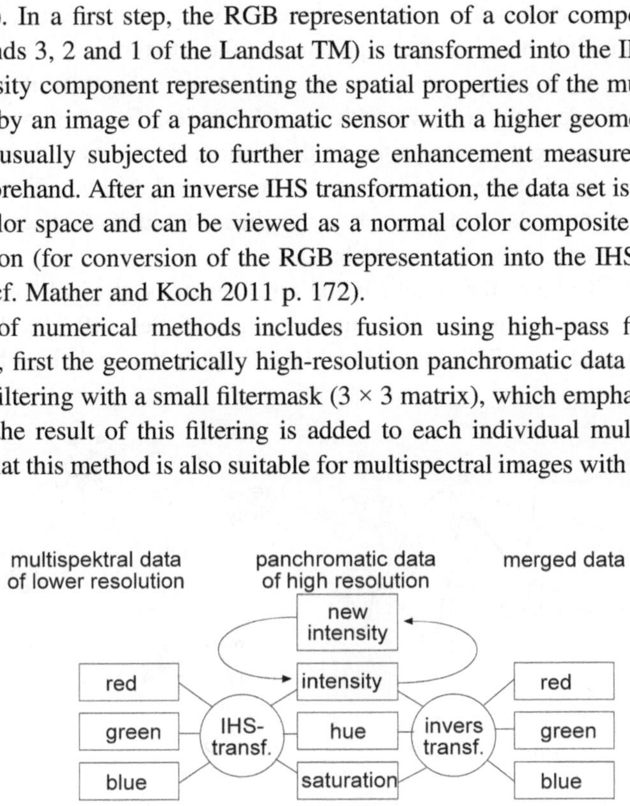

Fig. 10.25 Principle of the IHS transformation

bands. Overall, details in the multispectral data set become better recognizable, which ultimately increases interpretability.

Finally, it should be mentioned that image fusion has lost its importance. As previously mentioned, a primary purpose of image fusion is image sharpening (merging the high geometric resolution of a panchromatic dataset with the high spectral resolution of a multispectral dataset). This aspect has been pushed into the background due to the availability of high-resolution satellite data.

10.7 Classification

10.7.1 Pixel-Based Classification Methods: Principle

The general, civilian approach of remote sensing is to derive information from remote sensing data for environmental monitoring or for urban and environmental planning. This is primarily done with the help of the evaluation of images, which includes the classic, analog methods of photogrammetry and photo interpretation. Visual image interpretation identifies spatial objects based on color, brightness, texture, pattern, shape, size, position or shadow. This is mainly based on the experience of the interpreter.

Following the digital image pre-processing and image enhancement, the classic interpretation options are also possible with digital images. For this purpose, the color composites are printed out or interpreted on the monitor. In addition, there are procedures that analyses the digital images using numerical or statistical methods. The computer-aided classification is of particular importance here, i.e. the recognition of objects or features such as land cover types by analysing several bands.

Figure 10.26 illustrates the central initial considerations of pixel-based classification methods (for modern methods of pattern recognition cf. Sect. 10.7.9). The pixels to be classified are located in a multi-dimensional feature space that is defined by the recording

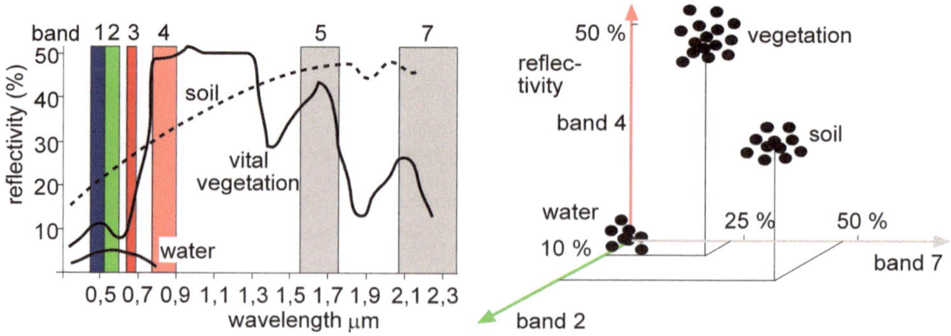

Fig. 10.26 Signature curves of three surfaces and representation of pixels for three land cover types in the three-dimensional space of three Landsat-TM bands

bands involved. Ideally, the pixels representing similar geoobjects are close together and can be separated. Using classification methods, these point clouds or clusters, i.e. pixels with similar characteristics, are identified and interpreted, for example, as different type of land cover. A distinction is made between two fundamentally different classification strategies.

Unsupervised classification does not require no prior information about the classes to be determined. The clusters are determined automatically by algorithms without any influence of the operator or scientist, i.e. unsupervised. The calculated clusters are only interpreted after the calculations have been completed.

In contrast, the *supervised classification* assumes that the land cover classes present in the satellite image are known. The operator or scientist knows that lakes are present, for example. In particular, some pixels can be clearly identified as water. Thus, supervised classification is fundamentally based on such external knowledge, which also includes the number of classes. The process is "supervised", i.e. on the basis of known land cover types of individual small test areas, so called *training areas*, within the entire digital image. In the next steps of the classification process, the remaining pixels of the image are automatically assigned to them, which represent different types of land cover.

A linguistically exact distinction must be noted. Strictly speaking, the classification methods can only derive land covers. How a body of water or a forest is used, for example as a water reservoir or for planting Christmas trees, can usually not be identified beyond doubt from a satellite image. However, the (linguistic) differences between *land cover* and *land use* are often blurred, which also depends, among other things, on the definition of the classes. For example, from the land cover "sealed settlement area" it is quite clear to conclude the use.

10.7.2 Pixel-Based Classification Methods: Implicit Assumptions

Classification methods are based on Cartesian coordinate systems and the distance from one pixel to another, which operationalize the similarity (cf. Fig. 10.26). The feature space is formed by the brightness values of the different bands, which already ensures that the order of magnitude of the objects to be classified is the same (e.g. 0–255 as 8-bit data).

The Euclidean distance or metric is by far the most important similarity measure for classification procedures (cf. Sect. 4.2.1). This distance measure is often used very carelessly, making it essential to discuss similarity measures.

The Euclidean distance is a special form of the more general so-called Minkowski-r distances or Minkowski-r metrics):

$$d_{ij} = \sqrt[r]{\sum_{s=1}^{m} \left| x_{is} - x_{js} \right|^{r}}$$

For $r = 1$, the so-called city-block or Manhattan metric results, which is of practical importance. This metric measures the distance between two objects by measuring parallel to the orthogonal coordinate axes, i.e. in the same way as in a rectangular street grid (cf. Manhattan). The city-block metric considers only the absolute differences, so it is less sensitive to extreme values. If there are many extreme values in a data set, it makes sense to use this measure.

For $r = 2$, the Euclidean distance is obtained, which is identical to the linear distance in the usual two-dimensional visual space (Pythagorean Theorem):

$$d_{ij} = \sqrt[2]{\sum_{s=1}^{2} |x_{is} - x_{js}|^2} = \sqrt{(x_{i1} - x_{j1})^2 + (x_{i2} - x_{j2})^2}$$

The calculation of the square root is often omitted, so that squared Euclidean distances are used in the classification procedures. If squares are used, then large differences in feature values are even more taken into account when determining the similarity of two objects. This artifact is not always useful.

More importantly, the use of Minkowski-r distances and therefore also the use of the Euclidean distance require uncorrelated features that generate the feature space and that define the feature axes. One can show that the cosine of the angle γ between two coordinate axes representing two variables (e.g. spectral bands) is equal to the correlation coefficient of these variables, that is: $r_{xy} = \cos\gamma$. Thus, for uncorrelated features: $r_{xy} = 0 = \cos 90°$, which means that the coordinate axes X and Y are orthogonal. Thus, for a correlation coefficient $r \neq 0$, there is an oblique feature space and strictly speaking, the Minkowski-r distances must not be computed. The Mahalanobis distance provides a solution, as it measures the distance between two points taking into account the covariances of the features, where the normalised covariances correspond to the correlations (cf. in more detail Steinhausen and Langer 1977 pp. 59 and Backhaus et al. 2016 pp. 248 and 511).

However, the prerequisite of uncorrelated features is (almost) never given in digital image processing. Adjacent spectral bands corresponding to coordinate axes are mostly highly correlated with each other. The requirement to exclude highly correlated bands often cannot be implemented. Sometimes a principal component analysis is carried out beforehand, which provides uncorrelated principal components, which are used for classification. Instead, it is necessary to ask, what influence correlated features (i.e. spectral bands) have on the classification. An important relation between cluster and principal component analysis is helpful:

Identical Euclidean distances and therefore also identical analysis results are obtained if, on the one hand, the Euclidean distance is calculated on the basis of z-normalised (correlated) original data and, on the other hand, the Euclidean distance is determined on the basis of principal components that are weighted with the square root of the associated eigenvalues (for the derivation and proof of this relationship cf. de Lange 2008). In view of this relationship, a classification of pixel values based on correlated spectral bands is quite possible, if the internal weighting of the features (spectral bands) is to be preserved and if this weighting can be detected by a principal component analysis and can be considered as meaningful.

10.7.3 Unsupervised Classification

The multivariate statistical methods of cluster analysis provide algorithms for performing *unsupervised classification*. In each case, the clusters (classes or groups) are formed based on the similarity of the objects or pixels, where the similarity is operationalised by proximity or distance in the feature space. A distinction is made between hierarchical methods, which are based on the analysis of a similarity matrix with $[n \cdot (n - 1)]/2$ values (n = number of pixels), and iterative methods (cf. de Lange and Nipper 2018 pp. 354). Hierarchical methods are not used in digital image processing, since, for example, almost $4 \cdot 10^{14}$ similarities would have to be analysed for one quarter of a Landsat scene. Instead, an iterative approach is used.

The iterated minimal distance method is fundamental to classification methods. Figure 10.27 illustrates the basic procedure, where simplified only 7 objects are given in a Euclidean coordinate system. In terms of digital scenes, "objects" are equivalent to "pixels". The general formulation assumes an initial decomposition of the n objects into k clusters. This decomposition or division occurs randomly or as in this example systematically since object 1 is in cluster 1, object 2 in cluster 2, object 3 in cluster 1 and so on. The black rings indicate objects in cluster 1 (cf. step 2 in Fig. 10.27). In the next step, the cluster centroids are determined as mean value vectors of the objects of the respective cluster. The centroids act as representatives of a cluster (cf. step 3 in Fig. 10.27). Then the check of the cluster membership and a possible reorganisation begin. All objects are assigned to the nearest centroid. In this example object 7 needs to be moved from cluster 1 to cluster

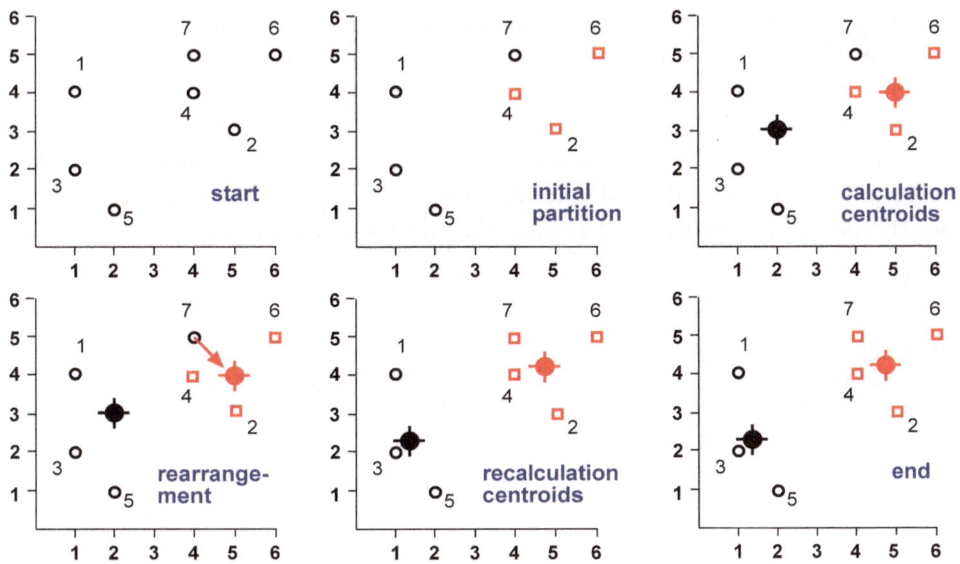

Fig. 10.27 Principle of the minimum distance method

2 (cf. step 4 in Fig. 10.27). Only after all objects have been checked and, if necessary, moved in one pass, the cluster centroids are recalculated (cf. step 5 in Fig. 10.27). Then the check of the cluster membership and any reorganisation starts again. The method ends at the latest when no object has changed a cluster. It is also possible to always calculate a homogeneity criterion in the iteration (e.g. the sum of all deviations from the cluster centroids). The method can then be terminated if the value of this criterion has not changed significantly between two steps.

In contrast to the iterated minimum distance method, the k-means algorithm directly tries to minimize the sum of squared distances of all objects in a cluster to the respective centroid (i.e. the "cluster variances" as a measure of homogeneity). After a pixel is moved from one cluster to another cluster, it is then immediately checked to see if the sum of squared distances has decreased. Ultimately, a reallocation is made to the cluster that provides the greatest improvement. After a shift has taken place, the cluster centroids are immediately recalculated. This is the main difference to the minimum distance method, in which the cluster centroids are only updated, when all objects have been checked and, if necessary, reassigned (cf. de Lange u. Nipper 2018 pp. 371). Such methods are relatively easy to implement and work relatively quickly. However, they require an initial decomposition that also assumes a predetermined number of clusters.

According to these statistical methods, the unsupervised classification requires no information about the clusters in an initial decomposition. The determination of the initial decomposition is done randomly, i.e. unsupervised. As in the example above, given k classes, every k^{th} object is assigned to class k. It can be seen that this method is highly dependent on the order of the objects (pixels).

In digital image processing, the so-called *isodata method* (also: iterative optimisation clustering or migrating means clustering) is often used, which is based on the k-means algorithm. However, the k-means algorithm is not able to resolve a cluster or to merge two clusters. This is exactly where the Isodata method goes further. For each cluster, the standard deviations of the classification variables (here spectral bands) and the Euclidean distances between the cluster centroids are calculated. If a cluster has one or more large standard deviations, it is divided with regard to these variables (cf. Fig. 10.28). If two cluster centroids have only a small distance, the two clusters are merged. In each case, user-

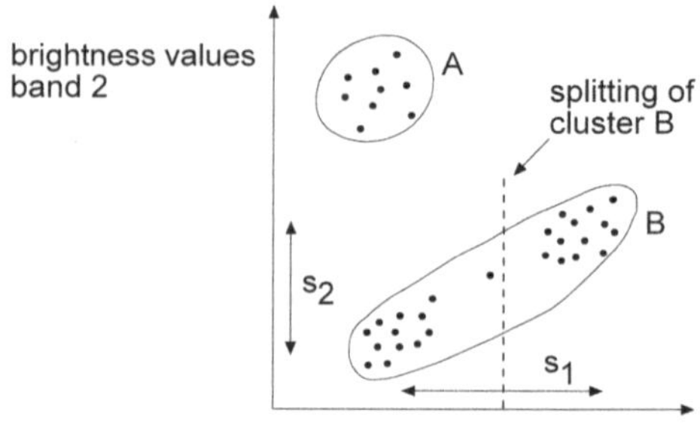

Fig. 10.28 Split and merge in the Isodata procedure (cf. Mather and Koch 2011 p. 235)

specific threshold values are specified for the standard deviations and centroid distances, which control this so-called split and merge process.

The approach of the k-means algorithm to minimize the squared distance of an object in a cluster to the respective centroid is also implemented in the minimum distance classifier in the supervised classification (cf. Sect. 10.7.5).

10.7.4 Determining Training Areas in the Supervised Classification

The *supervised classification* is based on external knowledge. An attempt is made to identify so-called *training areas* or (*training regions)* within the entire digital image to be classified, which are as homogeneous as possible in terms of land cover. All methods of supervised classification use samples of *training data*. Typically, training areas are identified through field work, maps, aerial photo or map interpretation. These areas or the selected pixels should be clearly recognizable, they act as a prototype or as a representative or as a model region for a known and previously defined type of land cover.

The parametric classification methods (cf. Sect. 10.7.5) also assume values for individual statistical parameters that ultimately define the classes or prototype of land cover. For example, the simple *parallelepiped method* requires estimates of the extreme values for each feature, the *minimum distance method* requires estimates for the multivariate means of the classes, which are taken to represent the classes, the *maximum likelihood method* requires estimates for the multivariate means and for the variance- covariances of the classes. These estimates are based on the data from the training areas. The more advanced methods, such as the algorithm based on neural networks, also work directly on the training data, but they do not impose strict requirements on the statistical distribution form of the variables (i.e. values of the spectral bands). However, it is evident that the result of a supervised classification depends on the selection as well as on the properties of the training regions.

The classification generally assumes that different land cover types each have their own characteristic reflectance in the various spectral bands. Using statistical characteristics of training areas identified as the prototype classes it is then possible to assign the remaining pixels to these classes. However, the signature curves for a specific land cover can change depending on the season or the state of the atmosphere, so that it is not possible to assume generally valid, standardised signature curves or statistical characteristics of, for example, a "forest" pixel. Within the supervised classification, the statistical characteristic values of the individual classes are always determined again, i.e. calibrated for a study area.

The forming of the sample areas, which are considered as prototypes of a special land cover type, is also achieved by so-called "region growing" algorithms. The basic principle is characterised by the fact that in the first step a starting pixel, i.e. a so-called "seed point", is set on a selected pixel of the input image, the meaning of which (i.e. land cover characteristics) is known to the user (cf. Fig. 10.29). As a rule, "seed points" are first set in water areas. From these single "seed points" larger areas grow up, which are very homogeneous with respect to the spectral values. It is checked whether an adjacent pixel

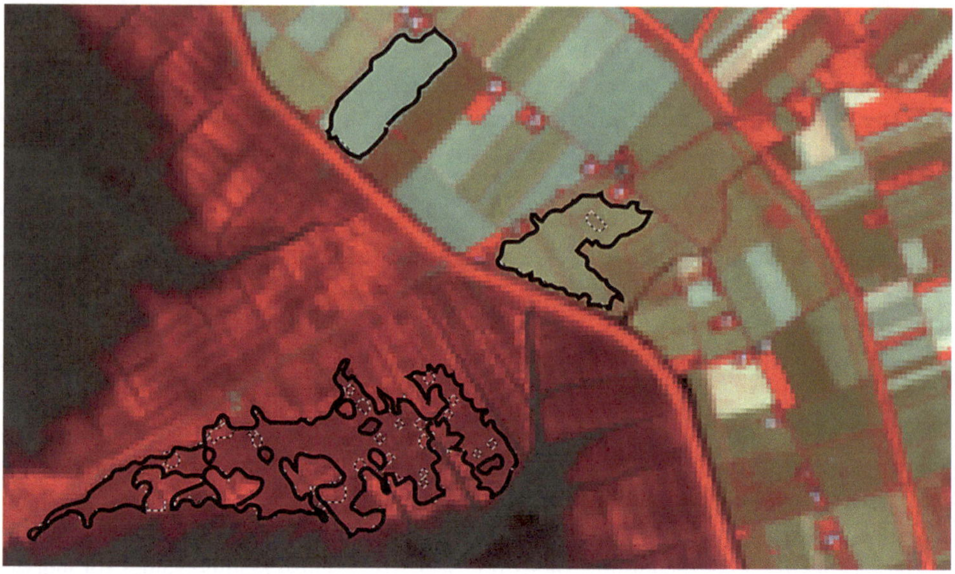

Fig. 10.29 Identification and labeling of training areas (i.e. prototypes): salt marsh in the dyke foreland and two croplands (Landsat 5 1999, West coast Schleswig-Holstein, channels 4-3-2)

fulfills a given homogeneity criterion. The spectral values of the new pixel must not differ much from those of the "seed point" in order to achieve the greatest possible homogeneity of the training area. The purpose of this preliminary work is not to cover the entire study area. Rather, pixels should be selected, whose importance is known beyond doubt. Therefore, the determination of the training areas requires skill and prior spatial knowledge of the operator or scientist. Following this quite complex preliminary work, supervised classification methods can usually achieve a good classification result.

10.7.5 Classification Based on Statistical Parameters

After determining training areas of land cover types with the associated spectral properties, the pixels of the overall scene are assigned to these classes in a second step. There are several methods for this assignment, which are also commonly used in cluster analysis:

- *parallelepiped classifier*
- *minimum distance classifier* (nearest neighbor procedure)
- *maximum likelihood classifier* (based on probabilities).

The simplest assignment method with very low computational effort is the parallelepiped classification. Each training area defines an n-dimensional cuboid for n spectral bands,

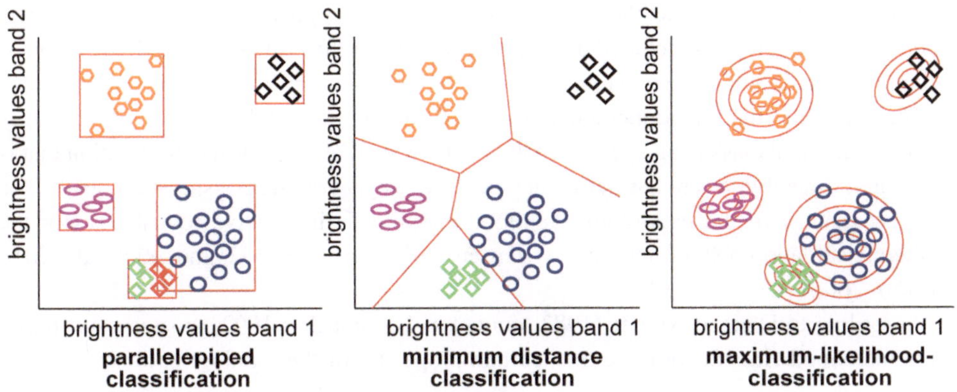

Fig. 10.30 Classification and assignment principles (cf. Lillesand et al. 2008 p. 550)

which is defined for each band from the minimum and maximum values of the respective brightness values.

For example, the brightness value limits of the spectral bands involved are determined by a training area that represents an asphalt surface, so that the land cover type "asphalt surface" is defined by a cuboid in the n-dimensional feature space. With a classification of only two spectral bands as in Fig. 10.30, this results in rectangular areas in the two-dimensional feature space. The pixels of the entire image, whose brightness value combinations fall within this cuboid are classified as "asphalt surfaces". However, the problem can arise that several classes (here: rectangles) overlap.

The minimum distance classifier first calculates the mean values of the brightness values based on the training areas for each prototype class and for each channel. Thus, cluster centroids are defined for the prototype classes in the feature space of the spectral bands involved. The pixels of the overall scene are then assigned to these crystallisation points, with the assignment being made to the nearest centroid. For a classification of only two channels as in Fig. 10.30, the feature space is decomposed into Thiessen polygons by this assignment rule (cf. Sect. 9.7.4). However, this strict decomposition using simple linear separation functions does not take into account the different scattering of the brightness values around the mean values of individual classes. This can lead to misclassifications.

In the case of the *maximum likelihood classification,* cluster centroids are determined for each prototype class according to the minimum distance classification. However, the assignment of the remaining pixels of the overall scene to these centroids is more complex. The probability that a pixel belongs to a specific cluster is calculated. It is assumed that the pixel values of each cluster are normally distributed around the cluster centroid. It must also be known how large the individual classes are or how likely a single class is. If these a priori probabilities for the occurrence of the individual classes are missing, all classes are assumed to be equally probable. However, an attempt can be made to estimate a priori probabilities based on visual interpretation or based on a previous classification in a corresponding study area with similar classes or prototypes. The different scattering of the cluster members around the cluster centroid, which represents a prototype (derived from the training area), is included in the determination of the probabilities. All in all, this is a computationally intensive procedure that usually delivers very good classification results.

The maximum likelihood method assumes a Gaussian normal distribution of the values of each individual class in all spectral channels. However, even a normal distribution that is only approximately present has hardly any negative impact on the classification result. The major advantage of the maximum likelihood classifier is its relative robustness (for the mathematical approach cf. especially Jensen 2015 pp. 398–402 and Richards 2013 pp. 250–260).

According to Dennert-Möller (1983, quoted in Schumacher 1992), a successful maximum likelihood classification requires certain properties of the training regions:

- The minimum number of training pixels sufficient to set up the calculations is n + 1 (cf. Swain and Davis 1978), where n represents the number of spectral bands. As a recommendation, Swain and Davis (1978) name a number of training pixels ranging from 10n to 100n, whereby these specifications have proven itself in practice for geometric resolutions such as Landsat (cf. Jensen 2015 p. 378, Lillesand et al. 2008 p. 559). Choosing multiple training regions across the image for each class is preferable to forming a single large training region.
- If possible, only the characteristic values of one prototype class should be recorded in a single training area.
- Although the training areas should be homogeneous, they should still reflect the entire variability within the spectral behavior of a prototype class (representativeness).
- The individual classes should be separated from each other as well as possible in order to keep the number of misclassifications low. The distributions around the cluster centroids of the different classes should overlap as little as possible.

Classification can be carried out using different methods, whereby a combination of supervised and unsupervised classification as well as a multi-stage procedure are possible. Such a strategy is always useful, if different classes can only be optimally identified by different procedures.

In a frequent approach, the different classes (e.g. land cover types) are separated one after the other in an iterative process. For example, water and land surfaces can first be separated relatively easily using the parallelepiped classificator. The water areas can then be masked out (for operations on raster data such as masking cf. Sect. 9.5.3). The classification then concentrates on the separation of different land cover types. Using a vegetation index (cf. Sect. 10.6.3.1), for example, it is then possible, to separate vegetation-covered surfaces from surfaces with little vegetation. Finally, maximum likelihood classification can fully focus on differentiating different vegetation formations. A multilevel classification strategy also makes sense, if additional initial information is available. For example, the building footprints as well as the traffic and water areas, which may come from digital cadastral maps, can be masked out so that the remaining pixels can be more easily classified with regard to vegetation differentiations.

Finally, it should be noted that Sects. 10.7.1, 10.7.2, 10.7.3, 10.7.4 and 10.7.5 deal with reliable and established classification methods, which have been developed for a long time mainly for images with a relatively coarse resolution (e.g. mainly Landsat). In the meantime, there are new methods of pattern recognition, especially for high-resolution images, which are based on methods of artificial intelligence (cf. Sect. 10.7.9).

10.7.6 Classification Accuracy

Site-specific procedures have become established for the quantitative assessment of the classification quality that not only allow a purely statistical statement about the proportion of correctly classified pixels, but also allow statements about the mutual dependencies of incorrect classifications between the different classes (cf. Jensen 2015 pp. 570). Such methods are based on the comparison between the classification result and reference data derived from sources assumed to be correct (e.g. own mapping, digital cadastral maps, biotope type mapping, topographic maps). The comparison is made using test areas, which must not have been used previously as training areas (for the selection of test areas cf. Congalton and Green 1999 pp. 22). In an analysis based on a stratified random sample, for example, a certain number of test pixels is randomly selected from the classification results for each class. Here, the number of test pixels per class depends on its areal extent. The minimum number of selected pixels per class should not fall below a value of 50 and should be between 75 and 100 pixels for more strongly represented classes (cf. Congalton 1991). It should be emphasised that these specifications apply to typical resolutions such as Landsat, but no longer to high-resolution satellite data.

The comparison is processed in so-called *confusion tables* (*confusion matrices*), in which the results of the classification are compared with the reference information (cf. Table 10.7). A confusion table allows the classification quality to be viewed from

Table 10.7 Example of a confusion table

	Reference data						
		Deciduous forest	Coniferous forest	Arable land	Meadow	Total	User's accuracy
Classified data	Deciduous forest	**139**	8	2	4	**153**	91%
	Coniferous forest	7	**101**	5	4	**117**	86%
	Arable land	5	4	**75**	2	**86**	87%
	Meadow	4	7	3	**45**	**59**	76%
	Total	**155**	**120**	**85**	**55**	**415**	
	Producer's accuracy	90%	84%	88%	82%		

several perspectives. The overall accuracy is estimated by the quotient of all correctly classified pixels (sum of the numbers in the main diagonal of the confusion table) and the total number of test pixels (here 360 / 415 = 87%). The accuracy from the user's point of view (so called user's accuracy) is calculated by dividing the number of correctly classified image elements by the row sum. It indicates the percentage or probability that a user of the classification result will actually come across the respective class in nature. In Table 10.7, for example, the user's accuracy for deciduous forest is 139 / 153 = 91%. The accuracy from the producer's point of view (so-called producer's accuracy) is calculated by dividing the correctly classified pixels by the column sum, is calculated. In Table 10.7, for example, the producer's accuracy for deciduous forest is 139 / 155 = 90%. Overall, the confusion table provides information on where difficulties arose in the clear class assignment and therefore indicates inhomogeneities in the training areas or even in the class definitions.

The correlation between the classified data and the reference data can be quantified with the help of contingency coefficients. In digital image processing, it is common, to calculate the *kappa coefficient*, which includes both the omissions (so-called errors of omission, number of pixels incorrectly not assigned to a class) and the additions (errors of commission, number of pixels incorrectly assigned to a class) in the assessment of the classification quality (cf. Jensen 2015 p. 570 and Richards 2013 p. 403):

$$\kappa = \frac{N \cdot \sum_{i=1}^{r} x_{ii} - \sum_{i=1}^{r} \text{row sum}_i \cdot \text{column sum}_i}{N \cdot N - \sum_{i=1}^{r} \text{row sum}_i \cdot \text{column sum}_i}$$

The values in Table 10.7 the result in:

$$\kappa = \frac{415 \cdot 360 - (155 \cdot 153 + 120 \cdot 117 + 85 \cdot 86 + 55 \cdot 59)}{415 \cdot 415 - 48310} = \frac{415 \cdot 360 - 48310}{415 \cdot 415 - 48310} = 0.82$$

Values of the kappa coefficient vary between 0 and 1. A value of 0 means no match and a value of 1 means complete match. Values of Kappa greater than 0.75 indicate a very good classification accuracy, values less than 0.4 indicate poor classification accuracy, but for such an assessment the requirements of the distribution form (random sample from a multinormal population) must be taken into account. However, it is much more important to assess the evaluation with regard to the classification objective. Thus, a classification with a kappa coefficient of 0.75 can be useless. It should also be noted that the reference data may be incorrect.

However, several papers highlight that the kappa coefficient is not the most appropriate measure to assess classification accuracy. Pontius et al. suggest that the kappa indices are misleading and/or erroneous for practical use, and rather suggest analysing the confusion matrix and considering the two parameters "quantity disagreement" and "allocation disagreement" (cf. Pontius and Millones 2011, Olofsson et al. 2014). Stehman and Foody 2019 review the current state of nearly 50 years of practice and argue strongly against the kappa coefficient and discuss further options (cf. also the more advanced parameters "quantity", "exchange" and "shift" in Pontius 2019).

10.7.7 Pixel-Based Classification Methods: Problems

The process strategies presented assume an unambiguous classification. However, this idealizes the actual situation. In practice some pixels often cannot be clearly assigned to a certain class. This can be the case, if several classes are relatively heterogeneous due to their statistical characteristics (variance of the brightness values) and a pixel can be assigned to several groups. In particular, the decomposition of the feature space of the spectral bands by linear separation functions as with the parallelepiped or minimum distance classifier is very rigid. Groups are sometimes not defined in such a simple or regular way and then cannot be clearly separated e.g. by Thiessen polygons. The maximum likelihood method brings improvements, it does not provide a solution in principle because, as a parametric method, it implies a certain form of distribution.

Mixed pixels, which represent different surface types in one image pixel due to the low geometric resolution pose greater problems (especially in transitional areas such as green areas and buildings on the outskirts of cities or green areas and water areas on the banks of rivers and lakes). Accordingly, in the case of low-resolution remote sensing data such as Landsat sensor data, a pixel cannot necessarily represent the spatial details of the earth's surface. The radiation values hitting the sensor are then mixed values of the emitted or reflected radiation on the ground. Clear assignment to a class is then no longer possible, which inevitably leads to misclassifications. The inclusion of non-spectral additional data such as cadastral data, which can be used for a pre-selection of the image data or a rough separation of image information (e.g. masking out settlement areas), is a possible solution. Furthermore, multi-temporal recordings can be used, in which e.g. the phenological development of the vegetation is used for clear differentiation. Methods of subpixel analysis (spectral mixture analysis), such as linear spectral unmixing (or spectral mixture analysis) can also be used (cf. Jensen 2015 pp. 480–484 and Mather and Koch 2011 pp. 258–263).

Another problem of the parametric methods is that each pixel is initially assigned to a land cover class, regardless of how low the probability of an actual class membership is. In this way, even the pixels that are not or insufficiently recorded by the previously defined training regions are assigned to the class whose properties they come closest to. This problem can be counteracted by defining a rejection threshold, which can be defined user-dependently and which rejects pixels with low membership probabilities as unclassifiable (cf. Richards 2013 pp. 253). As a result, those areas remain unclassified for which insufficient prototype classes were recorded in the training phase. This can also be the case, for example, for a class for which training areas are available, but which have different reflectance values at the unclassified locations (i.e. pixels) due to different topographic properties (illumination).

A so-called *salt-and-pepper effect* in the graphical presentation often occurs with parametric methods and especially with low-resolution images. The displayed classification result shows an unstable structure (sprinkles, incomprehensible pixel values), which complicates the evaluation. The use of a low-pass filter (cf. Sect. 10.6.4.1) can help to eliminate this effect. Median or modal filters are usually used for this.

Increasing resolution makes it much more difficult to find (larger) homogeneous image elements that represent a specific surface or land cover type. A single pixel can no longer be isolated, but must be considered as part of a larger object. For example, in the case of very high-resolution sensors, a pixel no longer detects building blocks, but rather individual parts of the building. Similarly, the smallest gaps or forest clearings can now be mapped in a forest, which are classified as open ground areas or farmland by a standard method such as maximum likelihood classification. The forest clearing can only be recognised by looking at the surroundings, i.e. by taking into account neighborhood information. For classification, pattern recognition and object formation, the consideration of the pixel neighborhood in particular now becomes essential. In addition to the neighborhood, texture and shape as well as color information also contribute to the cognitive process of a human interpreter.

10.7.8 Object-Oriented Image Segmentation and Classification

Image segmentation methods break away from the purely pixel-based view. Image segmentation, i.e. the extraction of segments or individual areas of an image, is basically not new. The simplest approaches include all methods with thresholding, i.e. (successive) masking of pixels from a certain value (e.g. NDVI >0.5). However, with regard to the classification of several different areas of an image, no satisfactory results are obtained.

In a further procedure, considerable prior knowledge is used. Boundaries of objects such as agricultural areas can be used. The pixels or their spectral values of a given area can be used to describe the area. However, this approach primarily classifies the areas rather than the pixels (so-called per-field approach versus a per pixel approach).

Object-oriented image segmentation is characterised by the fact that the algorithms include both spectral and spatial information (mostly based on the software eCognition, cf. Trimble 2023, for references to free software cf. Jensen 2015 p. 421). This makes them fundamentally different from pixel-based methods (for a broad overview of the development of object-based analyses cf. Blaschke 2010). The approach can generally be described as a *region-growing technique* (cf. determination of training areas in Sect. 10.7.3). The procedure starts with each pixel forming an object or region. At each step, objects are merged in pairs to form a larger object. The decision whether to merge two objects is based on a homogeneity criterion, which formalizes the similarity of adjacent image objects. A parameter that describes the spectral information (based on the standard deviations of the spectral values of all bands of the considered pixels) and another parameter that operationalizes the shape (based on the description of compactness and smoothness of the pixel clusters) are included in the calculation. The user specifies and weights the parameters (for a compilation cf. Jensen 2015 Table 9.12). He can determine whether the classification process is more clearly described by e.g. spectral properties. The merging of two adjacent pixels or objects is only done, if the extent of the match is less than a given value. A smaller value for this parameter "least degree of fitting" leads to a smaller number of mergers and therefore also to smaller objects. The procedure ends if no more mergers are possible. The user must also specify how the neighborhood of two pixels or objects is determined (e.g. N.4 neighborhoods, cf. Sect. 9.3.4).

The selection of the starting pixels for the formation of the regions and the determination of threshold values for the homogeneity criterion are critical points of the procedure. The operationalisation of the neighborhood also influences the object formation.

10.7.9 Modern Advanced Classification Approaches

Long-existing classification methods that belong to the field of machine learning have become the standard (for an introduction to the mindset of machine learning within classification methods in remote sensing cf. Hänsch and Hellwich 2017 pp. 604). These new classifiers, which belong to the supervised classification approaches, do not require any assumptions about the frequency distribution of the training data, as required by the maximum likelihood method, for example. The methods, which are therefore referred to as non-parametric methods, are not based on statistical information of the training regions (e.g., minimum, maximum, mean, variance-covariance of the features, i.e. bands, in a cluster). Due to these unnecessary assumptions, these methods are becoming increasingly popular for classification. Only the classification idea of the rather complex procedures can be reproduced here.

10.7.9.1 Classification with the Help of Neural Networks

Pattern recognition, e.g. classification, with neural networks follows a fundamentally different approach, compared to statistical methods. *Neural networks* are based on the way the human brain works, in which a huge number of nerve cells, neurons, are connected to one another (on this new field of computer science cf. Ertel 2016, Kruse et al. 2015, Nielsen 2015).

A neuron can be understood as a processing unit that receives weighted inputs from other neurons, adds them up, and sends a signal to other neurons, if the sum of the input values exceeds a threshold. The information is processed forward (feed-forward), starting from the first inputs to the last output (cf. Fig. 10.31). The (machine) learning capability is added to this very simple initial model. Before the model works reliably, it must be trained using training data (supervised learning). Initially, the weights are determined randomly. Since the output data determined from the input data using the weights and the thresholds do not initially match the corresponding classification of the training data, the weights and the threshold values must be adjusted. Mather and Koch compare this procedure with a

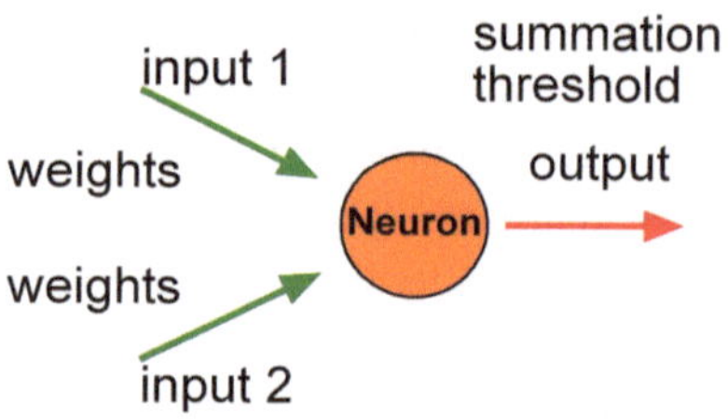

Fig. 10.31 Model of a neuron with two inputs and one output

child learning to read (cf. Mather and Koch 2011 p. 251). Reading skills are developed through repeated correction of mistakes and identification of letters combined with pronunciation.

In this way, for example, the logical "and" circuit could also be learned. In an iterative process, in which the two inputs are always either 0 or 1, the weights are changed until they both have the value 1/2 and the threshold is 1.

$$\text{sum} = \tfrac{1}{2} \cdot \text{input1 and } \tfrac{1}{2} \cdot \text{input2 and threshold} = 1$$

Compared to this simple model, which is called a single-layer perceptron, a more complex model consists of a multi-layer perceptron, which also assumes a more complex threshold function (continuous output values between 0 and 1) and a multi-layer arrangement of the neurons in layers. Instead of requiring discrete input values, the input values can be normalised (i.e., between 0 and 1). In the example of Fig. 10.32, the four neurons of the output layer provide outputs that serve as the basis for classification. In this example two neurons are in the middle, invisible layer, which are connected and weighted with all neurons of the input and output layers. These neurons of the so-called hidden layer implement the sums of the input values and the thresholding.

The model could be used to classify a pixel, each given three brightness values of visible light as input. The neural network could be trained such that if the first output neuron has a value close to 1 and the remaining output neurons have values close to zero, the input pixel is assigned to a first class. The pattern recognition or classification of an input pixel is therefore based on a known training pattern (training data pixel, e.g., brightness values 20, 12, 0 or normalised 0.08, 0.05, 0 for the first three spectral bands of Landsat 7) of which it is known that it belongs to a particular class (here water). For each input vector, the procedure returns a value for each neutron of the output layer. All neurons of the output layer almost always have a value of 0. Only for the known training pattern does the neuron denoting the associated class have the value 1. Learning an artificial neural network here means repeating the process again and again for input pixels and adjusting the weights or threshold values in such a way that training pixels, i.e. pixels with a known class membership, are ultimately assigned to exactly this class.

The multi-layer perceptron is trained using a complex learning rule called backpropagation. For each new pixel added to the training procedure, the output vector differs from the output vector of the uniquely classified training pixel by an error. Based on

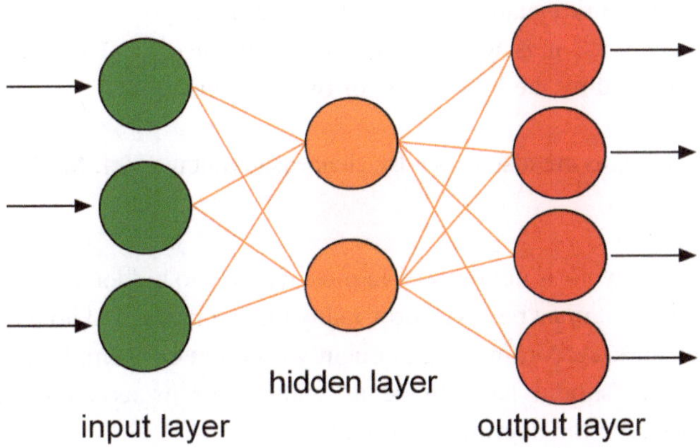

Fig. 10.32 Multilayer perceptron

this error, an attempt is made to readjust the weights. A transformation is therefore sought that maps input vectors to given output vectors, whereby no mathematical function being sought, but instead the mapping is operationalised by weights and threshold values. The quality of the mapping is described by an error function that must be minimised, but in general only a local minimum is found.

The backpropagation algorithm proceeds in a simplified way in three steps (for the mathematical formulation cf. e.g. Nielsen 2015 Chapter 2):

- The input values of a pixel are made available to the network at the input layer and processed forward by the neural network (feed forward, forward propagation).
- The output vector provided by the neural network is compared with the required output, since the class membership of the input pixel and therefore the allocation of the output neurons are known.
- The difference between the two vectors, i.e. the error of the network, is now returned to the input layer via the output layer (error back propagation, which explains the name of the method). The weights of the neuron connections and the threshold values are changed depending on their influence on the error.

A forward-directed neural network that uses backpropagation as a training method has several advantages (cf. Mather and Koch 2011 p. 252 f.):

- Regardless of their statistical frequency distribution, different input variables can be used for training.
- New input variables that have not yet been classified only have to be similar to the input data, with which the network was trained in order to produce a correct classification.

- Since a neural network usually consists of several layers of neurons that are linked by weighted connections, it is tolerant of noisy training data. The overall result is not significantly affected by the loss of one or two neurons.

However, the advantages are also offset by clear disadvantages (cf. Mather and Koch 2011 p. 253):

- The design of a network is not unique. In most cases, two hidden layers are sufficient. In principle, however, it must be asked how many hidden layers and how many neurons are needed in each case. According to a frequently used but only empirically available rule, the hidden layers should have twice as many neurons as input neurons. Are all connections between the neurons required?
- The training time of a neural network is long and can take several hours, because with many input and output neurons and a certain, predefined number of neurons of the hidden layer, there are many connections and extremely many weights have to be tested.
- The backpropagation algorithm may only be able to reach a local minimum of the error function or even oscillate cyclically.
- The classification results that can be achieved with a neural network depend strongly on the weights initially assigned to the connections between the neurons.
- The generalisation ability of neural networks depends on a complex interaction between the number of neurons on the hidden layers and the number of iterations during training. This can lead to an *overfitting* to the training data since the network was trained too much on properties of the training data. The neural network will then have difficulty assigning pixels to a class that are similar to a trained pattern but are not contained in the training data.

The forward directed, multi-layer perceptron is established in image processing, although it is not the only form of artificial neural network (cf. the special issue of the International Journal of Remote Sensing, which presented various approaches as early as 1997). In contrast to a simple neural network with only one hidden layer, deep learning neural networks are distinguished that use a large number of hidden layers. This also makes it possible to model complex structures between the input and the desired output (cf. in more detail Zhu et al. 2017).

Convolutional neuronal networks are the further development of neuronal networks that have reached their limits in image processing. There should be as many inputs as there are pixels with a huge number of connections between the neurons on the hidden layers. Convolutional neural networks are to be understood as optimisations of artificial neural networks that have numerous hidden layers that are "deeply" structured. A Convolutional neural network consists of different layers: the convolutional layer, the pooling layer and the fully meshed layer. The advantage is that the number of connections of the pooling and the convolutional layer is limited, even with a large number of input values, since there are only locally meshed subnets. The result is a lower memory requirement and a much shorter training time of the network (for an introduction cf. Goodfellow et al. 2016 and therein especially Chapter 9).

10.7.9.2 Classification with Support Vector Machine

Classification using a support vector machine represents a very optimal method, since it usually delivers more accurate classification results than other methods, and it requires quite little training data (for the beginnings in remote sensing cf. Huang et al. 2002). The classifier "*support vector machine*" represents a supervised, non-parametric statistical learning strategy that does not impose any requirement on the distributional form of the data.

Figure 10.33 illustrates the basic principle for the simplest case of a linear, binary classification that divides the pixels of a training area into one of two possible classes. As in all supervised classification methods, the basis is that the class membership of the pixels of the training domain is known. In Fig. 10.33, the pixels represent vectors in a two-dimensional vector space. The algorithm tries to decompose a set of objects into clusters in such a way that the widest possible section remains free between the objects. In general, an attempt is made to fit a hyperplane that separates the training objects. In mathematics, an (n-1) dimensional sub-vector space of an n-dimensional space is called a hyperplane in generalisation of a two-dimensional plane in the three-dimensional space of everyday life. In the example in Fig. 10.33, a hyperplane in the two-dimensional vector space is then a one-dimensional straight line.

In this example, two possible separation lines are drawn with their respective separation sections. Among all separators, one has an orientation for which the distance between the two nearest pixels (equivalent with vectors) becomes a maximum. These vectors are called support vectors (cf. the symbols marked in black in Fig. 10.33). The example clearly shows that the position of the separating line is only determined by the support vectors. The remaining objects, i.e. vectors, of the training data are ignored and do not influence the classification. Therefore, the classification remains stable even if the training data are changed, but the support vectors remain unchanged.

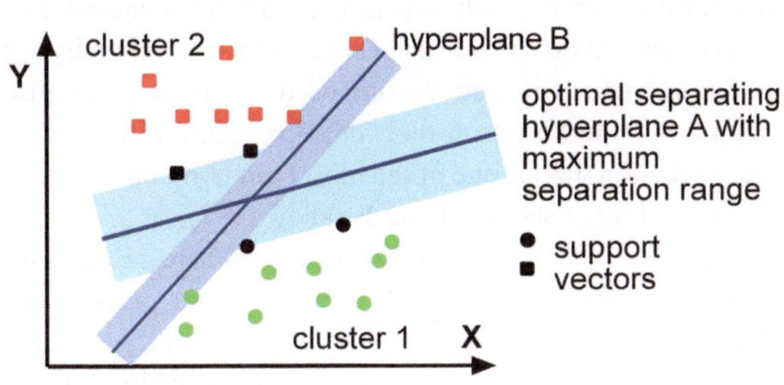

Fig. 10.33 The concept of support vectors

However, only a linear separation is possible with hyperplanes. But in most cases, the classes are usually not simply linearly separable in reality. The support vector machine approach solves this problem by mapping the training vectors into a sufficiently higher vector space in which a separating hyperplane can be determined. This is possible because in a vector space of sufficiently high dimension, every vector set is linearly separable. Several so-called kernel functions are available for this mapping, whereby the so-called radial basis function is frequently used. However, the parameters of this function are data-specific, so that they have to be re-determined for each classification.

The example only illustrates the simple situation that pixels must be assigned to exactly one of two classes. As a rule, several classes exist, so that the procedure must be modified (cf. Mather and Koch 2011 p. 268):

- The "one against the others" approach classifies one class against all other classes that are grouped into one class, this step being repeated for all classes.
- The "one against one" approach compares classes in pairs, with k classes resulting in (k · (k-1))/2 runs.
- Several, i.e. different, support vector machines can also be used for several classes.

Support vector machines have a high mathematically demand. However, they have become standard methods in machine learning. For digital image processing, they are characterised by the fact that, compared to other methods, they do not require large training regions, i.e., training pixels, and have simpler demands on the distributional form of the data (cf. further Mountrakis et al. 2011 and Steinwart and Christmann 2008). This makes them particularly well suited for the classification of hyperspectral data, since there are naturally only relatively small training regions to distinguish between many classes. However, the random-forest classifier offers more advantages, since it is less susceptible to overfitting (i.e. overfitting to the training data).

10.7.9.3 Classification with the Help of Decision Trees

Multi-level classification approaches can be organised in the form of decision trees. The *decision tree classifier* is a hierarchically organised, multi-level classification approach that progresses from top to bottom on a pixel basis. A decision tree generally serves as a tool for decision support, through which data are decomposed step by step into disjoint groups that are homogeneous as possible based on their characteristics values. The CART method (classification and regression tree) is one of the most common implementations (introduced by Breiman et al. in 1984, cf. also Friedl and Brodley 1997).

A binary tree represents the simplest decision tree, where at each stage the data are separated into only one of two possible classes. The complete classification does not take place on one level, but only after the complete decision tree has been traversed. Figure 10.34 illustrates the principle using three features for a scene in the Osnabrück area. The data set (Sentinel 2) is decomposed step by step on the basis of a splitting mechanism, which does not have to be the same for each stage.

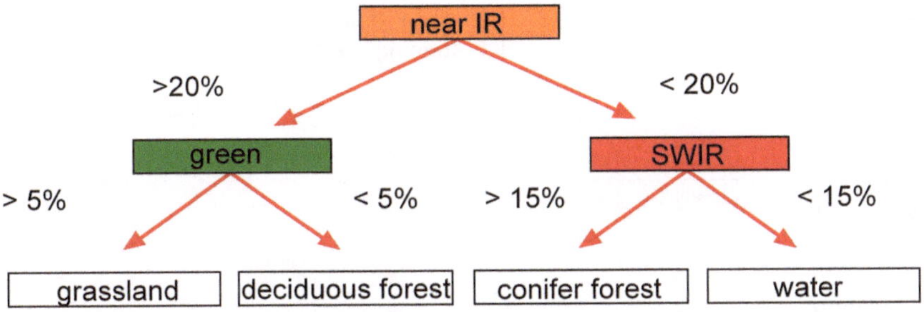

Fig. 10.34 Simple binary decision tree

When designing a decision tree, an optimal structure (i.e. depth of the tree), an optimal set of features at each node according to which the tree is further partitioned, and the decision rule for each node must be found. Since the number of possible tree structures can become very large even for a small number of classes, designing an optimal classifier is not easy. It is immediately obvious that the classification accuracy and efficiency is highly dependent on the tree structure. Therefore, various randomised methods for designing decisions have been developed (cf. Jensen 2015 p. 441 and Hänsch and Hellwich 2017 p. 610). In summary, decision trees combine many advantages:

- "Most approaches do not make any assumptions about the statistical distribution of the attributes or the target variables.
- You can deal with categorical, discrete and continuous features and their combinations.
- Missing measured values can be handled.
- They are robust to outliers and collinearities in the data, which have a strong (negative) impact on other parametric methods.
- Relationships between attributes can be discovered.
- Monotonic transformations of the attributes have no effect.
- They are not susceptible to the curse of "dimensionality" (Hänsch and Hellwich 2017 p. 611, translated).

A set of many decision trees that are generated and trained independently of one another is the basis for the *random forest classifier* (cf. Breiman 2001, cf. also Gislason et al. 2006). It is widely used mainly due to its excellent classification results and processing speed (for an overview cf. Belgiu and Drăgut 2016).

After the training phase of the random forest classifier, i.e. the different decision trees for the training pixels, a new pixel passes through all decision trees. Each tree provides a class assignment into one of k classes for this pixel. In the end, an evaluation of these (preliminary) assignments leads to the final classification (Fig. 10.35). The pixel is assigned to the class that was selected most frequently (majority criterion, for other assignments cf. Hänsch and Hellwich 2017 pp. 628).

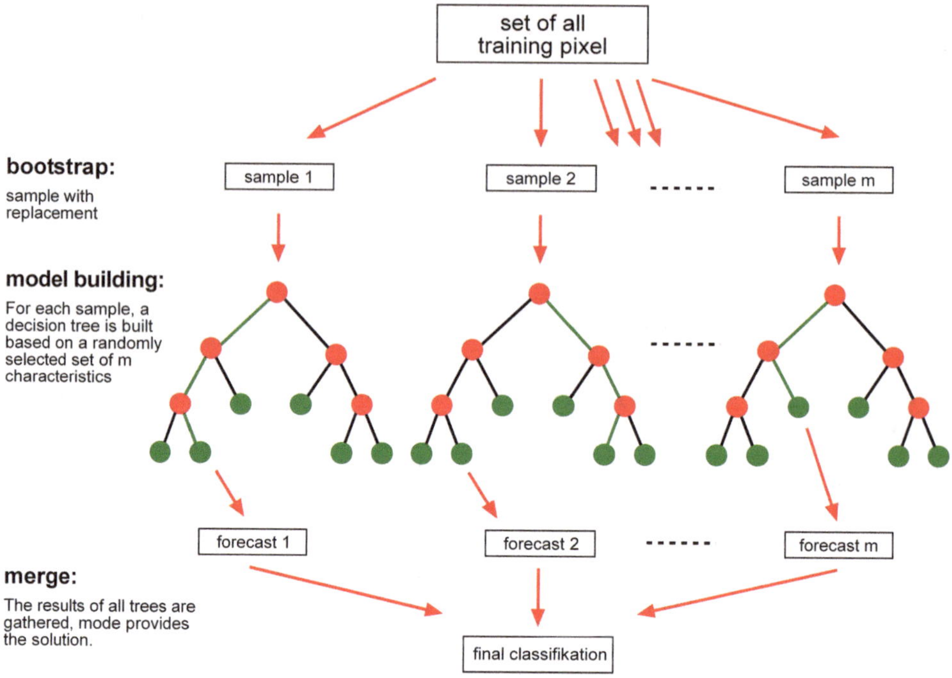

Fig. 10.35 Principle of the random forest classifier

The random forest classifier is built by a set of decision trees using the bagging algorithm. Here, "bagging" is an artificial word composed of the words "bootstrapping" and "aggregating". The individual trees are randomly selected according to a procedure that is referred to as "bootstrapping" in statistical methodology. In this process, random samples are repeatedly drawn from the set of all training pixels, whereby a pixel may appear in more than one sample or even in none. This is equivalent to draw a sample of numbered balls from a ballot-box and putting them back after the sample. The further procedure can be described by (cf. Belgiu and Drăgut 2016):

- About two-thirds of the training pixels (referred to as "in-the-bag samples") are used to train the trees. The remaining third (referred to as out-of-the-bag samples) is used to assess how good the resulting output is. This error estimate is referred to as the out-of-bag (oob) error.
- Each decision tree is created independently. Each node is partitioned based on multiple features, each randomly selected, with the number of features (mtry) determined by the user. This method is also called "feature bagging" and serves to eliminate existing correlations between the estimators.
- The number of trees (ntree) is also user-defined.

According to Breiman (2001), the algorithm generates trees that have high variance and low bias. Theoretical and empirical studies have shown that the classification accuracy is less sensitive to the parameter ntree than to mtry (cf. the literature review by Belgiu and Drăgut 2016). Very often, the ntree value is set at 500, since it seems that errors stabilize before this number of classification trees is reached. This common value could also be due to the fact that this is the default value in the R package. The square root of the number of features is usually taken as the value for the mtry parameter. Higher values increase the computing time considerably.

Overall, the random forest classifier gives good results (cf. Belgiu and Drăgut 2016). It is less sensitive to the quality of the training samples and to the problem of overfitting. This is mainly due to the large number of decision trees and their construction (random samples of the subsets of training patterns and the variables used to partition the nodes).

References

Achen, M. (1993): Untersuchungen über Nutzungsmöglichkeiten von Satellitenbilddaten für eine ökologisch orientierte Stadtplanung am Beispiel Heidelberg. Heidelberg: Selbstverlag des geographischen Institutes Heidelberg.

Airbus (2023a): Spot 6/7. https://www.intelligence-airbusds.com/en/8693-spot-67 (14.04.2023)

Airbus (2023b): Pléiades Neo. https://www.intelligence-airbusds.com/imagery/constellation/pleiades-neo/ (14.04.2023)

Airbus (2023c): Airbus Earth Oberservation Constellation. https://www.intelligence-airbusds.com/imagery/constellation/ (14.04.2023)

ASTRIUM (2023): SPOT 6 SPOT7 technical sheet. https://www.intelligence-airbusds.com/files/pmedia/public/r12317_9_spot6-7_technical_sheet.pdf (14.04.2023)

Albertz, J. (2009): Einführung in die Fernerkundung. Grundlagen der Interpretation von Luft- und Satellitenbildern. Darmstadt: Wiss. Buchgesellsch. 4. Ed.

Backhaus, K., Erichson, B., Plinke, W. u. R. Weiber (2016): Multivariate Analysemethoden: Eine anwendungsorientierte Einführung. Springer: Berlin 14. Ed.

Bannari, A., Morin, D. u. F. Bonn (1995): A Review of Vegetation Indices. In: Remote Sensing Reviews 13, S. 95–120.

Baret, F. (1995): Use of spectral reflectance variation to retrieve canopy biophysical characteristics. In: Danson, F.M. u. S. E. Plummer (Hrsg.): Advances in environmental remote sensing. Chichester: John Wiley.

Belgiu, M. u. L. Drăgut (2016): Random forest in remote sensing: A review of applications and future directions. In: ISPRS Journal of Photogrammetry and Remote Sensing 114, S. 24–31.

Blaschke, T. (2010): Object based image analysis for remote sensing. In: ISPRS Journal of Photogrammetry and Remote Sensing 65 (1), S. 2–16.

Bortz, J. u. C. Schuster (2010): Statistik für Human- und Sozialwissenschaftler. Lehrbuch mit Online-Materialien. Berlin: Springer. 7. Ed.

Breiman, L. u.a (1984): CART: Classification and Regression Trees. Wadsworth: Belmont, CA, 1984.

Breiman, L. (2001): Random Forests. In: Mach. Learning 45 (1), S. 5–32.

Chavez, P.S. (1996): Image-Based Atmospheric Corrections – Revisited and Improved. In: Photogrammetric Engineering and Remote Sensing 62, S. 1025–1036.

Cihlar, J., St-Laurent, L. u. J.A. Dyer (1991): Relation between the Normalized Difference Vegetation Index and Ecological Variables. In: Remote Sensing of Environment 35, S. 279–298.

Congalton, R.G. (1991): A Review of Assessing the Accuracy of Classifications of Remotely Sensed Data. In: Remote Sensing of Environment 37, S. 35–46.

Congalton, R.G. u. K. Green (1999): Assessing the Accuracy of Remotely Sensed Data. Principles and Practices. Boca Raton, Fl.: Lewis Publishers.

Copernicus (2023): Access to data. https://www.copernicus.eu/en/access-data (14.04.2023)

Crist, E.P. and R. C. Cicone (1984): A physically based transformation of Thematic Mapper data – The TM Tasseled Cap, IEEE Transactions on Geosciences and Remote Sensing, GE-22: 256–263

de Lange, N. (2008):): Notes on multispectral classification with correlated bands: effects on using euclidean distance. In: Schiewe, J. u. U. Michel (Hrsg.): Geoinformatics paves the highway to digital earth. GI-Reports 44. Osnabrück: Selbstverlag. S. 75–80.

de Lange, N. u. J. Nipper (2018): Quantitatice Methodik in der Geographie. Grundriss Allgemeine Geographie. Paderborn: Schöningh.

Dennert-Möller, E. (1983): Untersuchungen zur digitalen multispektralen Klassifizierung von Fernerkundungsaufnahmen mit Beispielen aus den Wattgebieten der deutschen Nordseeküste. Dissertation an der Universität Hannover, Fachrichtung Vermessungswesen.

DLR (2023a): Anwendungen und Projekte https://www.dlr.de/eoc/de/desktopdefault.aspx/tabid-5355/ (14.04.2023)

DLR (2023b): TanDEM-X https://www.dlr.de/rd/desktopdefault.aspx/tabid-2440/3586_read-16692 (14.04.2023)

DLR (2023c): Science Service System https://tandemx-science.dlr.de/ (14.04.2023)

DLR (2023d): RapidEye. https://www.dlr.de/rd/desktopdefault.aspx/tabid-2440/3586_read-5336 (14.04.2023)

DLR (2023e): Satellitendaten. https://www.dlr.de/eoc/desktopdefault.aspx/tabid-5356/ (14.04.2023)

Drury, S.A. (1990): A Guide to Remote Sensing. Interpreting Images of the Earth. Oxford: Oxford University Press.

DWD (2023): Wetter- und Klimalexikon. Absorption. https://www.dwd.de/DE/service/lexikon/Functions/glossar.html?lv2=100072&lv3=100160 (14.04.2023)

eoPortal (2023): Sharing Earth Oberservation Resources. https://eoportal.org/web/eoportal/home (14.04.2023)

EOS, Earth Observing System (2023a): SPOT 6&7. https://eos.com/find-satellite/spot-6-and-7/ (14.04.2023)

EOS, Earth Observing System (2023b): Pléiades 1. https://eos.com/find-satellite/pleiades-1/ (14.04.2023)

Ertel, W. (2016): Grundkurs Künstliche Intelligenz. Eine praxisorientierte Einführung. Springer Vieweg 4. Auflage, Heidelberg.

ESA (2023a): MetOP. https://www.esa.int/Applications/Observing_the_Earth/Meteorological_missions/MetOp (14.04.2023)

ESA (2023b): Europe's Copernicus programme. https://www.esa.int/Applications/Observing_the_Earth/Copernicus/Europe_s_Copernicus_programme (14.04.2023)

ESA (2023c): The sentinel missions. https://www.esa.int/Applications/Observing_the_Earth/Copernicus/The_Sentinel_missions (14.04.2023)

ESA (2023d): Introducing Sentinel 1. https://www.esa.int/Applications/Observing_the_Earth/Copernicus/Sentinel-1/Introducing_Sentinel-1 (14.04.2023)

ESA (2023e): Sentinel-2 im Dienste der Ernährung https://www.esa.int/Space_in_Member_States/Germany/Sentinel-2_im_Dienste_der_Ernaehrung (14.04.2023)

ESA (2023f): Introducing Sentinel-3. https://www.esa.int/Applications/Observing_the_Earth/Copernicus/Sentinel-3/Introducing_Sentinel-3 (14.04.2023)

ESA (2023g): Sentinel 4 and 5. https://www.esa.int/Applications/Observing_the_Earth/Copernicus/Sentinel-4_and_-5 (14.04.2023)

ESA (2023h): Sentinel 6. https://www.esa.int/Applications/Observing_the_Earth/Copernicus/Sentinel-6 (14.04.2023)

ESA (2023i): Planetscope instruments. https://earth.esa.int/eogateway/missions/planetscope (14.04.2023)

ESA (2023j): Skysat instruments. https://earth.esa.int/eogateway/missions/skysat (14.04.2023)

ESA (2023k): Sen2Cor. https://step.esa.int/main/snap-supported-plugins/sen2cor/ (14.04.2023)

EUMETSAT (2023a): Meteosat geostationary satellite series https://www.eumetsat.int/our-satellites/meteosat-series (14.04.2023)

EUMETSAT (2023b): Meteosat First Generation. https://www.eumetsat.int/meteosat-first-generation-retired (14.04.2023)

EUMETSAT (2023c): Meteosat Second Generation: https://www.eumetsat.int/meteosat-second-generation (14.04.2023)

EUMETSAT (2023d): Rapid scanning service https://www.eumetsat.int/rapid-scanning-service (14.04.2023)

EUMETSAT (2023e): Meteosat Third Generation: https://www.eumetsat.int/meteosat-third-generation (14.04.2023)

EUMETSAT (2023f): Meteosat series. https://www.eumetsat.int/our-satellites/meteosat-series?sjid=future (14.04.2023)

EUMETSAT (2023g): EUMETView https://view.eumetsat.int/productviewer?v=default (14.04.2023)

European Space Imaging (2023): Resources. https://www.euspaceimaging.com/resources/ (14.04.2023)

Friedl, M.A. u. C. E. Brodley, "Decision tree classification of land cover from remotely sensed data," Remote Sens. Environ., Vol. 61, No. 3, S. 399–409, Sep. 1997

Gislason, P.O., Benediktsson, J.A., Sveinsson, J.R., 2006. Random forests for land cover classification. Pattern Recognition Letters 27 (4), 294–300.

GOES-R (2023): GOES-R. SeriesNOAA's most advanced geostationary weather. https://www.goes-r.gov/education/docs/GOES-R_Overview_FS_FINAL.pdf (14.04.2023)

GOES-T (2023): GOES-T. https://www.goes-r.gov/education/docs/2021_GOES-T_MissionOverviewFS.pdf (14.04.2023)

Goodfellow I. et al. (2016): Deep Learning. Cambridge, Mass.: MIT Press. https://www.deeplearningbook.org/ (14.04.2023)

Hänsch, R. a. O. Hellwich (2017): Random Forests. In: Heipke, C. (Hrsg.) Photogrammetrie und Fernerkundung. S. 603–643. Handbuch der Geodäsie. (Hrsg. Freeden, W. u. R. Rummel). Berlin: Springer.

Harman, H.H. (1976) Modern Factor Analysis. Chicago: University of Chicago Press. 3rd ed.

Heipke, C. (2017a): Photogrammetrie und Fernerkundung – eine Einführung. In: Heipke, C. (Hrsg.) Photogrammetrie und Fernerkundung. S. 1–27. Handbuch der Geodäsie. (Hrsg. Freeden, W. u. R. Rummel). Berlin: Springer.

Heipke, C. (2017b): Photogrammetrie und Fernerkundung. Handbuch der Geodäsie. (Hrsg. Freeden, W. u. R. Rummel). Berlin: Springer.

Hildebrandt, G. (1996): Fernerkundung und Luftbildmessung für Forstwirtschaft, Vegetationskartierung und Landschaftsökologie. Heidelberg: Wichmann.

Huang, C. et al. (2002): An assessment of support vector machines for land cover classification, In: Int. J. Remote Sensing vol. 23, No. 4, S. 725–749.

Jähne, B. (2006): Digital Image Processing. Berlin: Springer. 6. Ed.

Jensen, J.R. (2015): Introductory Digital Image Processing. A Remote Sensing Perspective. 4. Ed. Glen View, Ill.: Pearson. Pearson series in geographic information science.

JPSS, Joint Polar Satellite System (2023a): JPSS Mission and Instruments https://www.nesdis.noaa.gov/current-satellite-missions/currently-flying/joint-polar-satellite-system/jpss-mission-and (14.04.2023)

JPSS, Joint Polar Satellite System (2023b): Visible Infrared Imaging Radiometer Suite (VIIRS) https://www.nesdis.noaa.gov/current-satellite-missions/currently-flying/joint-polar-satellite-system/jpss-mission-and-2 (14.04.2023)

Jürgens, C. (1997): The modified normalized difference vegetation index (mNDVI) – a new index to determine frost damages in agriculture based on Landsat TM data. In: International Journal of Remote Sensing 18, S. 3583–3594

Kauth, R.J. u. G.S. Thomas (1976): The tasseled cap -a graphic description of the spectral-temporal development of agricultural crops as seen as in Landsat. In: Proceedings on the Symposium on Machine Processing of Remotely Sensed Data, West Lafayette, Indiana, June 29 – July 1, 1976.

Kraus, K. (2012): Photogrammetrie: Geometrische Informationen aus Photographien und Laserscanneraufnahmen. Berlin De Gruyter. 7. Ed.

Kruse, R. et al. M. (2015): Computational Intelligence. Eine methodische Einführung in Künstliche Neuronale Netze, Evolutionäre Algorithmen, Fuzzy-Systeme und Bayes-Netze. Springer.

Lillesand, T. Kiefer R.W. u. J.W. Chipman (2008): Remote Sensing and Image Interpretation. Chichester: John Wiley. 6. Ed.

Löffler, E. (1994): Geographie und Fernerkundung – Eine Einführung in die geographische Interpretation von Luftbildern und modernen Fernerkundungsdaten. Stuttgart: Teubner. 2. Ed.

Louis, J. et al. (2016): SENTINEL-2 SEN2COR: L2A Processor for users. In: Proceedings of the Living Planet Symposium 2016. https://elib.dlr.de/107381/1/LPS2016_sm10_3louis.pdf (14.04.2023)

Markwitz, W. (1989): Vom Satellitensignal zur Bildkarte. Einsatz von Daten- und Informationstechniken zur Fernerkundung der Erde. In: Markwitz, W. u. R. Winter (Hrsg.): Fernerkundung: Daten und Anwendungen, S. 1–10. Karlruhe = Beiträge der Interessengemeinschaft Fernerkundung, Leitfaden 1.

Mather, P.M. u. M. Koch (2011): Computer Processing of Remotely Sensed Images – An Introduction. Chichester: Wiley-Blackwell. 4. Ed.

Mountrakis, G., J. Im u. C. Ogole (2011): Support vector machines in re-mote sensing: A review. ISPRS Journal of Photogrammetry and Remote Sensing 66, S. 247 – 259.

NASA (2023a): ECOSTRESS-Spectral Library https://speclib.jpl.nasa.gov/library (14.04.2023)

NASA (2023b): Landst sensors: pushbroom vs whiskbroom https://svs.gsfc.nasa.gov/12754 (14.04.2023)

NASA (2023c) The Worldwide Reference System https://landsat.gsfc.nasa.gov/about/the-worldwide-reference-system/ (14.04.2023)

NASA (2023d): Shuttle radar topographic mission. https://www2.jpl.nasa.gov/srtm/index.html (14.04.2023)

NASA (2023e): NASA's earth observing system. https://eospso.nasa.gov/content/nasas-earth-observing-system-project-science-office (14.04.2023)

NASA (2023f): TERRA. Theo EOS flagship. https://terra.nasa.gov/ (14.04.2023)

NASA (2023g): ASTER. https://asterweb.jpl.nasa.gov/characteristics.asp (14.04.2023)

Nielsen, M. A. (2015): Neural Networks and Deep Learning, Determination Press, 2015 (free online book http://neuralnetworksanddeeplearning.com/) (14.04.2023)

NOAA (2023a): Satellites https://www.noaa.gov/satellites (14.04.2023)

NOAA (2023b): Current Satellite Missions. https://www.nesdis.noaa.gov/current-satellite-missions/currently-flying (14.04.2023)

NOAA (2023c): NOAA's GOES-18 Satellite. https://www.nesdis.noaa.gov/current-satellite-missions/currently-flying/geostationary-satellites (14.04.2023)

NOAA (2023d): GOES-R Series Spacecraft & Instruments https://www.nesdis.noaa.gov/current-satellite-missions/currently-flying/geostationary-satellites/goes-r-series-spacecraft (14.04.2023)

Olofsson, P. et al. (2014): Good practices for estimating area and assessing accuracy of land change. Remote Sens. Environ. 148, 42–57.

Planet (2023a): Real-Time Satellite Monitoring with Planet. https://www.planet.com/products/monitoring/

Planet (2023b): High-resolution imagery with planet satellite tasking. https://www.planet.com/products/hi-res-monitoring/ (14.04.2023)

Pohl, C. u. J.L. van Genderen (1998): Multisensor Image Fusion in Remote Sensing: Concepts, Methods and Applications. In: International Journal of Remote Sensing 19, S. 823–854.

Pohl, C. u. J.L. van Genderen (1999): Multisensor Image Maps from SPOT, ERS und JERS. In: Geocarto International 14, S. 35–41.

Pontius, R.G. u. M. Millones (2011): Death to Kappa: Birth of quantity disagreement and allocation disagreement for accuracy assessment. In: Intern. Journ. of Remote Sensing 32(15):4407–4429.

Pontius, R.G. (2019): Component intensities to relate difference by categorie with difference overall. In. Int. Journal Appl. Earth Obs. Geoinformation 77, S. 94–99.

Richards, J.A. (2013): Remote Sensing Digital Image Analysis. 5. Ed. Berlin: Springer.

Richter, R. (1996): A spatially adaptive fast atmospheric correction algorithm. In: Intern. Journ. of Remote Sensing 17, S. 1201–1214.

Schowengerdt, R. A. (2006): Remote Sensing – Models and Methods for Image Processing. San Diego: Academic. 3. Ed.

Schumacher, H. (1992): Überwachte Klassifikation von Fernerkundungsaufnahmen. Oberpfaffenhofen. Forschungsbericht der DLR: DLR-FB 92 - 07.

SEOS (2023): Introduction to Remote Sensing. https://seos-project.eu/remotesensing/remotesensing-c01-p01.html (14.04.2023)

Shabanov, N. et al. (2002): Analysis of Interannual Changes in Northern Vegetation Activity Observed in AVHRR Data during 1981 to 1994. IEEE Transactions on Geoscience and Remote Sensing (download http://citeseerx.ist.psu.edu/viewdoc/download;jsessionid=8947D71FE6EDD401495D3C184E938F1B?doi=10.1.1.73.1217&rep=rep1&type=pdf) (14.04.2023)

SPOT IMAGE (2023): SPOT satellite technical data. https://www.intelligence-irbusds.com/files/pmedia/public/r329_9_spotsatellitetechnicaldata_en_sept2010.pdf (14.04.2023)

Stehman, S.V. u. G.M. Foody (2019): Key issues in rigorous accuracy assessment of land cover products. In: Remote Sensing of Environment 231, S.

Steinhausen, D. u. K. Langer (1977): Clusteranalyse. Einführung in Methoden und Verfahren der automatischen Klassifikation. Berlin: de Gruyter

Steinwart, I. u. A. Christmann (2008): Support Vector Machines Springer 2008

Swain, P.H. u. S.M. Davis (1978, Hrsg.): Remote Sensing: The Quantitative Approach. New York: McGraw-Hill.

Thenkabail, P.S., A.D. Ward, J.G. Lyon u. C.J. Merry (1994): Thematic Mapper Vegetation for De-termining Soybean and Corn Growth Parameters. In: Photogrammtric Engineering a. Remote Sensing 60, S. 437–442.

Trimble (2023): eCognition. https://geospatial.trimble.com/products-and-solutions/trimble-ecognition (14.04.2023)

Toth, C. u. B. Jutzi (2017): Plattformen und Sensoren für die Fernerkundung und deren Geopositionierung. In: Heipke, C. (2017): Photogrammetrie und Fernerkundung. S. 29–64. Handbuch der Geodäsie. (Hrsg. Freeden, W. u. R. Rummel). Berlin: Springer

USGS (2012): Landsat data continuity mission. https://pubs.usgs.gov/fs/2012/3066/fs2012-3066.pdf (14.04.2023)

USGS (2023a): Spectroscopy Lab https://www.usgs.gov/labs/spectroscopy-lab (14.04.2023)

USGS (2023b): Spectral library https://www.usgs.gov/labs/spectroscopy-lab/science/spectral-library (14.04.2023)

USGS (2023c): USGS EROS Archive – Advanced Very High Resolution Radiometer AVHRR https://www.usgs.gov/centers/eros/science/usgs-eros-archive-advanced-very-high-resolution-radiometer-avhrr (14.04.2023)

USGS (2023d): Landsat missions. https://www.usgs.gov/landsat-missions

USGS (2023e): Landsat – A global Land Imaging Mission. https://pubs.usgs.gov/fs/2012/3072/fs2012-3072.pdf (14.04.2023)

USGS (2023f): Landsat 7. https://www.usgs.gov/landsat-missions/landsat-7 (14.04.2023)

USGS (2023g): Landst satellite missions. https://www.usgs.gov/landsat-missions/landsat-satellite-missions (14.04.2023)

USGS (2023h): Landsat 8. https://www.usgs.gov/landsat-missions/landsat-8 (14.04.2023)

USGS (2023i): Landsat 9. https://www.usgs.gov/landsat-missions/landsat-9 (14.04.2023)

USGS (2023j): Landsat level-1 processing details. https://www.usgs.gov/landsat-missions/landsat-level-1-processing-details (14.04.2023)

USGS (2023k): Landsat levels of processing. https://www.usgs.gov/landsat-missions/landsat-levels-processing (14.04.2023)

USGS (2023l): USGS EROS Archive – Digital Elevation – Shuttle Radar Topography Mission (SRTM) 1 Arc-Second Global https://www.usgs.gov/centers/eros/science/usgs-eros-archive-digital-elevation-shuttle-radar-topography-mission-srtm-1 (14.04.2023)

USGS (2023m): Earth Resources Observation and Science (EROS) Center. Data. https://www.usgs.gov/centers/eros/data (14.04.2023)

USGS (2023n): Landsat Data Access. https://www.usgs.gov/landsat-missions/landsat-data-access (14.04.2023)

USGS (2023o): VIIRS Overview https://lpdaac.usgs.gov/data/get-started-data/collection-overview/missions/s-npp-nasa-viirs-overview/ (14.04.2023)

Vrabel, J. (2000): Multispectral Imagery Advanced Band Sharpening Study. In: Photogrammetric Engineering and Remote Sensing 66, S. 73–79.

Wessels, K. (2002): Integrierte Nutzung von Geobasisdaten und Fernerkundung für die kommunale Umweltplanung: dargestellt am Beispiel der Auswertung von Thermalscannerdaten. Osnabrücker Studien zur Geographie Bd. 21. Osnabrück: Rasch.

Zhu, X. X. et al. (2017): Deep Learning in Remote Sensing: A comprehensive review and list of resources. In: IEEE Geoscience and Remote Sensing Maga-zine 5; 4; S. 8–36.

Zink, M. et al. (2017): TanDEM-X. In: Heipke, C. (Hrsg.) Photogrammetrie und Fernerkundung. S. 525–554. Handbuch der Geodäsie. (Hrsg. Freeden, W. u. R. Rummel). Berlin: Springer.

Index

N. de Lange, *Geoinformatics in Theory and Practice*, Springer Textbooks in Earth Sciences, Geography and Environment,
https://doi.org/10.1007/978-3-662-65758-4